# Postharvest Physiology and Handling of Horticultural Crops

The increase in global population compels growers to use excessive fertilizers and pesticides to enhance agricultural production. This results in concerns with food safety and quality, as well as deterioration in soil and water quality. Furthermore, nearly 30% of fruits and vegetables are rendered unfit for consumption due to spoilage, quality loss or direct physical losses after harvest. This lowers the economic value of crops and forces the need for more production, which puts more strain on natural resources, mainly soil and water. Maintaining postharvest quality and reducing the losses are so crucial for ensuring sustainability in horticultural production. Excessive fertilizer use may also negatively affect the nutritional quality and preservation of horticultural products, reducing the shelf life and overall quality of fruits and vegetables. *Postharvest Physiology and Handling of Horticultural Crops* contains fundamental information that helps readers understand postharvest physiology of fresh fruits and vegetables and presents an in-depth analysis of the harmful impacts of agrochemicals. The book presents readers with eco-friendly, innovative techniques used to handle the fruit and vegetables during storage and through supply chains, helping to better preserve them.

## Features

- Describes available technologies to eliminate and minimize microbial infection for maintaining postharvest quality and safety of fresh produce.
- Explores and discusses approaches, technologies and management practices necessary to maintain products' storage quality by ensuring food safety and nutrition retention.
- Provides practical applications of latest developments in disinfection applications, smart packaging, nanoenabled applications, advances in fresh-cut products, light illumination and edible coatings.
- Presents an in-depth discussion of the harmful impacts of agrochemicals and aims to introduce readers to new, eco-friendly and innovative technologies.

With chapters written by experts in the field of postharvest fruit and vegetable preservation, this book provides information on the use of biomaterials in food preservation and provides practical information for students, teachers, professors, scientists, farmers, food packers and sellers, as well as entrepreneurs engaged in fresh food preservation industry.

# Postharvest Physiology and Handling of Horticultural Crops

Edited by
İbrahim Kahramanoğlu

CRC Press
Taylor & Francis Group
Boca Raton  London  New York

CRC Press is an imprint of the
Taylor & Francis Group, an **informa** business

First edition published 2024
by CRC Press
2385 NW Executive Center Drive, Suite 320, Boca Raton FL 33431

and by CRC Press
4 Park Square, Milton Park, Abingdon, Oxon, OX14 4RN

*CRC Press is an imprint of Taylor & Francis Group, LLC*

ISBN: 978-1-032-58945-9 (hbk)
ISBN: 978-1-032-58966-4 (pbk)
ISBN: 978-1-003-45235-5 (ebk)

DOI: 10.1201/9781003452355

Typeset in Times
by Apex CoVantage, LLC

# Contents

## SECTION I   Postharvest Physiology of Fruits and Vegetables

## SECTION II   Main Handling Practices for Fruits and Vegetables

## SECTION III   Plant- and Animal-derived Methods for Postharvest Quality Preservation

# Figures

# Figures

# Tables

# Preface

Water, oxygen, light and food are the most important sustainers of life on the earth. Humans need food to survive. The human population on earth is continuously increasing while the available soil and water resources are decreasing at a higher rate than the accelerating trend of human population. Therefore, the production of fruits and vegetables is becoming difficult. At the same time, nearly 30% of produced fruits and vegetables is never reaching the final consumers due to postharvest losses. Therefore, prevention of postharvest losses is highly important for sustainability on the earth, would reduce the pressure on soil and water for production and increase the availability of food for consumers. However, it is also of utmost importance to reduce postharvest losses by the application of new eco- and human-friendly technologies/applications.

Although several books are available in the market about the postharvest handling of fruits and vegetables, most of them do not cover these new methods. Therefore, this book is intended to discuss the postharvest physiology of fruits and vegetables, which is very important for the determination of the correct method of management, and also presents and discusses a selection of biomaterials used for the preservation of food quality during postharvest storage. It covers a description of the biomaterials, chemical constituents of the materials, application techniques, antimicrobial activities, food preservative characteristics, impacts on the food quality and mode of mechanism.

With chapters written by experts in the field of postharvest fruit and vegetable preservation, this book presents both new and innovative technologies as well as advancements in traditional technologies. It provides plenty of up-to-date information about the use of biomaterials in food preservation and provides practical information for students, teachers, professors, scientists, farmers, food packers and sellers, as well as entrepreneurs engaged in fresh food preservation industry. This book is comprised of three sections and 15 chapters:

- In Section I (Postharvest Physiology of Fruits and Vegetables), Chapter 1 describes the preharvest factors affecting the postharvest quality of fruits and vegetables in the context of maturity indices and postharvest physiology. Chapter 2 lists and discusses the biological and environmental factors affecting the postharvest quality of horticultural crops, and Chapter 3 talks about postharvest losses in the storage and supply chain.
- In Section II (Main Handling Practices for Fruits and Vegetables), Chapter 4 describes the concerns over agrochemicals, which still have an important place in controlling postharvest diseases, Chapter 5 discusses disinfection applications to fruits and vegetables, Chapter 6 concerns modified atmosphere packaging, Chapter 7 describes the importance of ethylene in postharvest physiology and methods of controlling it, Chapter 8 identifies and discusses nanoenabled applications in postharvest handling, Chapter 9 focuses on the integration of innovative technologies into postharvest fruit

storage systems, Chapter 10 presents recent technology and advances in fresh-cut products, and Chapter 11 elaborates on the most recent information and advantages of using light illumination in the postharvest handling of fruits and vegetables.

- In Section III (Plant- and Animal-derived Methods for Postharvest Quality Preservation), Chapter 12 describes the use of chitosan in the postharvest handling of fruits and vegetables, Chapter 13 covers the use of plant-based fixed oils in edible coating formulations, and Chapter 14 discusses the layer-by-layer applications of edible coatings in the postharvest handling of horticultural crops. Finally, Chapter 15 describes the recent advances in the improvement of the postharvest application of edible coatings.

Information is endless, and therefore the research and writing of a book are also endless. Therefore, I would appreciate receiving comments and criticism about the book, as well as new information to assist in future compilations. I am confident that this book will prove to be informative, practical, interesting and enlightening to its readers.

**İbrahim Kahramanoğlu**

# Acknowledgments

I would like to express my heartfelt thanks to the chapter authors for their valuable contributions. I also would like to share my gratitude to all the researchers whose studies/findings/comments contributed to postharvest knowledge, because this book would not be possible without their contributions. We have learned so much from them, and I hope that the readers will learn from us and continue this information flow.

Sincere thanks are due to Ms. Randy Brehm, Senior Editor of Life Sciences and Medicine at the CRC Press/Taylor & Francis Group LLC, without whose devoted help this book would not have been possible. I also thank all the other people who assisted us directly and indirectly during the preparation of this book.

Finally, I would like to express my gratitude to my mother Tezer Kahramanoğlu, father Mehmet Kahramanoğlu, sister Fatma Kahramanoğlu and cousins Fatma and Yücel for their support throughout my lifetime and during the preparation of this book.

# About the Editor

**İbrahim Kahramanoğlu, PhD,** was born in 1984 in Northern Cyprus. He has been an associate professor at the Department of Horticulture, Faculty of Agricultural Sciences and Technologies in the European University of Lefke (Northern Cyprus) since April 2018. He was the managing director of Alnar Pomegranates Ltd. from 2010 to 2018, which produced, packed and exported fresh pomegranate fruits and produced 100% natural and freshly squeezed pomegranate juice. Before that, he worked in the alternative crops project of USAID in Cyprus and managed more than 100 projects supported by national or international bodies including the European Commission. He has been consulting with farmers about Good Agricultural Practices (GAP) and the adaptation of modern technologies since 2008. Professor Kahramanoğlu is an expert in horticultural production, postharvest biology and technology, and GAP. His main studies are in postharvest physiology and handling of fruits; natural and novel technologies for handling and storage; digital and precision farming (agri 4.0) for sustainability; and adding value to horticultural crops. He has coauthored a monograph book with 16 chapters (*Pomegranate Production and Marketing*, CRC Press), edited three books (*Postharvest Handling, Modern Fruit Industry* and *Fruit Industry*), authored three book chapters, published more than 100 scientific publications related to his experience and attended more than ten international conferences and presented his research findings.

# Contributors

**Afiya John, PhD,** is Postharvest Technology and Postproduction Technology lecturer at The University of the West Indies in the Department of Food Production. She earned her PhD in biochemistry and molecular biology at the University of Chinese Academy of Sciences in China. Growing up in a rural, agricultural community in Trinidad and Tobago, she wasn't introduced to the term "postharvest technology" until she was required to complete a compulsory course while pursuing a BSc in general agriculture. After this course, she knew she was genuinely passionate and wanted to pursue a career in postharvest technology.

**Ayoub Mohammed, MPhil,** is a holder of a BSc (Hons) degree in general agriculture and an MPhil degree in agricultural engineering, both from the University of the West Indies. He also holds an MBA (Distinction) from Anglia Ruskin University. Apart from his academic qualifications, he earned a post-MBA from the Arthur LoK Jack Global School of Business. Over the years, Mr. Mohammed has received international training in postharvest technology, Good Agricultural Practices (GAP), Food Processing and Packaging, Hazard Analysis and Critical Control Points (HACCP), and Research Project Management. He was the recipient of an OAS fellowship, UWI scholarship and a Cochran fellowship. He has published some of his work in postharvest handling, food crop production and food safety. He joined the Ministry of Agriculture in 1996 where he worked at the Extension Training and Information Services Division (ETIS) as an agricultural officer 1. During the period 2009–2012, he worked at the National Agricultural Marketing and Development Corporation (NAMDEVCO) as manager of the Piraco Packinghouse facility. Mr. Mohammed has been a part-time lecturer at the University of the West Indies during the period 2014–2018 at the University of Trinidad and Tobago and the chief executive officer of NAMDEVCO during the period of July 2016–May 2017. Since February 2022, he acts as the manager of the University Field Station, Faculty of Food and Agriculture, of The University of the West Indies, Mt. Hope, Trinidad and Tobago.

**Bonga Lewis Ngcobo, PhD,** is a dedicated researcher with a PhD in Horticultural Sciences. He completed his postdoctoral fellowship at the University of Johannesburg and is currently working at Durban University of Technology. His research focuses on preharvest factors that influence postharvest performance of horticultural crops. Dr. Ngcobo has a passion for ensuring that the quality and safety of horticultural produce are maintained from the farm to the table. His work has been published in several reputable academic journals and communicated at international conferences.

**Chunpeng (Craig) Wan, PhD,** is working as Professor in the College of Agronomy, Jiangxi Agricultural University (JXAU), Nanchang, China. He earned his MS degree in traditional Chinese medicine (TCM) from Jiangxi University of Traditional Chinese Medicine and his PhD from Nanchang University, China. Furthermore,

Dr. Wan joined the Department of Biomedical and Pharmaceutical Sciences at the University of Rhode Island as a visiting scholar and postdoctorate fellow. Dr. Wan's research interests include phytochemistry and human health benefits, postharvest biology and technology of citrus fruits. His research continues to be funded by federal, state, and other agencies. Dr. Wan has edited various books and has authored over 150 publications. Dr. Wan also worked as editor-in-chief of *World Journal of Biological Chemistry*, as associate editor of *International Journal of Agriculture Forestry and Life Sciences*, as editor of *Frontiers in Nutrition, Evidence-Based Complementary and Alternative Medicine, Journal of Food Quality*, and as a Bentham Science Ambassador in 2019–2020.

**Cristina Arroqui, PhD,** was born in Pamplona, Spain in 1972. She earned a PhD in food engineering from the Public University of Navarre (UPNA) in 2001. She has worked in the Food Technology Department of the Public University of Navarra, where she has been tenured professor since 2010. Between 1998 and 2004, she made pre- and postdoctoral stages at the Universities of Lund (Sweden), Davis (California) and Leuven (Belgium). Currently, she works at the Innovation and Sustainability Food Research Institute of the UPNA. The research work carried out can be grouped into two lines of research: (1) food quality and process efficiency improvement using emerging technologies and (2) natural agents for the preservation of foods.

**Ersin Çağlar, PhD,** is currently a senior lecturer at the European University of Lefke, Northern Cyprus, via Mersin 10 Türkiye. He earned a PhD in management information systems (MIS) from Girne American University. He also earned his MBA in Business from the European University of Lefke (EUL). He teaches network security, computing, simulation and management information systems courses. His research interest is in the areas of management information systems, social media, network security, cloud computing, cryptocurrency and the Internet of Things (IoT).

**Hanifeh Seyed Hajizadeh, PhD**, is Associate Professor of Postharvest Physiology in the Department of Horticulture, Faculty of Agriculture, University of Maragheh, Maragheh, Iran. As initial research in postharvest physiology, he studied the effect of active modified atmosphere packaging on some fruits and vegetables (apples, mushrooms and grapes) and evaluated their shelf life. Recently, his research has been focused on plant stress and the role of antioxidants in enzymatic and non-enzymatic defense systems against abiotic stresses in fruits, vegetables and cut flowers after harvest. His upcoming book on harvest indices, postharvest disease and disorders, and the postharvest life of fruits, vegetables, and ornamentals has been compiled with regard to the produce fact sheets published by the UC Postharvest Technology Center.

**Mawande Hugh Shinga** is a promising young scholar pursuing a PhD in botany and plant biotechnology at the Postharvest Research Laboratory, Department of Botany and Plant Biotechnology, University of Johannesburg, South Africa. His research focuses on the regulation of postharvest fruit cell wall and cuticles using edible coating. Mr. Shinga is committed to developing innovative solutions to improve the quality and safety of horticultural produce. He is an excellent student and has received a number of awards and scholarships for his academic achievements.

**Menaka M.** has obtained her MSc degree in horticulture (PHT) and is currently pursuing her PhD from Division of Food Science and Postharvest Technology, ICAR–Indian Agricultural Research Institute, New Delhi, India. She is actively engaged in research on postharvest storage and fruit physiology. She has published five research papers in journals of international repute.

**Neela Badrie, PhD,** lectures in the areas of microbiology, biotechnology, food safety and risk analysis, food product development, epidemiology and foodborne diseases, international trade and food legislation to undergraduates and postgraduates at the Department of Food Production, Faculty of Food and Agriculture, The University of the West Indies, St. Augustine, Trinidad and Tobago, West Indies. She is also an attorney-at-law, admitted to practice law at the Supreme Court of Trinidad and Tobago. Professor Badrie is a fellow of the Caribbean Academy of Sciences (CAS) and The World Academy of Sciences (TWAS, Italy). She is the recipient of several awards such as the Fulbright Researcher Award, USA, Association of Commonwealth University, CARPIMS/EUROPEANUNION/ACP scholarship, European Union Lomé IV-CULP-UWI fellowship, United Nations University UNU/CFNI/PAHO/WHO fellowship and the Rudrunath Capildeo gold medal for applied science and technology, National Institute for Higher Education in Science and Technology.

**Nirmal Kumar Meena, PhD,** is working as Assistant Professor and Principal Investigator in the Department of Fruit Science at the College of Horticulture and Forestry, Jhalawar, Agriculture University, Kota, Rajasthan. He earned his MSc and PhD degrees in the discipline of postharvest technology (horticulture) from the prestigious, ICAR-Indian Agricultural Research Institute, New Delhi. He has guided four MSc students as a major advisor and seven students as member advisor. He is also Officer In-Charge at the Department of Postharvest Technology. He has published more than 33 research papers of international repute, 17 book chapters in Springer, Elsevier, and Taylor & Francis and 50-plus articles in magazines. He is on the editorial boards of several journals and magazines.

**Olaniyi Amos Fawole, PhD,** is an accomplished academic with a PhD in horticultural sciences, specializing in postharvest technology and agroprocessing. He is Full Professor and Founding Director of the Postharvest and Agroprocessing Research Centre (PARC), Department of Botany and Plant Biotechnology, at the University of Johannesburg, South Africa. Professor Fawole has led and managed over 30 research projects funded by various sources. He has published over 170 indexed articles and has successfully graduated 40 postgraduate students, with 24 currently under his supervision. He is a respected member of the academic community and has received numerous awards and honors for his contributions to the field of horticultural sciences.

**Olga Panfilova, Doctor of Agricultural Sciences**, is employed at the Russian Research Institute of Fruit Crop Breeding (VNIISPK), Zhilina, Russia, as a full senior researcher at the laboratory of berry breeding and variety study. Her academic and scientific opus is orientated toward horticulture, breeding and variety study, agricultural technology, plant physiology and biochemistry, plant anatomy and morphology, plant introduction, adaptivity of plants, drought resistance, heat resistance and winter hardiness. Dr. Panfilova is the author of two red currant cultivars: "Podarok Pobediteliam" (2019) and "Premiere" (2021); coauthor of the database number 2020621645 from 10.09.2020, "Sources and Donors of Economically Valuable Features of Red Currant Cultivars from VNIISPK Genetic Collection for Priority Directions of Breeding"; and author or coauthor of 90 publications in the field of horticulture.

**Oral Daley, PhD,** is a researcher and educator with professional experience in the areas of agricultural production, harvesting and postharvest management, food and quality management and agricultural education. He earned an ASc and BSc in agriculture, a postgraduate certificate in university teaching and learning, and a PhD in crop science. His PhD thesis was entitled "Ethnobotanical, Morphological, Physicochemical and Genetic Diversity Studies in Breadfruit [*Artocarpus altilis*, (Parkinson) Fosberg] in the Caribbean." Dr. Daley is a lecturer in the Department of Food Production, Faculty of Food and Agriculture, The University of the West Indies, St. Augustine Campus, Trinidad and Tobago. His current research focuses on crop germplasm characterization, evaluation and conservation with the aim of harnessing the genetic variation in underutilized tropical food crops for long-term genetic gain and diversity and to understand the genetic structure of complex traits that influences quality and yields in order to advance food and nutrition security in the tropics.

**Paloma Vírseda, PhD,** was born in February 1961 in Zaragoza, Spain. She has a degree in agricultural engineering (food industries specialty) from the Polytechnic University of Valencia in 1989, and in 1995 she earned her PhD in Food Engineering from the Public University of Pamplona (UPNA). Currently she is a member of the UPNA Research Commission, the Experts Committee of the National Center for Technology and Food Safety (CNTA) and the Technical Committee of the Spanish Technical Association for Air Conditioning and Refrigeration (ATECYR). She has participated in more than 20 research projects, supported by public and private funding, in the framework of two main research lines: improvement of food quality and process efficiency using emerging technologies and the use of natural products as food preservatives.

**Roghayeh Karimirad, PhD,** was born in 1988 in Iran. She was a talented student in her BSc, MSc and PhD courses. She completed her BSc degree in horticultural science in 2006–2010 at the University of Maragheh, Tabriz, Iran. She graduated with her MSc degree in physiology and breeding from the Olericulture Department in 2013 and with her PhD degree in the same department in 2017, both from the University of Mohaghegh Ardabili, Iran. Her PhD studies were on the nanoencapsulation of medicinal plant essential oil and their effects on postharvest quality of button mushroom (*Agaricus bisporus*). Dr. Karimirad has great interest, knowledge and experience in nanotechnology, release systems (as aspirin and medicinal plants) and food science. She was a teaching assistant in the field of nanotechnology, physiology and medicinal plants in the University of Mohaghegh Ardabili. To date, she has supervised about ten students for master's and doctorate degrees, has been the author of several chapters in various books and has published nearly ten articles in international journals.

**Sandra Horvitz, PhD,** was born in 1970 in Argentina and completed her agricultural engineering degree from the Universidad Nacional del Sur, Argentina. In 2009, she earned her PhD in food science and technology from the Public University of Pamplona, Spain. Currently she is Professor at the Food Technology Department and a member of the Innovation and Sustainability Food Research Institute of the Public University of Navarra. Her previous experience includes teaching and postharvest research at the Technical University of Ambato (Ecuador), a consultancy for the Iberoamerican Institute for Agricultural Cooperation (IICA) and more than ten years as a researcher in the Postharvest and Quality of Fresh Fruits and Vegetables Lab at the National Institute for Agricultural Technology (INTA, Argentina). She has authored several chapters in postharvest books and published research papers in both national and international journals.

**Vinod B. R.** is pursuing a PhD degree from the Division of Food Science and Postharvest Technology, ICAR-Indian Agricultural Research Institute, New Delhi, India, in postharvest management in horticulture.

**Wendy-Ann Isaac, PhD,** is Senior Lecturer in Crop Science in the Department of Food Production, Faculty of Food and Agriculture, The University of the West Indies, St. Augustine, Trinidad and Tobago. She earned her BSc, MSc, and PhD degrees from the UWI, St. Augustine and a MAppl Sci in agronomy from Lincoln University, New Zealand. Her research and teaching revolve around her area of expertise: sustainable vegetable production, integrated weed management, seed production technologies and innovative farming technologies, including protected agriculture and controlled environment agriculture. She has authored and coauthored several papers and chapters on these topics, which have been published in both regional and international peer-reviewed journals and has also coedited the book titled *Sustainable Food Production Practices in the Caribbean* and books on the topic of "Agricultural Development and Impacts of Climate Change on Food Production in Small Island Developing States (SIDS)".

# Abbreviations

| | |
|---|---|
| **1-MCP** | 1-methylcyclopropene |
| **AA** | Antioxidant activity |
| **AAS** | Atmospheric argon plasma |
| **AC** | Acemannan |
| **ACC** | 1-aminocyclopropane-1-carboxylic acid |
| **ADP** | Adenosine diphosphate |
| **AEW** | Acidic electrolyzed water |
| **Ag-NPs** | Silver nanoparticles |
| **AI** | Artificial intelligence |
| **AIB** | α-aminoisobutyric acid |
| **AO** | Aqueous ozone |
| **AOA** | 2-aminooxyacetic acid |
| **AOIB** | 2-aminooxyisobutyric acid |
| **AOP** | Advanced oxidative process |
| **AsA** | Ascorbic acid |
| **ATP** | Adenosine triphosphate |
| **AVG** | Aminoethoxyvinylglycine |
| **BCA** | Biocontrol agents |
| **CA** | Controlled atmosphere |
| **CAPP** | Cold atmospheric pressure plasma |
| **CF** | Chlorophyll fluorescence |
| **CFB** | Corrugated fiber boxes |
| **CFU** | Colony-forming unit |
| **CI** | Chilling injury |
| **CKs** | Cytokinins |
| **CMC** | Carboxymethyl cellulose |
| **CNCs** | Cellulose nanocrystals |
| **CNFs** | Cellulose nanofibers |
| **CNN** | Convolutional neural network |
| **CS** | Chitosan |
| **DACP** | Diazocyclopentadiene |
| **DCA** | Dynamic controlled atmosphere storage |
| **DDT** | Dichloro-diphenyl-trichloroethane |
| **DNA** | Deoxyribonucleic acid |
| **DPPH** | 2,2-Diphenyl-1-picrylhydrazyl |
| **EFE** | Ethylene forming enzyme |
| **EO** | Essential oil |
| **EOs** | Essential oils |
| **EPA** | Environmental Protection Agency |
| **ET** | Ethanol |
| **EU** | European Union |
| **EVA** | Ethylene vinyl acetate |

| | |
|---|---|
| **EVOH** | Ethylene-vinyl alcohol |
| **EW** | Electrolyzed water |
| **F&V** | Fruits and vegetables |
| **FA** | Fumaric acid |
| **FAO** | Food and Agriculture Organization of the United Nations |
| **FCFV** | Fresh-cut fruit and vegetables |
| **FDA** | The United States Food and Drug Administration |
| **FO** | Fixed oils |
| **FRAC** | Fungicide Resistance Action Committee |
| **GAPs** | Good Agricultural Practices |
| **GAs** | Gibberellins |
| **GC-MS** | Gas chromatography-mass spectrometry |
| **GRAS** | Generally recognized as safe |
| **HACCP** | Hazard Analysis and Critical Control Points |
| **HDPE** | High-density polyethylene |
| **HHP** | High hydrostatic pressure |
| **HNTs** | Halloysite nanotubes |
| **HPLC** | High-performance liquid chromatography |
| **HPP** | High pressure processing |
| **IAA** | Indole-3-acetic acid |
| **IoT** | Internet of things |
| **IPM** | Integrated pest management |
| **IR** | Ionizing radiation |
| **LAB** | Lactic acid bacteria |
| **LbL** | Layer-by-layer |
| **LD50** | Lethal dose 50 |
| **LDPE** | Low-density polyethylene |
| **LLDPE** | Linear low-density polyethylene |
| **LPM** | Low mercury lamp |
| **MA** | Modified atmosphere |
| **MAP** | Modified atmosphere packaging |
| **MAPS** | Microwave-assisted thermal pasteurization system |
| **MCP** | Momordica charantia polysaccharide |
| **MT** | Melatonin |
| **MVG** | Methoxyvinylglycine |
| **NBD** | 2,5-norbornadiene |
| **NFC** | Nanofibrillated cellulose |
| **NMs** | Nanomaterials |
| **NO** | Nitric oxide |
| **NPs** | Nanoparticles |
| **NTP** | Non-thermal plasma |
| **OG** | Ozone gas |
| **PA** | Peroxyacetic acid |
| **PAL** | Phenylalanine ammonia-lyase |
| **PAS** | Plasma-activated solution |

| | |
|---|---|
| **PAs** | Polyamines |
| **PDCA** | 2,4-pyridinedicarboxylic acid |
| **PEE** | Propolis ethanol extract |
| **PET** | Poly ethylene terephthalate |
| **PGR** | Plant growth regulator |
| **PL** | Pulsed light |
| **PLA** | Polylactic acid |
| **POD** | Peroxidase |
| **PP** | Polypropylene |
| **PPA** | Plasma-processed air |
| **PPD** | Postharvest physiological deterioration |
| **PPO** | Polyphenol oxidase |
| **PSO** | Pomegranate seed oil |
| **PVDC** | Polyvinylidene chloride |
| **PVOH** | Polyvinyl alcohol |
| **RH** | Relative humidity |
| **ROS** | Reactive oxygen species |
| **RQ** | Respiration quotient |
| **SA** | Salicylic acid |
| **SAEW** | Slightly acidic electrolyzed water |
| **SAM** | S-adenosyl-L-methionine |
| **SM** | Sodium metasilicate |
| **SOD** | Superoxide dismutase |
| **SSC** | Soluble solids concentration |
| **SSOPs** | Sanitation Standard Operating Procedures |
| **STHT** | Short time heat treatments |
| **TA** | Titratable acidity |
| **TCA** | Tricarboxylic acid |
| **TCA cycle** | Tricarboxylic acid cycle |
| **TCO** | Trans-cyclooctene |
| **TSS** | Total soluble solids |
| **TT** | Thermal treatment |
| **ULO** | Ultra-low oxygen |
| **US** | Ultrasound |
| **UV** | Ultraviolet radiation |
| **UV-C** | UltraViolet-C |
| **VCO** | Virgin coconut oil |
| **VPD** | Vapor pressure difference |
| **VUV** | Vacuum ultraviolet light |
| **WHO** | World Health Organization |
| **WSN** | Wireless sensor networks |
| **ZNPs** | Zein nanoparticles |
| **ZPLs** | ε-polylysine |

# Section I

Postharvest Physiology of
Fruits and Vegetables

# 1 Preharvest Factors, Maturity Indices and Postharvest Physiology of Fruits and Vegetables

*Oral Daley, Wendy-Ann Isaac,*
*Afiya John, and İbrahim Kahramanoğlu*

## 1.1 INTRODUCTION

Fruits and vegetables (F&V) are important for human nutrition, providing a diverse range of nutrients, minerals, antioxidants and fibers. Harvested F&V are living organisms, which continue to respire and transpire and so are perishable products. The senescence of F&V is highly diverse, similar to their morphological, physical, compositional and physiological characteristics, which result in varied requirements for postharvest management (Kader & Yahia, 2011). It is projected that, as the global population continues to expand, especially in urban areas, the demand for the packaging and transporting of fresh F&V will continue to increase. Furthermore, approximately 50% of fruits produced in tropical and subtropical regions are sold to the fresh fruit markets (Kader & Siddiq, 2012).

Some of the changes that occur in F&V after harvesting are desirable, while some are undesirable. These changes cannot be stopped but can be retarded or encouraged as necessary. The postharvest transportation, storage and processing of F&V are usually accompanied by quality deterioration, leading to the loss of commodity value (Wang et al., 2022). One of the major goals of postharvest management is to provide specific treatments that accelerate or decelerate the physiological changes and senescence of F&V. The final physiological processes that occur after harvesting are ripening and senescence, which are irreversible processes that lead to the breakdown and death of the F&V. Therefore, the postharvest quality of F&V is highly impacted and regulated by several biological and environmental factors that can accelerate or decelerate the deterioration of freshly harvested produce, thus reducing the shelf life and marketability. Respiration, transpiration and ethylene biosynthesis are the main biological factors where temperature, relative humidity, atmospheric composition and light are the main environmental factors impacting the postharvest quality of fresh F&V (Kahramanoğlu, 2017). Several genetic, environmental and cultural factors influence the development of crops in the field and their postharvest response. Therefore, an understanding of the physiological, environmental, cultural practices

DOI: 10.1201/9781003452355-2

and their interactions is necessary to develop successful treatments for maintaining quality and extending the postharvest life of F&V.

## 1.2 PREHARVEST FACTORS AFFECTING POSTHARVEST PHYSIOLOGY

Preharvest factors have a major influence on the postharvest quality attributes of F&V since postharvest technologies can only maintain but not improve quality attributes after harvest. The preharvest factors affecting the postharvest physiology of fruits and vegetables can be classified according to the following categories:

- Genetic factors
- Environmental/cultural factors
- Physicochemical factors
- Diseases

### 1.2.1 GENETIC FACTORS

The main genetic factors that influence postharvest physiology for any given crop is the choice of cultivar. Cultivars within a species show variations in quality attributes that will inevitably influence responses to postharvest treatments or technologies. One of the major areas of crop breeding and improvement is to develop or select cultivars that are of interest to producers and ultimately the consumers. F&V producers are generally interested in characteristics relating to yield, response to crop management, resistance or tolerance to diseases and adaptability to local growing conditions. Some of the most important quality attributes that influence consumers' choice of cultivars are size, color, shape, flavor, texture, firmness, juiciness and nutritional profile. The selection of the right cultivar or cultivars for production conditions, markets and consumers will significantly impact postharvest quality (Benkeblia et al., 2011). Therefore, producers select one cultivar or cultivars over others by finding a balance between the requirements and proposed end uses of the target market and production conditions.

Several studies have showed significant variation in the nutritional profile among cultivars of F&V. Carotenoids, vitamin A levels and sugar content vary among cultivars of banana (Englberger et al., 2006), breadfruit (Englberger et al., 2007) and pineapple (Ferreira et al., 2016). In a study with 21 breadfruit cultivars, significant differences were observed in instrument and sensory characteristics that are useful for cultivar identification and postharvest practices, such as maturity indices determination and cultivar selection for processing and packaging (Daley et al., 2016).

The genetics of rootstocks is also an important factor in those species that are propagated via grafting. Castle (1995) reported that rootstock used in citrus and other deciduous fruits can exert significant influence on fruit quality attributes such as fruit size, soluble solids concentrations (SSC), firmness, shelf life, as well as crop load and tree yield.

### 1.2.1.1 Classification of Fruits and Vegetables

Classification of the fruits and vegetables can be an important tool for understanding and managing the postharvest life of F&V. Botanically, fruit is the developed ovary and its associated parts. Fruits usually contain seeds, which are developed from the enclosed ovule after fertilization. However, some fruits, including bananas, develop without fertilization through parthenocarpy. Fruits can be classified according to different characteristics. One of the classifications is based on how the ovary and other flower organs are arranged and how the fruits develop. The three main groups are simple fruits, aggregate fruits and multiple (or composite) fruits (Ramaswamy, 2014).

1.  **Simple fruits** are derived from a single ovary. Simple fruits can be grouped as dry or fleshy; the ovary may be composed of one or more carpels, and the fruit may be dehiscent (splits open when mature) or indehiscent (does not split open). The *fleshy fruits* may be grouped as:
    a.  Drupe – stone fruits: developed from the single carpel of a single flower: e.g., cherry, peach, apricot, plum, olive, mango, etc.
    b.  Berry: a fleshy type of fruit that is derived from a compound ovary. it consists of a simple morphological structure with a thin skin enclosing a juicy flesh containing many seeds: e.g., tomato, banana, currants, papaya, etc.
    c.  Pome: derived from a flower with an inferior ovary. The flesh is an enlarged hypanthium, and the core is from the ovary: e.g., pomegranate, apple, pear, etc.
    d.  Hesperidium –citrus fruits: a modified form of berry, with a well developed peelable endocarp, called a hesperidium.
    e.  Pepo: includes fruits belonging to the cucumber family with a berry-like characteristic but with a hard outer layer developed from receptacle: e.g., melons, cucumber, etc.
    The *dry fruits* may be grouped as:
    a.  Achene: a dry, one-seeded fruit. The pericarp is easily separated from the seed coat: e.g., sunflower.
    b.  Grains/caryopsis/cereals: include dry, one-seeded, indehiscent fruit. It differs from the achene in that the pericarp and seed coat are firmly united all the way around the embryo: e.g., wheat, barley, rice, etc.
    c.  Nuts: consist of a hard nutshell protecting a kernel, which is usually edible: e.g., oak, chestnut, peanut, etc.
2.  **Aggregate fruits** are derived from many separate ovaries of a single flower, all attached to a single receptacle: e.g., strawberry.
3.  **Multiple (or composite)** fruits are the enlarged ovaries of several flowers grown together into a single mass: e.g., pineapple, fig.

Vegetables are soft edible parts of plants that are consumed by humans. The vegetables can be grouped as (1) bulky vegetative organs (roots, tubers and bulbs), (2) leafy succulent tissues (leafy, floral and stem) and (3) fruit vegetables (mature and immature) (Kader, 2002). Besides the botanical classification, fruits can be grouped, as climacteric and non-climacteric, according to their respiratory outputs and ethylene production. This type of classification is more important than the botanical one for improving the postharvest quality of F&V. Climacteric fruits can ripen after harvest, whereas non-climacteric fruits cannot. Fruit ripening is genetically programmed and is a complex process, which is generally associated with the loss of chlorophyll, softening, increase in soluble solids concentration, decrease in acidity, production of ethylene and development of odor and flavor (Kader, 1999). Maturity has two important definitions in postharvest knowledge:

- **Physiological maturity** refers to the development stage of the commodities when maximum growth and maturation have occurred but when the commodity has not reached marketable quality but has the capacity to ripen after harvest.
- **Horticultural maturity** refers to any development stage where the commodity has completed all the prerequisites for use by consumers and has developed marketable appearance and edibility.

Understanding the ripening and climacteric response of the F&V is very important because:

- Ripening is the beginning of senescence. So, since we aim to store the F&V for longer durations, we should harvest the climacteric F&V at physiological maturity and store them in controlled conditions to stop/delay ripening. On the other hand, we should harvest non-climacteric F&V that cannot ripen after harvest at horticultural maturity.
- F&V at physiological maturity generally have long storage life but are not horticulturally mature for marketing.
- Physiologically immature fruits cannot ripen or poorly ripen if detached from the plant, which results in an unacceptable commodity for consumers. These F&V are also highly susceptible to shriveling, bitter pit, superficial scald and decay.
- Harvesting climacteric fruits late in the maturation process when ripening has begun will limit storage life due to softening, low acidity and high susceptibility to chilling injury.

Climacteric fruit ripening is characterized by an increased rate of respiration and ethylene biosynthesis during ripening (Pech et al., 2008). Climacteric fruits show a slight decrease in respiration rate (preclimacteric minimum), shortly after harvest and an increase (climacteric rise) up to the climacteric maximum (respiratory peak: fully ripe, peak of edible ripeness) and a decrease again (postclimacteric decline: fruits are very susceptible to deterioration). Some important examples of climacteric fruits are apples, avocados, bananas, guava, kiwi, papaya, pears, peaches and tomato. The non-climacteric fruits have a decreasing rate of respiration and no or slight change in ethylene production. These fruits do not ripen after harvest and should be harvested at the right time, when they are fully ripe. Some important examples for this group of fruits are berries (raspberry, strawberry, cherry and blackberry), the citrus family, cucumber, eggplant, grape, loquat, pineapple and pomegranate. Some examples of climacteric and non-climacteric F&V are given in Table 1.1 (Paul et al., 2012; Osorio & Fernie, 2013; Watson et al., 2015).

Another important classification of fruits and vegetables, in terms of postharvest handling, is chilling sensitivity. Since harvested F&V are alive, respire and

**TABLE 1.1**

**Climacteric Characteristics and Chilling Sensitivity of Some Fruits and Vegetables**

| Chilling Sensitivity/ Climacteric Response | Climacteric | Non-climacteric |
|---|---|---|
| **Chilling-sensitive** | Avocado (*Persea americana*) | Bean (snap) (*Phaseolus vulgaris*) |
| | Banana (*Musa* spp.) | Citrus (*Citrus* spp.) |
| | Guava (*Psidium guajava*) | Cucumber (*Cucumis sativus*) |
| | Mango (*Mangifera indica*) | Eggplant (*Solanum melongena*) |
| | Papaya (*Carica papaya*) | Jujube (*Ziziphus* spp.) |
| | Passion fruit (*Passiflora edulis*) | Okra (*Abelmoschus esculentus*) |
| | Tomato (*Solanum lycopersicum*) | Olive (*Olea europaea*) |
| | | Pepper (*Capsicum* spp.) |
| | | Pineapple (*Ananas comosus*) |
| | | Pomegranate (*Punica granatum*) |
| | | Potato (*Solanum tuberosum*) |
| | | Pumpkin (*Cucurbita* spp.) |
| | | Squash (*Cucurbita* spp.) |
| | | Watermelon (*Citrullus lanatus*) |
| **Non-chilling-sensitive** | Apple (*Malus domestica*) | Artichoke (*Cynara cardunculus* var. *scolymus*) |
| | Apricot (*Prunus armeniaca*) | Asparagus (*Asparagus officinalis*) |
| | Fig (*Ficus carica*) | Bean (lima) (*Phaseolus lunatus*) |
| | Kiwifruit (*Actinidia* spp.) | Broccoli (*Brassica oleracea* var. *italica*) |
| | Nectarine (*Prunus persica* var. *nucipersica*) | Brussels sprouts (*Brassica oleracea* var. *gemmifera*) |
| | Peach (*Prunus persica*) | Cabbage (*Brassica oleracea* var. *capitata*) |
| | Pear (*Pyrus communis*) | Carrot (*Daucus carota*) |
| | Plum (*Prunus domestica*) | Cauliflower (*Brassica oleracea* var *botrytis*) |
| | | Cherry (*Prunus avium*) |
| | | Garlic (*Allium sativum*) |
| | | Grape (*Vitis vinifera*) |
| | | Lettuce (*Lactuca sativa*) |
| | | Onion (*Allium cepa*) |
| | | Radish (*Raphanus sativus*) |
| | | Spinach (*Spinacia oleracea*) |
| | | Strawberry (*Fragaria* × *ananassa*) |
| | | Sweet corn (*Zea mays*) |
| | | Turnip (*Brassica rapa* subsp. *rapa*) |

transpire, keeping them in cold rooms is very important for reducing both respiration and transpiration. However, some F&V are very sensitive to low but nonfreezing temperatures (0–12°C). Most crops of tropical and subtropical origin and some crops of temperate zones are sensitive or susceptible to chilling injury. The tissues of susceptible products weaken at these temperatures since they cannot continue their normal metabolic processes. Generally, the symptoms of chilling injury (CI) do not occur at these low temperatures but appear when removed from the chilling temperature to room temperature. The most important symptoms of chilling injury are surface pitting, discoloration, internal breakdown, failure to ripen, wilting, flavor loss and susceptibility to decay (Wang, 1989). Therefore, knowing the chilling sensitivity of the crops is very important for the correct selection of the postharvest handling practices. There are several methods for reducing/eliminating chilling injury in some fruits and vegetables. The details are discussed in other sections of this book. The names of some applicable methods according to Belwal et al. (2020) are:

- Temperature (low/high) conditioning,
- Intermitted warming,
- Waxing and/or coating,
- Controlled atmosphere storage,
- Modified atmosphere packaging,
- UV irradiation,
- Chemical applications, i.e., salicylic acid and methyl jasmonate

Some examples of chilling-sensitive and non-chilling-sensitive products, together with their climacteric characteristics, are given in Table 1.1.

### 1.2.1.2 Dynamics of Fruit Growth and Ripening

The main role of fruits is the production, protection and dispersal of seeds. Development of fruit can be divided into several stages after pollination: cell division, cell expansion, maturation, ripening and senescence. During ripening, the seeds become mature, while the fruit becomes soft. At that time, the soluble solids concentrate, and pigments and aroma volatiles increase while the acidity decreases (in general).

Hereafter, fruits become ripe and, if not harvested, over-ripe; cell structures deteriorate and make the fruit susceptible to pathogen infections (Atkinson et al., 2013). During the development of fruit, the ovary wall becomes a pericarp. The three morphologically distinct strata on fruits are exocarp (fruit skin), mesocarp (fruit flesh) and endocarp (inner cell layers). The exocarp has an important role in protecting fruit from external factors (pests and diseases), in which it develops a cuticle and/or hairs. The exocarp also helps to restrict the movement of gaseous and water vapor, which reduces respiration and transpiration and helps to improve the storability of the fresh fruits. The mesocarp tissues generally represent the fleshy part of fruit. The endocarps may develop as a dense hard layer around the seed in some fruits (i.e., apricot and peach).

From fertilization to maturity, the mass or volume of fruit can increase by more than 100-fold. The growth rate of a fruit (like plants) is not always the same. It is sometimes slow and sometimes rapid. This is a typical S-shaped curve (sigmoid growth curve). Some fruits, even from the same family, may exhibit a single sigmoid pattern (e.g., pear), while some others exhibit a double sigmoid pattern (e.g., peach and strawberry) (Pei et al., 2020). During the first weeks of fruit development, the growth is mainly stimulated by cell division, and flesh volume increases rapidly while the embryo volume remains small. Then the fruit growth slows down in many fruits until the end of the first sigmoid phase. At this time, the seed starts to form with the fertilized embryo and development of endosperm. After this first phase and the slowdown of the growth, a second phase begins with the growth and enlargement of the pericarp. This phase is generally driven by cell expansion. As the fruit ripens and becomes edible, several processes occur. This can be on- or off-tree (for climacteric fruits). Some of the well-known and general changes taking place in fruits during ripening are (Atkinson et al., 2013):

- Increase in soluble solids concentration (and so sweetness).
- Change in color (mainly loss of chlorophyll and increase in yellow, orange, red or purple pigments).
- Textural changes, mainly loss of firmness and increase in softening.
- Development and accumulation of aroma volatiles.
- Loss of acidity and accumulation of some organic acids.

### 1.2.2  ENVIRONMENTAL/CULTURAL FACTORS

Environmental factors affect F&V development through several processes including cell cycle duration, photosynthesis, respiration, transpiration, phloemic transport and metabolism (Benkeblia et al., 2011). The effect of the environment on these processes help to determine fruit size, external and internal fruit quality attributes and storability. Important environmental factors that affect fruit quality and storability include nutrients, soil, light, day length, temperature, relative humidity, irrigation, pruning, canopy position, rootstock and tree age.

### 1.2.2.1  Nutrients

Excess or deficiency of macro- and micronutrients affect the chemical composition of a crop, which also affects the F&V's quality and postharvest storability. For example, Lieten and Marcelle (1993) reported that the higher rates of nitrogen and potassium and increased ratio of nitrogen:calcium or potassium:calcium cause albinism (colorless fruits) in strawberry fruits. The decrease in the availability of nutrients in growing media results in higher physiological disorders in plants (Kumar & Kumar, 2016). Although there are variations in different plants, the general knowledge suggests that the increase in nitrogen content reduces the storability of the F&V (Thompson, 2008). Phosphorus, the other important macronutrient for plants, is also important for postharvest quality of F&V, which can affect membrane integrity and respiratory metabolism (Knowles et al., 2001). It was suggested that low phosphorus causes an increase in the respiration rate of cucumbers (Knowles et al., 2001). Potassium also positively impacts the storability of F&V, while some evidence suggests an increase in crop acidity as a response to high potassium rates and hence a decrease in crop storability (Thompson, 2008). A deficiency of calcium may also cause some disorders in stored fruits, i.e., bitter pits in apples (Atkinson et al., 2013). Calcium is involved in the stabilization of cell wall (binding to the pectin present in the cell walls) structures (Demarty et al., 1984), which improves the storability of the F&V.

### 1.2.2.2 Soil

The soil matrix is a living resource. For a healthy soil matrix, it is recommended to include about 45% mineral particles, 25% air, 25% water and 5% organic material. The mineral particles of soil are not living; however, due to the organic materials and living biodiversity in the matrix, the soil must be considered a living resource. The organic materials of the soil matrix, soil structure, water and nutrient content play a significant role in plant growth and development. For example, plant growth is restricted in compacted soil (due to the restrictions of water and nutrient uptake), which results in stunted growth, poor quality and higher susceptibility to postharvest disorders (Kumar & Kumar, 2016). Similar problems may occur in sandy and low-organic matter-containing soil types, which retain less water and nutrients as compared with fertile soils (Sainju et al., 2003). Soil also indirectly affects postharvest quality and storability of F&V by affecting nutrient, water and oxygen availability for roots.

### 1.2.2.3 Light

Light is the source of life on the earth. It warms our planet and has fundamental functions in several chemical, biochemical and physical processes, including the best known: photosynthesis. Therefore, light intensity and quality significantly affect plant growth and development and thus the quality of F&V due to the impact on photosynthesis rate and plant stress. Low light intensity reduces carbon fixation, which depletes the internal carbohydrate resources (Benichou et al., 2018). This may lead to abnormal cell metabolism and the loss of the postharvest life of F&V. On the other hand, high light intensity may also cause damage by causing plants to close stomata or generating some secondary metabolites that may result in cell damage (Kumar & Kumar, 2016). This can be associated with water stress or temperature stress. This type of stress increases the susceptibility of F&V to postharvest disorders. Sunlight also causes sunburn on several crops (i.e., pomegranate). The intense solar radiation alone or together with high temperatures cause sunburn on the pericarp (Kahramanoğlu & Usanmaz, 2016). Sunburn may cause damage in cell membranes and cell death, and it inhibits photosynthesis. Exposure to high sunlight causes an increase in the production of reactive oxygen species (ROS), which causes oxidative damage and induces senescence (Liu et al., 2022).

### 1.2.2.4 Day Length (Photoperiod)

The response of living organisms to changes in day length is photoperiodism, which enables plants to adapt to seasonal changes in their environment. Day length significantly impacts the flowering, bud dormancy, growth and development of F&V (Thomas, 2003). Different plants have evolved or have been bred for growing under certain day lengths. There are three categories of plants, depending on how they respond to day length: short day, long day and day neutral plants. A short-day plant requires less than 12 hours of sunlight (or more than 12 hours of uninterrupted darkness), and a long-day plant requires more than 12 hours of sunlight (or less than 12 hours of uninterrupted darkness) for flowering. Day neutral plants bloom flowers regardless of day length and the amount of light they receive. Besides the impacts on flowering, day length may also impact the correct maturation of the crops.

### 1.2.2.5 Temperature

Preharvest atmospheric temperature is a major factor influencing growth, development and productivity of F&V, mainly due to the impacts on photosynthesis rate. Photosynthesis has optimal limits for each plant and may be significantly impacted by lower or higher temperatures (Hikosaka et al., 2006). Quality characteristics such as shape, size and color are affected by temperature. Arpaia et al. (2004) reported that more rounded fruits were produced in avocado cultivar Harvest N4–5 in cooler environments of southern California when compared to the warmer environments of the San Joaquin Valley, California, which resulted in more elongated fruits. It is also known that the high temperature may cause stress on plants, which can be followed by the production and accumulation of heat shock proteins (Kotak et al., 2007). These heat shock proteins may be associated with susceptibility to postharvest disorders.

### 1.2.2.6 Relative Humidity

Relative humidity (RH) is the ratio of water vapor in a particular water–air mixture compared to the saturation humidity ratio at a given temperature. The rate of transpiration (loss of water vapor through the stomata of plants) significantly impacts the nutrient uptake of plants by enhancing the mass-flow of nutrients (Houshmandfar et al., 2018). Relative humidity has a significant impact on plant transpiration. Under low levels of RH, there is a higher driving factor for transpiration. If there is high relative humidity, then the driving factor for transpiration is reduced. Therefore, if the RH is too high, the transpiration and thus the nutrient removal from soil decrease.

   Low RH (high vapor pressure deficit) improves nutrient uptake to the leaves and fruit by the driving factor of transpiration. However, if the RH decreases too much, the leaf transpiration may be very high as compared to the fruit transpiration and may cause xylem sap backflow from fruit to leaves (de Freitas et al., 2014). This may impact the nutrient uptake, the fruit quality and postharvest storability. In addition, low or high RH may cause poor pollination, which can cause a reduction in fruit set and/or fruit quality (Sandhu, 2013; Sandhu et al., 2018).

### 1.2.2.7 Irrigation and Water Availability

Water is essential for plant growth and development for several reasons, including photosynthesis, uptake and transport of nutrients, transport of sugar, cooling and cell structural support. Hence water should be available during the growth and development of the plants. Scientific knowledge recommends that the water stress, mainly during fruit development, results in reduced fruit size (Barman et al., 2015). However, correct management of the water stress may sometimes be beneficial. For example, Crisosto et al. (1994) recommended that slight water stress before harvest may increase the soluble solids concentration of peach fruits, which may improve storability. Similar results were noted by Cui et al. (2008) for pear-jujube fruits, where the water deficit irrigation increased the SSC:TA ratio.

   Water scarcity and availability of freshwater are rapidly growing concerns around the globe (Kummu et al., 2016) and are among the most important limiting factors for crop production (Kahramanoğlu et al., 2020). Scientists are testing and adopting several irrigation practices to reduce water use and improve crop growth. Deficit irrigation is among the best known and adopted practices for improving water use efficiency (Ruiz-Sanchez et al., 2010). Deficits are generally applied when the sensitivity of plants to water stress is minimal. Most commonly, the deficit irrigation is applied after harvest, but it may also be applied before harvest to regulate some biochemical quality parameters of F&V. Two separate studies with Mollar de Elche pomegranate cultivar reported that sustained (32% evapotranspiration) deficit irrigation (Pena et al., 2013) and sustained (50% evapotranspiration) deficit irrigation (Laribi et al., 2013) significantly minimized water loss during storage as compared to fully irrigated fruits. Similar findings were noted by Romero-Trigueros et al. (2017) for grapefruit (cv. star ruby) under regulated (50% evapotranspiration) deficit irrigation. Deficit (50% evapotranspiration) irrigation was also noted to increase both SSC and TA contents of apricots and improve their storability (Pérez-Pastor et al., 2007). However, it must be kept in mind that the decrease in water availability during growth and development increases the susceptibility of F&V to the postharvest disorders (Kumar & Kumar, 2016). Studies by Bower et al. (1988) support this knowledge, where the water stress during the first 3 months after the fruit set of avocado fruits was reported to increase fruit browning potential and sensitivity to pathogens.

### 1.2.2.8   Pruning and Canopy Position

As explained in Section 1.2.2.3, light is very important for plant growth, development and the postharvest quality of fruits and vegetables. Therefore, the right pruning for good light penetration could help to improve fruit size, SSC content, phenolic content and reduce fruit TA. The fruits that develop in shaded places would have less flavor than the ones that receive more light. On the other hand, wrong pruning or the wrong timeline for pruning (for example, heavy pruning in summer) may reduce the energy of the tree and thus the carbohydrate level of the fruits. Studies of Agabbio et al. (1999) showed that the oranges located at the southern side of the tree canopy have higher SSC than the others. In a similar study, Nilsson and Gustavsson (2007) reported that the apples located at the outside of the canopy have higher SSC than the fruits located inside. This is associated with canopy position and light exposure.

### 1.2.2.9   Rootstock

A rootstock is a plant, with a root system, on which another compatible variety (scion) can be grafted or budded. Rootstocks play a very important role in horticultural practices. They help to combine the desirable characteristics of two different plants. The main advantages of rootstocks are improving plants' tolerance to biotic (pests, diseases, etc.) and abiotic (salinity, drought, flooding, contamination, high temperature, etc.) conditions and improving plants' growth and development by enhancing flowering and nutrition uptake and increasing yield and fruit quality (Nimbolkar et al., 2016). In short, rootstocks improve the adaptability of horticultural crops into different conditions. Therefore, the selection of the right rootstock is crucial for improving the harvest quality and storability of the fresh F&V.

### 1.2.2.10   Tree Age

Young trees are known to have bigger-sized and better-quality fruits. One good example was reported by Tahir et al. (2007), where young apple trees (less than 4 years old) were noted to have better flavor (SSC/TA) and color. Similar findings were noted for pomegranate fruits by Kahramanoğlu and Usanmaz (2016). The young trees were noted to have bigger size and better flavored fruits. The storability of the higher-quality fruits was also reported to be higher.

### 1.2.3 Crop Production Practices

Crop production practices such as soil fertility, propagation method, irrigation and pest management can influence the postharvest quality of fruits and vegetables (some already explained). Having the right balance of nutrients in the soil throughout the growing season is important to support optimum performance of crops, to prevent physiological disorders, diseases and to ensure quality produce. Researchers and producers are constantly investigating the appropriate types and quantities of fertilizers and times of application for crops to ensure that maximum yield and quality can be achieved. It is important to note that the level of soil fertility that contributes to improved postharvest quality may not be the same soil nutrient levels that produce the highest yields (Ladaniya, 2008; Sams, 1999).

### 1.2.4 Diseases (Preharvest Infections)

Many postharvest diseases are the result of pathogen infections that occur during crop cultivation. Therefore, preharvest treatments can be applied to prevent or minimize those diseases. Most of those treatments involve chemical applications or a combination of chemical and physical methods during the cultivation of the fruits or vegetables. However, several biological control strategies are emerging and show great promise in controlling postharvest diseases. The application of preharvest chitosan spray was reported to be effective during the postharvest storage of strawberry and to protect the fruits against infection and deterioration (Reddy et al., 2000). The postharvest rotting of sweet cherries was controlled by preharvest applications of *Aureobasidium pullulans* in combination with calcium chloride or sodium bicarbonate (Ippolito et al., 2005). Besides the control measures against pathogen infections, crop hygiene is also very important for the prevention of the infections. This can be done by removing infected plant parts and the effective control of pest hosts (i.e., weeds).

## 1.3 HARVESTING FACTORS

### 1.3.1 Quality Attributes

The quality attributes of F&V vary during growth and development. They refer to external attributes (color, size, shape, appearance and freedom from pathogens), internal attributes (color, flavor, internal defects, texture, juice content and number of seeds) and hidden attributes (safety and nutritional values) (Kahramanoğlu, 2017). Among these quality attributes, size and external color has been the most important for consumers for many years. However, the developments in science and the occurrence of health problems caused an increase in consumer awareness, which resulted in consumers paying more attention to safety and nutritional values. Therefore, the acceptability of F&V is now highly influenced by hidden quality attributes.

### 1.3.2 Maturity/Harvesting Indices

The stage of maturity at harvest greatly influences the composition, quality, losses and postharvest life of fresh F&V. To achieve the maximum postharvest life, the optimum harvest maturity must be met, but it also depends on the purpose for which the fruit or vegetable will be used. For example, the optimum maturity stage considered best for canning may not be best for dehydration, freezing or making jams or preserves. Generally, most fruits reach peak eating quality when harvested fully ripe, but some fruits (mainly the climacteric fruits) are usually harvested at the mature, unripe stage to decrease mechanical injury during postharvest handling. Fruits harvested at the immature stage are more prone to shriveling and mechanical damage and are of inferior quality when ripened while those harvested when overripe are likely to become soft and mealy with insipid flavor soon after harvest (Mishra & Gamage, 2007; Kader & Yahia, 2011). Maturity indices have been developed for many fruits and vegetables to help ensure that harvesting is done at the correct stage for best eating quality, balanced with the need to provide flexibility in marketing. Some important maturity indices used commercially include fruit size, shape, external and/or internal color, firmness, soluble solids concentration, titratable acidity (TA), juice content and SSC/TA ratio. The quality characteristics that need to be considered before harvest and their optimal limits are highly variable among the F&V species and among the different varieties/cultivars of same species. Some indicative values for selected F&V are given in Table 1.2.

### 1.3.3 Harvesting Methods

The postharvest quality and shelf life of fruits and vegetables are greatly affected by harvesting methods. When harvesting is not done properly, F&V get mechanical injuries such as bruising and surface abrasions, which leads to rapid decomposition. There are several factors that influence the harvesting method used for harvesting fruits and vegetables. These include maturity, delicacy of crop, climatic conditions, packing material, transportation facilities, and availability of skilled labor and size of the production operation. The two broad categories of harvesting methods used for fruits and vegetables are manual and mechanical harvesting. The choice of methods is influenced by several factors including the type of crop, size of operation and stage of maturity.

#### 1.3.3.1 Manual (Hand) Harvesting

Manual harvesting involves harvesting of F&V by hand, using fruit clippers, knives, or by pole mounted "cut and hold" picking shears. Manual harvesting is more feasible for produce that is consumed fresh and is particularly important for fruits that are harvested at various stages of maturity and need several pickings during the production season. Workers involved in harvesting should be trained to harvest and handle the fruits in such a way that minimum damage is done. Strawberry is among the most important examples for manual harvesting. Fruits bearing on trees, like citrus, apples, peach and mangoes, are also traditionally harvested by hand. For this reason, either a ladder should be carried from tree-to-tree or a picking platform can be used to enable the movement of harvesters among the trees. Although there is a mechanical

## TABLE 1.2
### Things to Consider Before Harvesting and Maturity Indices for Selected Fruits and Vegetables*

| Product | Things to Consider Before Harvesting | Maturity Indices | Additional Quality Indices |
|---|---|---|---|
| **Fruits** | | | |
| Apple (*Malus domestica*) | Visual appearance (skin color) Flesh firmness Flavor (SSC/TA) Starch content | In general, should be harvested before fully mature for longer storage Depending on the cultivar/variety, color ranges from green or yellow to red. Starch degradation should begin. Firmness (i.e., 53+ N for Golden Delicious, 80–98 N for Granny Smith) Acidity (i.e., 0.2–0.4% for Red Delicious, 0.8–1.2% for Granny Smith) SSC (i.e., 20%+ for Fuji) | Free from defects (decay, blemish, bruising, etc.) |
| Apricot (*Prunus armeniaca*) | Skin color Firmness Flavor (SSC/TA) | Skin color change from green to yellow Fruits should be firm SSC: ≥10% TA: 0.7–1.0% | Fruit size and shape Free from defects and decay |
| Avocado (*Persea americana*) | Dry matter and oil content | Dry matter may vary between 17–30% (depending on cultivar) Oil content: min. 8%, may be over 20% Keeping on tree may cause off-flavor when overmature If harvested in hot weather, again off-flavors may develop. | Fruit size and shape Skin color Free from sunburn, blemish and browning Free from diseases, mostly anthracnose |
| Banana (*Musa* spp.) | Skin color | Commercially harvested when mature-green, stored/transported and ripened under control conditions. Starch turns into sugar when ripe | Free from defects (i.e., scars, physical damage, insect injury) and decay |

**TABLE 1.2 (Continued)**
**Things to Consider Before Harvesting and Maturity Indices for Selected Fruits and Vegetables**

| Product | Things to Consider Before Harvesting | Maturity Indices | Additional Quality Indices |
|---|---|---|---|
| Cactus pear (*Opuntia* spp.) | Fruit size and fullness<br>Fruit firmness<br>Flavor (SSC/TA) | External color changes from green to cultivar color (red or yellow)<br>Abscission of the glochids (tufts of small spines)<br>SSC: 12–17%<br>TA: 0.03–0.12% | Intensity of the cultivar color<br>Uniformity<br>Size<br>Free from decay and defects |
| Cherry (*Prunus avium*) | Skin color<br>SSC | Cultivar-dependent: light-red color<br>SSC: 14–16% | Free from cracks, shriveling and decay<br>Green fleshy stems represent freshness |
| Date plum (*Diospyros lotus*) | Firmness<br>SSC<br>Tannin | Generally harvested when fully mature at Rutab and Tamar stages<br>High sugar (60–80%) and less tanning content | Fruit size<br>Free from sunburn, decay and insect |
| Fig (*Ficus carica*) | Skin color and flesh firmness | Fruits should be harvested when fully mature | Free from defects, sunburn, decay etc. |
| Grape (*Vitis vinifera*) | SSC | SSC: 14–17%<br>SSC/TA: ≥20 (might be used as a criterion) | Free from injury, decay, sunscald, cracking<br>Should appear fully turgid.<br>Rachis should be green. |
| Grapefruit (*Citrus × paradisi*) | Skin color<br>Minimum bitterness<br>SSC/TA | Should be harvested when fully mature<br>More than 60% of the skin should have yellow color<br>SSC/TA: min. 5.5 | Free from blemish<br>Peel should be turgid.<br>Size and shape<br>Color intensity and uniformity |
| Guava (*Psidium guajava*) | Skin color<br>SSC and TA | Color changes from dark to light green<br>Should be harvested at the mature-green stage<br>SSC and TA: 3% and 0.2% when green, changes to 10% and 1.5% when ripe, respectively | Color, size (9–12 cm) and shape<br>Free from defects, decay and insects<br>Less seeds, the better quality |
| Kiwi (*Actinidia* spp.) | SSC | Min. 6.5% at harvest and min. 14% at table, when ripe<br>Flesh firmness: 9–13 N | Free from cracks, bruises, scars, internal breakdown, and decay |

| | | | |
|---|---|---|---|
| Lemon (*Citrus limon*) | Juice content<br>Skin color | Juice content should be at least 25%.<br>If harvested at fully yellow, must be marked quickly and should be harvested at dark-green for longer storage. | Skin color<br>Size and shape<br>Free from decay and damage |
| Loquat (*Eriobotrya japonica*) | Skin color<br>Firmness<br>SSC<br>TA | Skin color change from green to yellow or orange color<br>SSC: ≥10%<br>TA: 0.3–0.6% | Free from defects and decay |
| Lychee (*Litchi chinensis*) | Skin color and size<br>Seed size and juiciness<br>SSC/TA | Do not ripen off-tree<br>Bright-red color is preferred.<br>Size should be minimum 25 mm in diameter<br>SSC/TA: ≥30 | Bright-red fruits<br>Free from browning, cracking and decay |
| Mandarin (*Citrus reticulata*) | SSC/TA | SSC/TA: ≥6.5<br>More than ¾ of the surface color should turn to cultivar color (yellow, orange or red) | Free from decay<br>Size and shape<br>Color<br>Firmness |
| Mango (*Mangifera indica*) | Skin color<br>Fruit shape | Shape changes, the cheeks become fullness<br>Skin color change from dark to light green<br>Flesh color change from greenish yellow to yellow | Free from defects and decay<br>Skin coloration<br>Uniformity in shape |
| Nectarine (*Prunus persica* var. *nucipersica*) | Skin color<br>Firmness<br>SSC | Skin color changes from green to yellow<br>Firmness: 9–13 N | Firmness less than 30 N<br>Free from decay |
| Olive (*Olea europaea*) | Size<br>Color | Right color for the cultivar and type (green or black) | Free from mechanical damage, shriveling, insect damage and decay<br>Oil content, right for the cultivar (12–30%) |
| Orange (*Citrus sinensis*) | SSC/TA<br>Color intensity and uniformity | SSC/TA: ≥8 if yellow-orange color exceeds ≥25%<br>SSC/TA: ≥10 if green-yellow color exceeds ≥25% | Free from decay, defect, and blemishes |

(Continued)

**TABLE 1.2 (Continued)**
**Things to Consider Before Harvesting and Maturity Indices for Selected Fruits and Vegetables**

| Product | Things to Consider Before Harvesting | Maturity Indices | Additional Quality Indices |
|---|---|---|---|
| Papaya (*Carica papaya*) | Skin color SSC | Skin color changes from dark to light green<br>Flesh color changes from green to yellow or red<br>SSC: ≥11.5% | Uniform size and color<br>Free from defects, sunburn, insect injury and decay |
| Pear (*Pyrus communis*) | Firmness | Juicy texture and aroma<br>Fruit firmness: 50–70 N depending on cultivars/varieties | Free from internal breakdown, insect damage and defects<br>Right firmness |
| Pineapple (*Ananas comosus*) | Skin color SSC TA | Shell color changes from green to yellow (at the base)<br>SSC: minimum 12%<br>TA: maximum 1% | Uniform size and shape<br>Free from decay, sunburn, internal breakdown and insect damage |
| Plum (*Prunus domestica*) | Skin color Firmness | Firmness: 9–13 N are best for eating. | Free from bruising and decay |
| Pomegranate (*Punica granatum*) | Skin color SSC TA | External red color (depending on the cultivar/variety)<br>SSC: between 8–21%<br>TA: less than 1% for sweet, 1–2% for sweet-sour and more than 2% for sour types | Free from cracks, sunburn and decay<br>Skin color and smoothness |
| Quince (*Cydonia oblonga*) | Skin color Firmness | Skin color changes from green to yellow<br>Fruits should be firm at harvest | Free from defects and decay |
| Strawberry (*Fragaria × ananassa*) | Skin color SSC TA | Minimum 2/3 of the fruit surface should be red or pink (depending on the cultivar)<br>SSC: minimum 7%<br>TA: maximum 0.8% | Free from decay |
| Watermelon (*Citrullus lanatus*) | Skin color Fruit sound SSC | Mature fruit sounds dull or hollow when thumped with the knuckles<br>Tendrils nearest the fruit may turn brown and dry during maturity | Well formed and uniform in shape, having a waxy, bright appearance<br>Free from decay, anthracnose, injury, bruising, sunburn, and scars |

**Vegetables**

| Vegetables | Maturity indices | Description | Quality characteristics |
|---|---|---|---|
| Artichoke (Globe) (*Cynara cardunculus* var. *scolymus*) | Globe size<br>Bracts color and size | The outer bracts should be tightly closed, firm, turgid and typical green. | Free from insect damages, defects and mechanical damages |
| Asparagus (*Asparagus officinalis*) | Spear height<br>Spear tip | Spears can be harvested after emergence but are generally picked at 10–25 cm length<br>Spear tips should be tightly closed. | Straight, tender and glossy spears are more preferred. |
| Beans (Snap) (*Phaseolus vulgaris*) | Pod color<br>Seed size<br>Days after flowering | Pod color should turn to a bright color and should be fleshy.<br>Seeds should be small and green.<br>Pods are generally ready for harvest, 8–10 days after flowering. | Should be straight, bright and firm.<br>Free from decay and chilling damage<br>Pod diameter is more important than length.<br>Beans should break easily when the pod is bent. |
| Broccoli (*Brassica oleracea* var. *italica*) | Head diameter and compactness | Heads should be dark and bright color at harvest.<br>Heads should be compact.<br>No discoloration on the stem and no yellow florets. | Dark or bright green color with closed flower buds is preferred.<br>Head should be firm. |
| Cabbage (*Brassica oleracea* var. *capitata*) | Head compactness and weight | Compact head, which can be only slightly compressed with moderate hand pressure<br>Very loose heads are immature, whereas the very firm heads are mature. | Free from decay<br>Leaves should be crisp and turgid. |
| Carrot (*Daucus carota*) | Length, diameter and size | Some markets prefer a diameter >1.8 cm. | Should be firm and straight.<br>Free from cracking or sprouting |
| Cauliflower (*Brassica oleracea* var. *botrytis*) | Head diameter and compactness | Generally, the heads are preferred to be >15 cm in diameter. | Heads should be firm and compact.<br>Free from decay, mechanical damage, browning and yellowing<br>Leaves, around the heads, should be turgid.<br>Absence of riceyness |

(*Continued*)

**TABLE 1.2 (Continued)**
**Things to Consider Before Harvesting and Maturity Indices for Selected Fruits and Vegetables**

| Product | Things to Consider Before Harvesting | Maturity Indices | Additional Quality Indices |
|---|---|---|---|
| Celery (*Apium graveolens*) | Stalk length | Desired stalk length is about 35–41 cm. Harvest should be performed before the development of pithiness at outer petioles. | Well formed petioles should be thick, compact, tender and light green in color. Free from defects, i.e., blackheart, seed stalks, cracks and insect damage |
| Cucumber (*Cucumis sativus*) | Size Firmness | Fruit firmness and external glossiness are the main indicators of harvest maturity. Size, depending on the cultivar, is important. | Dark green and firm, without pits or wrinkled ends Uniform shape and firmness are important. Free from decay and yellowing |
| Eggplant (*Solanum melongena*) | Seed development Size Firmness External glossiness | Fruits should be harvested immature, before hardening and enlargement of the seeds. | Free from defects and handling |
| Garlic (bulbs) (*Allium sativum*) | Above ground (top) plant parts | Fallen or dry tops are the most important indicator of the harvest time. SSC: above 35% | Cloves should be firm upon touch. |
| Lettuce (romaine) (*Lactuca sativa*) | Head development and number of leaves | Loose or compressible head is immature, while the firm head is over mature. | Bright to dark green color Free from insect damages, mechanical damages, and decay |
| Okra (*Abelmoschus esculentus*) | Pod size Color Days after flowering | Pods should be bright green and fleshy with small seeds during harvest. Typically, harvest maturity is reached 3–7 days after flowering. | Pod length between 5 to 15 cm is preferred. Pods should be flexible, bright-green and turgid. Free from blackening and bruising |
| Onion (dry) (*Allium cepa*) | Above ground (top) plant parts | When nearly 50–80% of tops are fallen over, buds are mature. | Free from mechanical damage, decay, sprouting, bruising and bottleneck. Bulb size |

| Crop | Maturity indices | Notes | Optimal values |
|---|---|---|---|
| Pea (*Pisum sativum*) | Pod size<br>Seed development | Frequent harvesting might be necessary.<br>Some cultivars/varieties should be harvested before seed development (snow peas), while some others (sugar snap peas) should be harvested after the development of visible seeds. | Bright green and fully turgid peas are preferred.<br>Free from defects and mechanical damage |
| Pepper (bell) (*Capsicum annuum*) | Color and size | Color peppers should have at least 50% coloration. | Free from decay, sunburn and cracks |
| Potato (*Solanum tuberosum*) | Sugar content<br>Size<br>Above ground (top) plant parts | Top senescence is used a preharvest quality indicator. | Tuber should be turgid and well shaped.<br>Free from adhering soil, mechanical damage, greening, sprouting and diseases |
| Radish (*Raphanus sativus*) | Size | Size should be checked regularly. Depending on the cultivar/variety, the minimum size should be 2 cm in diameter. | Radish should be tender, firm and crisp.<br>Free from harvest cuts, abrasions, insect damage, soil or other foreign material |
| Spinach (*Spinacia oleracea*) | Size and color | Spinach leaves are harvested at mid-maturity.<br>Older and yellowing leaves are avoided.<br>Regrowth of the spinach may require 3–4 weeks for second harvest. | Uniformly green and fully turgid leaves are preferred. |
| Squash (*Cucurbita* spp.) | Size<br>Seed development<br>Days after flowering | Depending on the cultivar/variety and temperature, days from flowering to harvest may be 45–60 for zucchini and 75+ for immature gourds.<br>Fruits can be harvested at desired size before seeds begin to enlarge and harden. | Tenderness and firmness<br>Surface should be shiny. |
| Tomato (*Solanum lycopersicum*) | Color<br>Shape<br>Appearance<br>Firmness | Depending on the marketing goals and cultivar/variety, tomato fruits can be harvested from physiological maturity (mature-green stage) through full-ripe. | Uniform shape and free from defects<br>Firm, turgid appearance and shiny color are preferred. |

* Some examples are shown for maturity indices, but it is cultivar/variety dependent, and right optimal values may vary. The information in this table was adopted from several references (Kitinoja & Kader, 2015; Thompson, 2008; Gross et al., 2016; de Freitas & Pareek, 2019) and authors' knowledge.

operation in the picking platforms, it is also considered manual harvesting because the fruit is removed from the tree by hand. Small vegetables are also harvested in a similar way to that of strawberries. Root crops, like potatoes and onions where the harvester needs to dig the soil, can be harvested by manual or mechanical means which can be performed by garden fork (Thompson, 2016).

### 1.3.3.2   Mechanical Harvesting

Mass removal of the F&V is highly important for quick harvesting, postharvest handling and marketing. Mechanical harvesting includes the systems designed to achieve this goal but is very limited for fresh fruits. Sometimes, if the F&V needs processing (apricot, tomato, grape for wine, peas, peaches and some leafy vegetables), mechanical ways can be followed. Powerful wind machines can be used to harvest oranges for processing. Tree shakers are commonly used for harvesting olives, which mostly goes for olive oil processing (Thompson, 2016). Tools for mechanical harvesting may damage the F&V and reduce the postharvest life of the crops. Thus, a proper selection of the methods is crucial for achieving longer storability. Sometimes, the mechanical harvesters may require support with chemical applications, which can loosen the mature fruits and make harvesting easier (Erkan & Dogan, 2019). Mechanical harvesting is common for root crops (including potato, garlic, onion, carrot, etc.) which are not sensitive to mechanical injuries. The advantages and disadvantages of mechanical harvesting follow:

### Advantages:

- Easily applied with less labor
- Ensuring harvesting of F&V in a short period of time
- Reduced running costs (The cost of equipment can be high in some countries.)

### Disadvantages:

- Possible damage to F&V and reduced postharvest storability and/or marketability
- No control on optimal maturity
- Not suitable for selective or multiple harvesting

According to Erkan and Dogan (2019), mechanical harvesting methods can be grouped into five categories:

- **Limb shaker** –mainly used for processing fruits. This method may cause bark and/or limb damages on trees. Abscission chemicals may need to be applied before using this equipment to improve harvest efficiency.
- **Canopy shaker** –works by vibrating trees and impacting the fruit branches.
- **Trunk shaker** – is mostly used for harvesting olives and nuts. They are mostly operated by a tractor. Trees and fruits may be damaged from this harvesting method.

- **Air blast** – a force-generated air blast used to remove the fruit from the tree.
- **Robotic harvester** – Although mechanization and automation are widely used in agriculture (automatic irrigation and fertigation, identification of pests, pesticide application, pruning, etc.), the use of mechanization and automation in harvesting is limited, mainly due to the complexity of growing environments and objectives of harvesting. Research is still continuing in the design and development of robotic harvesters to pick fruits automatically (Tang et al., 2020). Although there are several unsolved issues for visual harvesting robots, researchers believe that the success can be achieved.

## 1.4 POSTHARVEST PHYSIOLOGICAL PROCESSES

### 1.4.1 PLANT ONTOGENY

Plant ontogeny refers to the origin and development of plant and plant parts, which have a major influence on several physiological processes of fruits and vegetables. F&V undergo different stages of development, with different organs and tissue types developing at their own rates based on genetic and environmental factors affecting their metabolism and impacting their quality. The growth and maturation of F&V, collectively termed the development stage, only occur before harvesting. The five distinctive phases of F&V growth and development are (1) rapid cell division, (2) young or premature, (3) mature, (4) ripening and (5) senescence. The stage of development at which harvesting is done varies among fruits and vegetables with substantial variation in morphology and composition. Harvested F&V with tissues that are in the early stage of development or premature generally have relatively short shelf lives compared to those that are dormant or approaching dormancy (Blakey, 2011). Ripening is the result of several complex physiological and physical changes with distinct anabolic and catabolic processes that require large amounts of energy and prolonged membrane integrity (Bower et al., 1988; Wills et al., 1998). Senescence is the period when catabolic processes exceed anabolic processes, resulting in aging and necrosis (Wills et al., 1998).

### 1.4.2 RESPIRATION

Respiration is a metabolic process that allows living organisms to convert matter into energy. It involves enzymatic oxidation of various substrates such as carbohydrates, proteins, lipids and organic acids in the presence of atmospheric oxygen, resulting in the release of energy, carbon dioxide and water:

$$C_6H_{12}O_6 + 6O_2 \rightarrow 6CO_2 + 6H_2O + Energy$$

The process is generally expressed as the ratio of $CO_2$ production to $O_2$ consumed per unit weight of product per unit time and as heat liberated and substrate loss, as indicated by mass loss (Kandasamy, 2022; Mishra & Gamage, 2007). It also indicates the metabolic turnover and is proportional to the rate of deterioration in a product. Higher respiration rate is associated with faster deterioration, which causes accelerated senescence, mass loss, loss of food value and reduced flavor in the harvested product (Phan et al., 1975; Kays, 1991). However, respiration is necessary to maintain the vigor of plant tissues and provide resistance against spoilage in the postharvest life of fruits and vegetables. It is also responsible for some physiological disorders such as ripening, senescence, browning, molding, degradation of chlorophyll, decay and subsequent deterioration in their regular course of time (Raghavan & Gariepy, 1985; Kader et al., 1989).

Different substrates show differences in their degree of oxidation. The process is exothermic as a significant part of the energy produced is given off as heat, which causes an increase in the temperature of the harvested product (Mishra & Gamage, 2007). The release of heat is the result of many interrelated, simultaneous metabolic processes that occur in the product, some of which utilize respiratory energy (Kandasamy, 2022). Higher respiration rate also results in lower storability and shorter storage duration.

Respiration rate varies among crops and their maturity stage. In general, we aim to reduce the respiration rate of the harvested F&V to increase their storability. For this reason, the tools for us are temperature, atmospheric composition and mechanical damage, which are the three main factors accelerating/decelerating the respiration rate of F&V. For every 10°C increase in temperature there is a two- to threefold increase in biological reactions and respiration (Kahramanoğlu, 2017). Oxygen is among the main requirements of respiration. So the atmospheric composition is crucial to manage (Bovi et al., 2018). Reducing the oxygen concentration surrounding the F&V may reduce the respiration rate, but the elimination of oxygen causes anaerobic respiration and fermentation. In anaerobic fermentation, sugar is broken down into alcohol and carbon dioxide, which results in abnormal flavors in produce and promotes premature aging (Paltrinieri, 2014). Anaerobic fermentation mostly occurs at low oxygen or high carbon dioxide concentrations. The main metabolites of fermentation are ethanol, acetaldehyde and lactic acid, which are responsible for storage disorders, including necrotic or discolored tissues, off-odor and/or off-flavor. Under these conditions (low oxygen or high carbon dioxide), the tissue initiates anaerobic respiration, where the glucose is converted to pyruvate (by the glycolytic pathway) and then metabolized into ethanol, lactic acid or acetaldehyde (Kader et al., 1989). The limits for oxygen concentration varies among plant species and cultivars/varieties.

### 1.4.3 Transpiration and Water Stress

Transpiration is the loss of water or moisture in the form of vapor from living plant tissues. Movement of water from living plant tissue to outside occurs due to the humidity gradient between the internal and external atmospheres of F&V. It is a major postharvest issue because most F&V contain 80–95% of water by weight, and a 5–10% loss in moisture usually results in visible symptoms of deterioration (Kahramanoğlu, 2017; Kader & Yahia, 2011). Symptoms of deterioration in fresh fruits and vegetables due to moisture loss include shriveling, wilting, softening, poorer texture, loss in weight and lower quality. The rate of transpiration is affected by several environmental factors such as temperature, relative humidity, air movement and atmospheric pressure (Kader & Yahia, 2011). Also, internal or commodity factors make some F&V more vulnerable to moisture loss than others, including morphological and anatomical characteristics, surface:volume ratio and surface injuries (Kader & Yahia, 2011). Transpiration can be partially restrained or reduced by the fruit skin and/or the cuticle, which provides a barrier against water vapor loss. The structure, thickness and composition of the fruit cuticle affect this permeability and are highly variable among plant species and cultivars/varieties. Transpiration during postharvest can be controlled or reduced by storing above atmospheric pressure, maintaining low temperature and high relative humidity, reducing air movement, application of protective coverage including waxing and water-resistant coatings and application of protective packaging such as polyethylene film and modified atmosphere packaging (Kahramanoğlu, 2017; Mishra & Gamage, 2007).

### 1.4.4   RIPENING AND SENESCENCE

The life of F&V can be divided into three major physiological stages (with no clear distinctions) after pollination: growth, maturation and senescence. Growth and ripening patterns of climacteric and non-climacteric F&V significantly vary (see Figure 1.1 adapted from Wills and Golding (2016)). Growth and maturation can only occur on the plant, but ripening can be off the plant for climacteric fruits. Fruit ripening follows physiological maturity and is a highly coordinated, genetically programmed and irreversible phenomenon involving a series of physiological, biochemical and organoleptic changes that lead to the development of a soft and edible ripe fruit with desirable quality attributes (Prasanna et al., 2007).

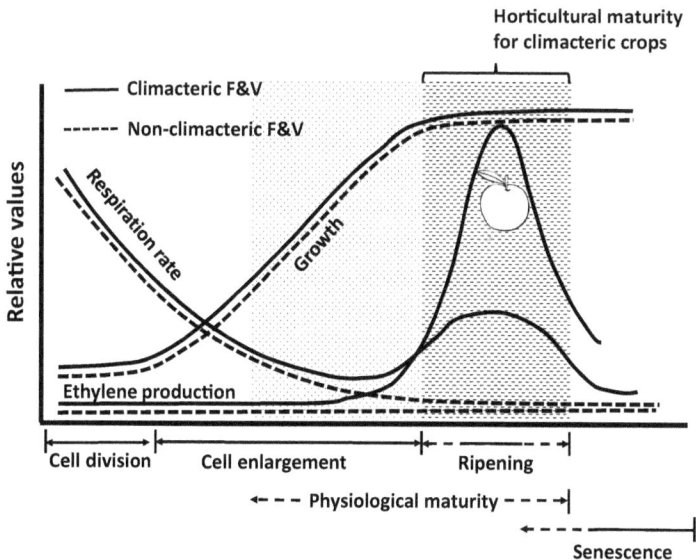

**FIGURE 1.1**   Growth and ripening patterns of climacteric and non-climacteric F&V [adapted from Wills and Golding (2016)].

The ripening process involves several structural, physical, chemical, nutritional, biochemical and enzymatic changes, which maybe degradative or synthetic (Mishra & Gamage, 2007). Degradative changes include chlorophyll breakdown, starch hydrolysis and cell wall degradation, while synthetic changes include formation of carotenoids and anthocyanin, aroma volatiles and ethylene formation. The ripening process is also associated with increased respiration and a transient increase in ethylene production.

The series of events leading to ripening are (1) thickening of cell wall and adhesion, (2) increased permeability of plasmalemma, (3) increased intercellular spaces contributing to softening, (4) changes in plastids, (5) transformation of chloroplasts into chromoplasts, (6) changes in color, (7) loss of texture, (8) formation of a visible and distinctive structure from epicuticular wax, (9) thickening of cuticle, (10) loss of epidermal hairs, and (11) lignification of endocarp (Mishra & Gamage, 2007).

Color changes that occur during ripening are the result of existing pigments becoming visible due to the degradation of chlorophyll, breakdown of the photosynthetic system and synthesis of different types of anthocyanin and their accumulation in vacuoles, and accumulation of carotenoids such as β-carotene, xanthophyll esters, xanthophyll and lycopene (Prasanna et al., 2007). The softening of fruit is the result of textural changes that are due to enzyme-mediated alteration in the structure and composition of the cell wall, partial or complete solubilization of cell wall polysaccharides such as pectin and cellulose, and hydrolysis of starch and other storage polysaccharides (Prasanna et al., 2007). The production of a complex mixture of volatile compounds such as ocimene and myrcene and degradation of bitter principles, flavanoids, tannins and related compounds all contribute to increased flavor during ripening (Prasanna et al., 2007). There is also a general increase in sweetness caused by the increased gluconeogenesis, hydrolysis of polysaccharides, especially starch, decreased acidity and accumulation of sugars and organic acids, resulting in an improved sugar/acid blend (Prasanna et al., 2007).

Some fruit species can continue to ripen after harvesting, known as climacteric fruits, while others, known as non-climacteric fruit, are incapable of ripening any further once detached from the plant. These two distinctive ripening patterns are also based on the respiratory patterns of fruits. Examples of climacteric fruits are avocado, banana, cherimoya, kiwifruit, mango, papaya and persimmon, while some non-climacteric fruits are the citrus species, pineapple and pomegranate. Climacteric fruits produce much larger quantities of ethylene in association with their ripening, and exposure to ethylene treatment will result in faster and more uniform ripening. However, non-climacteric fruits produce very small quantities of ethylene and do not respond to ethylene treatment except in terms of de-greening (removal of chlorophyll) in citrus fruits and pineapples. Ethylene is a hormone that is produced by all tissues of higher plants and some microorganisms. It regulates many aspects of plant growth, development and senescence and is physiologically active in trace amounts (less than 0.1 ppm) and induces fruit abscission, softening and some physiological disorders (Abeles et al., 2012). Both the beneficial and detrimental effects of ethylene hinge on its ability to alter or accelerate the natural processes of development, ripening and senescence.

Ripening is followed by senescence, which is the point where chemical synthesizing pathways stop and give way to degradative processes, leading to the aging and death of tissue. Fruits in storage are on a path to senescence, and several disorders can arise during that time. Senescence is controlled by genetic factors but can be induced by stressors such as injury, deficiency of nutrients and water during production, exposure to insects, pests and diseases, and adverse environmental conditions (Mishra & Gamage, 2007).

## 1.5 POSTHARVEST FACTORS

### 1.5.1 TEMPERATURE

Managing the postharvest temperature of F&V is the most important element of postharvest handling systems in maintaining the quality and safety of fresh produce. Different F&V have different optimum temperature ranges for successful postharvest handling and storage due to their different physicochemical properties. Generally, low temperatures decrease metabolic activity such as respiration and transpiration, reduce the incidence of pathogens and postharvest disease, and reduce insect activity, while high temperatures (up to a certain limit) accelerate all these deterioration factors (Kader & Yahia, 2011; Mishra & Gamage, 2007). According to Kader and Yahia (2011), the rate of deterioration of fruits and vegetables increase two- to fourfold for every increase of 10°C. Therefore, it is important to identify the specific optimal temperature range for each produce that will ensure the slowing down of deterioration and afford the opportunity to manipulate ripening during storage.

Postharvest managers are increasingly investing in precision temperature management tools, such as radio frequency identification (RFID) tags and time–temperature monitors to help them in their postharvest handling systems (Kader & Siddiq, 2012). Self-contained temperature and relative humidity (RH) monitors and recorders allow for data collection on produce by simple connecting these units to a personal computer with the appropriate software provided by the manufacturer (Kader & Siddiq, 2012). The use of infrared thermometers makes it easy to measure the surface temperature of fruits and vegetables for various postharvest processes and throughout the postharvest chain (Karabulut & Baykal, 2002). Electronic thermometers are used for measuring product temperature during cooling, storage and transport operations (Benichou et al., 2018).

### 1.5.2 RELATIVE HUMIDITY

Relative humidity (RH) can be considered the amount of moisture content in the atmosphere and is expressed as a percentage of the amount of moisture that can be retained by the atmosphere (Yahia, 2019). RH is closely related to temperature and is usually expressed at a given temperature and pressure without condensation. The optimum RH ranges for most fruits and vegetables during storage is 85–95% (Mishra & Gamage, 2007; Kader & Yahia, 2011).

Since most F&V contain a high percentage of water, they continue to lose water to the surrounding atmosphere after harvesting. This loss of moisture can be observed as shriveling, wilting, loss of crispness and toughness or mushiness, which makes F&V unacceptable to consumers or unsalable even at a weight loss of 5% (El-Ramady et al., 2015). Very high RH (above 95%) and water condensation on fruits can accelerate pathogen attack, impair the ripening process, cause decay and weaken packaging materials (Kader & Yahia, 2011; Kahramanoğlu, 2017; Mishra & Gamage, 2007). RH is usually monitored and measured along with temperature with devices such as self-contained temperature and relative humidity (RH) monitors and recorders.

### 1.5.3 ATMOSPHERIC GAS COMPOSITION

The postharvest physiological processes of F&V are strongly influenced by oxygen ($O_2$), carbon dioxide ($CO_2$), nitrogen ($N_2$) and ethylene ($C_2H_4$) in the postharvest atmosphere. The balance and concentrations of these gases in the postharvest environment can significantly affect the storage life of fruits and vegetables (Wu, 2010). Generally, a reduction of oxygen below 5% and increase of carbon dioxide above 3% delay deterioration of fresh fruits and vegetables due to a reduction in respiration rate of fresh produce, retardation of senescence and growth inhibition of many spoilage microorganisms (Wu, 2010; Kahramanoğlu, 2017). These conditions, along with maintaining low temperature, must be met in modified or controlled atmosphere storage, but it is highly dependent on the type of commodity, cultivar and maturity (Mishra & Gamage, 2007).

### 1.5.4 LIGHT

Light may also affect some postharvest physiological processes and cause some abnormal changes in F&V quality (Kahramanoğlu, 2017). First, it plays a role in controlling the synthesis/degradation of pigments responsible for color (chlorophyll and carotenoids). For example, the exposure of potatoes (*Solanum tuberosum*) to light during storage may result in the formation of chlorophyll, which appears as greening and the formation of solanine, which is toxic to humans (Patil et al., 1971). Light may also contribute to flavor development by catalyzing the oxidation of lipids, as well as degrading vitamins such as ascorbic acid and riboflavin (Mishra & Gamage, 2007). To prevent the adverse effect of light, light intensity should be minimized in the postharvest environment, or fruits and vegetables should be stored in the dark or packaged in materials that prevent light transmission.

### 1.5.5 MECHANICAL INJURY

Fruits and vegetables may experience mechanical injury in several ways, generally through poor handling practices. When fruits and vegetables have mechanical injuries, their internal tissue becomes exposed to contamination, increases the respiration rate, becomes susceptible to chemical and enzymatic reactions (i.e., browning), which allows the spread of decay microorganisms, and induces an overall quality decline (Mishra & Gamage, 2007). If injuries are caused in the early stages of development, some fruits and vegetables may repair and seal off the damaged area, but for most cases, the capacity of wound healing diminishes as the plant organs mature.

## 1.5.6 POSTHARVEST DISEASES OR INFECTIONS

Diseases cause significant postharvest losses in F&V, and these losses may occur at any time throughout the postharvest handling system. Postharvest diseases are often classified according to the three main ways that infections are initiated: (1) at the early stage of development when attached to the plant, (2) by direct penetration of certain fungi or bacteria through the intact cuticle or through wounds or natural openings in the surface and (3) through injuries in cut stems or damage to the surface (Mishra & Gamage, 2007; Coates & Johnson, 1997). Some postharvest diseases or infections can start as early as fruit set and continue until harvest and during storage. In fact, while environmental conditions influence the incidence and epidemiology of diseases in the field, many postharvest diseases are caused by latent infection that initiate in the field during the growing season (Michailides et al., 2009; Coates & Johnson, 1997). Latent infections play a major role in both the incidence and severity of postharvest diseases, and, when conditions are favorable, the incidence and severity of latent infections are higher and the risk for postharvest disease development increase and vice versa (Michailides et al., 2009). While most microorganisms can only invade damaged fruits and vegetables, a few are able to penetrate the skin of healthy produce. This usually starts with only one or a few pathogens invading and breaking down the tissues, followed by a broad-spectrum attack of several weak pathogens, resulting in the complete loss of the commodity due to the magnified damage (Mishra & Gamage, 2007; Singh & Sharma, 2018).

Most postharvest diseases of F&V are caused by fungi and bacteria. Fungi are very adaptable to a wide range of environments, which allows them to grow and flourish under various storage conditions. Many of the fungi that cause postharvest disease belong to the phylum Ascomycota, which consist of over 57,000 species. They are non-mobile, cellular organisms, whose structure is composed by threads called hyphae, and they reproduce by spores. However, the asexual stage (the anamorph) is usually encountered more frequently in postharvest diseases than the sexual stage (the teleomorph) (Coates & Johnson, 1997). Important genera of anamorphic postharvest pathogens include *Penicillium*, *Aspergillus*, *Geotrichum*, *Botrytis*, *Fusarium*, *Alternaria*, *Colletotrichum*, *Dothiorella*, *Lasiodiplodia* and *Phomopsis*. The genera *Phytophthora* and *Pythium* from the phylum Oomycota are important postharvest pathogens, causing several diseases such as brown rot in citrus (*Phytophthora citrophthora* and *P. parasitica*) and cottony leak of cucurbits (*Pythium* spp.) (Coates & Johnson, 1997). From the phylum Zygomycota the two most important genera in terms of postharvest diseases are *Rhizopus* and *Mucor*. *Rhizopus stolonifera* is one of the most common fungi in the world with global distribution, although it is more common in tropical and subtropical regions and affects a wide range of fruits and vegetables.

The most important postharvest bacterial pathogens belong to the genera *Erwinia*, *Pseudomonas*, *Bacillus*, *Lactobacillus*, *Xanthomonas* and *Pectobacterium*. Bacteria are usually a greater concern for vegetables than for fruit because they have low tolerance to the low pH associated with most fruits.

Different approaches can be used to manage postharvest diseases. The first approach should be to prevent infection by providing the right postharvest environment and proper handling and hygiene practices that help to main host resistance. Second, any sign of infection should be eradicated where possible to prevent spread. Where eradication is not practical or feasible, infections should be retarded by using fungicide or bactericides as appropriate.

## 1.6 CONCLUSION

Several factors affect the postharvest management and storage of fruits and vegetables. These factors can be grouped broadly into preharvest/production factors, harvesting factors and postharvest factors. The quality of the final produce is highly dependent on decisions made throughout the entire production and postharvest value chain. Correct selection of the handling practices, based on physiological knowledge, improves the storability of F&V.

## REFERENCES LIST

Abeles, F. B., Morgan, P. W., & Saltveit Jr., M. E. (2012). *Ethylene in plant biology.* Academic Press.

Agabbio, M., Lovicu, G., Pala, M., D'hallewin, G., Mura, M., & Schirra, M. (1999). Fruit canopy position effects on quality and storage response of "Tarocco" oranges. *Acta Horticulturae*, (485), 19–24. https://doi.org/10.17660/ActaHortic.1999.485.1

Arpaia, M.L., Van Rooyen, Z., Bower, J.P., Hofman, P.J., & Woolf, A.B. (2004). *Grower practices will influence postharvest fruit quality.* II International Avocado Conference Quillota, Chile, pp. 1–9. Retrieved October 29, 2022, from https://shorturl.at/myBX2.

Atkinson, R. G., David A. B., Jeremy N. B., Kevin J. P., & Robert J. S. (2013). Chapter 11 – Fruit growth, ripening and post-harvest physiology. In B. Atwell, P. Kriedemann, & C. Turnbull (Eds.), *Plants in action* (1st ed.). Macmillan Publisher. Retrieved October 29, 2022, from https://shorturl.at/bhnqI

Barman, K., Ahmad, M.S., & Siddiqui, M.W. (2015). Factors affecting the quality of fruits and vegetables: Recent understandings. In M.W. Siddiqui (Ed.), *Postharvest biology and technology of horticultural crops: Principles and practices for quality maintenance* (pp. 1–50). Apple Academic Press.

Belwal, P., Barman, K., & Yadav, N. (2020). Postharvest chilling injury in fruits and vegetables and its alleviation. *Agriculture and Food e-Newsletter*, 2(10), 171–172. Retrieved October 29, 2022, from https://shorturl.at/lrwJ8.

Benichou, M., Ayour, J., Sagar, M., Alahyane, A., Elateri, I., & Aitoubahou, A. (2018). Postharvest technologies for shelf-life enhancement of temperate fruits. In S.A. Mir, M.A. Shah, & M.M. Mir (Eds.), *Postharvest biology and technology of temperate fruits* (pp. 77–100). Springer. https://doi.org/10.1007/978-3-319-76843-4_4

Benkeblia, N., Tennant, D.P.F., Jawandha, S.K., & Gill, P.S. (2011). Preharvest and harvest factors influencing the postharvest quality of tropical and subtropical fruits. In E.M. Yahia (Ed.), *Postharvest biology and technology of tropical and subtropical fruits* (pp. 112–142e). Woodhead Publishing. https://doi.org/10.1533/9780857093622.112

Blakey, R.J. (2011). *Management of avocado postharvest physiology* (PhD Dissertation (Supervisor Prof. Dr. John P. Bower)). University of KwaZulu-Natal.

Bovi, G.G., Rux, G., Caleb, O.J., Herppich, W.B., Linke, M., Rauh, C., & Mahajan, P.V. (2018). Measurement and modelling of transpiration losses in packaged and unpackaged strawberries. *Biosystems Engineering*, *174*, 1–9. https://doi.org/10.1016/j.biosystemseng.2018.06.012

Bower, J.P., Cutting, J.G.M., & Wolstenholme, B. N. (1988). Effect of pre- and post-harvest water stress on the potential for fruit quality defects in avocado (Persea americana Mill.). *South African Journal of Plant and Soil*, 6(4), 219–222. https://doi.org/10.1080/025718 62.1989.10634516

Castle, W. S. (1995). Rootstock as a fruit quality factor in citrus and deciduous tree crops. *New Zealand Journal of Crop and Horticultural Science*, 23(4), 383–394. https://doi.org/10. 1080/01140671.1995.9513914

Coates, L., & Johnson, G. (1997). Postharvest diseases of fruit and vegetables. In J.F. Brown & H.J. Ogle (Eds.), *Plant pathogens and plant diseases* (pp. 533–548). Retrieved October 29, 2022, from www.appsnet.org/Publications/Brown_Ogle/

Crisosto, C.H., Johnson, R. S., Luza, J. G., & Crisosto, G. M. (1994). Irrigation regimes affect fruit soluble solids concentration and rate of water loss of O'Henry' peaches. *Hortscience*, 29(10), 1169–1171. https://doi.org/10.21273/HORTSCI.29.10.1169

Cui, N., Du, T., Kang, S., Li, F., Zhang, J., Wang, M., & Li, Z. (2008). Regulated deficit irrigation improved fruit quality and water use efficiency of pear-jujube trees. *Agricultural Water Management*, 95(4), 489–497. https://doi.org/10.1016/j.agwat.2007.11.007

Daley, O., Robert-Nkrumah, L.B., & Alleyne, A.T. (2016). Sensory and instrument assessment of colour and texture among breadfruit [*Artocarpus altilis* (Parkinson) Fosberg] cultivars. *Tropical Agriculture. Special Issue: International Breadfruit Conference*, 93(1), 92–108. Retrieved October 22, 2022, from https://shorturl.at/oJOR4.

de Freitas, S. T., McElrone, A. J., Shackel, K. A., & Mitcham, E. J. (2014). Calcium partitioning and allocation and blossom-end rot development in tomato plants in response to whole-plant and fruit-specific abscisic acid treatments. *Journal of Experimental Botany*, 65(1), 235–247. https://doi.org/10.1093/jxb/ert364

de Freitas, S.T., & Pareek, S. (2019). *Postharvest physiological disorders in fruit and vegetables*. CRC Press, Taylor & Francis Group. https://doi.org/10.1201/b22001

Demarty, M., Morvan, C., & Thellier, M. (1984). Calcium and the cell wall. *Plant, Cell and Environment*, 7(6), 441–448. https://doi.org/10.1111/j.1365-3040.1984.tb01434.x

El-Ramady, H.R., Domokos-Szabolcsy, É., Abdalla, N.A., Taha, H. S., & Fári, M. (2015). Postharvest management of fruits and vegetables storage. In E. Lichtfouse (Ed.), *Sustainable agriculture reviews*, 15 (pp. 65–152). Springer International Publishing. https://doi.org/10.1007/978-3-319-09132-7_2

Englberger, L., Alfred, J., Lorens, A., & Iuta, T. (2007). Screening of selected breadfruit cultivars for carotenoids and related health benefits in Micronesia. *Acta Horticulturae*, 757, 193–200. https://doi.org/10.17660/ActaHortic.2007.757.26

Englberger, L., Wills, R. B., Blades, B., Dufficy, L., Daniells, J. W., & Coyne, T. (2006). Carotenoid content and flesh color of selected banana cultivars growing in Australia. *Food and Nutrition Bulletin*, 27(4), 281–291. https://doi.org/10.1177/156482650602700401

Erkan, M., & Dogan, A. (2019). Harvesting of horticultural commodities. In E.M. Yahia (Ed.), *Postharvest technology of perishable horticultural commodities* (pp. 129–159). Woodhead Publishing. https://doi.org/10.1016/B978-0-12-813276-0.00005-5

Ferreira, E. A., Siqueira, H. E., Boas, E. V. V., Hermes, V. S., & Rios, A. D. O. (2016). Bioactive compounds and antioxidant activity of pineapple fruit of different cultivars. *Revista Brasileira de Fruticultura*, 38(3). https://doi.org/10.1590/0100-29452016146

Gross, K.C., Wang, C.Y., & Saltveit, M. (2016). *The commercial storage of fruits, vegetables, and florist and nursery stocks (No. 66)*. United States Department of Agriculture. Agricultural Research Service.

Hikosaka, K., Ishikawa, K., Borjigidai, A., Muller, O., & Onoda, Y. (2006). Temperature acclimation of photosynthesis: Mechanisms involved in the changes in temperature dependence of photosynthetic rate. *Journal of Experimental Botany*, 57(2), 291–302. https://doi.org/10.1093/jxb/erj049

Houshmandfar, A., Fitzgerald, G. J., O'Leary, G., Tausz-Posch, S., Fletcher, A., & Tausz, M. (2018). The relationship between transpiration and nutrient uptake in wheat changes under elevated atmospheric $CO_2$. *Physiologia Plantarum, 163*(4), 516–529. https://doi.org/10.1111/ppl.12676

Ippolito, A., Schena, L., Pentimone, I., & Nigro, F. (2005). Control of postharvest rots of sweet cherries by pre- and postharvest applications of *Aureobasidium pullulans* in combination with calcium chloride or sodium bicarbonate. *Postharvest Biology and Technology, 36*(3), 245–252. https://doi.org/10.1016/j.postharvbio.2005.02.007

Kader, A. A. (1999). Fruit maturity, ripening, and quality relationships. In *International symposium effect of pre-& postharvest factors in fruit storage* (Vol. 485, pp. 203–208). https://doi.org/10.17660/ActaHortic.1999.485.27

Kader, A.A. (2002). *Postharvest technology of horticultural crops* (3rd ed.). Regents of the University of California, Division of Agriculture and Natural Resources 94608.

Kader, A.A., & Siddiq, M. (2012). Introduction and overview. In *Tropical and Subtropical Fruits: Postharvest Physiology, Processing and Packaging* (pp. 1–16). https://doi.org/10.1002/9781118324097.ch1

Kader, A.A., & Yahia, E.M. (2011). Postharvest biology of tropical and subtropical fruits. In E.M. Yahia (Ed.), *Postharvest biology and technology of tropical and subtropical fruits* (pp. 79–111). Woodhead Publishing. https://doi.org/10.1533/9780857093622.79

Kader, A.A., Zagory, D., & Kerbel, E.L. (1989). Modified atmosphere packaging of fruits and vegetables. *Critical Reviews in Food Science and Nutrition, 28*(1), 1–30. https://doi.org/10.1080/10408398909527490

Kahramanoğlu, İ. (2017). Introductory chapter: Postharvest physiology and technology of horticultural crops. In İ. Kahramanoğlu (Ed.), *Postharvest handling* (pp. 1–5). https://doi.org/10.5772/intechopen.69466

Kahramanoğlu, I., & Usanmaz, S. (2016). *Pomegranate production and marketing.* CRC Press. https://doi.org/10.1201/b20151

Kahramanoğlu, İ., Usanmaz, S., & Alas, T. (2020). Water footprint and irrigation use efficiency of important crops in Northern Cyprus from an environmental, economic and dietary perspective. *Saudi Journal of Biological Sciences, 27*(1), 134–141. https://doi.org/10.1016/j.sjbs.2019.06.005

Kandasamy, P. (2022). Respiration rate of fruits and vegetables for modified atmosphere packaging: A mathematical approach. *Journal of Postharvest Technology, 10*(1), 88–102. Retrieved August 15, 2023, from https://shorturl.at/oEHM8

Karabulut, O.A., & Baykal, N. (2002). Evaluation of the use of microwave power for the control of postharvest diseases of peaches. *Postharvest Biology and Technology, 26*(2), 237–240. https://doi.org/10.1016/S0925-5214(02)00026-1

Kays, S. J. (1991). Science and practice of postharvest plant physiology. In *Postharvest physiology of perishable plant products* (pp. 1–22). Springer Publisher.

Kitinoja, L., & Kader, A. A. (2015). *Small-scale postharvest handling practices: A manual for horticultural crops.* University of California, Davis, Postharvest Technology Research. Retrieved October 29, 2022, from https://rb.gy/wonv0

Knowles, L., Trimble, M.R., & Knowles, N.R. (2001). Phosphorus status affects postharvest respiration, membrane permeability and lipid chemistry of European seedless cucumber fruit (*Cucumis sativus* L.). *Postharvest Biology and Technology, 21*(2), 179–188. https://doi.org/10.1016/S0925-5214(00)00144-7

Kotak, S., Larkindale, J., Lee, U., von Koskull-Döring, P., Vierling, E., & Scharf, K. D. (2007). Complexity of the heat stress response in plants. *Current Opinion in Plant Biology, 10*(3), 310–316. https://doi.org/10.1016/j.pbi.2007.04.011

Kumar, R., & Kumar, V. (2016). Physiological disorders in perennial woody tropical and subtropical fruit crops-A review. *Indian Journal of Agricultural Sciences, 86*, 703–717.

Kummu, M., Guillaume, J. H. A., de Moel, H., Eisner, S., Flörke, M., Porkka, M., Siebert, S., Veldkamp, T. I. E., & Ward, P. J. (2016). The world's road to water scarcity: Shortage and stress in the 20th century and pathways towards sustainability. *Scientific Reports*, 6(1), 1–16. https://doi.org/10.1038/srep38495

Ladaniya, M. S. (2008). Preharvest factors affecting fruit quality and post harvest life. In M. S. Ladaniya (Ed.), *Citrus fruit, biology, technology and evaluation* (pp. 79–102). Elsevier, Inc. Retrieved October 22, 2022, from https://shorturl.at/kGQRY.

Laribi, A. I., Palou, L., Intrigliolo, D. S., Nortes, P. A., Rojas-Argudo, C., Taberner, V., Bartual, J., & Pérez-Gago, M. B. (2013). Effect of sustained and regulated deficit irrigation on fruit quality of pomegranate cv. 'Mollar de Elche' at harvest and during cold storage. *Agricultural Water Management*, 125, 61–70. https://doi.org/10.1016/j.agwat.2013.04.009

Lieten, F., & Marcelle, R. D. (1993). Relationships between fruit mineral content and the "albinism" disorder in strawberry. *Annals of Applied Biology*, 123(2), 433–439. https://doi.org/10.1111/j.1744-7348.1993.tb04105.x

Liu, C., Su, Y., Li, J., Jia, B., Cao, Z., & Qin, G. (2022). Physiological adjustment of pomegranate pericarp responding to sunburn and its underlying molecular mechanisms. *BMC Plant Biology*, 22(1), 169. https://doi.org/10.1186/s12870-022-03534-8

Michailides, T. J., Morgan, D. P., & Luo, Y. (2009). Epidemiological assessments and postharvest disease incidence. In D. Prusky & M. L. Gullino (Eds.), *Postharvest pathology* (pp. 69–88). Springer. https://doi.org/10.1007/978-1-4020-8930-5_6

Mishra, V. K., & Gamage, T. V. (2007). Postharvest physiology of fruit and vegetables. In M. S. Rahman (Ed.), *Handbook of food preparation* (pp. 37–66). CRC Press. https://doi.org/10.1201/9780429091483

Nilsson, T., & Gustavsson, K. (2007). Postharvest physiology of "Aroma" apples in relation to position on the tree. *Postharvest Biology and Technology*, 43(1), 36–46. https://doi.org/10.1016/j.postharvbio.2006.07.011

Nimbolkar, P. K., Awachare, C., Reddy, Y. T. N., Chander, S., & Hussain, F. (2016). Role of rootstocks in fruit production – A review. *Journal of Agricultural Engineering and Food Technology*, 3(3), 183–188. Retrieved August 15, 2023, from https://shorturl.at/iqG37.

Osorio, S., & Fernie, A. R. (2013). Biochemistry of fruit ripening. *Molecular Biology and Biochemistry of Fruit Ripening*, 1–19. https://doi.org/10.1002/9781118593714.ch1

Paltrinieri, G. (2014). *Handling of fresh fruits, vegetables and root crops: A training manual for Grenada.* Food and Agriculture Organization of the United Nations.

Patil, B. C., Salunkhe, O. K., & Singh, B. (1971). Metabolism of solanine and chlorophyll in potato tubers as affected by light and specific chemicals. *Journal of Food Science*, 36(3), 474–476. https://doi.org/10.1111/j.1365-2621.1971.tb06391.x

Paul, V., Pandey, R., & Srivastava, G. C. (2012). The fading distinctions between classical patterns of ripening in climacteric and non-climacteric fruit and the ubiquity of ethylene – An overview. *Journal of Food Science and Technology*, 49(1), 1–21. https://doi.org/10.1007/s13197-011-0293-4

Pech, J. C., Bouzayen, M., & Latché, A. (2008). Climacteric fruit ripening: Ethylene-dependent and independent regulation of ripening pathways in melon fruit. *Plant Science*, 175(1), 114–120. https://doi.org/10.1016/j.plantsci.2008.01.003.

Pei, M. S., Cao, S. H., Wu, L., Wang, G. M., Xie, Z. H., Gu, C., & Zhang, S. L. (2020). Comparative transcriptome analyses of fruit development among pears, peaches, and strawberries provide new insights into single sigmoid patterns. *BMC Plant Biology*, 20(1), 108. https://doi.org/10.1186/s12870-020-2317-6

Pena, M. E., Artés-Hernández, F., Aguayo, E., Martínez-Hernández, G. B., Galindo, A., Artés, F., & Gómez, P. A. (2013). Effect of sustained deficit irrigation on physicochemical properties, bioactive compounds and postharvest life of pomegranate fruit (cv. 'Mollar de Elche'). *Postharvest Biology and Technology*, 86, 171–180. https://doi.org/10.1016/j.postharvbio.2013.06.034

Pérez-Pastor, A., Ruiz-Sánchez, M. C., Martínez, J. A., Nortes, P. A., Artés, F., & Domingo, R. (2007). Effect of deficit irrigation on apricot fruit quality at harvest and during storage. *Journal of the Science of Food and Agriculture, 87*(13), 2409–2415. https://doi.org/10.1002/jsfa.2905

Phan, C. T., Pantastico, E. B., Ogata, K., & Chachin, K. (1975). Respiration and respiratory climacteric. In E. B. Pantastico (Ed.), *Postharvest physiology, handling and utilization of tropical and subtropical fruit and vegetables* (pp. 86–102). AVI Publishing Company Inc.

Prasanna, V., Prabha, T. N., & Tharanathan, R. N. (2007). Fruit ripening phenomena–an overview. *Critical Reviews in Food Science and Nutrition, 47*(1), 1–19. https://doi.org/10.1080/10408390600976841

Raghavan, G., & Gariepy, Y. (1985). Structure and instrumentation aspects of storage systems. *Acta Horticulturae, 157*, 5–30. https://doi.org/10.17660/ActaHortic.1985.157.1

Ramaswamy, H. S. (2014). *Postharvest technologies of fruits and vegetables*. DEStech Publications, Inc.

Reddy, M. B., Belkacemi, K., Corcuff, R., Castaigne, F., & Arul, J. (2000). Effect of preharvest chitosan sprays on post-harvest infection by Botrytis cinerea and quality of strawberry fruit. *Postharvest Biology and Technology, 20*(1), 39–51. https://doi.org/10.1016/S0925-5214(00)00108-3

Romero-Trigueros, C., Parra, M., Bayona, J. M., Nortes, P. A., Alarcón, J. J., & Nicolás, E. (2017). Effect of deficit irrigation and reclaimed water on yield and quality of grapefruits at harvest and postharvest. *LWT – Food Science and Technology, 85*, 405–411. https://doi.org/10.1016/j.lwt.2017.05.001

Ruiz-Sanchez, M. C., Domingo, R., & Castel, J. R. (2010). Review. Deficit irrigation in fruit trees and vines in Spain. *Spanish Journal of Agricultural Research, 8*(S2), 5–20. https://doi.org/10.5424/sjar/201008S2-1343

Sainju, U. M., Dris, R., & Singh, B. (2003). Mineral nutrition of tomato. *Journal of Food, Agriculture and Environment, 1*(2), 176–183. https://doi.org/10.1234/4.2003.361

Sams, C. E. (1999). Preharvest factors affecting postharvest texture. *Postharvest Biology and Technology, 15*(3), 249–254. https://doi.org/10.1016/S0925-5214(98)00098-2

Sandhu, S. (2013). *Physiological disorders of fruit crops* (p. 189). New India Publishing Agency.

Sandhu, S., Singh, J., Kaur, P., & Gill, K. (2018). Heat stress in field crops: Impact and management approaches. In *Advances in crop environment interaction* (pp. 181–204). Retrieved August 15, from https://link.springer.com/chapter/10.1007/978-981-13-1861-0_7.

Singh, D., & Sharma, R. R. (2018). Chapter 1. Postharvest diseases of fruits and vegetables and their management. In M. W. Siddiqui (Ed.), *Postharvest disinfection of fruits and vegetables* (pp. 1–52). Academic Press. https://doi.org/10.1016/B978-0-12-812698-1.00001-7

Tahir, I. I., Johansson, E., & Olsson, M. E. (2007). Improvement of quality and storability of apple cv. Aroma by adjustment of some pre-harvest conditions. *Scientia Horticulturae, 112*(2), 164–171. https://doi.org/10.1016/j.scienta.2006.12.018

Tang, Y., Chen, M., Wang, C., Luo, L., Li, J., Lian, G., & Zou, X. (2020). Recognition and localization methods for vision-based fruit picking robots: A review. *Frontiers in Plant Science, 11*, 510. https://doi.org/10.3389/fpls.2020.00510

Thomas, B. (2003). Regulators of growth | Photoperiodism. In B. Thomas, B., D. J. Murphy, & B. G. Murray (Eds.), *Encyclopedia of applied plant sciences* (pp. 1077–1084). Academic Press.

Thompson, A. K. (2008). *Fruit and vegetables: Harvesting, handling and storage*. Blackwell Publishing.

Thompson, A. K. (2016). *Fruit and vegetable storage: Hypobaric, hyperbaric and controlled atmosphere*. Springer. https://doi.org/10.1007/978-3-319-23591-2

Wang, C. Y. (1989). Chilling injury of fruits and vegetables. *Food Reviews International*, *5*(2), 209–236. https://doi.org/10.1080/87559128909540850

Wang, J., Allan, A. C., Wang, W. Q., & Yin, X. R. (2022). The effects of salicylic acid on quality control of horticultural commodities. *New Zealand Journal of Crop and Horticultural Science*, *50*(2–3), 99–117. https://doi.org/10.1080/01140671.2022.2037672

Watson, J. A., Treadwell, D., Sargent, S. A., Brecht, J. K., & Pelletier, W. (2015). *Postharvest storage, packaging and handling of specialty crops: A guide for Florida small farm producers*. University of Florida.

Wills, R., & Golding, J. (2016). *Postharvest: An introduction to the physiology and handling of fruit and vegetables*. University of New South Wales Press.

Wills, R., McGlasson, B., Graham, D., & Joyce, D. (1998). *Postharvest: An introduction to the physiology and handling of fruit, vegetables and ornamentals* (4th ed). University of New South Wales Press.

Wu, C. T. (2010). *An overview of postharvest biology and technology of fruits and vegetables*. Technology on Reducing Post-harvest Losses and Maintaining Quality of Fruits and Vegetables Proceedings of 2010 AARDO Workshop.

Yahia, E. M. (2019). Chapter 1. Introduction. In E. M. Yahia (Ed.), *Postharvest technology of perishable horticultural commodities* (pp. 1–41). Woodhead Publishing. https://doi.org/10.1016/B978-0-12-813276-0.00001-8

# 2 Biological and Environmental Factors Affecting Postharvest Quality of Fruits and Vegetables

*Afiya John, Wendy-Ann Isaac, Oral Daley, and İbrahim Kahramanoğlu*

## 2.1 INTRODUCTION

The postharvest quality of fruits, vegetables, cereals and grains is highly impacted/regulated by several biological and environmental factors that can accelerate or decelerate the deterioration of freshly harvested products, thus reducing shelf life and marketability. Freshly harvested horticultural products and flowers are living organisms, in which respiration and transpiration occur. These two processes cause deterioration of the harvested products (Kahramanoglu, 2017). A wide range of biological agents affect fresh produce and include insects, fungi, viruses and bacteria (Farrell et al., 2002). Stored produce, fruits, seeds and tubers are dormant and are physiologically different from growing plants, and once they are kept in an unsterile environment, biological agents may attack them. In addition, several environmental factors accelerate or decelerate these biological factors (respiration, transpiration and pathogens) and can impact the quality of postharvest crops such as relative humidity, atmospheric composition and temperature. Postharvest crops contain essential organic acids, vitamins, sugars, minerals and antioxidants that constitute an essential component of daily human diets (Chen et al., 2021). Moreover, ethylene (the ripening hormone), which is biologically synthesized by almost all plants (from a negligible amount to very high concentrations) significantly impacts the postharvest quality of fresh produce. It can also be synthetically produced and exogenously applied to plants/products (Kahramanoglu, 2017).

Postharvest losses that result from the difficulty in preventing biological occurrences can be categorized as qualitative or quantitative. Freshly harvested crops are susceptible to attack by secondary pests and microorganisms due to the ease of entry. Crops with blemishes, insect attacks and discoloration are rejected by consumers and postharvest handlers, and they contribute to qualitative losses. Quantitative losses result from the direct consumption of the stored product by birds, rodents and insects as well as from the quick and widespread deterioration brought on by microbial

DOI: 10.1201/9781003452355-3

action. Since some insects show a preference for eating the seed's germ region, which can reduce the nutritional content of cereal grains and the viability of seeds, the feeding patterns of primary pests may result in quality losses (Farrell et al., 2002).

Fruits and vegetables (F&V) are living things, factors such as temperature, atmospheric composition, relative humidity during and after harvest, and the type and severity of microbial or insect infection all have a significant impact on marketability. These products lose moisture stored energy, nutrients and vitamins and become degraded due to pathogens, rodents; they also suffer physical losses from pest and disease attacks, as well as a loss in quality brought on by physiological disorders, fiber development, greening (potatoes), root growth, sprouting and seed germination during storage (Singh & Sharma, 2018).

## 2.2 BIOLOGICAL FACTORS

### 2.2.1 INTERNAL BIOLOGICAL FACTORS

#### 2.2.1.1 Respiration

Plants, fruits, and vegetables, even after harvesting (being detached from the plant), continue to respire throughout their postharvest life. Respiration together with transpiration are the two most important biological factors, limiting the storage duration and shelf life of fresh horticultural products after harvest. Respiration (biological oxidation) is the oxidative breakdown of stored carbohydrates, starch and/or organic acids, with the help of oxygen, into simpler end products: carbon dioxide, water and, most importantly, the energy (adenosine triphosphate, ATP) (Kader & Saltveit, 2002). During respiration, the stored carbohydrates are continually broken down into energy to keep the fresh products alive. Therefore, the breakdown of carbohydrates, "the food", leads to mass loss and negatively affects plant flavor, turgor and nutritional value. In other words, it can be said that the quality and storability of freshly harvested products have a negative correlation with the respiration rate. An increase in the respiration rate reduces the storability of fruits and vegetables. Together with energy, about 673 kcal of heat for each mole of sugar (180 g) is produced during respiration. This heat is the main reason for the requirements for refrigeration during storage and transport (Kader & Saltveit, 2002).

Respiration rate significantly varies among the different fruits and vegetables, where the ones with a faster respiration rates are more perishable. According to Kader and Saltveit (2002), dried fruits and nuts, garlic, onion, mature potato, taro and pumpkin have very low respiration rates (<5 ml $CO_2$ $kg^{-1}$ $h^{-1}$ at 5°C), whereas beet, cabbage, celery, cucumber, immature potato and tomato have a low rate of respiration (5–10 ml $CO_2$ $kg^{-1}$ $h^{-1}$ at 5°C). These products have higher storability than the others due to the low respiration rate. Several products are reported to have a moderate rate of respiration (11–20 ml $CO_2$ $kg^{-1}$ $h^{-1}$ at 5°C), i.e., cauliflower, eggplant, leaf lettuce and okra. On the other hand, some fruits and vegetables have high (21–30 ml $CO_2$ $kg^{-1}$ $h^{-1}$ at 5°C) (artichokes, beans, green onions, spinach, etc.) and extremely high (>30 ml $CO_2$ $kg^{-1}$ $h^{-1}$ at 5°C) (asparagus, broccoli, peas, parsley, sweet corn, etc.) respiration rates and low storability. Detailed information about the storage temperatures, rates of respiration, ethylene production and recommended atmospheric compositions are given in Table 2.1.

## TABLE 2.1
Recommended Optimum Temperature (°C), Relative Humidity (%), Oxygen (%) and Carbon Dioxide (%) Concentrations of Selected Fruits and Vegetables, Together with the Rates of Respiration (ml $CO_2$ $kg^{-1}$ $h^{-1}$) and Ethylene Production ($\mu l$ $C_2H_4$ $kg^{-1}$ $h^{-1}$)

| Product | Optimum Range for Storage Temperature °C | Optimum Relative Humidity (%) for Storage | Rates of Respiration at $x$ Temperature (ml $CO_2$ $kg^{-1}$ $h^{-1}$) | Rates of Ethylene Production at $x$ Temperature ($\mu l$ $C_2H_4$ $kg^{-1}$ $h^{-1}$) | Recommended $O_2$ and $CO_2$ Concentrations for CA Storage | Special Notes |
|---|---|---|---|---|---|---|
| **Fruits** | | | | | | |
| Apple (Gala) | −1–+1 | 90–95 | 6.5–8 at 0°C | 4–12 at 0°C | 1.5–2% $O_2$ and 1–2% $CO_2$ | Ethylene accelerates senescence. CA may extend storability up to 4–5 months. |
| Apple (Fuji) | −1–+1 | 90–95 | 4–6 at 0°C | 2–4 at 0°C | 1.5–2% $O_2$ and <0.5% $CO_2$ | Ethylene accelerates senescence. Late harvested products are not suitable for CA storage. |
| Apple (Golden Delicious) | −1–+1 | 90–95 | 3–6 at 0°C<br>7–12 at 10°C<br>15–30 at 20°C | 1–10 at 0°C<br>5–60 at 10°C<br>20–150 at 20°C | 1–3% $O_2$ and 1.5–3% $CO_2$ | CA may extend the storage duration from 6 to 10 months. |
| Apricot | −0.5–+0.5 | 90–95 | 2–4 at 0°C<br>6–10 at 10°C<br>15–20 at 20°C | <0.1 at 0°C | 2–3% $O_2$ and 2–3% $CO_2$ | 2–4 weeks without CA and 4–6 weeks with CA. Ethylene hastens ripening and encourages pathogen growth. |
| Avocado | +5–+13 for mature-green<br>+2–+4 for ripe | 90–95 | 10–25 at 0°C<br>25–80 at 10°C<br>40–150 at 20°C | >100 at 20°C (ripe fruits) | 2–5% $O_2$ and 3–10% $CO_2$ | Do not ripen on tree. Ethylene (100 ppm) enhance ripening. CA extends storage duration for unripe fruits, maybe up to 9 weeks. |
| Banana | +13–+14 for unripe storage<br>+15–+20 for ripening | 90–95 | 10–30 at 13°C<br>12–40 at 15°C<br>20–70 at 20°C | 0.1–2 at 13°C<br>0.2–5 at 15°C<br>0.3–10 at 20°C | 2–5% $O_2$ and 2–5% $CO_2$ | 100–150 ppm ethylene exposure is beneficial for quick ripening. CA delays ripening. Mature-green fruits to be stored 2–4 weeks in air and 4–6 weeks in CA. |

| | | | | | | |
|---|---|---|---|---|---|---|
| Cactus pear | +6–+8 | 90–95 | 15–20 at 20°C | <0.3 at 20°C | 2–3% $O_2$ and 2–5% $CO_2$ | Non-climacteric fruit. Have 2–4 weeks storage potential depending on cultivar and harvest maturity. CA may extend 4–8 weeks. |
| Cherry | −0.5–+0.5 | 90–95 | 3–5 at 0°C 15–17 at 10°C 22–28 at 20°C | – | 3–10% $O_2$ and 10–15% $CO_2$ | Have minimal response to ethylene. 1 week in air and 2–3 weeks in CA |
| Date plum | 0 for 6–12 months and −18 for longer durations | 70–75 | <25 at 20°C for Khalal stage <5 at 20°C for Tamar (ripe) stage | | | 20% $CO_2$ may extend the storability, but fruits should be consumed soon after CA storage. |
| Fig | −1–0 | 90–95 | 2–4 at 0°C 5–8 at 10°C 20–30 at 20°C | 0.4–0.8 at 0°C 1.5–3 at 10°C 4–6 at 20°C | 5–10% $O_2$ and 15–20% $CO_2$ | 1–2 weeks in air and 3–4 weeks in CA. Slightly sensitive to ethylene, mainly at temperatures above 5°C. |
| Grape | −1–0 | 90–95 | 1–2 at 0°C 5–8 at 10°C 12–15 at 20°C (together with stem, which has about 15-fold greater respiration than berries) | – | 2–5% $O_2$ and 1–5% $CO_2$ | Storage duration highly vary (2–5 months). $SO_2$ is so crucial for decay control. |
| Grapefruit | +12–+14 | 90–95 | 3–5 at 10°C 5–9 at 15°C 7–12 at 20°C | <0.1 at 20°C | 3–10% $O_2$ and 5–10% $CO_2$ (high $CO_2$ is toxic. CA is limited for grapefruit storage). | Can be stored for 4–12 weeks. 1–10 ppm ethylene exposure (1–3 days) to mature-green fruits at 20–25°C causes de-greening (loss of green and development of yellow color). |
| Guava | +8–+10 for mature-green +5–+8 for ripe | 90–95 | 4–30 at 10°C 10–70 at 20°C | 10 at 20°C (ripe fruits) | Limited research exist, 2–5% $O_2$ may delay ripening | Ethylene (100 ppm) accelerates ripening. Mature-green fruits have 2–3 weeks and fully-ripe fruits have 1 week storage potential. |

*(Continued)*

**TABLE 2.1** (*Continued*)
Recommended Optimum Temperature (°C), Relative Humidity (%), Oxygen (%) and Carbon Dioxide (%) Concentrations of Selected Fruits and Vegetables, Together with the Rates of Respiration (ml $CO_2$ $kg^{-1}$ $h^{-1}$) and Ethylene Production ($\mu l$ $C_2H_4$ $kg^{-1}$ $h^{-1}$)

| Product | Optimum Range for Storage Temperature°C | Optimum Relative Humidity (%) for Storage | Rates of Respiration at x Temperature (ml $CO_2$ $kg^{-1}$ $h^{-1}$) | Rates of Ethylene Production at x Temperature ($\mu l$ $C_2H_4$ $kg^{-1}$ $h^{-1}$) | Recommended $O_2$ and $CO_2$ Concentrations for CA Storage | Special Notes |
|---|---|---|---|---|---|---|
| Kiwi | 0–+1 | 90–95 | 1.5–2 at 0°C<br>5–7 at 10°C<br>15–20 at 20°C | <0.1 at 0°C (unripe)<br>0.1–0.5 at 20°C (unripe)<br>50–100 at 20°C (ripe) | 1–2% $O_2$ and 3–5% $CO_2$ | 7–10 days in air, 2–3 months in cold storage and 4–5 months in CA. Highly sensitive to ethylene. CA delays ripening. |
| Lemon | +12–+14 depending on maturity-ripeness | 90–95 | 5–6 at 10°C<br>7–12 at 15°C<br>10–14 at 20°C | – | 5–10% $O_2$ and 0–10% $CO_2$ | 1–10 ppm ethylene for 1–3 days at 20°C enhances de-greening. Ripening stage determines storability, can be up to 6–8 months. |
| Loquat | 0 depending on cultivar and maturity | 90–95 | 5–8 at 5°C<br>10–15 at 10°C<br>25–40 at 20°C | <0.5 at 20°C | 3–5% $O_2$ and 3–5% $CO_2$ | |
| Lychee | 5 (+1.5–+10) depending on cultivar. | 90–95 | 3–5 at 0°C<br>6–9 at 5°C | 0.1–0.3 at 0°C<br>0.2–0.6 at 5°C | – | Ethylene may enhance the loss of green color. |
| Mandarin | +5–+8 for 2–6 weeks | 90–95 | 2–4 at 5°C<br>3–5 at 10°C<br>10–15 at 20°C | – | 5–10% $O_2$ and 0–5% $CO_2$ | High $CO_2$ is toxic for mandarins. CA is limited. |

| Commodity | Temperature (°C) | RH (%) | Storage life | Respiration rate | CA (O₂ and CO₂) | Remarks |
|---|---|---|---|---|---|---|
| Mango | +13 for mature-green +10 for partially-/fully-ripe | 90–95 | 12–16 at 10°C 19–28 at 15°C 35–80 at 20°C | 0.1–0.5 at 10°C 0.3–0.4 at 15°C 0.5–0.8 at 20°C | 3–5% $O_2$ and 5–8% $CO_2$ | 2–4 weeks in air and 3–6 weeks in CA. Ethylene (100 ppm for ½–1 day at 20°C) accelerates ripening. CA delays ripening. |
| Nectarine | −1–0 | 90–95 | 2–3 at 0°C 8–12 at 10°C 32–55 at 20°C | 0.01–5 at 0°C 0.05–50 at 10°C 0.1–160 at 20°C | 6% $O_2$ and 17% $CO_2$ (recommended for reducing internal breakdown) | 1–2 weeks storage. There are numerous varieties/cultivars, where some may be stored for 7 weeks. |
| Olive (fresh green) | +5–+7.5 | 90–95 | 5–10 at 5°C 12–16 at 10°C 20–40 at 20°C | <0.1 at 20°C | 2–3% $O_2$ and 0–1% $CO_2$ (delays senescence) | 6–8 weeks in air and 10–12 weeks in CA. Moderately sensitive to ethylene. |
| Orange | +3–+8 | 90–95 | 2–4 at 5°C 6–12 at 10°C 11–17 at 20°C | <0.1 at 20°C | 5–10% $O_2$ and 0–5% $CO_2$ (high $CO_2$ is toxic). CA is limited for orange storage. | 3 months in air. 1–10 ppm ethylene for 1–3 days at 20°C enhance de-greening. |
| Papaya | +13 for mature-green +10 for partially ripe +7 for fully ripe | 90–95 | 3–5 at 7°C 4–6 at 10°C 15–35 at 15°C | 0.1–2 at 7°C 0.2–4 at 10°C 0.5–8 at 15°C | 3–5% $O_2$ and 5–8% $CO_2$ | 2–4 weeks in air and 3–5 weeks in CA. Ethylene (100 ppm) at 20°C for 1–2 days accelerates ripening. |
| Pear | −1–0 | 90–95 | 1–3 at 0°C 5–10 at 10°C 15–30 at 20°C | 2–5 at 0°C 5–15 at 10°C 40–80 at 20°C | 1–2% $O_2$ and 0–1% $CO_2$ | Ethylene enhance ripening. 2–3 months in air and 4–6 months in CA. |

*(Continued)*

**TABLE 2.1** (*Continued*)

**Recommended Optimum Temperature (°C), Relative Humidity (%), Oxygen (%) and Carbon Dioxide (%) Concentrations of Selected Fruits and Vegetables, Together with the Rates of Respiration (ml $CO_2$ kg⁻¹ h⁻¹) and Ethylene Production (µl $C_2H_4$ kg⁻¹ h⁻¹)**

| Product | Optimum Range for Storage Temperature°C | Optimum Relative Humidity (%) for Storage | Rates of Respiration at x Temperature (ml $CO_2$ kg⁻¹ h⁻¹) | Rates of Ethylene Production at x Temperature (µl $C_2H_4$ kg⁻¹ h⁻¹) | Recommended $O_2$ and $CO_2$ Concentrations for CA Storage | Special Notes |
|---|---|---|---|---|---|---|
| Pineapple | +10–+13 for partially ripe +7–+10 for fully ripe | 85–90 | 2–4 at 7°C 3–5 at 10°C 15–20 at 20°C | <0.2 at 20°C | 3–5% $O_2$ and 5–8% $CO_2$ | 2–4 weeks in air. CA delays senescence. Ethylene may induce de-greening. |
| Plum | –1–0 | 90–95 | 1–1.5 at 0°C 3–4 at 10°C 6–8 at 20°C | <0.01–5 at 0°C 0.04–60 at 10°C 0.1–200 at 20°C (higher values are for ripe plums) | 6% $O_2$ and 17% $CO_2$ | Storability is variety specific and is about 1–8 weeks. |
| Pomegranate | +5–+7 | 90–95 | 2–4 at 5°C 4–8 at 10°C 8–18 at 20°C | <0.1 at 10°C <0.2 at 20°C | 3–5% $O_2$ and 6–10% $CO_2$ | Up to 4–5 months in air and 5–6 months in CA |
| Quince | 0 | 90–95 | 2–5 at 0°C 10–14 at 10°C 20–40 at 20°C | 2–6 at 0°C 6–7 at 10°C 10–40 at 20°C | — | Has a storage potential of 2–3 months in air. Ethylene can stimulate uniform ripening. |

| Commodity | Temperature | RH (%) | Respiration rates | Ethylene production | CA conditions | Comments |
|---|---|---|---|---|---|---|
| Strawberry | 0–+0.5 | 90–95 | 6–10 at 0°C, 25–50 at 10°C, 50–100 at 20°C | 2 at 20°C | 10–15% $CO_2$ is recommended for reducing pathogen growth,= | 5–7 days in air and 7–14 days in CA |
| Watermelon | +15 (2 weeks) +7–+10 (3 weeks) | 85–90 | 3–4 at 5°C, 6–9 at 10°C, 17–25 at 20°C | 0.1–1 at 20°C | – | Ethylene, even at low concentrations, damages firmness. |
| **Vegetables** | | | | | | |
| Artichoke (Globe) | 0 | >95 | 8–22 at 0°C, 22–49 at 10°C, 67–126 at 20°C | – | 2–3% $O_2$ and 3–5% $CO_2$ | Quick cooling by hydrocooling or forced-air are recommended. Storage potential in air is about 2–3 weeks. |
| Asparagus | 0–+2 | 95–100 | 14–40 at 0°C, 45–152 at 10°C, 138–250 at 20°C | - | 5–10% $CO_2$ is recommended to reduce decay | Ethylene accelerates the lignification. Storage potential in air is 2–3 weeks at 2°C and 3–4 weeks at 0°C. |
| Beans (Snap) | +5–+7.5 | 95–100 | 10 at 0°C, 30 at 10°C, 65 at 20°C | – | 2–5% $O_2$ and 3–10% $CO_2$ (Higher $CO_2$ is detrimental.) | Storage potential is 10–12 days at recommended temperatures in air. Chilling injury occurs at temperatures below 5°C. Ethylene sensitive. |
| Broccoli | 0 | >95 | 10–11 at 0°C, 38–43 at 10°C, 140–160 at 20°C | Very low. | 1–2% $O_2$ and 5–10% $CO_2$ (Higher $CO_2$ is detrimental.) | Storage potential is 3–4 weeks at recommended temperature and decreases to 2 weeks at 5°C in air. Sensitive to ethylene. |
| Cabbage | 0–+2 | >95 | 2–3 at 0°C, 8–10 at 10°C, 14–25 at 20°C | Very low. | 2–5% $O_2$ and 3–6% $CO_2$ (Higher $CO_2$ is detrimental.) | Storage potential is 3–6 weeks at recommended temperatures in air. |

*(Continued)*

**TABLE 2.1 (*Continued*)**

**Recommended Optimum Temperature (°C), Relative Humidity (%), Oxygen (%) and Carbon Dioxide (%) Concentrations of Selected Fruits and Vegetables, Together with the Rates of Respiration (ml $CO_2$ $kg^{-1}$ $h^{-1}$) and Ethylene Production ($\mu l$ $C_2H_4$ $kg^{-1}$ $h^{-1}$)**

| Product | Optimum Range for Storage Temperature°C | Optimum Relative Humidity (%) for Storage | Rates of Respiration at $x$ Temperature (ml $CO_2$ $kg^{-1}$ $h^{-1}$) | Rates of Ethylene Production at $x$ Temperature ($\mu l$ $C_2H_4$ $kg^{-1}$ $h^{-1}$) | Recommended $O_2$ and $CO_2$ Concentrations for CA Storage | Special Notes |
|---|---|---|---|---|---|---|
| Carrot (mature roots) | 0 | 98–100 | 5–10 at 0°C<br>15–30 at 10°C<br>45–60 at 20°C | >0.1 at 20°C | >5% $CO_2$ increase spoilage. Moreover, low $O_2$ (<3%) is also not recommended. | Mature roots have storage potential up to 7–9 months in air. Exposure to ethylene may cause bitter flavor. |
| Cauliflower | 0 | 95–98 | 8–9 at 0°C<br>16–18 at 10°C<br>37–42 at 20°C | – | High $CO_2$ (>5%) and low $O_2$ (<2%) cause injury. | Highly sensitive to external ethylene. 3 weeks storage potential exist at recommended conditions in air. |
| Celery | 0–+2 | 98–100 | 3 at 0°C<br>12 at 10°C<br>32 at 20°C | – | 2–4% $O_2$ and 3–5% $CO_2$ High $CO_2$ (>10%) and low $O_2$ (<2%) cause injury. | Storage potential is 5–7 weeks in air. It is not very sensitive to ethylene. |
| Cucumber | +10–+12.5 | 95 | 12–15 at 10°C<br>12–17 at 15°C<br>10–26 at 20°C | – | 3–5% $O_2$ and <10% $CO_2$ | Highly sensitive to exogenous ethylene. Storage potential is 1–2 weeks in air. |

| | | | | | | |
|---|---|---|---|---|---|---|
| Eggplant | +7.5–+10 | 90–95 | 30–70 at 12.50°C depending on the cultivar | 0.1–0.7 at 12.5°C | 3–5% $O_2$ and <10% $CO_2$ | Sensitive to exogenous ethylene. Storage potential is 10–14 days in air. |
| Garlic (Bulbs) | –1–0 | 60–70 | 2–6 at 0°C<br>6–18 at 10°C<br>7–13 at 20°C | Very low | 3–5% $O_2$ and <10% $CO_2$ | Storage potential is 6–8 months at recommended conditions in air and 1–2 months at ambient conditions (20°C). Not sensitive to ethylene. |
| Lettuce (Romaine) | 0 | >95 | 9–12 at 5°C<br>15–20 at 10°C<br>30–40 at 20°C | Very low | 1–3% $O_2$ and 0–5% $CO_2$ | Storage potential is 2 weeks at 5°C and 3 weeks at 0°C in air. Sensitive to ethylene. |
| Okra | +7–+10 | 95–100 | 27–30 at 5°C<br>43–47 at 10°C<br>124–137 at 20°C | Very low | 3–5% $O_2$ and 4–10% $CO_2$ | Storage potential is 7–10 days at recommended conditions in air. |
| Onion (dry) | 0<br>(High temperatures (5–25) favors sprouting. | 75–80 | 3–4 at 5°C<br>27–29 at 25°C | Ethylene may encourage sprouting. | CA is not recommended. Low $O_2$ (<1%) damages the onions. | Heat curing (12–30 hours at 45°C) is beneficial. Mild types have up to 1 month and pungent ones up to 6–9 months storage potential in air. |
| Onion (green) | 0 | 98 | 5–16 at 0°C<br>18–31 at 10°C<br>40–90 at 20°C | Green onions are not sensitive to ethylene. | 2% $O_2$ and 5% $CO_2$ | Storage potential is about 4 months at recommended conditions in air. CA improves storability up to 6–8 weeks. |
| Pea | 0 | 95–98 | 15–24 at 0°C<br>34–59 at 10°C<br>123–180 at 20°C | Moderately sensitive to ethylene exposure | 2–3% $O_2$ and 2–3% $CO_2$ | Pods have up to 2 weeks storage potential at recommended conditions in air. CA would improve that duration. |
| Pepper (Bell) | +7.5–+10 | >95 | 3–4 at 5°C<br>5–8 at 10°C<br>18–20 at 20°C | 0.1 at 10°C<br>0.2 at 20°C | High $CO_2$ may damage the products. | Storage potential is 3–5 weeks in air. Non-climacteric. |

*(Continued)*

**TABLE 2.1 (Continued)**
**Recommended Optimum Temperature (°C), Relative Humidity (%), Oxygen (%) and Carbon Dioxide (%) Concentrations of Selected Fruits and Vegetables, Together with the Rates of Respiration (ml $CO_2$ kg$^{-1}$ h$^{-1}$) and Ethylene Production (μl $C_2H_4$ kg$^{-1}$ h$^{-1}$)**

| Product | Optimum Range for Storage Temperature°C | Optimum Relative Humidity (%) for Storage | Rates of Respiration at $x$ Temperature (ml $CO_2$ kg$^{-1}$ h$^{-1}$) | Rates of Ethylene Production at $x$ Temperature (μl $C_2H_4$ kg$^{-1}$ h$^{-1}$) | Recommended $O_2$ and $CO_2$ Concentrations for CA Storage | Special Notes |
|---|---|---|---|---|---|---|
| Potato | +4–+7 for table +10–+15 for frying +15–+20 for chipping | 95–98 | 6–8 at 5°C 7–11 at 10°C 9–23 at 20°C | Tubers are very sensitive to external ethylene | 5–10% $O_2$ and <10% $CO_2$ | Heat curing is beneficial (i.e. 10 days at 20°C). Storage potential is 3 weeks at 20°C, 4 months at 8°C and 8 months at 4°C in air (highly variable for types and maturity). |
| Radish | 0 | 95–100 | 2–4 at 0°C 6–7 at 10°C 19–26 at 20°C | Not sensitive to ethylene | 1–2% $O_2$ and 2–3% $CO_2$ | Storage potential is 1–2 weeks for bunched and 3–4 weeks for topped fruits in air. |
| Spinach | 0 | 95–98 | 9–11 at 0°C 41–69 at 10°C 89–143 at 20°C | Very sensitive to ethylene | 7–10% $O_2$ and 5–10% $CO_2$ | Storage potential is less than 2 weeks at recommended conditions in air. |
| Squash | +5–+10 | 95 | 6–7 at 0°C 17–18 at 10°C 42–48 at 20°C | 0.1–1 at 20°C | 3–5% $O_2$ and <10% $CO_2$ | Zucchini can be stored for 2 weeks at 5°C. Chilling injury increase at lower temperatures. |
| Tomato | +12.5–+15 for mature-green +7–+10 for firm-ripe | 90–95 | 7–8 at 10°C 12–15 at 15°C 12–22 at 20°C (for firm-ripe) | 1.2–1.5 at 10°C 4.3–4.9 at 20°C | 3–5% $O_2$ and 0–3% $CO_2$ | Storage potential is 2 weeks for mature-green and 8–10 days for firm-ripe at recommended conditions in air. |

*Source:* Information is adopted from several references (Kitinoja & Kader, 2003; Thompson, 2008; Watson et al., 2015; Gross et al., 2016; De Freitas & Pareek, 2019) and authors' knowledge.

Several internal and external factors impact the speed of respiration. Preharvest growing conditions, applications and climatic factors are important and determine the harvest quality of the products. The external factors of temperature and atmospheric composition are crucial factors. The higher the temperature, the higher the respiration rate. On the other hand, oxygen, which is among the main elements of respiration, significantly affects respiration (Bovi et al., 2018b). Therefore, lowering the temperature and modifying atmospheric composition are important practices for limiting respiration and improving storability.

Respiration continues until the reserves are all used; it is followed by aging, product decay, and senescence. Respiration is highly dependent on oxygen concentration, where the fresh air contains about 21% of the oxygen. However, eliminating the oxygen concentration around the fresh horticultural products is the ideal way for managing respiration, whereas anaerobic respiration (fermentation) takes place in the absence or limiting the supply of oxygen. In this process, sugar is broken down into alcohol and carbon dioxide, which results in abnormal flavors in the produce and promotes premature senescence (Paltrinieri, 2014).

## 2.2.1.2 Transpiration

Transpiration is the loss of water vapor from living tissues, including plants, flowers, fruits or vegetables. Water vapor tends to move from higher concentrations (inside the fruits and vegetables, 70–95% relative humidity) into lower concentrations (surrounding air, 50–60% relative humidity). Therefore, products lose water and shrink (lose mass) during transpiration. Harvested products do not have a connection to the plants and thus the roots, and they cannot replace this water loss, resulting in weight and quality loss (Bovi et al., 2016). The surface area of the living tissues and characteristics of the skin or peel significantly impacts the rate of transpiration. Likewise, the maturity stage of the product, skin injuries (if any), temperature, airflow, and relative humidity of the surrounding atmosphere all highly impact the rate of transpiration (Bovi et al., 2018a). Transpiration causes shriveling, wilting, softening, poorer texture, loss in weight and lower quality of fruits and vegetables. It can be reduced in storage by:

- Raising the relative humidity around fresh products,
- Reducing the air movements in the storage room,
- Lowering the air temperature in the storage room,
- Using protective coverage during packing, i.e., waxing, aloe vera gel coating, edible coatings or edible films, and/or
- Applying protective packaging, i.e., polyethylene film, modified atmosphere packaging, etc.

Besides reducing the transpiration, waxing and coating with edible materials have other advantages, i.e., improving appearance, protecting from microbiological pathogens, retarding wilting and shriveling, reducing physical losses, delaying respiration rate and maintaining product quality by reducing postharvest losses.

## 2.2.1.3 Ripening and Ethylene ($C_2H_4$) Biosynthesis

Fruit ripening involves numerous physiological, biochemical and molecular changes that are unique to plants, where the "green", unripe, acid-tasting fruits convert into "colorful", soft and sweet-tasting fruits. Fruit ripening also causes major changes in phenols composition. Ethylene synthesis plays a central role in fruit ripening. Ripening is followed by senescence (aging) with the help of respiration and transpiration (Brecht, 2019). The ethylene is biosynthesized from S-adenosyl-L-methionine (SAM) via 1-aminocyclopropane-1-carboxylic acid (ACC), which is catalyzed by ACC synthase and ACC oxidase (Adams & Yang, 1979).

Fruits generally fall into two groups – climacteric and non-climacteric fruits – according to their respiratory outputs and ethylene production. Climacteric fruits can ripen after harvest. These fruits show a slight decrease in respiration rate (pre-climacteric minimum), shortly after harvest and an increase (climacteric rise) up to the climacteric maximum (respiratory peak: fully ripe, the peak of edible ripeness) and a decrease again (postclimacteric decline: very susceptible to deterioration). Moreover, ethylene biosynthesis shows a burst during ripening in climacteric fruits. The production of ethylene shows an autocatalytic feature, where the initial exposure to ethylene causes an increase in ethylene biosynthesis. Some important examples of climacteric fruits are apples, avocados, bananas, guava, kiwi, melon, papaya, pears, peaches and tomato. However, some fruits, i.e., melon, guava and Japanese plum, show climacteric as well as non-climacteric behavior depending on the cultivar or genotype.

On the other hand, non-climacteric fruits have a decreasing rate of respiration and no or slight change in ethylene production. These fruits do not ripen after harvest and should be harvested at the right time, when they are fully ripe. Some important examples of this group of fruits are berries (raspberry, strawberry, cherry, blackberry), the citrus family, cucumber, eggplant, grape, loquat, pineapple and pomegranate (Paul et al., 2012; Osorio & Fernie, 2013; Watson et al., 2015). Non-climacteric plants, mainly the leafy vegetables, which cannot ripen after harvest, are sensitive to exogenous ethylene application/exposure and soften and rot if exposed.

Ethylene is a colorless gas, naturally produced by plants, and acts as a ripening hormone. Moreover, ethylene can be produced by a variety of other sources, including internal combustion engines, cigarette smoke, heaters and natural gas leaks (details are discussed in Section 3.5). Besides the genetic characteristics of species/genotypes, it is also well-known that wounding or other mechanical injury triggers ethylene production. Thus it should be kept in mind to gently harvest and handle fresh horticultural products.

Overall, ethylene can be either beneficial or detrimental for fruits and vegetables, depending on the type of product and storage aims. Although ethylene is required to ripen climacteric fruits, if the aim is to store the fruits for a longer period of time, it is important to harvest the produce before ripening and keep it away from ethylene because ripening is followed by senescence. Since ethylene accelerates ripening and de-greening, the ethylene should be controlled during storage. Several techniques, i.e., 1-methylcyclopropene (1-MCP), salicylic acid, nitric oxide, polyamines, edible coatings, controlled atmosphere, etc., have been developed to overcome the detrimental effects of ethylene. Selection of the appropriate method depends on the product type and aims in handling and marketing. Elimination of the harmful effects of ethylene can be achieved either by removing the ethylene or by inhibiting its biosynthesis/action (Lata & Sujayasree, 2021). Some methods follow:

- **Ethylene action inhibitors** – can be used to restrict the action of ethylene by blocking ethylene receptors. 1-MCP (known as Smart Fresh) and silver thio-sulfate are the two most important ethylene action inhibitors. 1-MCP binds with ethylene receptors with 10 times faster affinity. Use of 1-MCP effectively delays ripening, lignification, chilling injury, physiological disorders, pathogenic infections and softening in several horticultural crops (Xie et al., 2020)
- **Ethylene biosynthesis inhibitors** – are another important way of inhibiting the detrimental impacts of ethylene. Aminooxyacetic acid (AOA), aminoe-thoxyvinylglycine (AVG), polyamines (i.e., putrescine), nitric oxide (NO) and salicylic acid (SA) are some important inhibitors of ethylene biosynthe-sis (Ahmad & Ali, 2019; Baswal et al., 2020).
- Removal of ethylene
  - Avoidance of storing ethylene-sensitive products with climacteric fruits
  - Not using gas-powered equipment in cold rooms
  - Removing overripe, damaged or decayed fruit from storage rooms
  - Providing ventilation in storage rooms
  - Use of ethylene absorbers (Lata & Sujayasree, 2021)

### 2.2.2  EXTERNAL BIOLOGICAL FACTORS

### 2.2.2.1  Insects

Insects can cause significant damage to fruits and vegetables by introducing micro-organisms responsible for deterioration and losses by reducing the market potential. Fruit flies, fruit-feeding beetles and several Lepidoptera are examples of insects that feed internally in fruit and enter by a single orifice. These insects are challenging to identify and control (Yahia et al., 2019). The damage is caused by the larvae, which are hidden inside the developing fruit. The adult insect is in charge of identifying the host that the larvae will feed and develop on. The larvae can infest the fruits when the adult Lepidoptera lays eggs on or near the fruit, and the larvae penetrate the fruit after hatching (Yahia et al., 2019). The adult Coleoptera and Diptera utilize their spe-cialized mouthparts and ovipositors to pierce fruits and lay their eggs directly in the fruit tissue. Fruit flies, which infest a wide range of fruit crops and can either be host-specific or target a variety of fruit hosts, are the most prevalent postharvest pests. The female fruit fly uses her ovipositor to pierce the fruit's skin and to lay her eggs in the pulp, and the larvae start feeding and developing as soon as the adult fruit fly finds the fruit. When a larva reaches the last instar stage, it leaves the fruit, pupates in the ground and then emerges as an adult.

## Diptera

### *Tropical fruit flies*

Mediterranean fruit fly (Medfly): *Ceratitis capitata* (Wiedemann)
West Indian fruit fly: *Anastrepha obliqua* (Macquart)

Mexican fruit fly (Mexfly): *Anastrepha ludens* (Loew)
Sapote fruit fly: *Anastrepha serpentina* (Wiedemann)
Guava fruit fly: *Anastrepha striata* (Schiner)
South American fruit fly: *Anastrepha fraterculus* (Wiedemann)
Queensland fruit fly: *Bactrocera tryoni* (Froggatt)
Melon fly: *Zeugodacus cucurbitae* (Bragard et al., 2020)
Oriental fruit fly: *Bactrocera dorsalis* (HENDEL) (Loomans et al., 2019)
Papaya fruit fly: *Toxotrypana curvicauda* (Gerstaecker)

### Temperate fruit flies

Apple maggot: *Rhagoletis pomonella* (Walsh)
Eastern cherry fruit fly: *Rhagoletis cingulata* (Loew)
European cherry fruit fly: *Rhagoletis cerasi* (L.)

**Lepidoptera.** In Lepidoptera, the larval stage is the most destructive stage, and the larvae have biting and chewing mouthparts to feed on leaves, roots, fruits and stems. The adults have siphoning mouthparts.

Sweet potato moth: *Megastes grandalis*
Fall army worm: *Spodoptera frugiperda*
False codling moth: *Thaumatotibia (cryptophlebia) leucotreta* (Meyrick) (Tortricidae)
Avocado seed moth: *Stenoma catenifer* Walsingham (Elachistidae)
Codling moth: *Cydia pomonella* L. (Tortricidae)
Oriental fruit moth: *Grapholita (cydia) molesta* Busck (Tortricidae)
Diamond back moth: *Plutella xylostella*

**Coleoptera**. Insects in this order, which is comprised of beetles and weevils, typically have a pair of hardened front wings and chewing mouthparts. Coleoptera can cause damage through direct feeding or by transmitting various pathogens such as viruses, bacteria, fungi, and even nematodes. Many species also live and feed on stored feed and grain, causing significant postharvest losses. The rice weevil (*Sitophilus oryzae*), the lesser grain borer (*Rhyzopertha dominica)* and the flour beetles (*Tribolium* spp.) are a few examples. Fruit pests like beetles are fairly common. Weevils (Curculionidae) are particularly significant pests in postharvest circumstances because they typically oviposit internally and feed as larvae, and the oviposition opening frequently closes during fruit growth, making detection challenging. Beetle pupation frequently takes place in the pulp of seeds or fruits.

Mango seed weevil: *Sternochetus mangiferae* (Fabricius) (Bragard et al., 2018)
Large avocado seed weevil: *Heilipus lauri* (Boheman)
Guava weevil: *Conotrachelus dimidiatus* Champion
Plum curculio: *Conotrachelus nenuphar* (Herbst)
Sweet potato weevil: *Cylas formicarius*

**Hemiptera**. These insects have piercing and sucking mouthparts and feed on plant sap. Most hemipterans are phytophagous and affect many important agricultural crops. Damage can be caused directly from feeding or indirectly from the transmission of viruses or phytotoxic effects of injected saliva. Aphids can cause the growth of sooty mold due to the production of honeydew. Another significant insect pest is the pineapple mealybug, *Dysmicoccus brevipes* (Cockerell), which can also be detected on postharvest fruits, e.g., annona, banana, citrus and coffee. It is frequently discovered or intercepted in quarantine. Aside from the obvious results of feeding on this species, it also transmits pineapple wilt disease.

Pineapple mealybug: *Dysmicoccus brevipes* (Cockerell)
Papaya mealybug: *Paracoccus marginatus*
Pink hibiscus mealybug: *Maconellicoccus hirsutus*
Cottony cushion scale: *Icerya purchasi*
Cabbage aphid: *Brevicoryne brassicae*
Citrus aphids: *Toxoptera citricada*
White fly: *Aleurodicus disperus*

**Thysanoptera**. The adults and larvae collect in flowers, developing fruits, foliage and floral buds, inhabiting the tightly restricted and hidden spaces of plants (Reitz, 2009). The females deposit eggs into soft plant tissues (leaves, petioles, flower bracts, petals and developing fruit); after the eggs hatch, the larvae feed and develop. In the soil or on the plant, the pupal stage occurs before the adult emerges. The adult and larval stages feed by piercing the plant surface and sucking the content of the cells. This causes ghost spotting on tomato, sweet pepper and cucumbers. Viral diseases can spread through both direct and indirect damage, such as the tomato spotted wilt virus (Pupin et al., 2013).

Western flower thrip: *Frankliniella occidentalis*
Red-banded thrips: *Selenothrips rubrocinctus* (Giard)
Melon thrips: *Thrips palmi*

## 2.2.2.2 Arachnids

By consuming plant sap, larvae, nymphs and adults damage the host plant. Damage might include everything from feeding marks on leaves to bronzing, scarring and fruit distortion, depending on the species. The two-spotted mite (*Tetranychus urticae*) can cause webbing on leaves and fruit (Al-Shammery & Al-Khalaf, 2022). Feeding by the citrus bud mite (*Aceria sheldoni*) causes damage, such as fruit deformation, to developing fruit (Phillips & Walker, 1997).

## 2.2.2.3 Rodents

Only a small percentage of rodent species are pests. Rats and mice, which are commonly found in homes, make up the majority of pest species and have a long history of coexisting with people. *Rattus rattus* (ship rat), *Rattus norvegicus* (Norway rat), and *Mus musculus* (house mouse) are the three principal commensal species, and they can be found in both tropical and temperate climates. All food storage situations, from small-scale, on-farm store rooms to large-scale silos or warehouses, are susceptible to the problems that rodents can cause. Rodents in storage may consume an increased amount of food directly. They typically consume 10% of their body weight each day (Farrell et al., 2002).

In comparison to the direct loss of food, rodents damage and contaminate far greater amounts of food than they consume (Farrell et al., 2002). The main culprits behind the destruction are rodent hairs, urine and droppings. Grain contamination by rats is a particular issue because they are known carriers of numerous gastroenteric illnesses, including leptospirosis and toxoplasmosis (Gratz, 1988; Gorham, 1989). Food contamination may degrade its quality or render it unfit for human consumption.

### 2.2.2.4   Birds

Almost any bird that feeds on grains has the ability to invade grain storage. Typically, bird species that can roost or perch in the storage structure or facility and that are members of flocks that include pigeons and sparrows are thought to be potential postharvest pests. Through spilling and consumption, birds are extremely effective at inflicting significant losses to grain. Pecking birds can rip open bagged goods when they are stored, leading to significant spillage. Twenty adult pigeons can consume the same amount of food as one adult human; e.g., one adult pigeon can consume 35 g of grain every day (Farrell et al., 2002). Financially speaking, contamination is probably a bigger issue than actual grain loss. Birds carry germs and taint food that has been preserved. In addition to obvious contamination, bird excrement in the store can result in high zoonotic illness rates, such as Salmonella.

### 2.2.2.5   Fungi

Fungi pose a threat to the world's food supply, and efforts have been made to research effective, nontoxic methods of reducing postharvest infections. Synthetic fungicides are one of the most effective methods of controlling postharvest disease due to their ease of use and low cost (Chen et al., 2021). Fungi are organisms that resemble plants but lack chlorophyll, preventing them from using photosynthesis to create their own food. Fungi enzymatically hydrolyze complex substrates like proteins, lipids and carbohydrates and absorb the simpler substances through their cell walls (Farrell et al., 2002). Effective control is necessary to reduce fungi loss because losses can be up to 24%. Commercially important fruit disease losses are caused by pre- or postharvest fungal pathogen infections. They can grow and flourish under storage conditions due to their strong flexibility.

*Penicillium* rots, including *P. expansum*, *P. digitatum* and *P. italicum* (Errampalli, 2014; Palou, 2014), are examples of destructive fungal disease that affect postharvest crops. *Penicillium* is a wound pathogen that enters fruit tissue through injury caused by birds or insects (Errampalli, 2014). Postharvest losses in citrus fruits are caused by *P. digitatum* and *P. italicum* (Palou, 2014). *P. expansum* causes blue mold, which adversely affects orchard fruits like apples, pears, peaches, grapes, strawberries, and the spores can survive in organic matter and the soil (Errampalli, 2014). The fungus enters the fruit during harvest as a result of wounding. *P. expansum* may be present in the fruit at harvest, so preharvest treatment is necessary to lessen the pathogen's impact on postharvest degradation. *P. expansum* generates the mycotoxin patulin, a neurotoxic substance that is found in apples and apple products (Errampalli, 2014). *Penicillium* infection increases during optimal environmental conditions as a result of spores landing on fruit surfaces and infecting the fruit. Environmental factors, such as high inoculum levels, respiration rate, fruit age and wounding frequency, all influence the risk of *Penicillium* infection, which increases postharvest decay losses (Scholtz et al., 2017).

*Alternaria alternata* can infect a wide range of fruits and vegetables, e.g., avocado and mango. *A. alternata* is a pathogen that enters plant tissue through wounds and natural openings or by directly breaching the host cuticle, allowing the fungus to enter unripe tissue and lay dormant for weeks until the fruit ripens (Troncoso-Rojas & Tiznado-Hernández, 2014). The *Colletotrichum* genus of fungi is responsible for *Anthracnose* disease that affects numerous plant species, including bananas, mangoes, papayas and pome fruit (Pétriacq et al., 2018). A disease that is of economic importance is crown rot in banana, caused by *C. musae*, which is mainly dispersed by rainwater and colonizes on the decomposing leaves (Lassois & de Lapeyre de Bellaire, 2014). *C. musae*, the causative agent of anthracnose, establishes dormant infections in the field during the first month after banana flowering (De Lapeyre de Bellaire et al., 2000). *Colletotrichum musae*, a dormant fungus, produces anthracnose, causing patches and brown to black lesions during ripening, especially in immature fruits. If *C. musae* is not controlled early in the harvest, postharvest losses can reach up to 80%. During this time, control could delay the onset of symptoms for the duration of the storage period (Damasceno et al., 2019).

*Botrytis cinerea*, a necrotrophic fungal disease that causes gray mold (Romanazzi & Feliziani, 2014) that grows in the tissues of more than 200 plant species, including fruits and vegetables (Williamson et al., 2007). *B. cinerea* infections occur in the field and remain dormant until storage, when the pathogen proliferates due to higher relative humidity and low temperatures, which slow down host defenses (Romanazzi & Feliziani, 2014). *B. cinerea* may result in annual economic losses of more than $10 billion worldwide (Yahia et al., 2019).

### 2.2.2.6  Bacteria

*Pseudomonas*, *Xanthomonas* and *Erwinia* are important postharvest bacteria (though some *Erwinia* species have lately been transferred to the *Pectobacterium* genus) (Farrell et al., 2002). A gram-negative plant pathogen called *Pectobacterium carotovorum* causes blackleg in potatoes and soft rot disease in a range of plant hosts by degrading cell walls. Plant cell intercellular gaps are colonized by *P. carotovorum* cells, which use a type III secretion system to release potent effector chemicals (Aizawa, 2014). Fruit deterioration that occurs after harvest is caused by bacterial infections. There is a huge and diversified community of bacteria on the skin of fleshy fruits and vegetables. For example, infection with *Clavibacter michiganensis* subsp. *sepedonicus* causes tomato and potato bacterial wilt and canker, which can be very expensive in some regions and cultivars (Sen et al., 2015; Pétriacq et al., 2018). Bacteria can survive for a long time on tools, equipment, storage trays and other inert surfaces, making it difficult to control (Sen et al., 2015). Bacterial pathogens seem to have less of an effect on postharvest fruit degradation (Pétriacq et al., 2018).

### 2.2.2.7  Viruses

Viruses are less common than bacterial and fungal infections. They are typically not important for postharvest, but they can still lower the market value of agricultural produce. Through wounds, sap-sucking insects, mechanical abrasion, or harvesting, they may enter the plants. Since viruses are the main drivers of field crop disease and their effects are obvious, infected plants are typically removed from the food chain during harvest or sorting. Consequently, viruses are infrequently found in storage (Farrell et al., 2002). Instead of directly damaging the fruit, typical damage occurs in vegetative tissue during the growing season, impairing fruit growth and yield (Pétriacq et al., 2018).

## 2.3 ENVIRONMENTAL FACTORS

### 2.3.1 TEMPERATURE

Temperature is one of the most significant environmental factors that affect how long agricultural produce stays fresh and how it should be stored, with each product having an ideal storage temperature (see Table 2.1 for several selected horticultural products) based on its geographic origin (Fallik & Ilic, 2018). After harvest, all physiological processes, including respiration, transpiration and the supply of nutrients, continue since fresh commodities are still alive (Ahmad & Siddiqui, 2015; Duan et al., 2020). When the product is detached from the parent plant, the water supply is cut off. Produce degrades as a result of respiration, losing nutritional content, changing the weight, texture and flavor (Ahmad & Siddiqui, 2015). It is impossible to stop these processes, but it is possible to considerably slow down physiological and pathological invasion by properly controlling temperature and relative humidity during storage and transportation (Fallik & Ilic, 2018). Precooling is an essential step in regulating the temperature of postharvest fruit and vegetables. It slows down biological activities to preserve quality (Duan et al., 2020). Between various goods, the respiration rates vary greatly (see Table 2.1). Inadequate cold chain management can cause significant losses of fresh produce, especially if exposed to direct sunlight (high temperatures) after harvest or if the cold chain fails. The usage of low temperature is crucial (Duan et al., 2020). As a general rule, respiration rates and the development of decay organisms are slower at lower temperatures. For every 10°C drop in temperature, Van't Hoff's quotation (Q10) predicts that the rate of respiration will be halved and the shelf life will increase (Saltveit, 2019).

After harvest, cooling products as soon as feasible to extend shelf life and lower respiration and metabolic activity rates has considerable advantages (Ahmad & Siddiqui, 2015). Reduced respiration slows metabolic activity, and inhibition of microbial growth and development are some of the most efficient strategies to retain the quality of fresh fruits and vegetables (Brasil & Siddiqui, 2018; Duan et al., 2020). A 5°C temperature drop during cold storage has no influence on product quality when the temperature is above 20°C but has a substantial impact when the temperature is below 10°C. Proper temperature control is crucial due to the change in metabolic activity brought on by a rise in temperature, which exponentially increases the pace of physiological reactions.

Heat treatment of fruits and vegetables can also be used to quarantine insects, kill microorganisms and lessen the sensitivity of produce to cold. According to Koshita (2014), enzymes that code for color change are destroyed at temperatures above 30°C, along with the chances for chlorophyll breakdown, fast softening, wilting and dehydration. High-temperature curing is advantageous for a variety of plants, including pumpkins. To treat wounds sustained during harvesting, such as abrasions, bruises or scratches, pumpkins were kept at 28–30°C for 18 days (Mohammed et al., 2014).

Reduced temperatures are also linked to the development and proliferation of microorganisms that cause rotting and spoiling. Low temperatures inhibit the growth of microorganisms and slow the deterioration of fresh vegetables, which aids in the prevention of disease in stored products. Most pathogens that affect product postharvest flourish at temperatures ranging from 15 to 25°C; however, *Botrytis cinerea* can survive at low temperatures (Pandey et al., 2008). Although the majority of bacteria and fungi can live in low temperatures, their rate of growth is noticeably slower than under normal circumstances.

### 2.3.1.1  Freezing Injury

By maintaining temperatures around 0°C, the metabolism of harvested products is slowed, and their shelf life is increased (Saltveit, 2019). The concentration of dissolved solutes, like sugar, within the cells influences the temperature at which produce freezes. Water condenses into ice crystals inside and between cells when produce freezes, drying the cells. Cell barriers can also be penetrated by expanding ice crystals. The cells break down during the thawing process, giving the product the water-soaked look and structural integrity that are typical of freezing injury. The product is effectively "killed" by freezing because it can't resume its regular metabolic process. The likelihood of damage increases when the vegetables freeze gradually, and larger ice crystals start to develop. Perishable goods have large, heavily vacuolated cells and a high water content (75–95%). Their tissues have a rather high freezing point (varying from –3°C to –0.5°C), and the disturbance brought on by freezing typically causes the tissues to collapse and a complete loss of cellular integrity (Kader, 2013).

### 2.3.1.2  Chilling Injury

Chilling injury, which can result from exposing tropical or subtropical fruits or vegetables to low temperatures, can cause pitting, wilting, failure to ripen, water loss, rot, discoloration, and internal collapse (Wang, 1989). Damage might not be noticeable while the product is being stored, but it will be obvious once it is exposed to higher temperatures. The best way to preserve the quality of perishable horticultural commodities is by low-temperature storage, although depending on the product and the stage of maturity, this might be harmful to crops that are sensitive to chilling when held below a critical temperature (5–13°C) (Kader, 2013). Products can suffer from chilling injury if they are stored at temperatures that are higher than their freezing point but lower than those that cause physiological harm. All tissues, however, are harmed by freezing, but many horticultural products are physiologically harmed if kept at non-freezing temperatures below 10°C (Saltveit, 2019). The majority of products that are sensitive to chilling may tolerate brief exposure to temperatures below their typical temperature threshold, but as storage duration increases, the damage becomes more noticeable. Fruits and vegetables should be kept just above the freezing point to ensure postharvest quality for non-chilling sensitive goods. These temperatures are frequently cited as the best ones to use for storage and transportation.

### 2.3.1.3   Heat Injury

Because transpiration is essential to growing plants, exposing perishable goods to high temperatures (over 30°C) can be very hazardous (Kader, 2013). High temperatures have an impact on various aspects of fruit ripening, including softening, color development, respiration, and ethylene production. Plant parts that have been detached from the plant are not protected by transpiration, and exposure to direct sunlight can cause sunburn or sunscald by heating tissues above the thermal death point of their cells (Kader, 2013).

### 2.3.2   RELATIVE HUMIDITY

The amount of moisture (as water vapor) that the atmosphere can hold at any given temperature and pressure without condensation is known as relative humidity (RH), which is stated as a percentage. The capacity of air to contain moisture rises as the temperature rises. The vapor pressure difference (VPD) between a commodity and its surroundings is directly related to water loss. The relative humidity of the air surrounding the commodity and VPD are inversely correlated (Kader, 2013). Water makes up the majority of fruits and vegetables and water loss or dehydration which reduces the salable weight, affecting the appearance, firmness, crispness and texture. To reduce water loss, fruits and vegetables should be kept in storage at high relative humidity levels (Brasil & Siddiqui, 2018; Ahmad & Siddiqui, 2015). Humidifiers are used to create humidity in modern cold storage facilities (Ahmad & Siddiqui, 2015). If the relative humidity is above the recommended level, condensation can occur on the surface which can lead to the growth and development of microorganisms (Brasil & Siddiqui, 2018). Spore germination is reduced in dry conditions; however, if germination occurs, the exposed tissue might not be infected because it is too dry. Most fungi cannot develop if the relative humidity falls below 85–90%, but the low humidity is not ideal for produce predisposed to moisture loss, such as leafy vegetables.

In high rainfall areas, the incidence of pathogens is high. Severe damage to the fruits occurs if the rainy season increases during the ripening period of fruits. Moisture in the atmosphere would reduce moisture loss from produce; however, too much moisture encourages the growth of microorganisms, resulting in decay, reduced shelf life and product loss.

Condensation can occur on the packaging material or the surface of the fruit or vegetable if there are changes to the temperature during postharvest handling. The air's capacity to hold vapor decreases as the temperature drops. This can cause packaging materials to deteriorate during storage, shipping or handling. By lowering the transpiration coefficient ($k$) or the vapor pressure deficit (VPD), water loss is minimized (Holcroft, 2015; Duan et al., 2020). When produce is transported to ambient temperatures, the air surrounding them cools, and, depending on the relative humidity, the cooled air's water vapor will condense on the outside of the produce.

## 2.3.3  LIGHT

Both temperature and light help fruits attain the produce's overall eating quality. In some trees, utilization of light is responsible for the productivity and quality of fruits. The more leaf surface is exposed to light, the more assimilation of carbohydrates occurs within the fruit. The canopy of the fruit-bearing branch and the placement of the fruit also have an impact on how quickly plants synthesize oxygen, as the right canopy placements increase the capacity for photosynthetic activity in leaves, stems and fruit. For the development and growth of the buds, warm temperature and good light levels are needed to support photosynthesis. Another environmental aspect that has been investigated is the impact of light on the metabolism of edible plant tissues after harvest in order to maintain quality throughout storage. Light shortens the pre-climacteric stage and speeds up the natural ripening of banana fruits (Özdemir, 2016) and increases the lycopene content in tomatoes (Alba et al., 2000).

## 2.3.4  ATMOSPHERIC COMPOSITION

The atmosphere's gas composition can be changed/regulated to increase the shelf life of fruits and vegetables. In the package atmosphere or storage room, lowering the respiration rates and raising the carbon dioxide concentration can prevent the growth of microorganisms and insects (Brasil & Siddiqui, 2018). Oxygen ($O_2$) is consumed during respiration, and $CO_2$ is released. Significant respiration inhibition was seen when the $O_2$ concentrations were lowered to less than 5%. Carbon dioxide ($CO_2$), oxygen ($O_2$) and nitrogen ($N_2$) make up 400 ppm (or 0.04%) of air, so adjusting the gas concentrations in the air around fresh products can aid in maintaining quality and lengthen storage life (Fang & Wakisaka, 2021). Fruits and vegetables respond differently to altering environments.

### 2.3.4.1  Controlled Atmosphere (CA)

A controlled environment is one that deviates from the standard $CO_2$ and $O_2$ values for storing seeds, fruits, vegetables and tubers. A CA storage is an important handling practice for fresh horticultural products where the oxygen, carbon dioxide and nitrogen, as well as the temperature and relative humidity of the storage room, are measured and regulated continuously. The needed levels of $CO_2$ and $O_2$ are continuously checked to make sure the desired values are reached. By supplying fresh air to remove the excess $CO_2$ buildup, the levels are kept stable. The species, cultivar, stage of development and environmental and cultivation practices all play a role in choosing the best setting. Alcohol can develop, and physiological changes can occur in an anaerobic environment if the level of $CO_2$ rises or the level of oxygen falls (Ahmad & Siddiqui, 2015; Kubo, 2014). One probable use for CA is sea freight. Both $O_2$ and $CO_2$ can be reduced with nitrogen inside a closed shipping container. The container can be ventilated with air during travel in response to an $O_2$ sensor inside the unit to keep $O_2$ levels from getting too low.

### 2.3.4.2  Modified Atmosphere Packaging

In modified environment storage, the product's respiration is applied to lower $O_2$ and raise $CO_2$. The rate of product respiration, along with the surface area and permeability of the film, will all affect the atmosphere that forms. The objective is to provide a steady atmosphere where the rates of $O_2$ consumption and $CO_2$ production are equal to the rate of gas passage through the plastic film. In order to change the $O_2$ and $CO_2$ concentrations in the package atmosphere, actively respiring produce is sealed in polymeric film packages. These concentrations are attained via a natural interaction between the product's respiration rate and the transfer of gases through the packing material (Nasrin et al., 2022). A reduced respiration rate, which lowers the rate of substrate depletion and oxidation processes, is one of the main consequences of MAP (Nasrin et al., 2022). Understanding storage conditions and the typical range of respiration rates under those settings is necessary for designing a MAP system for any product. Understanding which environments will help the product and which will hurt it is equally crucial. Numerous items are harmed by high $CO_2$ levels.

As a result of gas passing through plastic film by dissolving into it and forming on the opposite side of the barrier, integral plastic films are equally permeable to both gases, in contrast to micro perforated films, which are more permeable to $CO_2$ than to $O_2$ (Renault et al., 1994). To make sure that the environment that forms inside a MAP is advantageous to the product, the film choice and microperforation level can be altered (Kader et al., 1989). In order to accomplish this effect at the anticipated storage temperature, films can be selected based on the optimal concentrations for extending storage life. The quantity of holes made during production affects the permeability of microperforated films. Due to this, packaging materials may be easily altered to suit the product and its uses (D'Aquino et al., 2016). This procedure is made possible by variations in gas concentration on opposite sides of the film. While $O_2$ has two equally sized atoms, the $CO_2$ molecule has three atoms and a little polar electrical charge. Due to this, $CO_2$ can dissolve into the plastic film more readily than $O_2$ (Mangaraj et al., 2014). Even more difficult than oxygen to dissolve through a membrane is nitrogen. The kind and thickness of plastic, manufacturing process and surface area all have an impact on how permeable a package is (Murmu & Mishra, 2017). The majority of plastics are 10 to 20 times more porous to $CO_2$ than to $N_2$ and three to four times more porous to $O_2$ than to $N_2$.

When the temperature rises, the rate at which the product respires typically increases faster than the rate at which the film permeates. This means that MAPs may become anaerobic at high temperatures (Exama et al., 1993). The fact that respiration rates increase significantly higher with temperature than does film permeability to $O_2$ and $CO_2$ is one of the main problems with MAP. The temperature increase of 20°C can double the permeability of polyethylene films. However, respiration can multiply by five or more across the same temperature range. The package will run out of oxygen if the environment is warmer than what it was intended to be because more oxygen will be consumed for breathing than can be transmitted via the film (become anaerobic). Packages, on the other hand, are unlikely to offer appreciable benefits (low $O_2$ and/or high $CO_2$) (Kader et al., 1989) during routine cold storage if they are made to provide an appropriate environment when the temperature is raised. Applications requiring strict temperature control are best suited for MAP.

### 2.3.5 ETHYLENE

Ethylene ($C_2H_4$), which is connected to growth, ripening and germination, is produced by F&V (Wei et al., 2020). In addition to speeding up ripening, ethylene can also cause overripening and even rot, which shortens shelf life and results in losses (Wei et al., 2020). Injuries or decay can also emit ethylene, which can have a variety of impacts, most of which are bad for the quality and longevity of storage. It is particularly connected to the ripening of some fruits, including tomatoes, bananas and mangoes. These fruits are referred to as climacteric due to the high levels of ethylene and the large increase in respiration that take place during ripening. By exposing climacteric fruit to ethylene while it is being stored, ripening can be hastened, and there are numerous commercial uses for this technique. Yellowing, shorter shelf life and a rise in disease incidence can all result from ethylene exposure. Ozone, venting and scrubbing systems can all be utilized to get rid of ethylene. The effects of ethylene are lessened at temperatures below 5°C.

Endogenous and exogenous ethylene are the two biological processes by which ethylene ripens fruits and vegetables. Exogenous ethylene comes from various sources, such as adjacent produce, auto exhaust, plastics and smoke, whereas endogenous ethylene is created by the plant through a biological mechanism. It is crucial to eliminate any adjacent sources of ethylene since fruits and vegetables will ripen independently of the source of ethylene. Because even a small amount of ethylene (0.1 μL $L^{-1}$) can have a major impact on plant growth and development, plants are extremely sensitive to ethylene (Chang, 2016). Many methods were created to lessen the effects of ethylene, including low-temperature storage and a controlled environment, to mention a couple (Wei et al., 2020). Plants might be sensitive to ethylene in different ways; however, it should be noted that ethylene has unfavorable effects on vegetables. Exposure to ethylene shortens storage life, increases sensitivity to chilling injury and can promote the spread of rot. It also induces chlorophyll breakdown (yellowing). Leafy vegetables can undergo unfavorable alterations in response to ethylene at low concentrations (0.1 ml $L^{-1}$, or 0.0001%). Some climacteric and non-climacteric fruits exhibit improved color consistency and consistent ripening due to commercially accessible ethylene.

## 2.4 CONCLUSION

The postharvest quality of fresh F&V is governed by many biological and environmental factors which, if not controlled/managed appropriately, may affect the rate of deterioration and spoilage. The management of these factors is critical in the quality assurance of fresh produce. Preharvest applications, correct maturity at harvest, harvesting gently, precooling, transportation conditions, sorting, sanitation, fungicide treatments, protective coverage, grading, sizing, packaging and storage are the basic steps in the postharvest handling of fresh fruits and vegetables (Kahramanoglu, 2017). The proper handling of fresh produce at all stages is therefore contingent on ensuring adequate education of all farmers, laborers, merchants and even consumers along the postharvest chain.

## REFERENCES LIST

Adams, D. O., & Yang, S. F. (1979). Ethylene biosynthesis: Identification of 1-aminocyclo-propane-1-carboxylic acid as an intermediate in the conversion of methionine to ethyl-ene. *Proceedings of the National Academy of Sciences of the United States of America, 76*(1), 170–174. https://doi.org/10.1073/pnas.76.1.170

Ahmad, A., & Ali, A. (2019). Improvement of postharvest quality, regulation of antioxidants capacity and softening enzymes activity of cold-stored carambola in response to poly-amines application. *Postharvest Biology and Technology, 148*, 208–217. https://doi.org/10.1016/j.postharvbio.2018.10.017

Ahmad, M. S., & Siddiqui, M. W. (2015). *Postharvest quality assurance of fruits.* Springer. https://doi.org/10.1007/978-3-319-21197-8

Aizawa, S.-I. (2014). Pectobacterium carotovorum – Subpolar hyper-flagellation. *Flagellar World, 58–59*. https://doi.org/10.1016/B978-0-12-417234-0.00018-9

Alba, R., Cordonnier-Pratt, M. M., & Pratt, L. H. (2000). Fruit-localized phytochromes regu-late lycopene accumulation independently of ethylene production in tomato. *Plant Physiology, 123*(1), 363–370. https://doi.org/10.1104/pp.123.1.363

Al-Shammery, K. A., & Al-Khalaf, A. A. (2022). Effect of host preference and micro habitats on the survival of Tetranychus urticae Koch (Acari: Tetranychidae) in Saudi Arabia. *Journal of King Saud University – Science, 34*(4). https://doi.org/10.1016/j.jksus.2022.102030

Baswal, A. K., Dhaliwal, H. S., Singh, Z., Mahajan, B. V. C., & Gill, K. S. (2020). Postharvest application of methyl jasmonate, 1-methylcyclopropene and salicylic acid extends the cold storage life and maintain the quality of "Kinnow" Mandarin. *Postharvest Biology and Technology, 161*, 111064. https://doi.org/10.1016/j.postharvbio.2019.111064

Bovi, G. G., Caleb, O. J., Herppich, W. B., & Mahajan, P. V. (2018a). Mechanisms and model-ing of water loss in horticultural products. In, Smithers, G., Trinetta, V., and Knoerzer, K. (Eds.), *Reference module in food science* (pp. 1–5). Elsevier. https://doi.org/10.1016/B978-0-08-100596-5.21897-0

Bovi, G. G., Caleb, O. J., Linke, M., Rauh, C., & Mahajan, P. V. (2016). Transpiration and moisture evolution in packaged fresh horticultural produce and the role of integrated mathematical models: A review. *Biosystems Engineering, 150*, 24–39. https://doi.org/10.1016/j.biosystemseng.2016.07.013

Bovi, G. G., Rux, G., Caleb, O. J., Herppich, W. B., Linke, M., Rauh, C., & Mahajan, P. V. (2018b). Measurement and modelling of transpiration losses in packaged and unpackaged strawberries. *Biosystems Engineering, 174*, 1–9. https://doi.org/10.1016/j.biosystemseng.2018.06.012

Bragard, C., Dehnen-Schmutz, K., Di Serio, F., Gonthier, P., Jacques, M., Jaques Miret, J. A. et al. (2018). Scientific opinion on the pest categorisation of Sternochetus mangiferae. *EFSA Journal, 16*(10), 5439. https://doi.org/10.2903/j.efsa.2018.5439

Bragard, C., Dehnen-Schmutz, K., Di Serio, F., Gonthier, P., Jacques, M., Jaques Miret, J. A., Justesen, A. F., Magnusson, C. S., Milonas, P., Navas-Cortes, J. A., Parnell, S., Potting, R., Reignault, P. L., Thulke, H., Van der Werf, W., Vicent Civera, A., Yuen, J., Zappalà, L., Bali, E. M.,. .. MacLeod, A. (2020). Pest categorisation of non-EU Tephritidae. *EFSA Journal, 18*(1). https://doi.org/10.2903/j.efsa.2020.5931

Brasil, I. M., & Siddiqui, M. W. (2018). Postharvest quality of fruits and vegetables: An overview. In M. W. Siddiqui (Ed.), *Preharvest modulation of postharvest fruit and vegetable qual-ity* (pp. 1–40). Academic Press. https://doi.org/10.1016/B978-0-12-809807-3.00001-9

Brecht, J. K. (2019). Chapter 14. Ethylene technology. In E. M. Yahia (Ed.), *Postharvest tech-nology of perishable horticultural commodities* (pp. 481–497). Woodhead Publishing. https://doi.org/10.1016/B978-0-12-813276-0.00014-6

Chang, C. (2016). Q&A: How do plants respond to ethylene and what is its importance? *BMC Biology, 14*(1), 7. https://doi.org/10.1186/s12915-016-0230-0

Chen, T., Ji, D., Zhang, Z., Li, B., Qin, G., & Tian, S. (2021). Advances and strategies for controlling the quality and safety of postharvest fruit. *Engineering*, *7*(8), 1177–1184. https://doi.org/10.1016/j.eng.2020.07.029

D'Aquino, S., Mistriotis, A., Briassoulis, D., Di Lorenzo, M. L., Malinconico, M., & Palma, A. (2016). Influence of modified atmosphere packaging on postharvest quality of cherry tomatoes held at 20°C. *Postharvest Biology and Technology*, *115*, 103–112. https://doi.org/10.1016/j.postharvbio.2015.12.014

Damasceno, C. L., Duarte, E. A. A., dos Santos, L. B. P. R., de Oliveira, T. A. S., de Jesus, F. N., de Oliveira, L. M., Góes-Neto, A., & Soares, A. C. F. (2019). Postharvest biocontrol of anthracnose in bananas by endophytic and soil rhizosphere bacteria associated with sisal (*Agave sisalana*) in Brazil. *Biological Control*, *137*, 104016. https://doi.org/10.1016/j.biocontrol.2019.104016

De Freitas, S. T., & Pareek, S. (2019). *Postharvest physiological disorders in fruit and vegetables*. CRC Press and Taylor & Francis Group. https://doi.org/10.1201/b22001

De Lapeyre de Bellaire, L., Chillet, M., Dubois, C., & Mourichon, X. (2000). Importance of different sources of inoculum and dispersal methods of conidia of *Colletotrichum musae*, the causal agent of banana anthracnose, for fruit contamination. *Plant Pathology*, *49*(6), 782–790. https://doi.org/10.1046/j.1365-3059.2000.00516.x

Duan, Y., Wang, G.-B., Fawole, O. A., Verboven, P., Zhang, X.-R., Wu, D., Opara, U. L., Nicolai, B., & Chen, K. (2020). Postharvest precooling of fruit and vegetables: A review. *Trends in Food Science and Technology*, *100*, 278–291. https://doi.org/10.1016/j.tifs.2020.04.027

Errampalli, D. (2014). *Penicillium expansum* (blue mold). In S. Bautista-Baños (Ed.), *Postharvest decay* (pp. 189–231). Academic Press. https://doi.org/10.1016/B978-0-12-411552-1.00006-5

Exama, A., Arul, J., Lencki, R. W., Lee, L. Z., & Toupin, C. (1993). Suitability of plastic films for modified atmosphere packaging of fruits and vegetables. *Journal of Food Science*, *58*(6), 1365–1370. https://doi.org/10.1111/j.1365-2621.1993.tb06184.x

Fallik, E., & Ilic, Z. (2018). Pre- and post-harvest treatments affecting flavor quality of fruits and vegetables. In M. W. Siddiqui (Ed.), *Preharvest modulation of postharvest fruit and vegetable quality* (pp. 139–168). Springer. https://doi.org/10.1016/B978-0-12-809807-3.00006-8

Fang, Y., & Wakisaka, M. A. (2021). A review on the modified atmosphere preservation of fruits and vegetables with cutting-edge technologies. *Agriculture*, *11*(10). https://doi.org/10.3390/agriculture11100992

Farrell, G., Hodges, R. J., Wareing, P. W., Meyer, A. N., & Belmain, S. R. (2002). Biological factors in PostHarvest Quality. In P. Golob, G. Farrell, & J. E. Orchard (Eds.), *Crop post-harvest: Science and technology volume 1: Principles and practice* (pp. 93–140). Blackwell Science. https://doi.org/10.1002/9780470751015.ch4

Gorham, J. R. (1989). HACCP and filth in food: The detection and elimination of pest infestation. *Journal of Environmental Health*, *52*, 84–86.

Gratz, N. G. (1988). Rodents and human disease: A global appreciation. In I. Prakash (Ed.), *Rodent pest management* (p. 69). CRC Press.

Gross, K. C., Wang, C. Y., & Saltveit, M. (2016). *The commercial storage of fruits, vegetables, and florist and nursery stocks (No. 66)*. United States Department of Agriculture. Agricultural Research Service.

Holcroft, D. (2015). Water relations in harvested fresh produce. In *The postharvest education foundation* (White Paper no. 15, 01). https://doi.org/10.1080/10408398909527490

Kader, A. A. (2013). Postharvest technology of horticultural crops – An overview from farm to fork. *Ethiopian Journal of Science and Technology* (Special Issue No.1), 1–8.

Kader, A. A., & Saltveit, M. E. (2002). Respiration and gas exchange. In J. A. Bartz & J. K. Br echt (Eds.), *Postharvest physiology and pathology of vegetables* (pp. 31–56). https://doi.org/10.1201/9780203910092

Kader, A. A., Zagory, D., & Kerbel, E. L. (1989). Modified atmosphere packaging of fruits and vegetables. *Critical Reviews in Food Science and Nutrition*, *28*(1), 1–30. https://doi.org/10.1080/10408398909527490

Kahramanoglu, I. (2017). Postharvest physiology and technology of horticultural crops. In I. T. Open (Ed.), *Postharvest handling (Kahramanoğlu, İ.)*, *1–5*. Rijeka. http://doi.org/10.5772/66538

Kitinoja, L., & Kader, A. A. (2003). *Storage of horticultural crops. Small-scale postharvest handling practices: A manual for horticultural crops* (4th ed.). Postharvest Horticulture Series No. 8E.

Koshita, Y. (2014). Effect of temperature on fruit colour development. In Y. Kanayama & A. Kochetov (Eds.), *Abiotic stress biology in horticultural plants* (pp. 47–58). https://doi.org/10.1007/978-4-431-55251-2_4

Kubo, Y. (2014). Ethylene, oxygen, carbon dioxide, and temperature in postharvest physiology. In Y. Kanayama & A. Kochetov (Eds.), *Abiotic stress biology in horticultural plants* (pp. 17–33). https://doi.org/10.1007/978-4-431-55251-2_2

Lassois, L., & de Lapeyre de Bellaire, L. (2014). Crown rot disease of bananas. In S. Bautista-Baños (Ed.), *Postharvest decay* (pp. 103–130). Academic Press. https://doi.org/10.1016/B978-0-12-411552-1.00003-X

Lata, D., & Sujayasree, O. J. (2021). Chapter 15. Significance of ethylene in postharvest technology. In S. Mitra (Ed.), *Postharvest management of horticultural crops* (pp. 250–262). JAYA Publishing House.

Loomans, A., Diakaki, M., Kinkar, M., Schenk, M., & Vos, S. (2019). Pest survey card on Bactrocera dorsalis. *EFSA Supporting Publications*, *16*(9). https://doi.org/10.2903/sp.efsa.2019.EN-1714

Mangaraj, S., Goswami, T. K., & Panda, D. K. (2014). Modeling of gas transmission properties of polymeric films used for MA packaging of fruits. *Journal of Food Science and Technology*, *52*(9), 5456–5469. https://doi.org/10.1007/s13197-014-1682-2

Mohammed, M., Isaac, W. A., Mark, N., St. Martin, C., & Solomon, L. (2014). Effects of curing treatments on physico-chemical and sensory quality attributes of three pumpkin cultivars. *Acta Horticulturae*, *1047*(1047), 57–62. https://doi.org/10.17660/ActaHortic.2014.1047.4

Murmu, S. B., & Mishra, H. N. (2017). Engineering evaluation of thickness and type of packaging materials based on the modified atmosphere packaging requirements of guava. *CV. Baruipur. LWT. Food Science and Technology*, *78*, 273–280. https://doi.org/10.1016/j.lwt.2016.12.043

Nasrin, T. A. A., Yasmin, L., Arfin, M. S., Rahman, M. A., Molla, M. M., Sabuz, A. A., & Afroz, M. (2022). Preservation of postharvest quality of fresh cut cauliflower through simple and easy packaging techniques. *Applied Food Research*, *2*(2), 100125. https://doi.org/10.1016/j.afres.2022.100125

Osorio, S., & Fernie, A. R. (2013). Biochemistry of fruit ripening. In S. Grahman, P. Mervin, G. James, & T. Gregory (Eds.), *The molecular biology and biochemistry of fruit ripening* (pp. 1–19). Blackwell Publishing Ltd. https://doi.org/10.1002/9781118593714.ch1

Özdemir, İ. S. (2016). Effect of light treatment on the ripening of banana fruit during postharvest handling. *Fruits*, *71*(2), 115–122. https://doi.org/10.1051/fruits/2015052

Palou, L. (2014). *Penicillium digitatum, Pencillium italicum* (green mold, blue mold). In S. Bautista-Baños (Ed.), *Postharvest decay* (pp. 45–102). Academic Press. https://doi.org/10.1016/B978-0-12-411552-1.00002-8

Paltrinieri, G. (2014). *Handling of fresh fruits, vegetables and root crops: A training manual for Grenada*. Food and Agriculture Organization of the United Nations.

Pandey, A. K., Jain, P., Podila, G. K., Tudzynski, B., & Davis, M. R. (2008). Cold induced Botrytis cinerea enolase (BcEnol-1) functions as a transcriptional regulator and is controlled by cAMP. *Molecular Genetics and Genomics, 281*(2), 135–146. https://doi.org/10.1007/s00438-008-0397-3

Paul, V., Pandey, R., & Srivastava, G. C. (2012). The fading distinctions between classical patterns of ripening in climacteric and non-climacteric fruit and the ubiquity of ethylene-an overview. *Journal of Food Science and Technology, 49*(1), 1–21. https://doi.org/10.1007/s13197-011-0293-4

Pétriacq, P., López, A., & Luna, E. (2018). Fruit decay to diseases: Can induced resistance and priming help? *Plants, 7*(4), 77. https://doi.org/10.3390/plants7040077

Phillips, P. A., & Walker, G. P. (1997). Increase in flower and young fruit abscission caused by citrus bud mite (acari: Eriophyidae) feeding in the axillary buds of lemon. *Journal of Economic Entomology, 90*(5), 1273–1282. https://doi.org/10.1093/jee/90.5.1273

Pupin, F., Bikoba, V., Biasi, W. B., Pedroso, G. M., Ouyang, Y., Grafton-Cardwell, E. E., & Mitcham, E. J. (2013). Postharvest control of western flower thrips (Thysanoptera: Thripidae) and California Red scale (Hemiptera: Diaspididae) with ethyl formate and its impact on citrus fruit quality. *Journal of Economic Entomology, 106*(6), 2341–2348. http://doi.org/10.1603/EC13111

Reitz, S. R. (2009). Biology and ecology of western flower thrips (Thysanoptera: Thripidae): The making of a pest. *Florida Entomologist, 92*(1), 7–13. https://doi.org/10.1653/024.092.0102

Renault, P., Souty, M., & Chambroy, Y. (1994). Gas exchange in modified atmosphere packaging. 1: A new theoretical approach for micro-perforated packs. *International Journal of Food Science and Technology, 29*(4), 365–378. https://doi.org/10.1111/j.1365-2621.1994.tb02079.x

Romanazzi, G., & Feliziani, E. (2014). *Botrytis cinerea* (gray mold). In S. Bautista-Baños (Ed.), *Postharvest decay* (pp. 131–146). Academic Press. https://doi.org/10.1016/B978-0-12-411552-1.00004-1

Saltveit, M. E. (2019). Respiratory metabolism. *Postharvest Physiology and Biochemistry of Fruits and Vegetables*, 73–91. https://doi.org/10.1016/b978-0-12-813278-4.00004-x

Scholtz, I., Siyoum, N., & Korsten, L. (2017). Penicillium air mycoflora in postharvest fruit handling environments associated with the pear export chain. *Postharvest Biology and Technology, 128*, 153–160. https://doi.org/10.1016/j.postharvbio.2017.01.009

Sen, Y., van der Wolf, J., Visser, R. G. F., & van Heusden, S. (2015). Bacterial canker of tomato: Current knowledge of detection, management, resistance, and interactions. *Plant Disease, 99*(1), 4–13. https://doi.org/10.1094/PDIS-05-14-0499-FE

Singh, D., & Sharma, R. R. (2018). Postharvest diseases of fruits and vegetables and their management. In M. W. Siddiqui (Ed.), *Postharvest disinfection of fruits and vegetables* (pp. 1–52). Springer. https://doi.org/10.1016/B978-0-12-812698-1.00001-7

Thompson, A. K. (2008). *Fruit and vegetables: Harvesting, handling and storage*. John Wiley & Sons.

Troncoso-Rojas, R., & Tiznado-Hernández, M. E. (2014). *Alternaria alternata* (black rot, black spot). In S. Bautista-Baños (Ed.), *Postharvest decay* (pp. 147–187). Academic Press. https://doi.org/10.1016/B978-0-12-411552-1.00005-3

Wang, C. Y. (1989). Chilling injury of fruits and vegetables. *Food Reviews International, 5*(2), 209–236. https://doi.org/10.1080/87559128909540850

Watson, J. A., Treadwell, D., Sargent, S. A., Brecht, J. K., & Pelletier, W. (2015). *Postharvest storage, packaging and handling of specialty crops: A guide for Florida small farm producers*. University of Florida.

Wei, H., Seidi, F., Zhang, T., Jin, Y., & Xiao, H. (2020). Ethylene scavengers for the preservation of fruits and vegetables: A review. *Food Chemistry, 337*, 127750. https://doi.org/10.1016/j.foodchem.2020.127750

Williamson, B., Tudzynski, B., Tudzynski, P., & Van Kan, J. A. L. (2007). Botrytis cinerea: The cause of grey mould disease. *Molecular Plant Pathology, 8*(5), 561–580. https://doi. org/10.1111/j.1364-3703.2007.00417.x

Xie, G., Feng, Y., Chen, Y., & Zhang, M. (2020). Effects of 1Methylcyclopropene (1-MCP) and ethylene on postharvest lignification of common beans (*Phaseolus vulgaris* L.). *ACS Omega, 5*(15), 8659–8666. https://doi.org/10.1021/acsomega.0c00151

Yahia, E. M., Neven, L. G., & Jones, R. W. (2019). Chapter 16. Postharvest insects and their control. In E. M. Yahia (Ed.), *Postharvest technology of perishable horticultural commodities* (pp. 529–562). Woodhead Publishing. https://doi.org/10.1016/ B978-0-12-813276-0.00016-X

# 3 Postharvest Losses during Storage and Supply Chain

*Ayoub Mohammed and Wendy-Ann Isaac*

## 3.1 INTRODUCTION

It is estimated that, by 2050, food production will need to increase by about 70% to meet the demand of feeding the world population which is expected to be around 9 billion people (Kiaya, 2014). Developing countries, which already find it difficult to increase its production of food to meet its growing food demand, also face the additional challenge of high postharvest losses of between 40–50% of its current production. Based on the estimates of the Food and Agricultural Organization of the United Nations (FAO), a 30% loss equals approximately 1.3 billion tonnes of food, which can feed up to 1.6 billion people annually. It goes without saying that food production is only half the battle in providing food for growing populations, and there is a need to implement effective strategies to reduce such losses, thus making more food available. Postharvest losses may be defined as the losses that occur from the point of harvest to the point of delivery to the final consumer (Yahia, 2019; GSARS, 2020). Reducing postharvest losses therefore is an important complementary strategy in increasing food production (Yahia, 2019).

A plethora of factors is known to impact postharvest losses including but not limited to preharvest or production factors, harvesting and field handling, packinghouse operations, packaging operations, predelivery storage operations, postprocessing delivery transportation and storage at the retail and wholesale facility (Hodges et al., 2011; Yahia, 2019). Magalhaes et al. (2021) identified 14 critical factors that lead to loss in the fresh produce supply chain:

1. Inadequate demand forecasting
2. Overproduction and excessive stock
3. Poor handling and operational performance
4. Storage at wrong temperature
5. Inadequate or defective packaging
6. Nonconformity with retail specifications
7. Sensorial or microbial deterioration
8. Short shelf life or expired products
9. Climate change and/or variability in weather conditions
10. Lack of postharvest storage facilities

DOI: 10.1201/9781003452355-4

11. Pricing strategies and promotion management
12. Lack of coordination and information management
13. Inadequate transport management
14. Inefficient in store storage management

From a supply chain perspective, the approach of Magalhães et al. (2021) brings together a deeper understanding of the multiplicity of factors that ought to be considered when managing a fresh produce business, as it discusses postharvest physiological factors and business consideration factors. Ramaswamy (2014) classified the main causes of postharvest losses into two categories: primary causes and secondary causes. Further, Ramaswamy (2014) listed six important primary causes: (1) biological and microbiological, (2) chemical and biochemical, (3) mechanical damage, (4) improper storage environments, (5) physiological factors and (6) psychological factors. Ramaswamy (2014) also listed six main secondary factors: (1) respiration, (2) ethylene production, (3) changes in composition, (4) growth and development considerations, (5) moisture loss due to transpiration and (6) physiological breakdown. While this classification is useful in understanding the main precursors of postharvest losses, these factors are not mutually exclusive, and very often an interplay of factors may set the stage for postharvest losses. From a fresh produce supply chain management perspective, it is important to note that often damage is latent and not visible until the problem leads to major damage at the point of the final consumer (Thompson, 2010). It becomes increasingly important therefore that loss reduction or mitigation strategies consider the underlying factors of both primary and secondary causes and mitigate against them along the postharvest supply chain. A simple diagrammatic representation of a postharvest supply chain is presented Figure 3.1.

A review of definitions of supply chain management is offered by Min et al. (2019). Despite the plethora of definitions, the supply chain is seen as a system that brings together people, activities, information and resources that are involved in moving products, in this case fresh produce from the point of production to final customers. Fresh produce supply chains are particularly challenging as the products are "alive" along the chain and are generally highly perishable, and the primary and secondary causes of deterioration mentioned earlier can result in the loss of the entire shipment from the point of production to the point of delivery to the final consumer. While the damage might be visible upon arrival of the final destination, the precursors for deterioration would have started long before. Mohammed (2005) demonstrated that hot peppers, harvested and air freighted to the United States of America from Trinidad, can result in 100% loss on arrival at US airports. Even a much shorter movement of produce from field to final consumer has resulted in significant qualitative and quantitative losses (Ramaswamy, 2014).

The highly perishable fresh produce supply chain can benefit from the principles involved in the four steps in the supply chains model often referred to as the supply chain operations (SCOR) model. The key elements of this model is planning, developing (sourcing), making, delivery and return. Planning involves determining the best logistics as well as resource and inventory planning, among other planning elements. At the sourcing stage, the supplier–buyer relationships are built and strengthened, with terms determined, finalized and agreed on for methods of

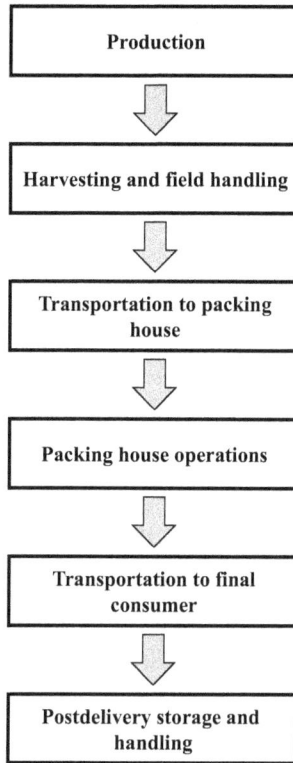

**FIGURE 3.1** Diagrammatic representation of postharvest supply chain.

delivery, produce-specific grade and standards, payment, shipping and delivery. Any delay, miscommunication or misunderstanding can lead to irrecoverable losses as produce needs to move quickly from supplier to buyer.

At the market stage, produce is harvested and prepared normally in a packing-house environment and readied for shipping to the buyer. While in other manufacturing arrangements involving nonperishables there may be ample time for returning defective products to the seller, fresh produce is much more challenging as deterioration can be very rapid once it begins, and therefore return and rework, while possible, are always challenging. Marrying the concepts of traditional supply chains, the processes involved in maintaining quality from the point of harvest can be considered as downstream processes, while those processes associated from the moment the products leave the packinghouse to its final destination as upstream. Fresh produce supply chains are becoming increasingly more complex in terms not merely of the distance fresh produce is now moving from the point of production but also of the increasing demand of consumers for higher-quality products in terms of qualitative and quantitative consumer expectations, as well meeting greater regulatory requirements in destination markets. As fresh produce supply chains become more complex, innovation and creativity, supply chain risk management, ethics and sustainability, cost reduction, and what is often referred to as the five rights (right price, right quality, right time, right quantity, right place) will play greater roles in the fresh produce industry. Notwithstanding the importance of understanding the complex postharvest requirement of fresh produce as it moves along the supply chain, there is also a need to manage other key processes: customer relationships, customer service management, demand management, order fulfilment, supplier relationship management (both for single-source and multiple-source) suppliers. Prior to harvesting, fresh produce obtains food and water from its "parent" plant. Once they are removed from this source of food, water and other vital nutrients, they are prone to deterioration, which can occur very rapidly. Much has been written on specific, individual postharvest requirements (Rees et al., 2012). Three important postharvest physiological considerations must be understood and managed if postharvest quality is to be maintained: respiration, ethylene production and transpiration.

Marketing channels for fresh produce vary in their complexities (Thompson, 2015). Thompson (2015) describe several types of supply chain arrangements that exists in fresh produce marketing, though there are myriads of permutations within each channel. In its simplest form, consumers may purchase directly from the farm. Products may be harvested on an as-needed basis or using a U-Pick model where customers are allowed to do their own harvesting operations. In other models, operators may purchase in bulk and move these products to be sold in wholesale markets. The time from harvesting to be sold on the wholesale market to then being purchased by retail market operators often leads to high postharvest losses, which is exacerbated by a number of factors including poor harvesting and field handling, poor field packaging, overstacking, delays in entering wholesale markets with produce spending upwards of 12–20 hours with very little protection from sunlight (Mohammed & Mujaffar, 2014). Studies conducted on tomato, hot pepper, pineapple, tomato, pumpkin, culantro and eggplant have all demonstrated that these marketing channels often lead to high postharvest losses that are often measured and expressed at the point of final sale (Mohammed & Di Chi, 2006). Studies done on hot peppers and pumpkin destined for exports markets have demonstrated large postharvest losses including loss of the entire shipment due to postharvest decay. Mohammed and Mujaffar (2014) shows that in hot peppers, oftentimes high levels of rejected fruits come into the packhouse, and, in the absence of strict adherence to a quality management system from farm to export, postharvest losses may render these businesses unprofitable. Other marketing channels also exists, such as when farmers offer their products directly for sale at municipal markets as well as at farmers markets (Thompson, 2015).

## 3.2 STRATEGIES TO REDUCE POSTHARVEST LOSSES IN FRESH PRODUCE SUPPLY CHAINS

### 3.2.1 Managing the Major Causes of Mechanical Crop Damage

Often overlooked, field handling and field packaging set the stage for either the maintenance of postharvest quality or the rapid deterioration of shelf life along the supply chain (Pathare & Al-Dairi, 2021). Bruising, more often than not, starts in the field, and these injuries become more evident as freshly harvested products move along the supply chain (Shewfelt, 1986; Lee, 2005). Bruising may affect freshly harvested horticultural products internally without any signs of external damage and may cause external blemishes or a combination of both external and internal damage (Prusky, 2011). Crop damage may also result in exacerbated increases in physiological processes leading to rapid produce deteriorations as fresh horticultural products move through the supply chain (Kumar et al., 2006). Crop damage will result in shorter shelf life as respiration and ethylene production are likely to increase; damaged produce also has more portals of entry for spoilage microorganisms (Prusky, 2011). Even if the damage is mild and does not affect internal quality, it will affect marketability and profitability as consumers may refuse to purchase these products based on damage to cosmetic appearance (Brosman & Sun, 2004; Shafie et al., 2017). Bruising may be affected by a number of factors including harvesting methods, time of harvesting, type of field packaging material, variation due to seasonality, time after harvest, maturity indices at the time of harvesting, precooling and final storage conditions (Hussein et al., 2019). Thompson (2015) identified six major causes of damage in fresh produce including cuts, scuffing, compression damage, impact bruising and vibration bruising. These various causes of mechanical damage can take place along the entire supply chain and often begin in the field (Ferreira et al., 2009). Compression bruising is defined as a force beyond above a certain critical value (threshold value) and may result in damage to fresh produce. While the relationship is dependent on a number of factors including time, the degree of compression and type of commodity involved, compression damage may not be immediately visible, especially in produce that are harvested mature-green and placed in storage to ripen. Overstacking of produce into poor field packaging containers on the field may be one of the primary contributors of compression and other types of crop bruising (Ferreira & Prokopets, 2009). The use of polypropylene sacks in field packaging of delicate commodities such as hot pepper and tomato has been linked to poor shelf life and quality losses in hot peppers, sweet peppers and cassava. The use of stackable nesting crates may provide a simple long-term solution that can significantly reduce the incidence of compression bruising. Impact bruising is attributed to the sudden fall of produce or an object hitting the produce (Toivonen et al., 2007). Table 3.1 provides a summary of the types of field packaging and their advantages and disadvantages. Impact may not affect the external appearance of the damage at the time the bruising occurs, but it may affect the internal tissues (Shewfelt, 1986). Avocadoes exported from Mexico to European markets are often impacted by two major causes of postharvest losses. Anthracnose caused by the fungus *Collectotrichum gloeosporiodes* and internal blackening, which has been attributed to harvest and postharvest impact bruising (Zamora-Magdaleno et al., 2001). Scuffing damage is often observed when fresh horticultural products keep rubbing against a hard surface. This may occur when produce is kept in tight spaces and their outer peel rubs against one another or against another surface such as packaging material. From a postharvest management standpoint, steps should be taken to prevent scuffing as the damage to the peel can lead to greater damage as cells become damaged due to abrasion. It was found that in citrus fruits, scuffing damage was responsible for the greatest loss in postharvest shelf life (Tariq et al., 2001). In managing the fresh produce supply chain, it is important to understand from the point of harvest to final display, fresh horticultural

**TABLE 3.1**

**Summary of the Types of Field Packaging and Their Advantages and Disadvantages**

| Type of Field Harvesting Container | Advantages | Disadvantages |
|---|---|---|
| Wooden crates | Offers protection to some products<br>Offers ventilation<br>Stable when stacked<br>Reparable<br>Relatively durable<br>Convenient to handle<br>Labels can be attached.<br>Reusable | High purchase cost<br>Water damage can cause rotting.<br>Difficult to clean<br>Disposal problem<br>Poor construction can lead to puncture and bruises to delicate produce.<br>Can harbor pests--weevils, termites |
| Woven basket | Offers some protection of produce<br>Has ventilation<br>Reusable | Low compression strength; therefore does not offer adequate protection for delicate fresh produce.<br>Difficult to clean and sanitize<br>Does not utilize space efficiently<br>Difficult to stack<br>Has rough surfaces which can cause cuts and puncture to fresh produce. |
| Corrugated fiber board boxes | Lightweight<br>Easy to handle<br>Smooth surfaces<br>Good ventilation<br>Labels can be easily attached.<br>Offers good protection<br>Stable when stacked<br>Easy to palletize<br>Available in many different styles<br>Available in different wall thickness | Not suitable for field packing of fresh produce as field conditions will cause the packaging to be easily damaged.<br>Cold temperatures and high relative humidity will cause the packaging to lose its strength. |
| Polypropylene sacks | Low cost<br>Easy availability<br>Can be used for products such as dried coconuts, dried cocoa beans, etc.<br>Unsuitable for most fresh fruits and vegetables | Offers little or no protection for most fresh fruits and vegetables.<br>Inadequate ventilation<br>Allows heat buildup<br>Difficult to stack<br>Cause serious damage to delicate perishable crops |
| Ventilated high-density polyethylene crates | Durable<br>Offers good protection<br>Excellent ventilation<br>Stable when stacked<br>Low cargo volume when empty—nestable<br>Reusable<br>Easy to clean<br>Comes with UV protection<br>Made from recyclable plastic<br>Can be used for a variety of crops. | Higher initial capital cost, but think of it as an investment. |

products go through a complex number of handling steps which need to be carefully designed, planned and executed in order to maximize revenue, income and customer loyalty.

### 3.2.2  TEMPERATURE MANAGEMENT

Managing the cold chain temperature is the single most important management strategy required to reduce postharvest losses in fruits and vegetables (Kitinoja & Thompson, 2010). Management of temperature starts at the time of harvest and continues all the way to delivery to the final customer. At the level of field handling, it is best to harvest fresh produce early in the morning after it has benefitted from cooler nights and thus lower transpiration and respiration rates. After harvesting, it is best to put in the appropriate field packaging material, keeping the produce in a cool place, thus preventing the buildup of field heat. As soon as possible, produce should be removed and taken to a packinghouse facility to complete the prestorage treatments and post-treatment at a safe low temperature storage (Ahmad & Siddiqui, 2015). Simply keeping fresh produce in a shaded area on the field can positively impact produce. Rickard and Coursey (1979) reported that fresh produce kept in a shaded area on the field has pulp temperatures that were 3–10°C when compared to unshaded produce. Kitinoja (2010) reported that tomato and melongene, which were kept in the open sun had pulp temperatures that were around 15°C higher when compared to tomato and eggplant kept in a shaded area on the field. These simple field temperature management strategies are important since the lower the temperature maintained prior to precooling, the less the refrigeration load and the faster the cooling takes place during precooling operations (Thompson & Singh, 2008). Maximum benefit is obtained when fruits and vegetables are stored at temperatures just above their freezing temperatures. Fruits and vegetables of tropical and subtropical origin are prone to a physiological disorder at temperatures above 0°C and as high as 17°C. The phenomenon of chilling injury has been reviewed, but what is critical to the supply chain is that these products need to be stored at refrigerated temperatures above 10–12°C. Even under the best refrigerated storage conditions, the postharvest shelf life of these commodities is much shorter than those of temperate fruits and vegetables, which can be stored at temperatures closer to 0°C without any deleterious effects. If tropical and subtropical commodities are stored at chilling temperatures, the physiological expression of the disorder will result in devastating postharvest losses. Chilling injury has a time–temperature relationship, and beyond a certain threshold, the physiological impairment and, more importantly from a supply chain perspective, the expression of the disorder can result in high economic losses. There are several reviews of the lowest safe temperatures of chilling sensitive commodities (Wang & Zhu, 2017). Symptomatic expressions of chilling injury include pitting, failure to ripen, wilting, loss of color and flavor, decay and internal browning, to name a few. One of the challenging considerations of chilling injury is that symptoms are more visible when produce is removed from chilling to non-chilling temperatures, which has implications for a supply chain management perspective. Along the supply chain, mechanisms need to be put in place to ensure that the refrigerated temperatures of chilling-sensitive products are always kept outside the chilling temperature zone. Individual products may express more specific symptoms of chilling damage over others, and this is important as it gives operators a framework of what symptoms they need to pay close attention to for specific types of products.

### 3.2.2.1   Role of Precooling in Temperature Management of Fresh Produce

One of the most important practices that is used to reduce field heat prior to final refrigerated storage of fresh produce is precooling (Kitinoja & Thompson, 2010). Defined as the rapid cooling of fresh produce to reduce field heat, this practice has been shown to reduce microbial spoilage and to reduce respiration rate, ethylene production and moisture loss in the long-term storage of fresh produce (Ferreira et al., 1994; Reina et al., 1995), all of which helps to extend the postharvest shelf life and quality of fresh produce. Many systems are employed in the precooling of fresh produce, including room cooling, forced air cooling, vacuum cooling, top icing and hydrocooling. These methods have been reviewed and described by Brosnan and Sun (2001) and Thompson (2008). A brief summary of the advantages and disadvantages of these methods is summarized Table 3.2.

While the benefits of precooling in maintaining quality is well-known, small producers and suppliers are often confronted with the difficult challenge of finding appropriate technologies that are cost-effective but that ensure that the quality of the

### TABLE 3.2
### Advantages and Disadvantages of Commonly Used Precooling Methods in Fresh Produce

| Precooling Method | Advantages | Disadvantages |
|---|---|---|
| Room Cooling | Low-cost relative to other methods<br>Best suited for less perishable commodities<br>Effective for chilling sensitive crops<br>Cooling rates can be increased by improving spaces between stacks. | Slow method of cooling since cooling is primarily by conduction. |
| Hydrocooling | Can be used for a wide range of products.<br>Can accommodate a wide range of bulk containers.<br>There is no moisture loss during the cooling process.<br>The method can slightly rehydrate slightly wilted products.<br>Energy efficient compared to some other methods | Can lead to spread of decay organisms, affecting produce if the sanitation of the cooling water is not properly managed.<br>Water must be drained and equipment cleaned daily to prevent food spoilage and contamination by human pathogens. |
| Top icing | Effectively cools products with high respiration rates and high perishability.<br>Heat transfer occurs rapidly, leading to fast cooling of products. | High initial capital cost<br>Packaging should withstand contact with ice and water.<br>Leads to higher freight cost as icing adds weight to the shipment.<br>In transit, melting of ice can damage other products in the container. |

fruits and vegetables they offer in the marketplace meets the demanding standards of the markets they intend to supply (Kitinoja & Thompson, 2010). To meet the needs of these suppliers, Kitinoja and Thompson (2010) describes a range of options that are available especially and that can easily be adapted for small-scale commercial fresh produce operations. These include mobile forced air cooling systems; batch-type mobile and stationery coolers that can precool batches from as little as 1500 kg per batch with an efficacy that gets to the target temperature between 1.5 and 4 hours; evaporative cooling units, which are also called desert coolers or swamp coolers; passive evaporative cooling systems that do not require electricity; the use of ice, package ice and ice banks; solar chilling systems; structures that are supported by cool night temperatures, to name a few.

### 3.2.2.2 Importance of Postharvest Pretreatments in Preventing Postharvest Losses

A packinghouse is an important part of the fresh produce supply chain and serves as the demarcation point between inbound logistics and outbound logistics. On arrival at the packinghouse facilities, produce may be subject to a sampling procedure that may be used to generate a quality report. Mohammed and Mujaffar (2014) described the use of a quality reporting system for hot peppers destined for export markets. Similar reporting systems have been developed for many other products (Mohammed & De Chi, 2006). Apart from the critical role of precooling, other important product-specific pretreatments are also done in a packinghouse facility (Thompson, 2015). Pretreatments include cleaning, sorting, washing, chlorination, waxing, curing, fungicidal dips, heat treatment and fumigation (Rees et al., 2012).

Cleaning may be done either as a dry-cleaning operation, as some products can become damaged if they are cleaned with water; products that are not susceptible to water damage are wet-cleaned. The main objective of cleaning is to assist in reducing the microbial load of the produce. In systems where potable water is recycled during the wet cleaning process, a number of sanitizing agents may be used, with chlorine being the most common. Sorting and grading is another crucial operation done at packinghouses. This may be achieved through automation for large-scale operations or through small-scale to manual sorting for small-scale packinghouse operations. While the term sorting is often used to describe the removal of undesirable produce, such as that with visible signs of mechanical damage, grading is often used to define the categorizing of produce based on weight size or some other defined grade and standards to meet specific market requirements.

Coating fresh produce with edible waxes is a well established practice in the fresh produce trade. Following waxing, a thin layer of wax remains on the surface of the fruits. It has been used in passion fruits, apples, peaches, pineapples, sweet potatoes, tomato and cassava. Edible coatings may be derived from lipids, proteins and polysaccharides. A review of edible coatings and their application has been reviewed by Baldwin (1994) and Raghav et al. (2016). Edible coatings positively impact fresh produce and aids in shelf life extension and quality by a number of mechanisms including reducing moisture loss, preventing or delaying the onset of postharvest physiological deterioration (PPD) in cassava, reducing the rate of respiration and, in some cases, delaying the onset of physiological disruption in fresh produce. Thompson (2015) describes the waxing aids in maintaining crispiness and freshness as produce moves the supply chain. The use of edible waxes therefore plays an important role in reducing losses along the outbound logistics of the supply chain. Recent advances in wax coatings have allowed waxes to be used as carriers and antimicrobial agents, thus further reducing the risks of pathogen invasion during storage. Curing is a well-known strategy used to reduce the moisture content of some commodities, thus extending their shelf life. While the practice is done mostly to hardy vegetables such as onions, garlic roots and tuber crops, pumpkin and squash, curing plays an important role in healing small surface wounds, thus preventing microbial invasion and spoilage of these commodities. In onions and garlic, for example, curing starts with leaving them in the field to dry, which tightens the neck and starts the process of drying the outer sheath leaves. The process of drying is then completed in storage specifically designed for these commodities.

## 3.3  PACKAGING FOR WHOLESALE AND RETAIL DISPLAY

Once products are pretreated, they are generally packaged for the wholesale and retail markets. The role of field packaging has been discussed. Final packaging prior to delivery is equally important as the wrong packaging and handling can also lead to damage along the sometimes long and complicated journey to the final consumer. Apart from the primary functions of packaging, fresh producer operators often need to consider the recycling and biodegradability characteristics of the packaging material, sales and consumer appeal, packaging requirements for bulk wholesale customers versus retail marketers, the shape of the produce, economic considerations, and the impact of the packaging on shelf life and information that may be important to the potential buyer. Fresh produce packaging is classified as consumer (unit packaging), transport packaging and unit load packaging. Consumer packing consists of low-density polyethylene bags, netted bags, trays and sleeves. Corrugated fiber boxes (CFB) dominates the market for the transport of fresh produce. CFBs offer numerous advantages as they are smooth and will not cause abrasion damage to produce. They can also be designed to facilitate ventilation as well as waxed to accommodate produce that may need to be transported on ice and may also allow produce to be transported loose or separated by dividers to act as a second layer of cushioning. Corrugation offers good cushioning of the produce contained within and meets the requirements for recyclability and reuse. Palletizing of produce prior to shipping has been discussed by Kitinoja and Kader (2015), who summarized the benefits of palletization as reducing operational cost, reducing labeling as labels can be placed on the palletized load, maximizing storage space, making it easy to move as it can facilitate forklift operations and reduced mechanical stress and mechanical damage of produce. Palletization can make other operations such as precooling much more effective. Palletized loads are also easy to unitize, which makes transportation much more effective.

Modified and controlled atmosphere packaging have been shown to extend the shelf life of a range of products and have been reviewed by numerous authors (Kitinoja & Kader, 2015). Kitinoja and Kader (2015) described some of the trends in packaging that are all aimed at reducing losses: packaging impregnated with antimicrobial agents that reduce decay, carbon dioxide and oxygen scavengers and ethylene scrubbers that are aimed at preventing physiologically exacerbated produce deterioration during storage.

## 3.4  MANAGEMENT OF POSTHARVEST PHYSIOLOGICAL PROCESSES

Unmanaged respiration, ethylene production and transpiration can lead to significant qualitative and quantitative losses along the fresh produce supply chain. Fruits are often classified as being either climacteric or non-climacteric. A graphical representation of the classification is given in Figure 3.2. Sometime after harvest, climacteric fruits and vegetables show a marked spike in respiration rate concomitant with a marked increase in ethylene production, which results in the physical and biochemical changes associated with fruit ripening. Ethylene is a plant hormone associated with many physiological changes in plants. In the postharvest storage environment, it triggers the ripening process in climacteric fruits. Ripening is associated with a plethora of compositional and metabolic changes in fruits and vegetables. Managing respiration and ethylene are two key aspects that need to be managed during the ripening process in climacteric fruits. Non-climacteric fruits and vegetables show a

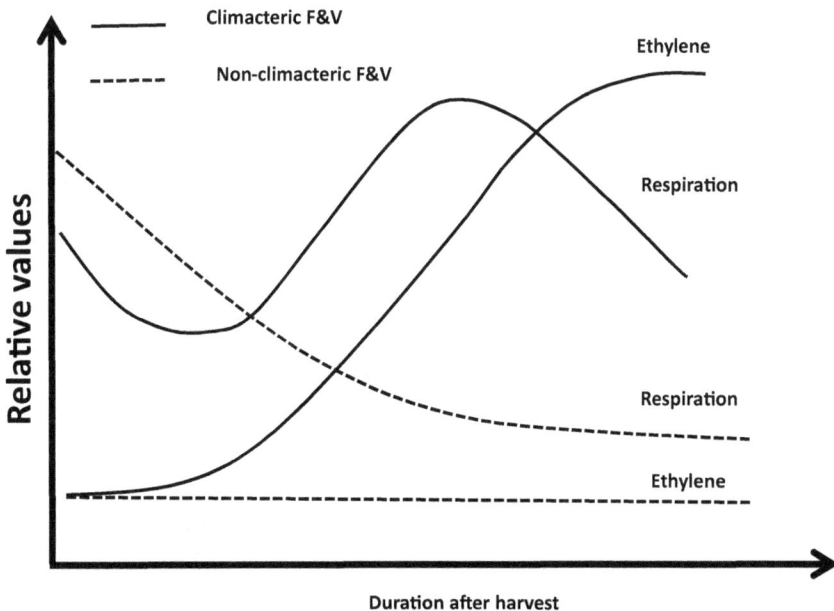

**FIGURE 3.2**  Ethylene production and respiration patterns of climacteric and non-climacteric crops [adapted from Capino & Farcuh, 2021)].

markedly different physiological response after harvest as there is a decline in both respiration rate and ethylene production (Kadar & Saltveit, 2003).

These differences are important as each category of horticultural products requires different management along the supply chain. Non-climacteric fruits will not undergo any major changes in ripening after harvest, and therefore careful attention must be paid to the optimum stage of maturity at the point of harvest. Unlike climacteric horticultural products, which can be harvested at the mature-green stage and allowed to complete ripening, non-climacteric commodities must be harvested at optimum maturity since ripening cannot be completed in storage Kader and Saltveit (2002). Ethylene can have both beneficial and detrimental effects in the postharvest environment, and this must be understood and managed in order to ensure that its benefits are maximized and that its deleterious effects do not result in postharvest and financial losses. The response of climacteric fresh produce to ethylene is referred as autocatalytic Ramaswamy (2015). Once climacteric commodities are exposed to ethylene, whether it is generated from the fruits or vegetables themselves or from an external source, the biochemical processes associated with ripening are initiated, and the irreversible changes associated with ripening begin. An understanding of this phenomenon is important since in many commercial activities where climacteric fruits are being transported to distant markets, they are generally harvested at the mature-green stage and are ripened on arrival at their intended market. If such commodities are exposed to ethylene or precursors of ethylene during storage, then the ripening process can be triggered in transit, resulting in overripe fruits and subsequent losses. Apart from its autocatalytic behavior, exposure to ethylene can have deleterious effects on some horticultural products including leaf senescence, wilting of flower, stimulation of sprouting in tomato, lignification and toughening of asparagus spears and russet spotting in lettuce, peel discoloration and yellowing and softening of cucumbers. If ethylene goes unchecked in the storage environment, it can lead to rapid deterioration and high postharvest losses. Ethylene exerts its effects, whether they are beneficial or damaging, in very minute quantities and often in parts per million, making its management even more important Kader (2002). The principal strategies used to manage ethylene in the postharvest environment fall into the following categories:

1. Eliminating the sources of ethylene
2. Ventilation
3. Use of ethylene absorbers
4. Use of compounds that alter the biosynthesis of ethylene

Strategies that are often used to eliminate ethylene in storage environments include removal of the sources of external ethylene that may trigger damaging effects such as the use of electric over fossil-based-fuel forklifts since the exhaust fumes of the latter can trigger ethylene; removal of ripening fruits or decaying fruits from storage areas which are meant to store mature-green climacteric fruits; and separation of loading trucks from cold storerooms that contain fresh produce. Ventilation is aimed at removing ethylene, which may be produced by fruits while held in storage. Applying ethylene absorbers is a well established practice in the commercial fresh produce trade. They are important in removing ethylene which may be produced by fruits and vegetables while in storage or during transit to destination markets. Potassium permanganate and charcoal are well established in the commercial fresh produce trade as ethylene absorbers, and lastly is the development of 1-MCP, which prevents ethylene biosynthesis by binding to receptors that produce ethylene thus delaying ripening Kader (2002).

Unmanaged water loss along the supply chain translates into economic loss to a fresh produce business (Ramaswamy, 2014). Transpiration refers to the loss of water by evaporation (Ramaswamy, 2014; Bartz & Brecht, 2003). After harvest, the edible portion of the plant is now cut off from a supply of nutrients and water from the parent plant. As the edible portion enters the supply chain, it will continue to transpire, but the goal is to manage the rate of transpiration such that the produce remains alive without significantly impacting the quality of the product when presented at the final consumer to the point that it is rejected or the price has to be renegotiated downward. Transpiration, like other physiological factors, are affected by a number of factors (Bartz & Brecht, 2003), including the nature of the produce, genetics, field production practices, maturation and ripening, respiration environmental factors and postharvest handling practices (Bartz & Brecht, 2003).

Water can be lost through stems, peels, stomata and other natural openings and through damage (Cantwell et al., 2010). The water content in fresh produce ranges from around 65% for products such as garlic to as high as 95% for leafy greens such as lettuce. A compilation of the rate of moisture loss on a fresh weight basis is presented in Bartz and Brecht (2003); there is wide variation in water loss among different produce. Surface:volume ratio is an important factor that fresh produce operators need to pay particular attention to. Burton (1982) showed that leafy greens with high surface:volume ratios have markedly higher water loss than produce with dense heads. Large, dense fruits that have extremely high volume:surface areas lose less water than fruits with smaller volume:surface ratios.

According to Bartz and Brecht (2003), the seminal work that examined the relationship between ripening and water loss was conducted by Wardlaw and Leonard (1936, 1939, 1940). They demonstrated that in mangoes and bananas, the respiration rates remain low at 29°C and a relative humidity of 85% during the preclimacteric phase but rose sharply at the onset of the climacteric, followed by achieving steady-state moisture loss at the climacteric peak. Unlike mangoes and bananas, there was marked increase in moisture loss in avocado and apples that coincided with the ripening. These observations are important as they indicate product specificity, which is important factor when planning postharvest management strategies for individual products and which, while there are generally agreed principles in terms of how fresh produce supply chains may be developed and handled, mean that specific products will behave differently and must be taken into account.

Moisture loss is well correlated with shriveling and decrease in firmness in fresh produce, which negatively impacts consumer acceptability. It is also noteworthy that a relatively small amount of water loss can result in a large decrease in fruit firmness. Cantwell et al. (2010) demonstrated that a 4% reduction in the moisture content of broccoli results in a 30% decrease in firmness in broccoli heads. The fundamental principle of managing moisture loss is reducing the water vapor pressure deficit between the produce and the environment. Key postharvest management strategies to manage transpiration is given by Ramaswamy (2014), Bartz and Brecht (2003). Maintaining a low temperature difference between the produce and the cooling coils by optimizing the refrigeration systems and through humidification of the air either through misting, wet floor, use of postharvest coating and plastic films.

Unmanaged moisture loss is linked to senescence and curtailment of shelf life in some produce since water stress has been linked to increased electrolyte leakage, decline in water potential at the cellular level in bell peppers, followed by softening, which is associated with produce senescence. Water stress is believed to increase the concentration of free radicals at a cellular level and deterioration observed at a cellular level. In some climacteric fruits, increased water stress is correlated with early ripening and senescence.

## 3.5 MANAGING POSTHARVEST DISEASES ALONG THE SUPPLY CHAIN

If left unmanaged, postharvest diseases can cause losses of entire shipments of fresh fruits and vegetables. Infection, colonization and growth of bacterial and fungal organisms are the main causes of microbial damage in postharvest environments, though yeasts and molds may also be involved. A systems approach employing multiple strategies from production through delivery and storage is the foundation of managing postharvest diseases. The main fungal organisms involved in postharvest disease are *Alternaria, Botrytis, Colletotrichum, Diplodia, Penicillium, Phytophthora* and *Rhizopus* spp. (Mudaliar et al., 2023). A systems approach offers the best strategy for managing these diseases since infection can occur in the field or greenhouses, on harvesting tools, in harvesting totes and during packinghouse operations and storage. While fruits and vegetables do have some protection of the outer layers due to a waxy cuticle and closely packed outer cells, they are still largely delicate, and, even if they do not suffer from mechanical bruising, they are still susceptible to attack by microorganisms. Some microorganisms interact with specific commodities while not having any effect on other species. As a case in point, *Penicillium digitatum* attacks citrus, while *Penicillium expansum* attacks apples and pears but not citrus. Attacks by microorganisms may occur through natural openings Kader (2002) or through cuts and other mechanical wounds. Physiological damage, such as chilling injury during ripening and senescence when the peel is soft and least resistant, can result in microbial infection. Some fungal infections such as *Colletotrichum* spp. occur in the field. Fungal spores enter through the flower, form a germ tube and remain quiescent until the conditions are right before displaying symptoms of decay at which time it is too late as the damage is irreversible, resulting in quantitative and financial losses. Other fungal organisms only need minor damage of the fruit surface, which may not be visible to the naked eye, leading to microbial growth and rotting of the infected fruit or vegetable. During the infection and early stages of colonization of the peel of the produce, there are few visible symptoms of the disease by way of rotting. Several factors are responsible for early disease development: temperature, humidity and produce defense mechanisms Kader (2002).

Strategies to control postharvest decay may be curative, aimed at preventing or suppressing growth of fungal organisms or aimed at changing the microclimate in which the pathogens survive. From a systems handling approach, the field strategies should include proper field sanitation, tree pruning, drainage and use of fungicides. From a practical application standpoint, cleaning and sanitizing harvesting totes with chlorine-based sanitizers have been shown to reduce the microbial load. Cleaned and sanitized tools, as well chlorination of postharvest wash water and the use of hot water dips in approved fungicidals have all been shown to arrest the development of postharvest disease in produce Kader (2002).

## 3.6   CONCLUSION

Several factors affect postharvest/food loss during storage and through the stages in the food supply chain. This chapter examined some of these postharvest losses and discussed mitigation strategies that should be implemented to reduce food losses along the supply chain. It examined the management of mechanical crop damage and temperature. It also looked at the role of precooling in the temperature management of fresh produce and pretreatments in preventing postharvest losses. Mitigation strategies include innovative technologies for packaging and precooling, all aimed at managing postharvest physiological processes and diseases along the supply chain.

## REFERENCES LIST

Ahmad, M., & Siddiqui, M. (2015). Factors affecting postharvest quality of fresh fruits. *Postharvest Quality Assurance of Fruits*, 7–32. https://doi.org/10.1007/978-3-319-21197-8_2

Baldwin, E. A. (1994). Edible coatings for fresh fruits and vegetables: Past, present, and future. In J. M. Krochta, E. A. Baldwin, & M. O. Nisperos-Carriedo (Eds.), *Edible coatings and films to improve food quality* (pp. 25–64). Technomic Publishing.

Bartz, J. A., & Brecht, J. K. (2003). *Postharvest physiology and pathology of vegetables*. CRC Press. https://doi.org/10.1201/9780203910092

Brosnan, T., & Sun, D. W. (2001). Precooling techniques and applications for horticultural products-a review. *International Journal of Refrigeration*, *24*(2), 154–170. https://doi.org/10.1016/S0140-7007(00)00017-7

Brosnan, T., & Sun, D. W. (2004). Improving quality inspection of food products by computer vision: A review. *Journal of Food Engineering*, *61*(1), 3–16. https://doi.org/10.1016/S0260-8774(03)00183-3

Burton, W. G. (1982). *Postharvest physiology of food crops*. Longman Group Ltd.

Cantwell, M., Hong, G., & Nie, X. (2010). Using tissue ammonia and fermentative volatile concentrations as indicators of beneficial and stressful modified atmospheres for leafy and floral vegetables. *Acta Horticulturae*, (876), 165–172. https://doi.org/10.17660/ActaHortic.2010.876.20

Capino, A., & Farcuh, M. (2021). Ethylene and the regulation of fruit ripening. *Vegetable and Fruit News*, *12*(1). Retrieved September 20, 2022, from https://extension.umd.edu/resource/ethylene-and-regulation-fruit-ripening

Ferreira, M. D., Brecht, J. K., Sargent, S. A., & Aracena, J. J. (1994). Physiological responses of strawberry to film wrapping and precooling methods. *Proceedings of the Florida State Horticultural Society*, *107*, 265–269.

Ferreira, J., & Prokopets, L. (2009). Does offshoring still make sense? Supply chain. Management Review, 13(1), 2027.

Grolleaud, M. (2002). Postharvest losses: Discovering the full story. In *Overview of the phenomenon of losses during the post-harvest*. Retrieved August 23, 2022, from https://agris.fao.org/agris-search/search.do?recordID=XF2016055548

GSARS. (2020). *Guidelines on the measurement of harvest and postharvest losses*. GSARS Guidelines. Retrieved August 23, 2022, from http://www.fao.org/documents/card/en/c/CB1554EN/

Hodges, R. J., Buzby, J. C., & Bennett, B. (2011). Postharvest losses and waste in developed and less developed countries: Opportunities to improve resource use. *Journal of Agricultural Science, 149*(Suppl. 1), 37–45. https://doi.org/10.1017/S0021859610000936

Hussein, Z., Fawole, O. A., & Opara, U. L. (2019). Determination of physical, biochemical and microstructural changes in impact-bruise damaged pomegranate fruit. *Journal of Food Measurement and Characterization, 13*(3), 2177–2189. https://doi.org/10.1007/s11694-019-00138-z

Kader, A. A. (2002). Postharvest technology of horticulture crops. *Oakwood University of California Agriculture and Natural Resources Publication, 3311*, 535.

Kader, A. A., & Saltveit, M. E. (2002). Respiration and gas exchange. In J. A. Bartz & J. K. Brecht (Eds.), *Postharvest physiology and pathology of vegetables* (pp. 31–56). https://doi.org/10.1201/9780203910092

Kiaya. (2014). *Postharvest losses and strategies to reduce them*. Retrieved July 21, 2022, from http://www.actioncontrelafaim.org/wp-content/uploads/2018/01/technical_paper_phl__.pdf

Kitinoja, L. (2010). *Identification of appropriate postharvest technologies for improving market access and incomes for small horticultural farmers in sub-Saharan Africa and South Asia. WFLO Grant Final Report to the Bill & Melinda Gates Foundation, March 2010.318p*. Retrieved July 22, 2022, from http://ucce.ucdavis.edu/files/datastore/234-1847.pdf

Kitinoja, L., & Kader, A. A. (2015). *Small-scale postharvest handling practices: A manual for horticultural crops* (5th ed). Postharvest Horticulture Series No. 8E. University of California, Davis, Postharvest Technology Research and Information Center.

Kitinoja, L., & Thompson, J. F. (2010). Pre-cooling systems for small-scale producers. *Stewart Postharvest Review, 6*(2), 1–14. https://doi.org/10.2212/spr.2010.2.2

Kumar, D. K., Basavaraja, H., & Mahajanshetti, S. B. (2006). An economic analysis of post-harvest losses in vegetables in Karnataka. *Indian Journal of Agricultural Economics, 61*, 134–146.

Lee, E. (2005). Quality changes induced by mechanical stress on roma-type tomato and potential alleviation by 1-methylcyclopropene (Doctoral Dissertation). University of Florida.

Magalhães, V. S. M., Ferreira, L. M. D. F., & Silva, C. (2021). Using a methodological approach to model causes of food loss and waste in fruit and vegetable supply chains. *Journal of Cleaner Production, 283*, 124574. https://doi.org/10.1016/j.jclepro.2020.124574

Min, S., Zacharia, Z. G., & Smith, C. D. (2019). Defining supply chain management: In the past, present, and future. *Journal of Business Logistics, 40*(1), 44–55. https://doi.org/10.1111/jbl.12201

Mohammed, A. (2005). *Postharvest handling and food safety issues for hot pepper production, seminar*. Proceedings of the on Opportunities in Hot Pepper Production, RCLRC Couva.

Mohammed, A., & De Chi, W. (2006). *Production and good agricultural practices for pumpkin, IICA*. Retrieved July 17, 2022, http://www.namistt.com/DocumentLibrary/Production%20Profiles/GAP%20PUMPKINS.pdf

Mohammed, A., & Mujaffar, S. (2014). Postharvest handling of hot peppers. *Researchgate Presentations*. https://doi.org/10.13140/RG.2.1.4866.7286

Mudaliar, K., Sharma, V., Agnihotri, C., Agnihotri, S., Deora, A., & Singh, B. P. (2023). Microbiological impact and control strategies to monitor postharvest losses in fruits and vegetables. In B. P. Singh, S. Agnihotri, G. Singh, V. K. Gupta (Eds.), *Postharvest Management of Fresh Produce* (pp. 113–147). Academic Press. https://doi.org/10.1016/B978-0-323-91132-0.00003-4

Pathare, P. B., & Al-Dairi, M. (2021). Bruise damage and quality changes in impact-bruised, stored tomatoes. *Horticulturae, 7*(5), 113. https://doi.org/10.3390/horticulturae7050113

Prusky, D. (2011). Reduction of the incidence of postharvest quality losses, and future prospects. *Food Security, 3*(4), 463–474. https://doi.org/10.1007/s12571-011-0147-y

Raghav, P. K., Agarwal, N., & Saini, M. (2016). Edible coating of fruits and vegetable: A review. *International Journal of Scientific Research and Modern Education, 1*(1), 188–204.

Ramaswamy, H. S. (2014). *Postharvest technologies of fruits and vegetables.* DEStech Publications, Inc.

Rees, D., Farrell, G., & Orchard, J. (2012). Introduction. In D. Rees, G. Farrell, & J. Orchard (Eds.), *Crop postharvest: Science and technology, 3: Perishables* (p. 464). Wiley-Blackwell.

Reina, L. D., Fleming, H. P., & Humphries, E. G. (1995). Microbial control of cucumber hydro-cooling water with chlorine dioxide. *Journal of Food Protection, 58*(5), 541–546. https://doi.org/10.4315/0362-028X-58.5.541

Rickard, J. E., & Coursey, D. G. (1979). The value of shading perishable produce after harvest. *Appropriate Technology, 6*(2), 18–19.

Shafie, M. M., Rajabipour, A., & Mobli, H. (2017). Determination of bruise incidence of pomegranate fruit under drop case. *International Journal of Fruit Science, 17*(3), 296–309. https://doi.org/10.1080/15538362.2017.1295416

Shewfelt, R. L. (1986). Postharvest treatment for extending the shelf life of fruits and vegetables. *Food Technology, 40,* 70–89.

Tariq, M. A., Tahir, F. M., Asi, A. A., & Pervez, M. A. (2001). Effect of curing and packaging on damaged citrus fruit quality. *Journal of Biological Sciences, 1,* 13–16. https://doi.org/10.3923/jbs.2001.13.16

Thompson, A. K. (2008). *Fruit and vegetables: Harvesting, handling and storage.* John Wiley & Sons.

Thompson, A. K. (2010). *Controlled atmosphere storage of fruits and vegetables* (2nd Rev. ed.). CABI Publishing. Retrieved August 22, 2022, from http://www.cabidigitallibrary.org/doi/book/10.1079/9781845936464.0000

Thompson, A. K. (2015). *Postharvest transport.* CABI Compendium. https://doi.org/10.1079/cabicompendium.98019896.

Thompson, J. F., & Singh, R. P. (2008). *Status of energy use and conservation technologies used in fruit and vegetable cooling operations in California.* California Energy Commission, PIER Program. CEC-400–1999–005.

Toivonen, P. M. A., Hampson, C., Stan, S., McKenzie, D. L., & Hocking, R. (2007). Factors affecting severity of bruises and degree of apparent bruise recovery in a yellow-skinned apple. *Postharvest Biology and Technology, 45,* 276280. https://doi.org/10.1016/j.postharvbio.2007.01.018

Wang, B., & Zhu, S. (2017). Pre-storage cold acclimation maintained quality of cold-stored cucumber through differentially and orderly activating ROS scavengers. *Postharvest Biology and Technology, 129,* 1–8. https://doi.org/10.1016/j.postharvbio.2017.03.001

Wardlaw, C. W., & Leonard, E. R. (1936). Studies in tropical fruits. I. Preliminary observations on some aspects of development, ripening and senescence, with special reference to respiration. *Annals of Botany, 50*(3), 621–653. https://doi.org/10.1093/oxfordjournals.aob.a090608

Wardlaw, C. W., & Leonard, E. R. (1939). Studies in tropical fruits. IV. Methods in the investigation of respiration with special reference to the banana. *Annals of Botany, 3*(1), 27–42. https://doi.org/10.1093/oxfordjournals.aob.a085056

Wardlaw, C. W., & Leonard, E. R. (1940). Studies on tropical fruits. IX. The respiration of bananas during ripening at tropical temperatures. *Annals of Botany, 4*(2), 269–315. https://doi.org/10.1093/aob/4.2.269

Yahia, E. M. (2019). Chapter 1. Introduction. In E. M. Yahia (Ed.), *Postharvest technology of perishable horticultural commodities* (pp. 141). Woodhead Publishing. https://doi.org/10.1016/B978-0-12-813276-0.00001-8

Zamora-Magdaleno, T., Cardenas-Soriano, E., & Cajuste-Bontemps, J. (2001). Anatomy of damage by friction and by *Colletotrichum gloesporioides* penz. in avocado fruit "Hass". *Agrocencia, 35,* 237–244.

# Section II

---

*Main Handling Practices
for Fruits and Vegetables*

# 4 Postharvest Challenges, Chemical Control of Postharvest Diseases and Concerns over Agrochemicals

*İbrahim Kahramanoğlu*

## 4.1 INTRODUCTION

Postharvest diseases cause huge losses of fruits and vegetables. Feliziani and Romanazzi (2013) reported that the losses of stored fruits may exceed 50% in a single year as a result of the infections of *Penicillium italicum* and *Penicillium digitatum*. In addition to the losses by deterioration, microbial toxins produced by fungi and/or bacteria cause unsafe food (Hsieh & Gruenwedel, 1990). One of the most important examples of microbial toxins is the aflatoxins produced by *Aspergillus flavus* and *A. parasiticus*, which are found mostly on peanuts, tree nuts, wheat, cotton and maize (Severns et al., 2003). Aflatoxins are reported to cause serious threats to human health through respiratory, mucous or cutaneous routes, which may result in hepatotoxicity and immunotoxicity (Roze et al., 2013). It is also reported to be a potent carcinogen (Ellis et al., 1991).

Effective and consistent control of postharvest diseases requires integrated control. Until the end of the 1980s, the most important elements of integrated control strategies included the selection of diseases-resistant cultivars, correct nutrition and irrigation practices, correct determination of the harvest maturity, disinfecting of foods, sanitation of packing and storing areas, storage of products under unsuitable conditions for the growth of pathogens and the application of preharvest and postharvest agrochemicals. However, among these methods, agrochemicals took the lead and became the most widely used method, due to their rapid effect and easy application. These are the main reasons of the selection of agrochemicals in horticultural fields. However, according to today's knowledge, it is an indisputable fact that the excessive misuse of agrochemicals causes serious problems in the environment and for human health (Ruffo Roberto et al., 2019). At this point, fungicides are like a double-edged sword. Chemicals used for the prevention of postharvest diseases and foodborne illness may again create a food safety problem by leaving chemical residues.

Food safety refers to routines in the production, harvesting, handling and storing of food products in a way to prevent chemical residues, mechanical injury and postharvest diseases during the journey through the supply chain. Fungicides are not the only tools for the effective control of postharvest and foodborne diseases, where there are several important and widely used alternatives: waxing, edible coatings (Riva et al., 2020), modified atmosphere packaging (Kahramanoğlu & Wan, 2020), light irradiation (Papoutsis et al., 2019) and hot water treatments (48–55°C for 2–5 min) (Kahramanoğlu et al., 2020). Similarly to agrochemicals, plastics are also under discussion in the public media and among scientists due to their long-lasting effects on nature (Ferreira et al., 2016). Therefore, both the scientific world and public media have turned their gaze to the eco-friendly alternatives of agrochemicals, where the edible films and edible coatings emerge as the most important alternatives (Chen et al., 2019; Riva et al., 2020). These environmentally friendly materials have high biodegradability and are mostly safe for human beings (Nor & Ding, 2020). Widespread use of these methods is believed to reduce postharvest losses in a sustainable and healthy way and to reduce the pressure on horticultural production, which would then contribute to achieve the balance between food supply and demand (Marangoni Júnior et al., 2020).

In line with this information, this chapter aims to emphasize the importance of eco-friendly alternatives of agrochemicals by highlighting the history of agrochemicals, their negative impacts on human and environmental health, and resistant problems.

## 4.2 CHEMICAL CONTROL OF DISEASES: HISTORY AND DEVELOPMENTS

Any substance or mixture of substances that is used for controlling pests (insects, bacteria, fungi, weeds, etc.) and that causes yield or quality loss during production and/or storage of the crops is considered a pesticide (Zacharia, 2011b). Pesticide is the broad term of agrochemicals, and specific pests have different groups of pesticides. Thus pesticides can be classified based on target pests: acaricides against mites, bactericides against bacteria, fungicides against fungi, insecticides against insects, herbicides against weeds, nematicides against nematodes, molluscicides against snail, rodenticides against rodents, etc. Moreover, pesticides are classified according to the mode of entry to the target organisms. This can either be systemic or non-systemic (contact). Systemic pesticides are absorbed by the target organisms and moved to tissues that are not treated. Contact pesticides have their impact only by contact with the pest. Pesticides can also be classified according to their origin as organic, inorganic or biological. Organic and biological pesticides are more host specific, while inorganic pesticides are nonspecific with a wide range of target groups and non-target organisms (Abubakar et al., 2020). Since ancient times, human beings have aimed to control pests rapidly and effectively, both for production and for storing of their food products. During that time, elemental sulfur had been used for removing pests from plants. One of the oldest known documents about the management of pests is the *Papyrus Ebers*, which is an Egyptian compilation of medical texts dated about 1550 b.c.e. Although the document is more about herbal knowledge, it includes such information about the techniques for controlling insects on foods (Hallmann-Mikołajczak, 2004). Since then up to the pesticide era (1940s and later), mercury and arsenic emerged and were used against several pests (Abubakar et al., 2020; Fishel & Ferrell, 2013).

The need for food and the human population have a positive correlation; thus an increase in human population causes an increase in the need for food. The human population was about 1.6 billion in 1900 and has increased to about 7.95 billion today (Roser et al., 2021; Anonymous, 2022). Thus the need for food is increased. Food plays an important role in feeding and hence in the growing global population. However, production of horticultural crops is limited by the natural resources (mainly soil, water and weather conditions) and by pests, diseases and weeds (Stoytcheva, 2011). Throughout history, human beings have been combating pests to increase the yield of horticultural crops.

One of the most important developments in horticultural production was the development of modern pesticides. The pesticide era began with the discovery of dichloro-diphenyl-trichloroethane (DDT) by Paul Muller in 1939, which later earned him the Nobel Prize in Medicine due to the decrease in the health-related problems like malaria, typhus and the other insect-borne human diseases. This did not last long, and in 1962, Rachel Carson published the book *Silent Spring* and elaborated the negative impacts of DDT on human health. Thereafter, states began to ban the use of DDT.

The United States Environmental Protection Agency (EPA) ordered the cancellation of the DDT in 1972 due to its potential human health risks and negative impacts on the environment. DDT is known to be very persistent in nature, highly accumulative in fatty tissues, and able to travel way into the atmosphere. It is also nowadays classified as a probable human carcinogen by the United States and international authorities (EPA, 2022). Since then, both the public media and scientific world have fixed their gaze on the impacts of agrochemicals. Thus not only DDT but almost all agrochemicals are being discussed. Although there is a widespread public concern about the dangers of the misuse or/excessive use of pesticides, due to the exponential growth of the population and need to increase agricultural production, the global production and use of pesticides are also growing (Fishel & Ferrell, 2013). Besides these impacts on human health and environment, another big problem is raised mainly from the misuse of agrochemicals as pesticide-resistant genotypes (Buchel, 1983; Fishel & Ferrell, 2013). The main reasons of the selection of pesticides by farmers, households, scientists, states and industries are (1) effective and quick control of pests, (2) improving the yield by controlling the pests and (3) controlling the spread of human disease vectors (FAO, 2004; Ross, 2005). However, at the point reached today, it has become inevitable to evaluate the advantages and disadvantages of pesticides together and to establish a controlled system that can maximize benefits while minimizing their harm.

Pesticides have been used to protect plants (including both flowers and horticultural crops) from specific pests and are an undeniable part of modern life. Without the benefits of pesticides, the crop yield would decrease (nearly half of present levels), which might cause a decrease in food supply, an increase in prices, a food crisis and an increase in undernourished people. Not only the production of the crops but also the protection of the harvested fruits and vegetables are significantly important, which are subject to attack by fungi and other pests after harvesting.

The group of pesticides that are mainly used in the postharvest storage of the fruit and vegetable products are the fungicides, which are used to inhibit the growth and/or the development of fungi or fungal spores. Newly developed fungicides are mostly focused on inhibiting the growth of the pathogens for a period of time rather than killing them (Latijnhouwers et al., 2000; Rouabhi, 2010). Similar to pesticides, fungicides can be either contact or systemic. Residues of fungicide are generally becoming problematic when used as postharvest applications (Brooks & Roberts, 1999). Fungicides can be classified in two groups according to their origins: biologically based (biofungicides) and chemically based fungicides. The biofungicides may contain living microorganisms, such as *Bacillus licheniformis*, where the chemically based fungicides are mostly synthesized from organic and inorganic chemicals (Rouabhi, 2010). Fungal diseases are among the most important threats for horticultural crop production (Fisher et al., 2012) and crop storage and global food security (Strange & Scott, 2005; Kahramanoğlu, 2017). Fungicide sales, based on mass, accounts for the highest share (nearly 46%) in the pesticide market of the European Union (Eurostat Pesticide Sales, 2022). After the application, fungicides may enter into soil, groundwater and aquatic ecosystems and can be harmful to target and/or non-target organisms. These harms are discussed in the following sections.

## 4.3 IMPACTS OF FUNGICIDES ON HUMAN HEALTH

Fungal pathogens are a serious concern for horticultural commodities during growing, harvest, postharvest storage, transport and marketing. In addition to the yield loss during crop growing, fungal pathogens lead in quality and mass loss after harvest. Fungicides are among the main means of combating these issues. In recent years, it has been seen that the excessive and incorrect use of fungicide can cause negative effects on human health. However, on the other hand, it is known that uncontrolled fungi can also cause undesirable toxic effects on human health. A large number of fungi may derive carcinogenic mycotoxins and mutagenic secondary metabolites (Klich, 2007). One of the well-known examples is the saprophytic soil fungi *Aspergillus flavus*, which produces carcinogenic secondary metabolite aflatoxin. This aflatoxin cause human diseases (including liver cancer) if enough is consumed during the consumption of contaminated food (Hedayati et al., 2007).

The increased use of agrochemicals has associated long-term impacts on human health (Bhanti & Taneja, 2007). The contamination of foods with agrochemicals may negatively impact human health. Agrochemicals may cause short-term (acute) negative impacts on health, as well as long-term (chronic) negative impacts on health. Acute health impacts include stinging eyes, headaches, dizziness, nausea and diarrhea, while the chronic impacts may be asthma, birth defects, nervous system disorders, cancer, diabetes, endocrine disruption, immune system disorders and sexual dysfunction (Berrada et al., 2010; Beyond Pesticides, 2022). The hazard, or damage, caused by agrochemicals is determined by the combination of toxicity and exposure. Toxicity is generally measured and considered for single exposure as acute toxicity, usually measured as the dose of chemical required to kill 50% of the animals in a test population. The lethal dose 50 (LD50) is used to express this toxicity in milligrams of chemical per kilogram of body weight (mg kg$^{-1}$) of the test animal. The lower the LD50 value of a chemical is, the greater the toxicity will be, so that the pesticides with high LD50s are the least toxic to humans.

According to Beyond Pesticides (2022), which was established in 1981 as a non-profit organization in Washington, D.C., there is a scientifically confirmed link between pesticides and several diseases. Of course, this link does not exist under correct and controlled conditions but may appear when pesticides are used incorrectly and excessively. Moreover, it is well-known that some people, like infants and young children, are more vulnerable than others to pesticide impacts. In addition, the duration of exposure is important, which makes farm workers more vulnerable due to high and long exposures to agrochemicals. Therefore, it can be said that exposure to the dose is the determinant of the toxicity ("the dose makes the poison", Paracelsus). At this point, it is important to know the limits of each pesticide. The maximum residue limit (MRL) is the maximum allowed level of residue that can remain in the tissue or product that is legally tolerated in or on food. It is an important trading standard set by national and international authorities with the help of scientific studies to ensure that residues are controlled, but it is not a toxicological parameter (Shaw & Vannoort, 2001). The joint cooperation of Food and Agriculture Organization (FAO) of the United Nations and the World Health Organization (WHO) created the Codex Alimentarius in 1963. Nowadays, the codex MRLs are internationally agreed standards, but small modifications may exist in different countries. Scientific studies reveal that food products containing the residues of pesticides above MRL can cause health hazards (Bhanti & Taneja, 2005).

In the light of this information, a short research was carried out by choosing the most used fungicides today or used a lot in the past and on which there are many scientific studies. The potential harms of the active ingredients of these fungicides on human health are listed in Table 4.1 by scanning various Internet resources. It should not be forgotten that while the damages in this list may not occur in the correct use of the agrochemicals in question, they may cause significant problems in excessive and misuse.

## 4.4   IMPACTS OF FUNGICIDES ON ENVIRONMENT

Despite the positive impacts of agrochemicals on crop yield and quality, the increased use of agrochemicals has resulted in contamination of the environment (Bhanti & Taneja, 2007). Agrochemicals released into the nature may have negative impacts on the environment, both short-term and/or long-term, which may damage the normal functioning of the nature (Zacharia, 2011a). Pesticides, which are sometimes limited to test organisms (selective in their mode of action), generally are capable of damaging different forms of life, but more are non-selective and damage all forms of the organisms rather than just target organisms (Zacharia, 2011a).

Pesticides, once released into the environment, may have many different outcomes. Once the pesticides are applied to plants and/or soil, they can move through the air (through spray drift due to wind and/or air currents), soil (through washing and leaching) and through water (via runoff), which may end up in the non-targeted parts of the nature. With these ways, pesticides may reach to lower soil layers and

**TABLE 4.1**
**Potential Hazards of Some Selected Active Ingredients on Human Health [retrieved from Beyond Pesticides (2022)]**

| Active Ingredient | Birth and Development Defects | Cancer | Endocrine Disruption | Kidney/ Liver Damage | Nervous System Disorders | Reproductive Effects | Sensitizer/ Irritant | Reference |
|---|---|---|---|---|---|---|---|---|
| Azoxystrobin | ND* | ND | ND | ND | ND | ND | Yes | (EPA, 2022a) |
| Captan | Yes | Possible | ND | ND | Yes | Yes | Yes | (Fitzmaurice et al., 2014; EPA, 2005) |
| Chlorothalonil | ND | Likely | ND | Yes | Yes | Yes | Yes | (Riley & Sherrerd, 2000; Coscolla et al., 2017; EPA, 2022a) |
| Cyprodinil | ND | ND | ND | ND | Possible | ND | ND | (Coleman et al., 2012) |
| Difenoconazole | ND | Possible | suspected | ND | ND | ND | ND | (PesticideInfo, 2022) |
| Fludioxonil | Possible | Possible | likely | Yes | Possible | Possible | Yes | (Orton et al., 2011; Teng et al., 2013; Brandhorst & Klein, 2019; Ko et al., 2019) |
| Imazalil | Yes | Likely | ND | Yes | ND | Yes | Yes | (EPA, 2022a) |
| Iprodione | ND | Likely | Yes | Yes | ND | ND | Yes | (Colborn et al., 1996; EPA, 2022a) |
| Mancozeb | Yes | Yes | Yes | ND | ND | ND | Yes | (Colborn et al., 1996; EPA, 2022a) |
| Mandipropamid | ND | Not likely | ND | Yes | ND | ND | Yes | (EPA, 2022b) |
| Metalaxyl | ND | ND | Likely | Yes | ND | ND | Yes | (Lerro et al., 2021; EPA, 2022a) |
| Propiconazole | ND | Possible | ND | Yes | ND | ND | ND | (EPA, 2022a) |
| Pyrimethanil | ND | Possible | Yes | Yes | ND | ND | ND | (Colborn et al., 1996; Coleman et al., 2012) |
| Sulfur (elemental) | ND | ND | ND | ND | ND | ND | Yes | (EPA, 2022a) |
| Tebuconazole | ND | Possible | Yes | ND | ND | ND | ND | (Hass et al., 2012; EPA, 2022a) |
| Thiabendazole | ND | Yes | ND | Yes | ND | ND | ND | (EPA, 2022a) |
| Thiram | Yes | ND | Yes | Yes | Yes | Yes | Yes | (Briggs, 2018; Colborn et al., 1996; EPA, 2022a) |

*ND: Not documented. This does not mean that the active ingredient in question does not have a harmful effect. However, no such damage was found in the sources examined.

groundwater (Harrison, 1990). The movement of pesticides in the environment is very complex and is significantly impacted by the characteristics (solubility, adsorption, persistence and volatility) of the pesticide (UKDE, 2016). Pesticide compounds break down into simpler and often less toxic chemicals by several processes, including chemical degradation (reaction with water), microbial action (by soil microorganisms) and photodegradation (reaction with sunlight). This process may be rapid for some pesticides (occurring in even hours), and others may take years or more (UKDE, 2016). In the field applications, several methods need to be followed for reducing the pesticide residues (Gill & Garg, 2014). Some of them are listed here:

- First, crop selection should be site specific, according to the soil, water, weather and other ecological conditions of the area, for being able to reduce the pest problems and the need for pesticides.
- Second, using non-chemical control methods (cultural, physical, mechanical and biological) is very important for integrated pest management and for reducing the need for agrochemicals
- Finally, if pesticides still need to be used, attention should be paid to the following issues:
  - Choosing the least toxic and pest-specific (not broad-spectrum) chemicals when possible
  - Right identification of the pest problems
  - Applying pesticides only when they are required
  - Applying pesticides under correct weather conditions
  - Right nozzle and pressure selection for the prevention of spray drift
  - Calibrating pesticide application equipment regularly
  - Using recommended doses of pesticides
  - Identifying vulnerable areas and protecting from pesticides
  - Handling pesticides safely

Although some of the agrochemicals used in the control of postharvest diseases are used during crop production and/or before harvest, postharvest disease control is mostly carried out in controlled conditions (packhouses, etc.) after harvest. However, such applications can also reach the environment through air and water and cause various types of damage. Some of the best known damages by pesticides (not for postharvest application but for crop production in general) are damage to:

- **Species diversity of the ecosystems**. Incorrect use of pesticides may eliminate a species that might have known or unknown functions for the entire community and ecosystem (Zacharia, 2011a).
- **Pollinators, such as honeybees and butterflies**. Pesticides can also reduce crop yield (Whitehorn et al., 2012). This may also negatively impact the natural ecosystem and plant growth.
- **Soil microorganisms**. Soil microorganisms have several important functions, such as biomass decomposition, nutrient cycle, degradation of contaminants and maintenance of soil structure. Thus the damage to soil microorganisms may significantly impact nutrient cycle and reduce soil fertility (Dutta et al., 2010).

- **Soil erosion**. Pesticides may increase the vulnerability of the soil against soil erosion. This might be due to the loss of soil organisms and/or soil vegetation due to the uncontrolled use of pesticides (Zacharia, 2011a).
- **Quality of groundwater**. Pesticides may reach groundwater or other water sources, i.e., river, and several researchers have reported that pesticide concentrations exceeded allowable limits for drinking water in some samples (Singh & Mandal, 2013).
- **Birds**. Pesticides may directly (by residues in insects and/or plants) or indirectly (by eliminating the insects and/or plants) kill or damage birds (Guerrero et al., 2012). Some birds in the higher levels of food chains, i.e., eagles and owls, may be highly impacted from pesticide residues (Zacharia, 2011a).
- **Fish or other aquatic organisms**. This is among the major problems associated with the pesticide contamination and was reported to cause widespread mortality of fish and aquatic organisms due to residues. As with birds, this negative impact on fish can also be direct and indirect (Scholz et al., 2012).
- **The natural balance between pests and beneficial insects**. Incorrect selection of the pesticides, mostly the broad-spectrum pesticides, may damage the population of beneficial insects, and this may cause an increase in the pest populations and make it difficult to control the pests (Talebi et al., 2011).

As previously explained, the application of fungicides for the prevention of postharvest pathogens are mostly applied after harvest but are also applied before harvest to prevent pathogen infections. Thus the impacts of fungicidal compounds on non-target organisms and on the environment should be considered. These compounds may have side effects and significantly damage the non-target soil microorganism, bees, birds and/or aquatic organisms (Yang et al., 2011). Bees provide valuable ecological services, including the pollination of wild flowering plants and cultivated crops (Woodcock et al., 2019). So they are very critical for agricultural systems in relation to food and fiber production (Meeus et al., 2018). There are increasing concern and public awareness on this issue, i.e., reducing bees' exposure to toxic agrochemicals. The negative impacts of insecticides on bees has been extensively investigated since the end of the 20th century, but comparatively little is known about the fungicides (Belsky & Joshi, 2020). Detailed information has been summarized in Table 4.2. The field applications of the fungicides can come in contact with the *Apis* and non-*Apis* bee species (Heller et al., 2020), and these exposures are reported to be linked with some physiological changes at the molecular and the genetic levels in bees (Mao et al., 2017). In a recent study, Fisher et al. (2017) reported that the iprodione application alone or in combination with boscalid, pyraclostrobin and/or azoxystrobin significantly decreases the forager survival of *Apis mellifera* (honey bees) in almond orchards. On the other hand, Wade et al. (2019) also noted that the propiconazole or iprodione + chlorantraniliprole consumption by bee larvae significantly reduced adult emergence. Fungicides may, on the other hand, have indirect impact on bees by increasing the susceptibility to other pesticides. Findings of Robinson et al. (2017) support this knowledge, where the propiconazole application increased the susceptibility of *Osmia bicornis* against clothianidin toxicity.

**TABLE 4.2**

**Potential Hazards of Selected Active Ingredients on the Environment**

| Active Ingredient | Detected in Groundwater | Toxic to Birds | Toxic to Aquatic Organisms | Toxic to Bees | References |
|---|---|---|---|---|---|
| Azoxystrobin | Yes | ND | Yes | ND | (Rodrigues et al., 2013; EPA, 2022b) |
| Captan | ND | Yes | Yes | ND | (Mineau et al., 2001; EPA, 2022a) |
| Chlorothalonil | Yes | ND | Yes | ND | (EPA, 2022a) |
| Cyprodinil | ND | ND | Yes | ND | (EPA, 2022b) |
| Difenoconazole | ND | ND | Yes | ND | (PesticideInfo, 2022) |
| Fludioxonil | Possible | Possible | Yes | Yes | (Lopez-Antia et al., 2016; EPA, 2022b) |
| Imazalil | ND | Yes | Yes | ND | (EPA, 2022a) |
| Iprodione | ND | ND | Yes | Yes | (Fisher et al., 2017; EPA, 2022a) |
| Mancozeb | ND | ND | Yes | Yes | (Tew, 1996; EPA, 2022a) |
| Mandipropamid | ND | ND | Yes | ND | (EPA, 2022b) |
| Metalaxyl | ND | Yes | ND | ND | (Mineau et al., 2001) |
| Propiconazole | ND | ND | Yes | ND | (EPA, 2022a) |
| Pyrimethanil | ND | ND | Yes | ND | (EPA, 2022a) |
| Sulfur (elemental) | ND | ND | ND | ND | (EPA, 2022a) |
| Tebuconazole | ND | ND | ND | ND | (EPA, 2022a) |
| Thiabendazole | ND | ND | Yes | ND | (EPA, 2022a) |
| Thiram | ND | Yes | Yes | Yes | (Briggs, 2018) |

*Source:* Retrieved from Beyond Pesticides (2022).

*ND:* Not documented (this does not mean that the active ingredient in question does not have such a harmful effect. However, no such damage was found in the sources examined)

## 4.5 DEVELOPMENT OF FUNGICIDE-RESISTANT GENOTYPES

The ability of a pest, disease or weeds to withstand an agrochemical is called pesticide resistance. It is a predictable consequence of the repeated use of pesticides with the same mode of action. However, according to the Insect Resistance Action Committee (IRAC, 2022), the true resistance should have an inheritable change whereby the pests, diseases or weeds must be able to pass on that resistance ability to their offspring. This means that the resistant species are new genotypes with small modifications from their original ancestors. Thus it can be said that the resistance needs some mutations in the genetic makeup of the organism. According to IRAC (2022), there might be several different mechanisms of the development of resistance, as summarized here:

- **Target site resistance**. This is the most common type of the resistance development. Generally, each pesticide has an action site on the target organisms, which is usually a receptor protein. Repeated applications of the same pesticide on the same action site may cause mutations in the genetic code for the receptor protein. This can modify the shape of the protein and thus prevent the pesticide to interact with the action site. Thus the pesticide becomes ineffective due to being unable to interact with the pest.

- **Metabolic resistance**. This may occur with the increased amount of the specific enzymes that break down unwanted molecules (so that the agro-chemicals) and so the organisms may become more efficient at breaking down the pesticide and survive.
- **Physical adaptation**. This resistance/adaptation may occur again with the help of some mutations, where the organisms may develop some physical protection (i.e., a waxy cuticle) against pesticides.
- **Behavioral adaptation**. This type of adaptation is not so common. However, it may take place with the alterations in the natural behavior of the pests, again due to mutations, which may reduce the exposure of the pests to the pesticides.

Resistance to pesticides is a growing and serious problem threatening agricultural production and productivity. Although there are several other methods for control-ling pests, the pesticides have been the most useful tools against the pests due to their easy application and quick effects. However, as previously explained, the resistant organisms are very well adapted against pesticides, can pass their ability to their offspring and survive. This makes agricultural production more difficult.

There might be several reasons for the development of fungicide resistance, which might be due to (1) fungicide application (wrong dose, ineffective/poor spray cover-age, incorrect application time, antagonism, etc.), (2) environmental and plant devel-opment conditions (too wet or dry conditions, loss of fungicide due to wash-off, etc.) and (3) incorrect identification of the problem (Buhler, 2022). The development of resistance in fungicides can be managed with several simple management strategies, summarized here FRAC (2022a):

- Fungicide rotation (applying the fungicides not exclusively but as a mixture with one or more active ingredients of different type of actions or in rotation with alternative active ingredients with different type of actions)
- Reducing the number of treatments
- Following the product guidelines and recommended doses
- Avoiding the use of eradicant or broad-spectrum active ingredients
- Following the simple rules of integrated disease/pest management (IPM)

According to FRAC Code List 2022 (FRAC, 2022b), several fungi, in different places in the world, were noted to develop resistance against several active ingre-dients of fungicides. Grey mold (caused by the necrotrophic and airborne fun-gus, *Botrytis cinerea*) is among the most important postharvest problems of pome fruits, stone fruits, grapes and berries (Wan et al., 2021). *B. cinerea* was reported to have a high capacity to develop specific resistance to single-site fungicides. For example, previous publications noted specific resistance of *B. cinerea* against ben-omyl, thiophanate-methyl and carbendazim (Yourman & Jeffers, 1999), iprodione

(Northover & Matteoni, 1986; Yourman & Jeffers, 1999), cyprodinil and pyrimethanil (Myresiotis et al., 2007) and boscalid (Leroux et al., 2010).

*Alternaria* spp. is a broad genus of fungi, which mainly cause *Alternaria* rot on pome fruits and stone fruits, was also reported to develop some resistance against fungicides. Studies of Avenot et al. (2016) noted a cross-resistance to difenoconazole, propiconazole and tebuconazole, whereas similar findings were noted by Yang et al. (2019) against mancozeb and difenoconazole. The other most important postharvest fruit damaging pathogens are blue mold (*Penicillium italicum*) and green mold (*Penicillium digitatum*), which significantly impact the storability of citrus fruits, pome fruits, stone fruits, grapes and berries (Wan et al., 2021). An important study with the thiabendazole (TBZ) active ingredient was conducted in Taiwan. A total of 40 isolates of *P. digitatum* were tested, and more than 97% of test isolates were noted to be resistant to TBZ (Lee et al., 2011). In a recent study by Xu et al. (2022), it was noted that the high sporulation capacity and the great diversity in fitness of *P. digitatum* make it very powerful for developing resistance against boscalid, where some populations were noted as resistant in the research mentioned. Blue mold (rot) caused by *Penicillium expansum* is also very common in apples and pears. It was reported that some populations of *P. expansum* developed resistance against imazalil (Baraldi et al., 2003) and pyrimethanil (Caiazzo et al., 2014).

## 4.6  CONCLUSION

Nowadays, in the year 2022, a decrease in food supply has occurred in the world mostly due to the impacts of the COVID-19 pandemic, the impacts of climate change (extreme heat, flooding and drought), shortages of natural gas supply, an increase in fuel and transport prices and war between Russia and Ukraine. All this has caused an increase in food prices and food insecurity throughout the world. Until today, globalization has had an increasing trend among humanity, and world has become more interdependent, wherein the fate of one state is linked to the fate of another state. However, the COVID-19 pandemic and food crisis have caused a new shift from globalization to nationalism across the globe. In both the globalized and nationalized worlds, food is the most important requirement of human beings where its production and storage had to be kept high enough to feed the world.

Since the beginning of agriculture, there has been a need for protecting crops (and harvested products) against pests, diseases and weeds. Numerous agrochemicals have been produced and used for these purposes. Due to the wrong agricultural practices (monoculture, wrong site selection for crops, failure to follow ecological principles in production, wrong identification of the problems, excessive and inappropriate use of agrochemicals), reduction in soil fertility, decrease in water quality and quantity, climate change, etc., current agricultural systems cannot be successful without agrochemicals. However, it is also clear that agrochemicals have been associated with several negative impacts on human health, the environment and agricultural systems.

Just at this point, as mentioned in this chapter, agrochemicals are like a double-edged sword. Agrochemicals, which provide fast and effective solutions for the problems encountered during production and postharvest storage, can cause irreversible and difficult problems when used incorrectly and/or excessively. Therefore, it is extremely important to disseminate the eco-friendly and innovative methods mentioned in this and the other chapters of the book.

## REFERENCES LIST

Abubakar, Y., Tijjani, H., Egbuna, C., Adetunji, C. O., Kala, S., Kryeziu, T. L. et al. (2020). Pesticides, history, and classification. In C. Egbuna & B. Sawicka (Eds.), *Natural remedies for pest, disease and weed control* (pp. 29–42). Academic Press. https://doi.org/10.1016/B978-0-12-819304-4.00003-8

Anonymous. (2022). *Worldometer web page.* Retrieved June 8, 2022, from http://www.worldometers.info

Avenot, H. F., Solorio, C., Morgan, D. P., & Michailides, T. J. (2016). Sensitivity and cross-resistance patterns to demethylation-inhibiting fungicides in California populations of *Alternaria alternata* pathogenic on pistachio. *Crop Protection, 88*, 72–78. https://doi.org/10.1016/j.cropro.2016.05.012

Baraldi, E., Mari, M., Chierici, E., Pondrelli, M., Bertolini, P., & Pratella, G. C. (2003). Studies on thiabendazole resistance of *Penicillium expansum* of pears: Pathogenic fitness and genetic characterization. *Plant Pathology, 52*(3), 362–370. https://doi.org/10.1046/j.1365-3059.2003.00861.x

Belsky, J., & Joshi, N. K. (2020). Effects of fungicide and herbicide chemical exposure on Apis and non-Apis bees in agricultural landscape. *Frontiers in Environmental Science, 8*, 81. https://doi.org/10.3389/fenvs.2020.00081

Berrada, H., Fernández, M., Ruiz, M. J., Moltó, J. C., Mañes, J., & Font, G. (2010). Surveillance of pesticide residues in fruits from Valencia during twenty months (2004/2005). *Food Control, 21*(1), 36–44. https://doi.org/10.1016/j.foodcont.2009.03.011

Beyond Pesticides. (2022). *Website of beyond pesticides.* Retrieved July 10, 2022, from http://www.beyondpesticides.org/resources/pesticide-gateway

Bhanti, M., & Taneja, A. (2005). Monitoring of organochlorine pesticide residues in summer and winter vegetables from Agra, India-A case study. *Environmental Monitoring and Assessment, 110*(1–3), 341–346. https://doi.org/10.1007/s10661-005-8043-6

Bhanti, M., & Taneja, A. (2007). Contamination of vegetables of different seasons with organophosphorous pesticides and related health risk assessment in northern India. *Chemosphere, 69*(1), 63–68. https://doi.org/10.1016/j.chemosphere.2007.04.071

Brandhorst, T. T., & Klein, B. S. (2019). Uncertainty surrounding the mechanism and safety of the post-harvest fungicide fludioxonil. *Food and Chemical Toxicology, 123*, 561–565. https://doi.org/10.1016/j.fct.2018.11.037

Briggs, S. A. (2018). *Basic guide to pesticides: Their characteristics and hazards.* CRC Press. https://doi.org/10.1201/9781315138107

Brooks, G. T., & Roberts, T. R. (1999). *Pesticide chemistry and bioscience – The food-environment challenge. A volume in Woodhead publishing series in food science, technology and nutrition.* Woodhead Publishing.

Buchel, K. H. (1983). *Chemistry of pesticides.* John Wiley & Sons, Inc.

Buhler, W. (2022). *Fungicide resistance module in pesticide environmental stewardship.* Retrieved July 15, 2022, from https://pesticidestewardship.org/resistance/fungicide-resistance/

Caiazzo, R., Kim, Y. K., & Xiao, C. L. (2014). Occurrence and phenotypes of pyrimethanil resistance in *Penicillium expansum* from apple in Washington state. *Plant Disease, 98*(7), 924–928. https://doi.org/10.1094/PDIS-07-13-0721-RE

Chen, J., Shen, Y., Chen, C., & Wan, C. (2019). Inhibition of key citrus postharvest fungal strains by plant extracts in vitro and in vivo: A review. *Plants, 8*(2), 26. https://doi.org/10.3390/plants8020026

Colborn, T., Dumanoski, D., & Myers, J. P. (1996). *Our stolen future: Are we threatening our fertility, intelligence and survival? – A scientific detective story*. Dutton.

Coleman, M. D., O'Neil, J. D., Woehrling, E. K., Ndunge, O. B. A., Hill, E. J., Menache, A., & Reiss, C. J. (2012). A preliminary investigation into the impact of a pesticide combination on human neuronal and glial cell lines in vitro. *PLoS One*, *7*(8), e42768. https://doi.org/10.1371/journal.pone.0042768

Coscollà, C., López, A., Yahyaoui, A., Colin, P., Robin, C., Poinsignon, Q., & Yusà, V. (2017). Human exposure and risk assessment to airborne pesticides in a rural French community. *Science of the Total Environment*, *584–585*, 856–868. https://doi.org/10.1016/j.scitotenv.2017.01.132

Dutta, M., Sardar, D., Pal, R., & Kole, R. K. (2010). Effect of chlorpyrifos on microbial biomass and activities in tropical clay loam soil. *Environmental Monitoring and Assessment*, *160*(1–4), 385–391. https://doi.org/10.1007/s10661-008-0702-y

Ellis, W. O., Smith, J. P., Simpson, B. K., & Oldham, J. H. (1991). Aflatoxins in food: Occurrence, biosynthesis, effects on organisms, detection, and methods of control. *Critical Reviews in Food Science and Nutrition*, *30*(4), 403–439. https://doi.org/10.1080/10408399109527551

EPA. (2005). EPA weight-of-evidence category, Group B2 – probable human carcinogen. Office of pesticide programs. In *List of chemicals evaluated for carcinogenic potential*. Retrieved May 10, 2005, July 10, 2022, from http://www.epa.gov/pesticides/carlist/

EPA. (2022). *Fact sheets on new active ingredients*. Retrieved July 10, 2022, from http://web.archive.org/web/20120107215849/http%3A//www.epa.gov/opprd001/factsheets/index.htm

EPA. (2022a). *Pesticide registration status*. Retrieved July 10, 2022, from https://archive.epa.gov/pesticides/reregistration/web/html/status.html

EPA. (2022b). *DDT – A brief history and status*. Retrieved June 7, 2022, from http://www.epa.gov/ingredients-used-pesticide-products/ddt-brief-history-and-status#:~:text=DDT%20(dichloro%2Ddiphenyl%2Dtrichloroethane,both%20military%20and%20civilian%20populations

Eurostat Pesticide Sales. (2022). Retrieved June 8, 2022, from http://appsso.eurostat.ec.europa.eu/nui/show.do?dataset=aei_fm_salpest09

Feliziani, E., & Romanazzi, G. (2013). Preharvest application of synthetic fungicides and alternative treatments to control postharvest decay of fruit. *Stewart Postharvest Review*, *9*(3), 1–6. https://doi.org/10.2212/spr.2013.3.4

Ferreira, A. R., Alves, V. D., & Coelhoso, I. M. (2016). Polysaccharide-based membranes in food packaging applications. *Membranes*, *6*(2), 22. https://doi.org/10.3390/membranes6020022

Fishel, F. M., & Ferrell, J. A. (2013). *Managing pesticide drift. Agronomy department*. PI232, University of Florida. Retrieved June 7, 2022, from https://edis.ifas.ufl.edu/pi232

Fisher, A., Coleman, C., Hoffmann, C., Fritz, B., & Rangel, J. (2017). The synergistic effects of almond protection fungicides on honey bee (Hymenoptera: I) forager survival. *Journal of Economic Entomology*, *110*, 802–808. https://doi.org/10.1093/jee/tox031

Fisher, M. C., Henk, D. A., Briggs, C. J., Brownstein, J. S., Madoff, L. C., McCraw, S. L., & Gurr, S. J. (2012). Emerging fungal threats to animal, plant and ecosystem health. *Nature*, *484*(7393), 186–194. https://doi.org/10.1038/nature10947

Fitzmaurice, A. G., Rhodes, S. L., Cockburn, M., Ritz, B., & Bronstein, J. M. (2014). Aldehyde dehydrogenase variation enhances effect of pesticides associated with Parkinson disease. *Neurology*, *82*(5), 419–426. https://doi.org/10.1212/WNL.0000000000000083

Food and Agriculture Organization. (2004). *Pesticides residue committee report*. Retrieved July 7, 2022, from http://www.fao.org/fileadmin/templates/agphome/documents/Pests_Pesticides/JMPR/Reports_1991-2006/report2004jmpr.pdf

FRAC. (2022a). *How does fungicide resistance evolve? A publication of fungicide resistance action committee.* Retrieved July 15, 2022, from http://www.frac.info/fungicide-resistance-management/background

FRAC. (2022b). *FRAC code list 2022.* Retrieved July 15, 2022, from http://www.frac.info/docs/default-source/publications/frac-code-list/frac-code-list-2022--final.pdf?sfvrsn=b6024e9a_2

Gill, H. K., & Garg, H. (2014). Pesticide: Environmental impacts and management strategies. In M.L. Larramendy & S. Soloneski (Eds.), *Pesticides.* IntechOpen. https://doi.org/10.5772/57399

Guerrero, I., Morales, M. B., Oñate, J. J., Geiger, F., Berendse, F., de Snoo, G., Eggers, S., Pärt, T., Bengtsson, J., Clement, L. W., Weisser, W. W., Olszewski, A., Ceryngier, P., Hawro, V., Liira, J., Aavik, T., Fischer, C., Flohre, A., Thies, C., & Tscharntke, T. (2012). Response of ground-nesting farmland birds to agricultural intensification across Europe: Landscape and field level management factors. *Biological Conservation, 152,* 74–80. https://doi.org/10.1016/j.biocon.2012.04.001

Hallmann-Mikołajczak, A. (2004). Ebers papyrus. The book of medical knowledge of the 16th century B.C. Egyptians. *Archiwum Historii i Filozofii Medycyny, 67*(1), 5–14.

Harrison, S. A. (1990). *The fate of pesticides in the environment, agrochemical fact Sheet # 8.* Penn.

Hass, U., Boberg, J., Christiansen, S., Jacobsen, P. R., Vinggaard, A. M., Taxvig, C., Poulsen, M. E., Herrmann, S. S., Jensen, B. H., Petersen, A., Clemmensen, L. H., & Axelstad, M. (2012). Adverse effects on sexual development in rat offspring after low dose exposure to a mixture of endocrine disrupting pesticides. *Reproductive Toxicology, 34*(2), 261–274. https://doi.org/10.1016/j.reprotox.2012.05.090

Hedayati, M.T., Pasqualotto, A.C., Warn, P.A., Bowyer, P., & Denning, D.W. (2007). *Aspergillus flavus*: Human pathogen, allergen and mycotoxin producer. *Microbiology, 153*(6), 1677–1692. https://doi.org/10.1099/mic.0.2007/007641-0

Heller, S., Joshi, N. K., Chen, J., Rajotte, E. G., Mullin, C., & Biddinger, D. J. (2020). Pollinator exposure to systemic insecticides and fungicides applied in the previous fall and pre-bloom period in apple orchards. *Environmental Pollution, 265*(A), 114589. https://doi.org/10.1016/j.envpol.2020.114589

Hsieh, D. P., & Gruenwedel, S. H. (1990). Microbial toxins. In C. Winter, J. Seiber, & C. Nuckton (Eds.), *Chemicals in the human food chain* (pp. 239–267). Van Nostrand Reinhold.

IRAC. (2022). *Insecticide resistance training basic module: Crop protection.* Retrieved July 15, 2022, from Insect Resistance Action Committee. https://irac-online.org/about/irac/

Kahramanoğlu, İ. (2017). Introductory chapter: Postharvest physiology and technology of horticultural crops. In İ. Kahramanoğlu (Ed.), *Postharvest handling* (pp. 1–5), Intech Open. https://doi.org/10.5772/intechopen.69466

Kahramanoğlu, I., Okatan, V., & Wan, C. (2020). Biochemical composition of propolis and its efficacy in maintaining postharvest storability of fresh fruits and vegetables. *Journal of Food Quality, 2020,* 8869624. https://doi.org/10.1155/2020/8869624

Kahramanoğlu, İ., & Wan, C. (2020). Determination and improvement of the postharvest storability of little mallow (*Malva parviflora* L.): A novel crop for a sustainable diet. *Hortscience, 55*(8), 1378–1386. https://doi.org/10.21273/HORTSCI15179-20

Klich, M.A. (2007). *Aspergillus flavus*: The major producer of aflatoxin. *Molecular Plant Pathology, 8*(6), 713–722. https://doi.org/10.1111/J.1364-3703.2007.00436.X

Ko, E. B., Hwang, K. A., & Choi, K. C. (2019). Effects of fludioxonil on cardiac differentiation of mouse embryonic stem cells. *Endocrine Abstracts, 63,* 323. https://doi.org/10.1530/endoabs.63.P323

Latijnhouwers, M., de Wit, P.J., & Govers, F. (2000). Oomycetes and fungi: Similar weaponry to attack plants. *Trends in Microbiology, 11*(10), 462–469. https://doi.org/10.1016/j.tim.2003.08.002

Lee, M. H., Pan, S. M., Ngu, T. W., Chen, P. S., Wang, L. Y., & Chung, K. R. (2011). Mutations of β-tubulin codon 198 or 200 indicate thiabendazole resistance among isolates of *Penicillium digitatum* collected from citrus in Taiwan. *International Journal of Food Microbiology*, *150*(2–3), 157–163. https://doi.org/10.1016/j.ijfoodmicro.2011.07.031

Leroux, P., Gredt, M., Leroch, M., & Walker, A. S. (2010). Exploring mechanisms of resistance to respiratory inhibitors in field strains of Botrytis cinerea, the causal agent of gray mold. *Applied and Environmental Microbiology*, *76*(19), 6615–6630. https://doi.org/10.1128/AEM.00931-10

Lerro, C. C., Beane Freeman, L. E. B., DellaValle, C. T., Andreotti, G., Hofmann, J. N., Koutros, S., Parks, C. G., Shrestha, S., Alavanja, M. C. R., Blair, A., Lubin, J. H., Sandler, D. P., & Ward, M. H. (2021). Pesticide exposure and incident thyroid cancer among male pesticide applicators in agricultural health study. *Environment International*, *146*, 106187. https://doi.org/10.1016/j.envint.2020.106187

Lopez-Antia, A., Feliu, J., Camarero, P. R., Ortiz-Santaliestra, M. E., & Mateo, R. (2016). Risk assessment of pesticide seed treatment for farmland birds using refined field data. *Journal of Applied Ecology*, *53*(5), 1373–1381. https://doi.org/10.1111/1365-2664.12668

Mao, W., Schuler, M. A., & Berenbaum, M. R. (2017). Disruption of quercetin metabolism by fungicide affects energy production in honey bees (*Apis mellifera*). *Proceedings of the National Academy of Sciences of the United States of America*, *114*(10), 2538–2543. https://doi.org/10.1073/pnas.1614864114

Marangoni Júnior, L., Cristianini, M., & Anjos, C. A. R. (2020). Packaging aspects for processing and quality of foods treated by pulsed light. *Journal of Food Processing and Preservation*, *44*(11). https://doi.org/10.1111/jfpp.14902

Meeus, I., Pisman, M., Smagghe, G., & Piot, N. (2018). Interaction effects of different drivers of wild bee decline and their influence on host–pathogen dynamics. *Current Opinion in Insect Science*, *26*, 136–141. https://doi.org/10.1016/j.cois.2018.02.007

Mineau, P., Baril, A., Collins, B. T., Duffe, J., Joerman, G., & Luttik, R. (2001). Reference values for comparing the acute toxicity of pesticides to birds. *Reviews of Environmental Contamination and Toxicology*, *170*, 13–74.

Myresiotis, C. K., Karaoglanidis, G. S., & Tzavella-Klonari, K. (2007). Resistance of *Botrytis cinerea* isolates from vegetable crops to anilinopyrimidine, phenylpyrrole, hydroxyanilide, benzimidazole, and dicarboximide fungicides. *Plant Disease*, *91*(4), 407–413. https://doi.org/10.1094/PDIS-91-4-0407

Nor, S. M., & Ding, P. (2020). Trends and advances in edible biopolymer coating for tropical fruit: A review. *Food Research International*, *134*, 109208. https://doi.org/10.1016/j.foodres.2020.109208

Northover, J., & Matteoni, J. A. (1986). Resistance of *Botrytis cinerea* to benomyl and iprodione in vineyards and greenhouses after exposure to the fungicides alone or mixed with captan. *Plant Disease*, *70*(5), 398–402. https://doi.org/10.1094/PD-70-398

Orton, F., Rosivatz, E., Scholze, M., & Kortenkamp, A. (2011). Widely used pesticides with previously unknown endocrine activity revealed as in vitro antiandrogens. *Environmental Health Perspectives*, *119*(6), 794–800. https://doi.org/10.1289/ehp.1002895

Papoutsis, K., Mathioudakis, M. M., Hasperué, J. H., & Ziogas, V. (2019). Non-chemical treatments for preventing the postharvest fungal rotting of citrus caused by *Penicillium digitatum* (green mold) and *Penicillium italicum* (blue mold). *Trends in Food Science and Technology*, *86*, 479–491. https://doi.org/10.1016/j.tifs.2019.02.053

PesticideInfo. (2022). *Pesticide action network pesticide database*. Retrieved July 10, 2022, from http://www.pesticideinfo.org/Search_Chemicals.jsp

Riley, B., & Sherrerd, J. (2000). *Northwest coalition for alternatives to pesticides*. Retrieved July 10, 2022, from http://www.pesticide.org/pesticide-factsheets

Riva, S. C., Opara, U. O., & Fawole, O. A. (2020). Recent developments on postharvest application of edible coatings on stone fruit: A review. *Scientia Horticulturae, 262*, 109074. https://doi.org/10.1016/j.scienta.2019.109074

Robinson, A., Hesketh, H., Lahive, E., Horton, A. A., Svendsen, C., Rortais, A., Dorne, J. L., Baas, J., Heard, M. S., & Spurgeon, D. J. (2017). Comparing bee species responses to chemical mixtures: Common response patterns? *PLoS One, 12*(6), e0176289. https://doi.org/10.1371/journal.pone.0176289

Rodrigues, E. T., Lopes, I., & Pardal, M. Â. (2013). Occurrence, fate and effects of azoxystrobin in aquatic ecosystems: A review. *Environment International, 53*, 18–28. https://doi.org/10.1016/j.envint.2012.12.005

Roser, M., Ritchie, H., & Ortiz-Ospina, E. (2021). *World population growth*. Retrieved June 8, 2022, from https://ourworldindata.org/world-population-growth

Ross, G. (2005). Risks and benefits of DDT. *Lancet, 366*(9499), 1771–2; author reply 1772. https://doi.org/10.1016/S0140-6736(05)67722-7

Rouabhi, R. (2010). Introduction and toxicology of fungicides. In O. Carisse (Ed.), *Fungicides*. IntechOpen. https://doi.org/10.5772/12967

Roze, L. V., Hong, S. Y., & Linz, J. E. (2013). Aflatoxin biosynthesis: Current frontiers. *Annual Review of Food Science and Technology, 4*, 293–311. https://doi.org/10.1146/annurev-food-083012-123702

Ruffo Roberto, S., Youssef, K., Hashim, A. F., & Ippolito, A. (2019). Nanomaterials as alternative control means against postharvest diseases in fruit crops. *Nanomaterials, 9*(12), 1752. https://doi.org/10.3390/nano9121752

Scholz, N. L., Fleishman, E., Brown, L., Werner, I., Johnson, M. L., Brooks, M. L., Mitchelmore, C. L., & Schlenk, D. (2012). A perspective on modern pesticides, pelagic fish declines, and unknown ecological resilience in highly managed ecosystems. *BioScience, 62*(4), 428–434. https://doi.org/10.1525/bio.2012.62.4.13

Severns, D. E., Clements, M. J., Lambert, R. J., & White, D. G. (2003). Comparison of Aspergillus ear rot and aflatoxin contamination in grain of high-oil and normal oil corn hybrids. *Journal of Food Protection, 66*(4), 637–643. https://doi.org/10.4315/0362-028X-66.4.637

Shaw, R. F., & Vannoort, R. (2001). Pesticides. In D. Watson (Ed.), *Food chemical safety* (1st ed.). Contaminants (Woodhead Publishing Series in Food Science, Technology and Nutrition). Woodhead Publishing.

Singh, B., & Mandal, K. (2013). Environmental impact of pesticides belonging to newer chemistry. In A. K. Dhawan, B. Singh, M. Brar-Bhullar, & R. Arora (Eds.), *Integrated pest management* (pp. 152–190). Scientific Publishing.

Stoytcheva, M. (2011). Pesticides – Formulations, effects, fate. In M. Shokrzadeh & S. S. Saeedi Saravi (Eds.), *Pesticides in agricultural products* (pp. 225–242). Intech Open, Rijeka, Croatia. https://doi.org/10.5772/13419

Strange, R. N., & Scott, P. R. (2005). Plant disease: A threat to global food security. *Annual Review of Phytopathology, 43*, 83–116. https://doi.org/10.1146/annurev.phyto.43.113004.133839

Talebi, K., Hosseininaveh, V., & Ghadamyari, M. (2011). Ecological impacts of pesticides in agricultural ecosystem. In M. Stoytcheva (Ed.), *Pesticides in the modern world–risks and benefits* (pp. 143–168). Intech Open, Rijeka, Croatia. https://doi.org/10.5772/949

Teng, Y., Manavalan, T. T., Hu, C., Medjakovic, S., Jungbauer, A., & Klinge, C. M. (2013). Endocrine disruptors fludioxonil and fenhexamid stimulate miR-21 expression in breast cancer cells. *Toxicological Sciences, 131*(1), 71–83. https://doi.org/10.1093/toxsci/kfs290

Tew, J. E. (1996). *Protecting honeybees from pesticides. Ohio state University cooperative extension*. Retrieved July 11, 2022, from http://web.archive.org/web/20031123075324/http://beelab.edu/factsheets/sheets/2161.html

UKDE. (2016). *Pesticides in the environment*, by Kentucky Pesticide Education Program of the University of Kentucky Department of Entomology. Retrieved July 12, 2022, from http://www.uky.edu/Ag/Entomology/PSEP/pdfs/6environment.pdf

Wade, A., Lin, C. H., Kurkul, C., Regan, E. R., & Johnson, R. M. (2019). Combined toxicity of insecticides and fungicides applied to California almond orchards to honey bee larvae and adults. *Insects*, *10*(1), 20. https://doi.org/10.3390/insects10010020

Wan, C., Kahramanoğlu, İ., & Okatan, V. (2021). Application of plant natural products for the management of postharvest diseases in fruits. *Folia Horticulturae*, *33*(1), 203–215. https://doi.org/10.2478/fhort-2021-0016

Whitehorn, P. R., O'Connor, S., Wackers, F. L., & Goulson, D. (2012). Neonicotinoid pesticide reduces bumble bee colony growth and queen production. *Science*, *336*(6079), 351–352. https://doi.org/10.1126/science.1215025

Woodcock, B. A., Garratt, M. P. D., Powney, G. D., Shaw, R. F., Osborne, J. L., Soroka, J., Lindström, S. A. M., Stanley, D., Ouvrard, P., Edwards, M. E., Jauker, F., McCracken, M. E., Zou, Y., Potts, S. G., Rundlöf, M., Noriega, J. A., Greenop, A., Smith, H. G., Bommarco, R.,. . . Pywell, R. F. (2019). Meta-analysis reveals that pollinator functional diversity and abundance enhance crop pollination and yield. *Nature Communications*, *10*(1), 1481. https://doi.org/10.1038/s41467-019-09393-6

Xu, Q., Luo, C., Fu, Y., & Zhu, F. (2022). Risk and molecular mechanisms for Boscalid resistance in *Penicillium digitatum*. *Pesticide Biochemistry and Physiology*, *184*, 105130. https://doi.org/10.1016/j.pestbp.2022.105130

Yang, C., Hamel, C., Vujanovic, V., & Gan, Y. (2011). Fungicide: Modes of action and possible impact on nontarget microorganisms. *ISRN Ecology*, *2011*, 1–8. https://doi.org/10.5402/2011/130289

Yang, L. N., He, M. H., Ouyang, H. B., Zhu, W., Pan, Z. C., Sui, Q. J., Shang, L. P., & Zhan, J. (2019). Cross-resistance of the pathogenic fungus *Alternaria alternata* to fungicides with different modes of action. *BMC Microbiology*, *19*(1), 205. https://doi.org/10.1186/s12866-019-1574-8

Yourman, L. F., & Jeffers, S. N. (1999). Resistance to benzimidazole and dicarboximide fungicides in greenhouse isolates of *Botrytis cinerea*. *Plant Disease*, *83*(6), 569–575. https://doi.org/10.1094/PDIS.1999.83.6.569

Zacharia, J. T. (2011a). Ecological effects of pesticides. In M. Stoytcheva (Ed.), *Pesticides in the modern world* (pp. 1–18). Intech Open. https://doi.org/10.5772/20556

Zacharia, J. T. (2011b). Identity physical and chemical properties of pesticides. In M. Stoytcheva (Ed.), *Pesticides in the modern world* (pp. 1–18). Intech Open. https://doi.org/10.5772/17513

# 5 Eco-friendly, Non-thermal Postharvest Disinfection Applications to Fruits and Vegetables
## *Produce Safety and Quality*

*Neela Badrie*

## 5.1 INTRODUCTION

### 5.1.1 Postharvest Losses

It is reckoned that about one-third of all fruits and vegetables globally produced are never consumed by humans (Shinde et al., 2019). Fresh produce is perishable after the harvest of fruits and vegetables due to continuing respiration and transpiration, subject to senescence and decay and postharvest damage during postharvest handling (Aslam et al., 2020; Kahramanoğlu, 2017). Several factors may contribute to postharvest losses in fruits and vegetables, such as physiological and biochemical processes, extrinsic stressors, preharvest cultivation practices, poor postharvest handling infrastructures and unacceptable produce quality attributes at harvest (Shinde et al., 2019). Microbial contamination is a major cause for food loss in the fresh produce industry and accounts for more than 25% before the food reaches the consumer's plate (Varalakshmi, 2021). Hence, appropriate postharvest handling practices and disinfection treatments are required to minimize the losses, to assure optimum quality and to maximize the storage life of fresh produce (Mathushika & Gomes, 2021). Postharvest management involves a set of postproduction practices of which disinfection is one of the postharvest steps (El-Ramady et al., 2015). Chemical pesticides took the lead and became the most widely used practices due to their quick impact and easy application. However, scientific confirmation of the negative impacts of excessive and misuse of pesticides on human and environment health have led humanity to look for alternatives (Ruffo Roberto et al., 2019). The global trend is toward eco-friendly alternatives, and non-thermal disinfection methods have been developed to inactivate or destroy microorganisms and retain quality in the postharvest handling of fresh produce (Jeong & Jeong, 2018).

DOI: 10.1201/9781003452355-7

## 5.1.2 DISINFECTION/SANITIZATION

Fresh fruits and vegetables harbor microorganisms that cause decay and losses and increase the risks of foodborne illness to consumers (Chinchkar et al., 2022; Praeger et al., 2018). Sanitization/disinfection is a step in the postharvest handling of fresh produce to inactivate or destroy pathogenic organisms for public health. Numerous disinfection techniques are applied throughout the food chain, including conventional disinfectants such as chlorine and chlorine compounds (Xuan & Ling, 2019). Some non-thermal disinfection technologies offer non-hazardous and eco-friendly methods for decontaminating microbes in food (Chinchkar et al., 2022), such as chlorine dioxide, ozonation, electrolyzed water, UV-C, pulsed light, ultrahigh pressure, cold atmospheric pressure plasma, ultrasound and nanotechnology.

## 5.1.3 ALTERNATIVES TO CHLORINE

Washing and disinfection are critical steps in removing dirt and pesticide residues and in reducing the risk of foodborne pathogen contamination and unintended cross-contamination during postharvest handling of perishables (Fan & Song, 2020; Ali et al., 2018). The use of either chlorine or chlorine-related compounds as an effective sanitizing agent in water offers a barrier against cross-contamination and controls spoilage and pathogens in fresh produce (De Corato, 2020; Gadelha et al., 2019; Ali et al., 2018). However, chemical disinfection techniques such as chlorine applied as hypochlorous acid and hypochlorite may leave chemical residues, possess low inactivation efficacy and have a negative effect on produce quality (Meireles et al., 2016).

Chlorine is the most common disinfectant applied in the fresh produce supply chain due to its reliability, availability, high efficacy against a wide spectrum of microorganisms, low cost and ease of use (De Corato, 2020; Praeger et al., 2016; Ali et al., 2018), but the disinfection of by-products can be a risk (Vivek et al., 2019; Ali et al., 2018; Praeger et al., 2018; Meireles et al., 2016). The chlorine-derived disinfection of by-products such as trihalomethanes and chlorates can be absorbed by washed vegetables (Gadelha et al., 2019). It is a practical issue in the produce industry to maintain an ideal free chlorine concentration that could prevent the risk of contamination within the wash system (Abnavi et al., 2019). Flow cytometry has offered a quick method to distinguish the different physiological states of bacteria after the disinfection procedures and has been applied to evaluate the disinfection effect on *Escherichia coli* in lettuce (Teixeira et al., 2020).

Several alternative "green" postharvest disinfection technologies, such as electrolyzed water (EW), ultraviolet radiation (UV), pulsed light (PL), high-pressure processing (HPP), ultrasound (US) and cold atmospheric pressure plasma (CAPP), are improved alternatives to traditional chlorine (Deng et al., 2020a; Ali et al., 2018) without the harmful effects to or compromise of the quality of fresh-cut produce. Many chemical disinfection technologies, such as chlorine dioxide, ozone, electrolyzed water, essential oils, carbon dioxide, and organic acids, are alternatives to the common chlorine-based sanitizers without the production of harmful by-products (Deng et al., 2020a). Several combined hurdles may have synergistic lethal effects against microorganisms and minimize the undesirable quality effects on produce (Varalakshmi, 2021; Deng et al., 2020a; Zhang & Jiang, 2019)

A set of methods used for disinfection/inactivation of microorganisms in food storage is called hurdle technology. Edible coatings and films with the inclusion of antimicrobials, such as natural plant-based products and improved packaging, can delay the senescence of fruits and vegetables and resist pathogenic diseases (Wan et al., 2021; Kahramanoğlu et al., 2020; Vivek et al., 2019).

## 5.2 ECO-FRIENDLY, NON-THERMAL ALTERNATIVE DISINFECTION APPLICATIONS

There has been a shift toward environment friendly and non-thermal disinfection technologies as possible alternatives that have antimicrobial effects with the advantage of maintaining the sensory and nutritional qualities of fresh produce (Varalakshmi, 2021; Srivastava & Mishra, 2021; Praeger et al., 2018). The non-thermal chemical disinfection technologies presented in this chapter include chlorine dioxide ($ClO_2$), ozone ($O_3$), electrolyzed water (EW), and the physical disinfection technologies are ultraviolet light (UV-C), pulsed light (PL), high hydrostatic pressure (HPP), cold atmospheric pressure plasma (CAPP), ultrasound (US) and nanotechnology. Their mechanisms of disinfection, antimicrobial activity, effects on food safety and quality, and the applications to fruits and vegetables are all included and discussed in the following sections.

### 5.2.1 CHEMICAL DISINFECTION TECHNOLOGIES

#### 5.2.1.1  Chlorine dioxide ($ClO_2$)

Chlorine dioxide ($ClO_2$), a reddish to yellowish-green gas at room temperature, dissolves in water and can kill microbes due to its strong oxidizing activity, being about 2.5 times that of Cl (Takaki et al., 2021; Praeger et al., 2018). The effectiveness of $ClO_2$ gas against pathogens depends on the gas concentration, exposure time, temperature and relative humidity (Vivek et al., 2019; Praeger et al., 2016). Chlorine dioxide is applied in both aqueous and gaseous forms. However, unlike Cl, it does not react with organic compounds to produce toxic by-products but can be explosive at higher concentrations and could be toxic to humans at more than 1000 ppm (Lacombe et al., 2020; Praeger et al., 2018; Pillai et al., 2009). The application of aqueous $ClO_2$ to fruits and vegetables is permissible to a maximum concentration of 3 ppm residual $ClO_2$ in the United States (Praeger et al., 2018). The microbicidal effects of $ClO_2$ at a low concentration of 0.1 ppm on produce surface and wash water relate to the disruption effects on cell membranes, oxidization of cell-surface proteins and increase in potassium efflux due to changes in cell membrane permeability (Praeger et al., 2016).

Gaseous $ClO_2$ was shown to be effective against presumptive coliforms. Cold storage provided an additional decontamination step before the distribution as a waterless postharvest intervention for blueberry processors (Lacombe et al., 2020).

The slow release of $ClO_2$ gas formulations in the sanitization of tomatoes and cantaloupes resulted in the minimal deposition of perchlorate and chlorate residues

(Smith et al., 2015). The mechanism of antifungal activity of $ClO_2$ against green mold *Penicillium digitatum* in citrus fruits was identified using flow cytometry and scanning electron microscopy (Liu et al., 2020c). Chlorine dioxide inhibited fungal growth by causing disruption to the cell membrane. The increase of $ClO_2$ concentration resulted in an increase in malondialdehyde and nucleic acid leakage of *P. digitatum* and damage to the mycelium structure.

### 5.2.1.2  Electrolyzed Water (EW)

Electrolyzed water (EW) is an approved organic sanitizer as a replacement for the conventional Cl in washing water by highly concentrated NaOCl (Yan et al., 2021; Ignat et al., 2016; Tomas-Callejas et al., 2012). As an antimicrobial and antibiofilm agent, EW has several advantages such as on-the-spot application, eco-friendliness, low cost, and safety for human health as a promising disinfection treatment for vegetables (Yan et al., 2021).

Applications of non-thermal, chemical disinfection technologies and the variations in electrolyzed water (EW) treatments and ozonation for different fruits and vegetables are summed in Table 5.1. Acidified electrolyzed water (AEW) suppressed disease in longan by delaying pulp softening through the suppression of cell wall disassembly, down-regulated the degrading-related genes of the longan pulp cell wall and retained higher levels of longan pulp cell wall polysaccharides, covalent and ionic soluble pectin, cellulose and hemicellulose (Sun et al., 2022). In other research, the EW, especially AEW with a pH of 3.66 and Cl of 230 mg $L^{-1}$ decontaminated 90% of broccoli sprouts after washing for 20 sec (Puligundla et al., 2018). The AEW has minimized decay and retained the physicochemical quality of nectarine fruit compared to Cl water treatment (Belay et al., 2021). In another study, the slightly acidic electrolyzed water (SAEW) with fumaric acid and CaO proved to be an effective sanitizer, which improved the quality of apple, mandarin and tomato (Chen et al., 2019). Moreover, the study of Li et al. (2019) reported that the SAEW promoted metabolism, which induced the accumulation of bioactive sulforaphane in broccoli sprouts.

### 5.2.1.3  Ozonation

Ozone technology is an environmentally friendly and non-thermal disinfection for extending the storage life of fresh agricultural crops (Botondi et al., 2021; Mohd Aziz & Ding, 2018). It is an approved antimicrobial agent that is generally recognized as safe (GRAS) (Chuwa et al., 2020; Perry & Yousef, 2011). It can be applied either as a gas (OG) or as in aqueous form (AO) to enhance the quality and safety of produce, to prolong postharvest shelf life and to sanitize food production plants (Öztekin, 2019; Shah et al., 2019; Mohd Aziz & Ding, 2018; Tzortzakis & Chrysargyris, 2017).

Ozone has higher oxidizing capacity with an efficiency of 1.5 times higher compared to Cl and about 3000 times higher than the oxidation capacity of HOCl (Chuwa et al., 2020; Mahajan et al., 2014). The contact periods for antimicrobial action are typically four to five times less than that for Cl (Chuwa et al., 2020; Mahajan et al., 2014). The ozone, being a strong oxidative agent, allows for quick decomposition to oxygen, leaves no residue traces, has rapid action and is effective against a broad spectrum of microorganisms; its reactions do not contain toxic halogenated

**TABLE 5.1**
**Applications of Non-thermal Chemical Disinfection Technologies to Fruits and Vegetables**

| Disinfection Technology | Fruits and Vegetables | Investigation/Treatment | Main Effects | References |
|---|---|---|---|---|
| Acidic electrolyzed water (AEW) | Longan | Effects of treatment on disease occurrence and resistant enzyme activity of longans during postharvest storage | Disinfection fruit disease resistance and ROS scavenging capacity | (Sun et al., 2022) |
| Alkaline electrolyzed water (AEW) | Nectarine | Effects of AEW, sodium hypochlorite on quality and decay incidence of fresh nectarine stored at −0.5 °C for 31 days | AEW (200 mg $L^{-1}$, 10 min) retain freshness and reduce surface decay compared to chlorine-treated and control nectarine. | (Belay et al., 2021) |
| Combined electrolyzed acidic water (AEW) with fumaric acid (FA) and calcium oxide (CaO) | Apple, mandarin and tomato | Investigate efficacy of combined treatment on disinfection by microbial reduction from fruit surfaces | Synergistic properties of the combine treatment reduced microbial risk and did not affect sensory quality. | (Chen et al., 2019) |
| Slightly acidic electrolyzed water (SAEW) | Broccoli sprout | Evaluate SAEW with available chlorine concentration on bioactive compounds sulforaphane content | Compared to control, SAEW treatment improve the sulforaphane of broccoli sprouts by 61.2%, and increase in glucoraphanin concentration. Reduce microbial concentration. | (Li et al., 2019) |
| Electrolyzed water (EW), acidic electrolyzed water (AEW) | Broccoli sprout | Evaluate EW and AEW for effectiveness of disinfection washing for broccoli sprout | AEW treatment of pH of 3.66 had no negative effects on sensory and physicochemical quality, AEW reduced 90% microbial after washing for 20 sec. | (Puligundla et al., 2018) |

| | | | | |
|---|---|---|---|---|
| Aqueous ozone (AO) | Fresh-cut apple | Investigate AO treatment on textural quality and enzymatic activity in fresh-cut apple during cold storage | Delayed softening, degradation of cell wall polysaccharides and enzyme activities of β galactosidase and α-arabinofuranosidase. | (Liu et al., 2021b) |
| Ozone gas (OG) | Tomato | Evaluate $O_3$ treatment by dose and time on antimicrobial reduction of *Escherichia coli* O157:H7 and *Listeria monocytogenes* on tomato | High dose of 3 µg $O_3$ $g^{-1}$ of fruit reduced most the microbial survival dependence on time. Low to moderate dose (1–2 µg $O_3$ $g^{-1}$ of fruit) had minor reduction of microbes. In first hour of application, more genes were downregulated at high OG dose. | (Shu et al., 2021) |
| Aqueous ozone (AO) | Broccoli waxy leaf | Test antibacterial effect of AO on broccoli waxy leaf surfaces with surfactant | No detection of *Salmonella typhimurium* on waxy leaf surfaces after washing with AO with surfactant. Lower attachment of *S. typhimurium* attached to waxy leaf surfaces compared to glossy and pesticide-treated waxy leaf surfaces. | (Song et al., 2021) |
| Aqueous ozone (AO) | Tomato | Investigate combined treatment of calcium chloride, chitosan, hydrogen peroxide and AO on tomato quality stored at 10°C for 28 days. | All treatments regulate ripening and delay senescence of postharvest tomato under cold storage conditions. Treatment with either chitosan or calcium chloride was most effective to maintain fruit quality attributes. | (Shehata et al., 2021) |

*(Continued)*

**TABLE 5.1 (Continued)**
**Applications of Non-thermal Chemical Disinfection Technologies to Fruits and Vegetables**

| Disinfection Technology | Fruits and Vegetables | Investigation/Treatment | Main Effects | References |
|---|---|---|---|---|
| Sonolytic-ozonation, ozone gas (OG) and ultrasound (US) | Spinach leaves | Evaluate combined treatment at different times on microorganisms, pesticide residues and quality comparison to water | Combined OG and US for 10 min decontaminated the microbial and chemical in spinach leaves without affecting quality up to 1 week at 5°C. | (Siddique et al., 2021) |
| Ozone gas (OG) | Cantaloupe melon | Analyze the transcriptome of cantaloupe (peel and pulp) to treatment using high-throughput RNA sequencing approach | Delayed postharvest decay and improved intrinsic quality of cantaloupe on OG application. Identified genes in cantaloupe peel (570) and pulp (313), correlated to pectin metabolism ethylene and flavonoid. | (Zhang et al., 2021) |
| Ozone gas (OG) | Cantaloupe | Examine effects of different OG doses on quality of cantaloupes during storage | An OG dose of 15.008 mg m$^{-3}$ improved and extended the postharvest fruit quality. | (Chen et al., 2020) |
| Aqueous ozone (AO) | Parsley leaves | Effects of AO and chlorine for 5 min washes on *Escherichia coli* and *Listeria innocua* and quality of fresh parsley leaves. | Higher microbial reduction with AO than with distilled water. More effective reduction against *E. coli* with Cl than AO and with no major reduction difference for *L. innocua* counts with Cl or AO. No negative effect on pigments and antioxidant compounds of parsley leaves on washing. | (Karaca & Velioglu, 2020) |

| Treatment | Commodity | Objective | Findings | Reference |
|---|---|---|---|---|
| Aqueous ozone (AO) and ultrasound (US) | Strawberry | Investigate combine efficacy of AO and US on microbial pathogen and pesticide residues | Combined treatment for 3 min reduced bacteria and various pesticides on strawberry compared to control. Combined AO and US treatment for 3 min reduced fungal decay and weight loss maintain higher catalase, peroxidase, and superoxide dismutase in fruit tissue during cold storage for 12 days. | (Maryam et al., 2020) |
| Sonozonation, aqueous ozone (AO) and ultrasound (US) | Cherry tomato | Investigate the combined application of AO and US washing on microorganisms and nutritional quality of cherry tomato | Reduced spoilage microorganisms and enhanced quality of fruit. Recommend single washing technique for microbial safety. | (Mustapha et al., 2020) |
| Aqueous ozone (AO) combines with sodium metasilicate (SM) | Cabbage | Evaluate combined treatment on fresh-cut cabbage quality at 4°C and 90% RH for 12 days | Reduce $E.\ coli$ in comparison to control (water-treated) samples after 12 days. No negative sensory quality, ascorbic acid content, total phenol and carotenoid effects on treated cabbage. | (Nie et al., 2020) |
| Ozone gas (OG) | Raspberries | Effect of treatment on microbiological and antioxidant quality of raspberries stored at 20–25°C | An OG dose application of 8–10 ppm applied for 30 min every 12 hours over for 3 days reduced growth of aerobic mesophilic bacteria and fungi. Total eradication of $Botrytis\ cinerea$ with OG and 12% infection for control fruit after storage. | (Piechowiak et al., 2019) |

compounds, and its application is allowed in food industrial use with organic certification when compared to other chemical alternatives (Botondi et al., 2021; Shahi et al., 2021; Horvitz & Cantalejo, 2014; Miller et al., 2013). However, exposure to elevated levels of OG could harm the health of workers (Mohd Aziz & Ding, 2018; Horvitz & Cantalejo, 2014).

The efficacy of ozone disinfection treatment is influenced by the microbial population, concentration, temperature and contact time between AO and the fresh produce during the washing process, vegetable and fruit surfaces, packaging material, and the materials used in equipment design and kinetic modeling (Öztekin, 2019; Botondi et al., 2021; Aslam et al., 2020; Sachadyn-Król & Agriopoulou, 2020; Glowacz & Rees, 2016). The contact time for application by OG is usually longer than for AO (Botondi et al., 2021). The $O_3$ is an unstable triatomic molecule that autodecomposes suddenly and rapidly into oxygen atoms at ambient temperature, and hence it requires continuous generation in situ (Botondi et al., 2021; Carletti et al., 2013).

An ozonation procedure by a liquid whistle reactor produced hydrodynamic cavitation for the disinfecting water (Ghernaout & Elboughdiri, 2020). Aqueous ozone could be produced by the generation of OG by an ionizer in water. The ozone-generated disinfectant has high oxidation and reduction potential and is soluble in water (Sachadyn-Król & Agriopoulou, 2020). Raw materials such as produce for disinfection can be exposed to the modified atmosphere enriched with OG (Piechowiak et al., 2019).

Ozone is effective against most microorganisms within a short disinfection contact time; however, its bactericidal activity could be brief, due to changes in the ozone in the presence of oxygen molecules after reacting on the organic materials (Takaki et al., 2021). Ozone inactivates microbes on fresh produce due its operative antimicrobial properties, high penetrability and reactivity (Vivek et al., 2019). Its microbicidal action is on cellular constituents such as proteins, nucleic acids in the cytoplasm, cell walls, spore coats and virus capsids (Mahajan et al., 2013; Miller et al., 2013).

Applications of ozonation and their impacts on fruits and vegetables as chemical disinfection technology are summed in Table 5.1. Ozone has disinfecting properties to enhance the staying quality and storage life of fresh produce (Aslam et al., 2020). According to the research of Shu et al. (2021), a high dose (3 µg $O_3$ g$^{-1}$ of fruit) resulted in a major reduction in bacteria survival in a time-dependent manner (see Table 5.1 for details). A short exposure to $O_3$ at a moderate (2 µg $O_3$ g$^{-1}$ of fruit) dose resulted in major reduction of *Listeria monocytogenes*. The main defenses in *Escherichia coli* were attenuated after exposure to a low dose (1 µg $O_3$ g$^{-1}$ of fruit) as demonstrated by transcriptome profiling by RNA sequence.

In another research, sonozonation, a combined treatment of AO and US, was applied in washing techniques to cherry tomato (Mustapha et al., 2020). As a separate application, US required elevated power and longer application time to inactivate microorganisms as compared to US combined with other disinfection technologies. The hurdle treatment resulted in a ≥3 log reduction in spoilage microorganisms. There was greater antimicrobial efficiency of AO at higher concentrations and exposure times, but these treatment factors affected the quality of the treated cherry tomato. In another study, a combined treatment of calcium chloride, chitosan, hydrogen peroxide and AO applied to tomato were noted to regulate fruit ripening and delayed senescence during storage at 10°C (Shehata et al., 2021).

The effects of combined treatment of ozone and packaging application were investigated on the quality and the storage life of tomato fruit. The effect of packaging delayed microbiological contamination and preserved fruit freshness at room temperature for 12 days (Zainuri et al., 2018). The combined effects of ozone and a perforated polyethylene bag resulted in the best quality tomato fruit. The effects of sequentially forced OG and nonstop advanced oxidative process (AOP) evaluated the inactivation of *Listeria monocytogenes* on and within Empire apples (Murray et al., 2018). A 20 min $O_3$ application into an airstream through the apple bed reduced *L. monocytogenes* with no significant effect of apple position within the bed. The continuous AOP in combination with other disinfection methods could reduce internal microbial populations.

An AO treatment of 1.4 mg $L^{-1}$ for 5 min applied to fresh-cut apple was noted to reduce fruit softening by delaying the increase in water-soluble pectin (Liu et al., 2021b). An AO contact application for 5 min removed pesticide residues and prolonged storage. The study of Chen et al. (2020) suggested that an ozone dose of 15.008 mg $m^{-3}$ maintained the firmness of cantaloupes by lowering respiration rate and ethylene production. The transcriptomes of cantaloupe peel and pulp in response to ozone treatment have been identified by RNA sequencing and revealed that the genes regulated metabolism of ethylene flavonoid and pectin (Zhang et al., 2021). Also, ozone preserved the firmness of the cantaloupe through changes in pectin metabolites and reduced ethylene production by regulation of the relevant genes, especially in the peel.

Two-mode continuous application of AO and US to strawberry inhibited microbial spoilage, reduced pesticide residues and retained postharvest physical and nutritional quality of strawberry fruits stored at 2°C (Maryam et al., 2020). The application of the preservation hurdles for 3 min extended the marketability of strawberry fruits for 6 more days under cold storage. An ozonation dose of 8–10 ppm for 30 min for 36 hours was effective in reducing storage diseases for raspberries at 20–25°C and had higher antioxidant content due to reduced loss of the polyphenols (Piechowiak et al., 2019). The control fruit was completely infected with *Botrytis cinerea* compared to the ozone-treated fruits of 12%.

An AO with an addition of a surfactant improved the antibacterial activity on the waxy produce surfaces of broccoli (Song et al., 2021). The AO washing was less effective on waxy surfaces than that on glossy broccoli surfaces. However, the addition of a surfactant to AO resulted in lower levels of *Salmonella typhimurium* being attached to the broccoli waxy leaf than on glossy leaf. Hence, an emulsified pesticide with a surfactant promoted the attachment of *S. typhimurium* on waxy broccoli leaf.

In another study, an AO (1.4 mg $L^{-1}$) treatment for 5 min was found to be most effective to eliminate pesticide residues and extend the storage of fresh-cut cabbage (Liu et al., 2021a). In addition, the AO application promoted initial respiratory metabolism and inhibited ethylene production during 12 days of storage. Research of Nie et al. (2020) with a combined treatment of OG with sodium metabisulfite was found to be more effective than that for individual treatments in controlling both pathogens and inherent microorganisms on cabbage. The hurdle application extended the storage up to 12 days at 4°C with no adverse effects on sensory characteristics, vitamin C, total phenols and carotenoids.

The combined hurdles of ozone and US treatment (sonolytic-ozonation) reduced microorganisms and pesticides in spinach, without influencing its quality (Siddique et al., 2021). The best storage of pre-air-dried spinach leaves was with 10 min contact time. The 10 min and 15 min sonolytic-ozonation treatments resulted in optimal reduction of the microbial counts. The combined ozone and US disinfection treatment proved to be effective and eco-friendly/green decontamination technique for spinach. the effects of ozone and Cl washes on microbial (*Escherichia coli* and *Listeria innocua*) quality and shelf life of fresh parsley leaves were studied by Karaca and Velioglu (2020). In comparison to water, AO was more effective in reducing the microbial load of parsley leaves, while Cl was more effective against *E. coli* than AO. The washing treatments did not adversely affect chlorophyll, ascorbic acid, total phenols and antioxidant activity. Disinfection conducted with OG and application of 0–5 mg $L^{-1}$ and with AO and application of 0–10 mg $L^{-1}$ had no quality effects on carrots such as firmness, weight loss and color of the carrots (de Souza et al., 2018). The OG extended the shelf life of carrots by preventing an increase in total soluble solids for 5 days at 18°C at RH of 80%.

### 5.2.2 PHYSICAL DISINFECTION TECHNOLOGIES

The physical disinfection applications include ultraviolet light (UV-C), ionizing radiation (IR), pulsed light (PL), cold atmospheric pressure plasma (CAPP), high hydrostatic pressure (HPP), high-intensity ultrasound (US) and nanotechnology (see Table 5.2). The research drift is to apply two or more hurdle technologies to increase the synergistic lethal effects against microorganisms and to reduce the negative effects of individual treatments (Varalakshmi, 2021; Deng et al., 2020a, 2020b). Also, edible coating materials such as chitosan, methylcellulose, carrageenan and alginate are eco-friendly and serve as carriers for antimicrobials and antioxidants in preserving fruit and vegetable surfaces after harvesting (Vivek et al., 2019; Pérez-Pérez et al., 2016).

#### 5.2.2.1 Ultraviolet Light (UV)

Ultraviolet light (UV) illumination is a cold disinfection technology used in fresh-cut produce for the disinfection of water and surfaces (Ochoa-Velasco et al., 2020). It is easy to use, lethal to most microbial types (Khaire et al., 2021) and has no chemical residues. The optimum germicidal effect of UV-C light is within range 200–280 nm, which is related to the bacterial DNA absorbing UV photons at approximately 260 nm (Green et al., 2020; Collazo et al., 2018). The efficacy of UV-C radiation for the decontamination of fruits and vegetables depends on the disinfection systems, mode of application, pulse intensity, morphology, type and location of microorganisms on fruits and vegetables and the distance from the irradiation source (Darré et al., 2022; Esua et al., 2020). The short-wave ultraviolet light of UV-C, 254 nm, is effective in reducing the microbial load in air, in potable water and on non-porous surfaces free from organic residues, unlike long wave UV light (UV-A, >320 nm),

**TABLE 5.2**
**Applications of Non-thermal Physical Disinfection Technologies to Fruits and Vegetables**

| Disinfection Technology | Fruits and Vegetables | Treatment | Main Effects | References |
|---|---|---|---|---|
| Ultraviolet light-C (UV-C) photolysis and vacuum ultraviolet light (VUV) Combined UVC and VUV-assisted titanium dioxide (UVC-TiO$_2$, (VUV-TiO$_2$) | Carrots | Investigate the effects of treatments on degradation of pesticides and inactivation in solution *Escherichia coli* and *Saccharomyces cerevisiae* and on fresh carrot surface | Treatments of combined UV-C and VUV treatments and combined use of UV-C-TiO$_2$ and VUV-TiO$_2$ treatments showed higher removal of residual pesticides on fresh-cut carrots than water washing. UV-TiO$_2$ treatment showed slightly higher microbial inactivation effects than combined UV. | (Choi et al., 2020) |
| Water-assisted ultraviolet light-C (WUV) and peroxyacetic acid | Broccoli | Test WUV doses single 0.5 kJ m$^{-2}$, 0.3 kJ m$^{-2}$) or combined with peroxyacetic acid (50 mg L$^{-1}$) on decontamination of fresh-cut broccoli | Both WUV and combined with peroxyacetic acid reduced natural aerobic mesophilic microorganisms and *Listeria innocua* without affecting physical quality. | (Collazo et al., 2018, 2019) |
| UV-C LEDs and low mercury lamp (LPM) | Apple and lettuce | Compare efficacy of UV-C LEDs (277 nm) to LPM (253.7 nm) for germicidal activity on foodborne pathogens on *Escherichia coli* O157:H7 lettuce leaves and *Listeria monocytogenes* on apple skin | UVC LEDs and LPM lamps had comparable germicidal efficacy, and both treatments showed similar microbial reductions on lettuce leaves and on apple skin Microbial log reduction for UV-C LEDs was similar at 4°C and 25°C. | (Green et al., 2020) |

*(Continued)*

**TABLE 5.2 (Continued)**
**Applications of Non-thermal Physical Disinfection Technologies to Fruits and Vegetables**

| Disinfection Technology | Fruits and Vegetables | Treatment | Main Effects | References |
|---|---|---|---|---|
| Pulsed light (PL) | Chinese kale | Evaluate influence of PL intensity on removal of microbial load, pesticide residue and quality changes of Chinese kale | Eliminate pathogenic bacteria, such as *Escherichia coli*, *Staphylococcus aureus* and pesticide substances at PL intensity 8.4 J cm $^{-2}$. | (Minh, 2021) |
| Pulsed light (PL), with mild thermal (TT) | Strawberries | Effect of single PL or TT and combinations on fungal strawberry physical microstructure stored at 5°C | TT with 2.5 and 5.0 min in water at 46°C was extra active than PL fluence (11.9 J cm $^{-2}$) to reduce fungal decay. PL and TT with 5 min improved firmness compared to control fruit. | (Contigiani et al., 2020) |
| Ultrasound-assisted (US)-osmotic dehydration (OD) | Peach | Investigate the effectiveness of US-OD on reducing foodborne pathogens. Apply US waves (50 and 75% amplitudes) and inoculate peach with pathogens. Immerse samples in 70% sucrose for 4, 8 and 12 hours for osmotic dehydration | Highest reduction of microbial pathogens at US amplitude of 75% after 12 hours storage in osmotic solution. Lower moisture, total phenolics and antioxidant scavenging radicals with OD treatment. | (Hashemi & Jafarpour, 2021) |
| High-pressure processing (HPP)/microwave-assisted thermal pasteurization system (MAPS) | Green beans | Investigate effects of HPP or MAPS on quality attributes and control of *Listeria innocua* for green beans | More effective control of *L. innocua* with MAPS than HPP, but both treatments had similar effects on quality of green beans. Major decrease in vitamin content for HPP-treated beans compared to MAPs-treated beans. | (Inanoglu et al., 2021) |

| Treatment | Commodity | Objective | Findings | Reference |
|---|---|---|---|---|
| Cold atmospheric pressure plasma (CAPP) | Mulberries | Effects of CAPP different time and strength on fungal rot (*Botrytis cinerea*) mulberries | CAPP destroyed the structure and affected metabolism of *B. cinerea* but did not affect the quality of mulberries. | (Yinxin et al., 2022) |
| Cold atmospheric pressure plasma (CAPP) (coplanar surface barrier discharge) | Red currants | Evaluate treatment on microbial load and quality of red currants at 7°C for 10 days | Nitrogen generated CAPP reduced total aerobic mesophiles and air-generated CAPP reduced yeast and mold load. Quality attributes of red currants were unaffected by CAPP. | (Limnaios et al., 2021) |
| Non-thermal plasma (NTP) | Button mushroom | Inactivate *Pseudomonas tolaasii* on inoculated button mushroom by NTP at 4°C for 21 days | NTP reduced *P. tolaasii* and microbial contamination and reduced firmness of mushroom in storage. | (Pourbagher et al., 2021) |
| Plasma-processed air (PPA) | Chinese bayberries | Evaluate PPA on *Penicillium citrinum* of intact and damage Chinese bayberries over 3 days storage at 20°C and pretreatment on metabolites | PPA pretreatment reduced decay index and enhanced antioxidant activity. PPA inhibited mycelial growth of *P. citrinum* and reduced the decay indices. | (Shen et al., 2021) |
| Plasma-activated solution (PAS) | Lettuce | Effect of PAS incubation on cell integrity of *Staphylococcus aureus* for and quality of fresh lettuce leaves | PAS reduced *S. aureus* population increase cell membrane leakage. Increased reactive oxygen species and affected cellular DNA content but had no negative effects on produce quality. | (Zhao et al., 2021b) |
| Atmospheric argon plasma (AAS) | Almond | AAS application at different contact time to decontaminate almond slices surface | AAS reduced microorganisms after 20 min, with further reduction with longer contact time. Quality attributes of color, peroxide value and sensory attributes were unaffected by CAPP. | (Shirani et al., 2020) |

which is limited in its microbiocidal properties (Khaire et al., 2021; Bintsis et al., 2000).

The UV-A at an irradiation wavelength of 365 nm for different microbial types was found to enhance the bactericidal activity of ferulic acid. The hurdle application increased the oxidative change with further disruption of the bacterial cell membrane of *Escherichia coli* (Shirai & Yasutomo, 2018). The disinfection activity of ferulic acid and UV-A was due to cytotoxicity rather than genotoxicity effects on microorganisms.

The germicidal efficacy of UVC LEDs at 277 nm was just as effective a germicidal wavelength compared to low-pressure mercury lamps at 253.7 nm against *Escherichia coli* O157:H7 on lettuce leaves and *Listeria monocytogenes* on apple skin (Green et al., 2020) but resulted in higher browning rates. The UVC light as a dry disinfection application was a suitable alternative to aqueous sanitizers for the disinfection of both lettuce leaves and apple skin.

A single application of water-assisted UV-C technology (0.5 kJ m$^{-2}$) reduced the microbial load of fresh-cut broccoli and wash water without affecting produce quality (Collazo et al., 2018). The combination of 50 mg L$^{-1}$ peracetic acid and water-assisted UV-C of 0.3 kJ m$^{-2}$ was effective in reducing mesophiles by 2 log$_{10}$ in organic broccoli. The effects of UV photolysis and UV-TiO$_2$ photocatalysis under vacuum ultraviolet light (VUV) and UV-C sources on the degradation of pesticides and microorganisms on fresh carrot surface were investigated (Choi et al., 2020). The UV-TiO$_2$ was effective in the removal of residual pesticides from fresh-cut carrot surfaces, while the UVC, VUV, UVC-TiO$_2$ and VUV-TiO$_2$ did not vary in microbial inactivation effects. These disinfection technologies are non-chemical and residue-free on fresh produce.

The microbial load of tomatoes was reduced by the combined instantaneous application of UV-C radiation and ultrasonic energy (Esua et al., 2018). The total aerobic bacteria decreased with an increase in UV-C dosage. The combined UV-C radiation and ultrasonic energy input increased the total phenols of tomatoes. The dosage treatment is significant for the disinfection of tomato and to increase its total phenolic content and antioxidant activity.

### 5.2.2.2   Pulsed light (PL)

Pulsed light (PL) or high-intensity light pulse (HILP) is comprised of short and high peak pulses of broad spectrum of electromagnetic wavelengths (100–1100 nm) (Varalakshmi, 2021; Mahendran et al., 2019; Palgan et al., 2011). It is an eco-friendly cold sterilization postharvest treatment for fruits and vegetables and is an alternative to UV technology (Fang et al., 2020; Urban et al., 2018; Kramer et al., 2017). The UV-C component (200–280 nm) of the PL spectrum is associated with the microbial inactivation (Kramer et al., 2017). If the wavelength range should be below 320 nm, the UV is ineffective, and PL will not be as lethal on microorganisms (Fang et al., 2020). Further, the mechanism of UV decontamination is associated with the formation of thymine dimer in microbial cell DNA, which blocks DNA synthesis, inhibits microbial cell replication and results in the death of microbial cells (Varalakshmi, 2021; Khandpur & Gogate, 2016). Pulsed light (PL) eradicated the microbial and pesticide residues of green vegetables and retained the physicochemical properties and phytochemical components (Minh, 2021). The modes of action against micro-organisms by pulse light treatment relate to its direct effect on the specific microbial cells and to the barrier mechanism of the produce medium (Duarte de Sousa et al., 2019; Pongener et al., 2018; Urban et al., 2018; Duarte-Molina et al., 2016). The germicidal effect of PL is related to its photothermal and photochemical consequences (Deng et al., 2020a; Elmnasser et al., 2007). The cold sterilization effect of PL against microorganisms depends on the intensity of light, the number of pulses, pulse width, microbial species and produce characteristics (Deng et al., 2020a; Fang et al., 2020). The sterilization effect of PL increases with higher light intensity. The antimicrobial action of PL is due to its highly reactive oxidizing species (photodynamic inactivation), which cause oxidative disruption to the lipids and proteins in the cell membranes of microbes (Varalakshmi, 2021). The PL technology is considered non-thermal; however, it could cause the heating of produce if a longer treatment time is applied (Vorobiev & Lebovka, 2019). The inactivation of microbes by PL is due to the increased release of instantaneous energy (Khaire et al., 2021; Deng et al., 2020a). After the PL application, the enzymatic repair of DNA does not occur due to severe cell damage (Fang et al., 2020). The PL effects on microbial reduction depend on the produce matrix properties (Huang & Chen, 2014) and on sample transparency (Yousefi et al., 2021). The advantage of PL processing is that a short application time could deliver the minimum dosage required for microbial inhibition (Salehi, 2022).

A proteomic study assessed PL damage on the formation of dimers on DNA for *Listeria innocua* (Aguirre et al., 2018). The cyclobutane pyrimidine dimers and the pyrimidine photoproducts were detected in the microbial cells. Also, the PL treatment induced stress proteins and other proteins. The proteome differences in survivors may have accounted for the adaptation of *L. innocua* after PL treatment. PL technology allows for the rapid surface decontamination of vegetables using high-intensity light pulses for noticeably short times (Bhavya & Hebbar, 2017). The advantages of PL for the cold disinfection of fruits and vegetables relate to its sanitization efficiency, minimal adverse quality effects at low doses, continuous process, and no harmful residues (Deng et al., 2020b). The PL limitations relate to the low extent of penetration, thermal damage and the negative impact on produce quality at high doses (Deng et al., 2020a). A combination of mild thermal treatment (TT) and PL was just as effective for fungal inhibition than individual TT application on storage of strawberries (Contigiani et al., 2020). A TT application contact time for 5 min delayed the onset of infection for 6 days and resulted in the least infected fruits during storage.

Pulsed light treatment was used to decontaminate *Cryptosporidium parvum* oocysts on raspberries to achieve oocyst reductions (Le Goff et al., 2015). In other research, PL application was applied as swift light pulses for microbial and pesticide removal on Chinese kale leaf (Minh, 2021). At a PL intensity 8.4 J cm$^{-2}$ resulted in the removal of the microbes and pesticides. Hence, the Chinese kale was safer with better retention of quality attributes on employment of PL. In another study, a combination of pulsed light (PL) with an antimicrobial wash hindered microbial growth on cherry tomato for 3 weeks in storage (Leng et al., 2020). This joint application demonstrated synergistic inactivation of *Salmonella enterica* on the stem scars of cherry tomato. Other research showed that PL fluence of 9 J cm$^{-2}$ limited ethylene production due to an increase in polyamines, thereby resisting postharvest quality changes in Cantaloupe melons (Duarte de Sousa et al., 2019). The study of Moreira et al. (2017) also showed that the PL-treated, pectin-coated, fresh-cut apple pieces demonstrated higher antioxidant activity than fresh and PL-control samples. The combined treatment resulted in the microbial count reduction of 2 log CFU $g^{-1}$ toward the end of storage.

### 5.2.2.3   Ultrasound (US)

Power ultrasound (US) or high-intensity ultrasound is an eco-friendly and nontoxic disinfection technology (Majid et al., 2015). The combination of US with other disinfection methods intensified the US-based effects (Khaire et al., 2021). US is suggested for surface disinfection of fresh produce within the contact times between 1 and 10 min at a low frequency of 20–100 kHz (Fan et al., 2021; Dai et al., 2020; Vivek et al., 2019).

Acoustic cavitation is the main antibacterial mechanism in the different applications of US. This induces mechanical stress on bacterial cells and produces radicals (Dai et al., 2020), which have further impact by increasing the permeability of the microbial cell membrane, oxidative damage to the bacterial cell and eventually bringing about cell death (Khaire et al., 2021; Majid et al., 2015).

Ultrasound (US) has limited disinfection efficacy in the control of microbial growth and is often combined with other disinfection technologies to improve its disinfection efficacy (Wang & Wu, 2022; Khaire et al., 2021). The combined US application with one or more disinfection methods in hurdle technologies result in the synergistic effects of improving microbial inactivation without undesirable effects on produce quality (Khaire et al., 2021; Deng et al., 2020b; Pou & Raghavan, 2020). The single application of US has minimal effect on microbial growth.

A combination of hurdles such as US with PL of wide spectrum between 100 and 1000 nm and brief time pulse (100–400 μs) was effective for microbial reduction (Khaire et al., 2021). Also, an application of US with liquid chemical sanitizers resulted in the deeper penetration of the sanitizers in the hydrophobic surface of fruits and vegetables (Vivek et al., 2019). An US-assisted osmotic dehydration was effective in reducing foodborne pathogens on pears, which improved with an increase in US amplitude (Hashemi & Jafarpour, 2021). The highest pathogen reduction was noted at 75% amplitude after 12 hours' storage in the osmotic solution.

A study compared the impact of US application and malic acid, as individual treatments and in combination, on the inactivation of foodborne pathogens in sweet lemon juice (Hashemi & Jafarpour, 2020). The combination of US and malic acid was more effective on microbial inactivation than for separate treatment. The synergistic bactericidal action of the combined treatment was related to the damage of the cell membrane and inactivation of enzymes.

### 5.2.2.4   Cold Atmospheric Pressure Plasma (CAPP)

Cold atmospheric pressure plasma (CAPP) or cold plasma (CP) is a non-thermal plasma technology and environmentally friendly approach with advantages over other commercial decontamination methods (Ansari et al., 2022; Yinxin et al., 2022). The CAPP is a "green" technology that requires negligible water handling and on-time generation of reactive agents with no residues (Bovi et al., 2019; Pignata et al., 2017; Thirumdas et al., 2015).

Plasma-activated water (PAW) was generated by the transfer of chemical reactivity and energy from CAPP to water with no chemical residues or environmental pollution (Soni et al., 2021; Zhou et al., 2020). The CAPP is composed of ionized gases and reactive species that are dissociated by electrical energy input to acquire antimicrobial properties (Vivek et al., 2019). The CAPP is applied in the disinfection of surfaces and for food sterilization (Zhao et al., 2021a; Pankaj & Keener, 2017). It is an appropriate disinfection for heat-sensitive foods such as fresh produce due to its generation at ambient temperature and atmospheric pressure (Baier et al., 2015; Bußler et al., 2017).

The CAPP is applied for water disinfection due to its antimicrobial mechanism. The application of a 5 min plasma treatment to yeast cells in water resulted in cell reduction and cell death by apoptosis and necrosis subject to the contact time (Xu et al., 2020). The cell fatality depended on the level of induced oxidative stress. Cold plasma-activated solution (PAS), which was generated by dielectric barrier discharge plasma treatment, reduced the bacterial load on fresh lettuce. It affected the microbial cells by leakage, changed the ratio of unsaturated to saturated fatty acids of cell membrane and increased the reactive oxygen species in cells (Zhao et al., 2021b).

Plasma-activated hydrogen peroxide solution prepared using dielectric-barrier-discharge plasma served as an anti-biofilm technology that reduced *Staphylococcus aureus* in biofilm (Zhao et al., 2021a). Also, the PAS had no negative effects on the quality of fresh leafy vegetables (Zhao et al., 2021b). The argon jet plasma process increased food safety in almond slices without affecting color, peroxide value and sensory attributes (Shirani et al., 2020). However, the textural hardness was affected by the plasma treatment; hence the first 10 min application is recommended for consumer satisfaction. The CAPP has been applied to many berry fruits that have tender skin and are susceptible to injury and pathogen invasion (Bovi et al., 2019). The CAPP has effectively inhibited *Botrytis cinerea*, the gray mold rot in mulberries. Yinxin et al. (2022) reported that CAPP damaged the structure of the mold and caused a redox imbalance that supports the control of the mold.

In this study, currants were treated with air and nitrogen-generated diffuse coplanar surface barrier discharge CAPP at various times (0–10 min) and stored at 7°C for 10 days (Limnaios et al., 2021). The air-generated plasma reduced the yeast and mold count, and the nitrogen-generated plasma reduced the aerobic mesophilic total viable count. Both treatments did not impact most quality attributes. The microorganism, *Pseudomonas tolaasii*, causes blotch disease in button mushroom. In vivo antibacterial effects of CAPP were found to improve microbial contamination and enhance the qualitative properties of button mushroom (Pourbagher et al., 2021). Also, CAPP could delay mushroom softening. As reported from the several different studies on different fruits, there was a need to identify the optimum process parameters of CAPP to apply to each produce (Bovi et al., 2019).

In another study by Shen et al. (2021), inoculated Chinese bayberries with *Penicillium citrinum* and treated with plasma-processed air (PPA). This pretreatment of PPA caused oxidative stress of the fruit resulting in antioxidative compounds and antioxidant activity. The *Momordica charantia* polysaccharide (MCP) as a nanofiber matrix served in active food packaging. CAPP was used to treat the nanofiber membranes and to increase their application performance by Cui et al. (2020). Results suggested that phlorotannin (PT) has antibacterial and antioxidant properties and was encapsulated in the MCP nanofibers. The CAPP improved the release of PT from nanofibers. The antibacterial and antioxidant activities of the PT/MCP nanofibers were improved.

### 5.2.2.5 High Hydrostatic Pressure (HHP)

High hydrostatic pressure (HHP) or ultra-high-pressure processing (UHP) or high-pressure processing (HPP) operates at higher pressures (100–700 MPa) with or without external heat (Murray et al., 2017). Postharvest HHP application aided in pathogen reduction on fresh produce (Varalakshmi, 2021); however, tissue injury can be a major limitation (Agregán et al., 2021). There are many variations of the application of HPP, such as by vacuum packaging of the product and then pressurizing by the medium, usually water. The advantages of HHP for the disinfection of fruits and vegetables include in-package treatments and no harmful residues while there is the limitation of possible thermal damage depending on treatment conditions (Deng et al., 2020a). The HHP was more effective against vegetative microbes than bacterial spores (Jongman et al., 2021). The HHP disinfection treatment inactivated bacterial vegetative cells by multiple mechanisms, which included DNA inhibition synthesis, rupture of cell membrane, induced protein denaturation and enzyme inactivation (Huang et al., 2017; Lou et al., 2015).

Current knowledge suggests that the microwave-assisted thermal pasteurization is more effective than HPP to control *Listeria innocua* in green beans, but the effects on quality were similar for both treatments (Inanoglu et al., 2021).

### 5.2.2.6 Nanotechnology

Nanotechnology is an alternative disinfection technology to manage postharvest diseases and to maintain the quality of fruits and vegetables (González-Estrada et al., 2019; Liu et al., 2020b; Ruffo Roberto et al., 2019). Nanoparticles, nanocomposites and their by-products can release target molecules at specific and non-targeted organisms, reduce potential toxicity and maintain the functional agents (Alghuthaymi et al., 2020). The nanoparticles target the cell walls and cell membrane, lipopolysaccharide layer and phospholipid bilayer of microorganisms in disinfection (Rikta, 2019). The nanohybrid antifungal is a synergistic approach to manage fungal pathogens (Alghuthaymi et al., 2020). Nanotechnology applications for fruits and vegetables include nanocoatings/nanoemulsions, thin-film packaging and nanosensors for labeling of fresh products (Alghuthaymi et al., 2021). Nanocomposites have aided in the postharvest management of diseases and infections, leading to produce loss (Mathushika & Gomes, 2021).

A nanofilm is a very thin layer, less than 100 nm. It is applied as a fruit coating and as an external film for the retention of acceptable quality (Li et al., 2021). A chitosan-nano-silica-sodium alginate composite film lengthened the postharvest storage of winter jujube for about 1 month over untreated fruits (Kou et al., 2019).

A UV/ozone-treated low-density polyethylene (LDPE) film was coated with polyethyleneimine and polyacrylic acid polymer solution with loaded antimicrobial silver nanoparticles (Ag-NPs) (Azlin-Hasim et al., 2016). The film inclusion with Ag-NPs resulted in better antimicrobial properties. A chitosan-nano-$SiO_2$ film minimized the changes in acidity, anthocyanin and growth of mesophilic microorganisms on fresh blueberry fruits (Li et al., 2021).

## 5.3  SAFETY ISSUES OF DISINFECTION TECHNOLOGIES

Several eco-friendly alternatives to the conventional use of chlorine for disinfection for fruits and vegetables are generally recognized as safe. Excessive exposure of ozone could have deleterious effects not only on human health but on the physico-chemical quality of fruits and vegetables. Some critical factors in the applications of ozone are the concentration, contact time, sensitivity of the produce and worker's safety. The U.S. Food and Drug Administration (FDA) has recommended that workers shall not be exposed to more than 0.05 ppm of ozone for 8 hours (Ali et al., 2018; Miller et al., 2013). Also, the system design and process must comply with the established health standards recommended by the FDA for use of ozone in the fresh-cut industry (Ali et al., 2018). An issue of ozone safety is that it remains in the output air treatment system; however, it can be removed by the use of an appropriate photocatalyst (Szeto et al., 2020). The ozone treatment of commodities with strong aroma should be avoided as their volatile compounds could be affected (Mohd Aziz & Ding, 2018). The safety concerns of HPPP stem from the use of high pressures in applications (Deng et al., 2020b).

The safety attributes of nanomaterials have not been thoroughly researched (Liu et al., 2020b) as there could be a migration of nanoparticles into produce through absorption, dissolution and diffusion. The higher levels of reactive oxygen species of nanoparticles (NPs) have been linked to the genotoxicity of metal NPs (Mortezaee et al., 2019). Predictive modeling research could relate the effects of nanotechnology on the estimated produce quality and shelf life (Liu et al., 2020b).

## 5.4  CONCLUSION AND FUTURE FOCUS

Freshly harvested and minimally processed fruits and vegetables are perishable and are vulnerable to physiological and microbial contamination throughout the various postharvest processing stages. Emerging advanced postharvest disinfection technologies could delay microbial decay and maximize quality, thereby minimizing produce losses and prolonging the storage life of the fresh produce. An ideal disinfectant should be effective against foodborne pathogens, non-hazardous and eco-friendly. Chlorine dioxide application is limited to the microorganisms on the produce surface but was effective in preventing cross-contamination in the washing step. However, its antimicrobial efficacy could be reduced in the presence of organic material.

Electrolyzed water (EW) application is an environmentally friendly, cheap and safe method for human health and serves as an effective antimicrobial and antibiofilm agent. Variations in EW treatments such as slightly acidic to acidic electrolyzed water and alkaline electrolyzed water were applied to broccoli sprouts, longan, nectarine, apple mandarin and tomato. Pulsed light (PL) was applied to remove microorganisms and pesticide residues of green vegetables. Its sterilization effects against microorganisms depended on the light intensity, number of pulses, microbial species and produce properties.

The ultrasound (US) application has limited disinfection efficacy; hence it is recommended that US should be combined with other disinfection application for fresh produce. US has been combined with ozone, acids, pulsed light, osmotic dehydration with synergistic antimicrobial effects and better retention of produce quality.

Cold atmospheric pressure plasma (CAPP) has been applied to many tender berries, being susceptible to pathogen attack. It is critical to maximize the process parameters of CAPP such as the generation source, gas type setup, and process parameters for each produce (Bovi et al., 2019).

The antimicrobial efficacy of ozone was increased with its higher concentration for cherry tomatoes but was influenced by temperature and contact time. In ozonation, ozone can be applied as gaseous oxygen (OG) or aqueous ozone (AO) for the disinfection of such fresh produce as cherry tomatoes, cantaloupes, fresh-cut apples, strawberry, cabbage, spinach, fresh parsley, etc.

Nanotechnology has various applications in the control of postharvest diseases such as by alternative nanofilms, nanopackaging, nanobiosensors, nanodisinfectants and nanoantifungal agents. There is the increasing use of biopolymer nanofilm such as chitosan with the incorporation of silicon and titanium dioxides in the commercial storage of produce.

There is a growing trend of combining several disinfection hurdle applications as the resultant synergistic effects are lethal to microorganisms and could minimize the deterioration of produce.

Fruit and vegetable processors should implement preventative decontamination programs such as Good Agricultural Practices (GAPs), Sanitation Standard Operating Procedures (SSOPs), Hazard Analysis and Critical Control Points (HACCP).

## REFERENCES LIST

Abnavi, M. D., Alradaan, A., Munther, D., Kothapalli, C. R., & Srinivasan, P. (2019). Modeling of free chlorine consumption and *Escherichia coli* O157: H7 cross-contamination during fresh-cut produce wash cycles. *Journal of Food Science, 84*(10), 2736–2744. https://doi.org/10.1111/1750-3841.14774

Agregán, R., Munekata, P. E. S., Zhang, W., Zhang, J., Pérez-Santaescolástica, C., & Lorenzo, J. M. (2021). High-pressure processing in inactivation of Salmonella spp. in food products. *Trends in Food Science and Technology, 107*(14), 31–37. https://doi.org/10.1016/j.tifs.2020.11.025

Aguirre, J. S., García de Fernando, G., Hierro, E., Hospital, X. F., Espinosa, I., & Fernández, M. (2018). Characterization of damage on *Listeria innocua* surviving to pulsed light: Effect on growth, DNA and proteome. *International Journal of Food Microbiology, 284*, 63–72. https://doi.org/10.1016/j.ijfoodmicro.2018.07.002

Alghuthaymi, M. A., Abd-Elsalam, K. A., Paraliker, P., & Rai, M. (2020). Mono and hybrid nanomaterials: Novel strategies to manage postharvest diseases. *Multifunctional Hybrid Nanomaterials for Sustainable Agric.-Food and Ecosystems*, 287–317. https://doi.org/10.1016/B978-0-12-821354-4.00013-3

Alghuthaymi, M. A., Rajkuberan, C., Rajiv, P., Kalia, A., Bhardwa, J. K., Bhardwaj, P., Abd-Elsalam, K. A., Valis, M., & Kuca, K. (2021). Nanohybrid antifungals for control of plant diseases: Current status and future perspectives. *Journal of Fungi, 7*(1), 48. https://doi.org/10.3390/jof7010048

Ali, A., Yeoh, W. K., Forney, C., & Siddiqui, M. W. (2018). Advances in postharvest technologies to extend the storage life of minimally processed fruits and vegetables. *Critical Reviews in Food Science and Nutrition, 58*(15), 2632–2649. https://doi.org/10.1080/10 408398.2017.1339180

Ansari, A., Parmar, K., & Shah, M. (2022). A comprehensive study on decontamination of food-borne microorganisms by cold plasma. *Food Chemistry. Molecular Sciences, 4,* 100098. https://doi.org/10.1016/j.fochms.2022.100098

Aslam, R., Alam, M. S., & Saeed, P. A. (2020). Sanitization potential of ozone and its role in postharvest quality management of fruits and vegetables. *Food Engineering Reviews, 12*(1), 48–67. https://doi.org/10.1007/s12393-019-09204-0

Azlin-Hasim, S., Cruz-Romero, M. C., Cummins, E., Kerry, J. P., & Morris, M. A. (2016). The potential use of a layer-by-layer strategy to develop LDPE antimicrobial films coated with silver nanoparticles for packaging applications. *Journal of Colloid and Interface Science, 461,* 239–248. https://doi.org/10.1016/j.jcis.2015.09.021

Baier, M., Ehlbeck, J., Knorr, D., Herppich, W. B., & Schlüter, O. (2015). Impact of plasma processed air (PPA) on quality parameters of fresh produce. *Postharvest Biology and Technology, 100,* 120–126. https://doi.org/10.1016/j.postharvbio.2014.09.015

Belay, Z. A., Botes, W. J., & Caleb, O. J. (2021). Effects of alkaline electrolyzed water pretreatment on the physicochemical quality attributes of fresh nectarine during storage. *Journal of Food Processing and Preservation, 45*(10), e15879. https://doi.org/10.1111/jfpp.15879

Bhavya, M. L., & Umesh Hebbar, H. U. (2017). Pulsed light processing of foods for microbial safety. *Food Quality and Safety, 1*(3), 187–202. https://doi.org/10.1093/fqsafe/fyx017

Bintsis, T., Litopoulou-Tzanetaki, E., & Robinson, R. K. (2000). Existing and potential applications of ultraviolet light in the food industry – A critical review. *Journal of the Science of Food and Agriculture, 80*(6), 637–645. https://doi.org/10.1002/(SICI)10970010(20000501)80:6<637:AID-JSFA603>3.0.CO

Botondi, R., Barone, M., & Grasso, C. A. (2021). A review into the effectiveness of ozone technology for improving the safety and preserving the quality of fresh-cut fruits and vegetables. *Foods, 10*(4), 748. https://doi.org/10.3390/foods10040748

Bovi, G. G., Fröhling, A., Pathak, N., Valdramidis, V. P., & Schlüter, O. (2019). Safety control of whole berries by cold atmospheric pressure plasma processing: A review. *Journal of Food Protection, 82*(7), 1233–1243. https://doi.org/10.4315/0362-028X. JFP-18-606

Bußler, S., Ehlbeck, J., & Schlüter, O. K. (2017). Pre-drying treatment of plant related tissues using plasma processed air: Impact on enzyme activity and quality attributes of cut apple and potato. *Innovative Food Science and Emerging Technologies, 40,* 78–86. https://doi.org/10.1016/j.ifset.2016.05.007

Carletti, L., Botondi, R., Moscetti, R., Stella, E., Monarca, D., Cecchini, M., & Massantini, R. (2013). Use of ozone in sanitation and storage of fresh fruits and vegetables. *Journal of Food, Agriculture and Environment, 11*(3&4), 585–589.

Chen, C., Zhang, H., Zhang, X., Dong, C., Xue, W., & Xu, W. (2020). The effect of different doses of ozone treatments on the postharvest quality and biodiversity of cantaloupes. *Postharvest Biology and Technology, 163,* 111124. https://doi.org/10.1016/j.postharvbio.2020.111124

Chen, X., Tango, C. N., Dalir, E. B.-M., Oh, S.-Y., & Oh, D.-H. (2019). Disinfection efficacy of slightly acidic electrolyzed water combined with chemical treatments on fresh fruits at the industrial scale. *Foods, 8*(10), 497. https://doi:10.3390/foods8100497. https://doi.org/10.3390/foods8100497

Chinchkar, A. V., Singh, A., Singh, S. V., Acharya, A. M., & Kamble, M. G. (2022). Potential sanitizers and disinfectants for fresh fruits and vegetables: A comprehensive review. *Journal of Food Processing and Preservation, 46*(10). https://doi.org/10.1111/jfpp.16495

Choi, S. W., Shahbaz, H. M., Kim, J. U., Kim, D.-H., Yoon, S., Jeong, S. H., Park, J., & Lee, D.-U. (2020). Photolysis and TIO₂ photocatalytic treatment under UVC/VUV irradiation for simultaneous degradation of pesticides and microorganisms. *Applied Sciences, 10*(13), 4493. https://doi.org/10.3390/app10134493

Chuwa, C., Vaidya, D., Kathuria, D., Gautam, S., Sharma, S., & Sharma, B. (2020). Ozone (O₃): An emerging technology in the food industry. *Food and Nutrition Journal, 5*(2), 224. https://doi.org/10.29011/2575-7091.10012410

Collazo, C., Charles, F., Aguiló-Aguayo, I., Marín-Sáez, J., Lafarga, T., Abadias, M., & Viñas. I. (2019). Decontamination of Listeria innocua from fresh-cut broccoli using UV-C applied in water or peroxyacetic acid, and dry-pulsed light. *Innovative Food Science & Emerging Technologies, 52*, 438–449. https://doi.org/10.1016/j.ifset.2019.02.004

Collazo, C., Lafarga, T., Aguiló-Aguayo, I., Marín-Sáez, J., Abadias, M., & Viñas, I. (2018). Decontamination of fresh-cut broccoli with a water–assisted UV-C technology and its combination with peroxyacetic acid. *Food Control, 93*, 92–100. https://doi.org/10.1016/j.foodcont.2018.05.046

Contigiani, E. V., Jaramillo Sánchez, G. M., Castro, M. A., Gómez, P., L., & Alzamora, S. M. (2020). Efficacy of mild thermal and pulsed light treatments individually applied or in combination, for maintaining postharvest quality of strawberry cv. Albion. *Journal of Food Processing and Preservation, 45*(1), e15095. https://doi.org/10.1111/jfpp.15095

Cui, H., Yang, X., Abdel-Samie, M. A., & Lin, L. (2020). Cold plasma treated phlorotannin/*Momordica charantia* polysaccharide nanofiber for active food packaging. *Carbohydrate Polymers, 239*(1), 116214. https://doi.org/10.1016/j.carbpol.2020.116214

Dai, J., Bai, M., Li, C., Cui, H., & Lin, L. (2020). Advances in the mechanism of different antibacterial strategies based on ultrasound technique for controlling bacterial contamination in food industry. *Trends in Food Science and Technology, 105*(4), 211–222. https://doi.org/10.1016/j.tifs.2020.09.016

Darré, M., Vicente, A. R., Cisneros-Zevallos, L., & Artés-Hernández, F. (2022). Postharvest ultraviolet radiation in fruit and vegetables: Applications and factors modulating its efficacy on bioactive compounds and microbial growth. *Foods, 11*(5), 653. https://doi.org/10.3390/foods11050653

De Corato, U. (2020). Improving the shelf-life and quality of fresh and minimally processed fruits and vegetables for a modern food industry: A comprehensive critical review from the traditional technologies into the most promising advancements. *Critical Reviews in Food Science and Nutrition, 60*(6), 940–975. https://doi.org/10.1080/10408398.2018.1553025

de Souza, L. P., Faroni, L. R. D. A., Heleno, F. F., Cecon, P. R., Gonçalves, T. D. C., da Silva, G. J., & Prates, L. H. F. (2018). Effects of ozone treatment on postharvest carrot quality. *LWT, 90*(9), 53–60. https://doi.org/10.1016/j.lwt.2017.11.057

Deng, L. Z., Mujumdar, A. S., Pan, Z., Vidyarthi, S. K., Xu, J., Zielinska, M., & Xiao, H. W. (2020a). Emerging chemical and physical disinfection technologies of fruits and vegetables: A comprehensive review. *Critical Reviews in Food Science and Nutrition, 60*(15), 2481–2508. https://doi.org/10.1080/10408398.2019.1649633

Deng, L.-Z., Tao, Y., Mujumdar, A. S., Pan, Z., Chen, C., Yang, X.-H., Liu, Z.-L., Wang, H., & Xiao, H.-W. (2020b). Recent advances in non-thermal decontamination technologies for microorganisms and mycotoxins in low-moisture foods. *Trends in Food Science and Technology, 106*, 104–112. https://doi.org/10.1016/j.tifs.2020.10.012

Duarte de Sousa, A. E., de Almeida Lopes, M. M., Renan Moreira, A. D., Nunes Macedo, J. J., Herbster Moura, C. F., Souza de Aragão, F. A., Zocolo, G. J., Alcântara de Miranda, M. R., & de Oliveira Silva, E. (2019). Induction of postharvest resistance in melon using pulsed light as abiotic stressor. *Scientia Horticulturae, 246*, 921–927. https://doi.org/10.1016/j.scienta.2018.11.066

Duarte-Molina, F., Gómez, P. L., Castro, M. A., & Alzamora, S. M. (2016). Storage quality of strawberry fruit treated by pulsed light: Fungal decay, water loss and mechanical properties. *Innovative Food Science and Emerging Technologies*, *34*, 267–274. https://doi.org/10.1016/j.ifset.2016.01.019

Elmnasser, N., Guillou, S., Leroi, F., Orange, N., Bakhrouf, A., & Federighi, M. (2007). Pulsed-light system as a novel food decontamination technology: A review. *Canadian Journal of Microbiology*, *53*(7), 813–821. https://doi.org/10.1139/W07-042

El-Ramady, H. R., Domokos-Szabolcsy, É., Abdalla, N. A., Taha, H. S., & Fári, M. (2015). Postharvest management of fruits and vegetables storage. In E. Lichtfouse (Ed.), *Sustainable agriculture reviews* (pp. 65–152). Springer Nature. https://doi.org/10.1007/978-3-319-09132-7_2

Esua, O. J., Chin, N. L., & Yusof, Y. A. (2018). Simultaneous UV-C and ultrasonic energy treatment for disinfection of tomatoes and its antioxidant properties. *Journal of Advanced Agricultural Technologies*, *5*(3), 209–214. https://doi.org/10.18178/joaat.5.3.209-214

Esua, O. J., Chin, N. L., Yusof, Y. A., & Sukor, R. (2020). A review on individual and combination technologies of UV-C radiation and ultrasound in postharvest handling of fruits and vegetables. *Processes*, *8*(11), 1433. https://doi.org/10.3390/pr8111433

Fan, K., Wu, J., & Chen, L. (2021). Ultrasound and its combined application in the improvement of microbial and physicochemical quality of fruits and vegetables: A review. *Ultrasonics Sonochemistry*, *80*, 105838. https://doi.org/10.1016/j.ultsonch.2021.105838

Fan, X., & Song, Y. (2020). Advanced oxidation process as a postharvest decontamination technology to improve microbial safety of fresh produce. *Journal of Agricultural and Food Chemistry*, *68*(46), 12916–12926. https://doi.org/10.1021/acs.jafc.0c01381

Fang, W., Xue, S., & Yue, Y. (2020). Progress of pulsed light sterilization technology in the food field. In L. Yang & Z. Xu (Eds.), *E3S Web of Conferences International Conference on Energy*. Environment and Bioengineering (ICEEB 2020), Xi'an, China.185, (4pp), id.04072. https://doi.org/10.1051/e3sconf/202018504072

Gadelha, J. R., Allende, A., López-Gálvez, F., Fernández, P., Gil, M. I., & Egea, J. A. (2019). Chemical risks associated with ready-to-eat vegetables: Quantitative analysis to estimate formation and/or accumulation of disinfection byproducts during washing. *EFSA Journal*, *17*(Suppl. 2), e170913. https://doi.org/10.2903/j.efsa.2019.e170913

Ghernaout, D., & Elboughdiri, N. (2020). Towards enhancing ozone diffusion for water disinfection – Short notes. *Open Access Library Journal*, *7*, e6253. https://doi.org/10.4236/oalib.1106253

Glowacz, M., & Rees, D. (2016). The practicality of using ozone with fruit and vegetables. *Journal of the Science of Food and Agriculture*, *96*(14), 4637–4643. https://doi.org/10.1002/jsfa.7763

González-Estrada, R. R., Blancas-Benitez, F. J., Moreno-Hernández, C. L., Coronado-Partida, L., Ledezma-Delgadillo, A., & Gutiérrez-Martínez, P. (2019). Nanotechnology: A promising alternative for the control of postharvest pathogens in fruits. In D. Panpatte & Y. Jhala (Eds.), *Nanotechnology for agriculture: Crop production and protection* (pp. 323–337). Springer. https://doi.org/10.1007/978-981-32-9374-8_15

Green, A., Popović, V., Warriner, K., & Koutchma, T. (2020). The efficacy of UVC LEDs and low-pressure mercury lamps for the reduction of *Escherichia coli* O157:H7 and *Listeria monocytogenes* on produce. *Innovative Food Science and Emerging Technologies*, *64*, 102410. https://doi.org/10.1016/j.ifset.2020.102410

Hashemi, S. M. B., & Jafarpour, D. (2020). Ultrasound and malic acid treatment of sweet lemon juice: Microbial inactivation and quality changes. *Journal of Food Processing and Preservation*, *44*(11), e14866. https://doi.org/10.1016/j.ultsonch.2020.105261

Hashemi, S. M. B., & Jafarpour, D. (2021). Antimicrobial and antioxidant properties of Saturn peach subjected to ultrasound-assisted osmotic dehydration. *Journal of Food Measurement and Characterization*, *15*(3), 2516–2523. https://doi.org/10.1007/s11694-021-00842-9

Horvitz, S., & Cantalejo, M. J. (2014). Application of ozone for the postharvest treatment of fruits and vegetables. *Critical Reviews in Food Science and Nutrition, 54*(3), 312–339. https://doi.org/10.1080/10408398.2011.584353

Huang, H.-W., Wu, S.-J., Lu, J.-K., Shyu, Y.-T., & Wang, C.-Y. (2017). Current status and future trends of high-pressure processing in food industry. *Food Control, 72*(A), 1–8. https://doi.org/10.1016/j.foodcont.2016.07.019

Huang, Y., & Chen, H. (2014). A novel water-assisted pulsed light processing for decontamination of blueberries. *Food Microbiology, 40*, 1–8. https://doi.org/10.1016/j.fm.2013.11.017

Ignat, A., Manzocco, L., Maifreni, M., & Nicoli, M. C. (2016). Decontamination efficacy of neutral and acidic electrolyzed water in fresh-cut salad washing. *Journal of Food Processing and Preservation, 40*(5), 874–881. https://doi.org/10.1111/jfpp.12665

Inanoglu, S., Barbosa-Cánovas, G. V., Patel, J., Zhu, M.-J., Sablani, S. S., Liu, F., Tang, Z., & Tang, J. (2021). Impact of high-pressure and microwave assisted thermal pasteurization on inactivation of *Listeria innocua* and quality attributes of green beans. *Journal of Food Engineering, 288*(7), 110162. https://doi.org/10.1016/j.jfoodeng.2020.110162

Jeong, M.-A., & Jeong, R.-D. (2018). Applications of ionizing radiation for the control of postharvest diseases in fresh produce: Recent advances. *Plant Pathology, 67*(1), 18–29. https://doi.org/10.1111/ppa.12739

Jongman, M., Carmichael, P., Loeto, D., & Gomba, A. (2021). Advances in the use of biocontrol applications in preharvest and postharvest environments: A food safety milestone. *Journal of Food Safety, 42*(2), e12957. https://doi.org/10.1111/jfs.12957

Kahramanoğlu, I. (2017). Introductory chapter: Postharvest physiology and technology of horticultural crops. In I Kahramanoğlu (Ed.), *Postharvest handling* (pp. 1–5). Intech Open Science. https://doi.org/10.5772/intechopen.69466

Kahramanoğlu, I., Okatan, V., & Wan, C. (2020). Biochemical composition of propolis and its efficacy in maintaining postharvest storability of fresh fruits and vegetables. *Journal of Food Quality, 2020*, article ID 8869624, 9 pages. https://doi.org/10.1155/2020/8869624

Karaca, H., & Velioglu, Y. S. (2020). Effects of ozone and chlorine washes and subsequent cold storage on microbiological quality and shelf life of fresh parsley leaves. *LWT, 127*, 109421. https://doi.org/10.1016/j.lwt.2020.109421

Khaire, R. A., Thorat, B. N., & Gogate, P. R. (2021). Applications of ultrasound for food preservation and disinfection: A critical review. *Journal of Food Processing and Preservation, 2021*, e16091. https://doi.org/10.1111/jfpp.16091

Khandpur, P., & Gogate, P. R. (2016). Evaluation of ultrasound-based sterilization approaches in terms of shelf life and quality parameters of fruits and vegetable juices. *Ultrasonics Sonochemistry, 29*, 337–353. https://doi.org/10.1016/j.ultsonch.2015.10.008

Kou, X., He, Y., Li, Y., Chen, X., Feng, Y., & Xue, Z. (2019). Effect of abscisic acid (ABA) and chitosan/nano-silica/sodium alginate composite film on the color development and quality of postharvest Chinese winter jujube (*Zizyphus jujuba* Mill. cv. Dongzao). *Food Chemistry, 270*(1), 385–394. https://doi.org/10.1016/j.foodchem.2018.06.151

Kramer, B., Wunderlich, J., & Muranyi, P. (2017). Recent findings in pulsed light disinfection. *Journal of Applied Microbiology, 122*(4), 830–856. https://doi.org/10.1111/jam.13389

Lacombe, A., Antosch, J. G., & Wu, V. C. H. (2020). Scale-up model of forced air-integrated gaseous chlorine dioxide for the decontamination of lowbush blueberries. *Journal of Food Safety, 40*(4), e12793. https://doi.org/10.1111/jfs.12793

Le Goff, L., Hubert, B., Favennec, L., Villena, I., Ballet, J. J., Agoulon, A., Orange, N., & Gargala, G. (2015). Pilot-scale pulsed UV light irradiation of experimentally infected raspberries suppresses *Cryptosporidium parvum* infectivity in immunocompetent suckling mice. *Journal of Food Protection, 78*(12), 2247–2252. https://doi.org/10.4315/0362-028X.JFP-15-062

Leng, J., Mukhopadhyay, S., Sokorai, K., Ukuku, D.O., Fan, X., Olanya, M., & Juneja, V. (2020). Inactivation of Salmonella in cherry tomato stem scars and quality preservation by pulsed light treatment and antimicrobial wash. *Food Control*, *110*(3), 107005. https://doi.org/10.1016/j.foodcont.2019.107005

Li, L., Song, S., Nirasawa, S., Hung, Y.C., Jiang, Z., & Liu, H. (2019). Slightly acidic electrolyzed water treatment enhances the main bioactive phytochemicals content in broccoli sprouts via changing metabolism. *Journal of Agricultural and Food Chemistry*, *67*(2), 606–614. https://doi.org/10.1021/acs.jafc.8b04958

Li, Y., Rokayya, S., Jia, F., Nie, X., Xu, J., Elhakem, A., Almatrafi, M., Benajiba, N., & Helal, M. (2021). Shelf life, quality, safety evaluations of blueberry fruits coated with chitosan Nano material films. *Scientific Reports*, *11*(1), 55. https://doi.org/10.1038/s41598-020-80056-z

Limnaios, A., Pathak, N., Grossi Bovi, G.G., Fröhling, A., Valdramidis, V.P., Taoukis, P.S., & Schlüter, O. (2021). Effect of cold atmospheric pressure plasma processing on quality and shelf life of red currants. *LWT*, *151*, 112213. https://doi.org/10.1016/j.lwt.2021.112213

Liu, C., Chen, C., Jiang, A., Zhang, Y., Zhao, Q., & Hu, W. (2021a). Effects of aqueous ozone treatment on microbial growth, quality, and pesticide residue of fresh-cut cabbage. *Food Science and Nutrition*, *9*(1), 52–61. https://doi.org/10.1002/fsn3.1870

Liu, C., Chen, C., Zhang, Y., Jiang, A., & Hu, W. (2021b). Aqueous ozone treatment inhibited degradation of cell wall polysaccharides in fresh-cut apple during cold storage. *Innovative Food Science and Emerging Technologies*, *67*, 1466. https://doi.org/10.1016/j.ifset.2020.102550

Liu, W., Zhang, M., & Bhandari, B. (2020b). Nanotechnology – A shelf-life extension strategy for fruits and vegetables. *Critical Reviews in Food Science and Nutrition*, *60*(10), 1706–1721. https://doi.org/10.1080/10408398.2019.1589415

Liu, X., Jiao, W., Du, Y., Chen, Q., Su, Z., & Fu, M. (2020c). Chlorine dioxide controls green mold caused by *Penicillium digitatum* in citrus fruits and the mechanism involved. *Journal of Agricultural and Food Chemistry*, *68*(47), 13897–13905. https://doi:10.1021/acs.jafc.0c05288

Lou, F., Neetoo, H., Chen, H., & Li, J. (2015). High hydrostatic pressure processing: A promising nonthermal technology to inactivate viruses in high-risk foods. *Annual Review of Food Science and Technology*, *6*(1), 389–409. https://doi.org/10.1146/annurev-food-072514-104609

Mahajan, P.V., Caleb, O.J., Singh, Z., Watkins, C.B., & Geyer, M. (2014). Postharvest treatments of fresh produce. *Philosophical Transactions. Series A, Mathematical, Physical, and Engineering Sciences*, *372*, 20130309. http://doi.org/10.1098/rsta.2013.0309

Mahendran, R., Ramanan, K.R., Barba, F.J., Lorenzo, J.M., López-Fernández, O., Munekata, P.E.S., Roohinejad, S., Sant'Ana, A.S., & Tiwari, B.K. (2019). Recent advances in the application of pulsed light processing for improving food safety and increasing shelf life. *Trends in Food Science and Technology*, *88*(2), 67–79. https://doi.org/10.1016/j.tifs.2019.03.010

Majid, I., Nayik, G.A., & Nanda, V. (2015). Ultrasonication and food technology: A review. *Cogent Food and Agriculture*, *1*(1), 1071022. https://doi.org/10.1080/23311932.2015.1071022

Maryam, A., Anwar, R., Malik, A.U., Raheem, M.I.U., Khan, A.S., Hasan, M., U., Hussain, Z., & Siddique, Z. (2020). Combined aqueous ozone and ultrasound application inhibits microbial spoilage, reduces pesticide residues, and maintains storage quality of strawberry fruits. *Journal of Food Measurement and Characterization*, *15*(2), 1437–1451. https://doi.org/10.1007/s11694-020-00735-3

Mathushika, J.M., & Gomes, C. (2021). Emerging concepts and practices in post-harvest management of horticultural crops revisited. *International Journal of Current Science Research and Review*, *4*(8), 859–876. https://doi.org/10.47191/ijcsrr/V4-i8-04

Meireles, A., Giaouris, E., & Simões, M. (2016). Alternative disinfection methods to chlorine for use in the fresh-cut industry. *Food Research International*, *82*, 71–85. https://doi.org/10.1016/j.foodr es.2016.01.021

Miller, F. A., Silva, C. L. M., & Brandão, T. R. S. (2013). A review on ozone-based treatments for fruit and vegetables preservation. *Food Engineering Reviews*, *5*(2), 77–106. https://doi.org/10.1007/s12393-013-9064-5

Minh, N. P. (2021). Postharvest treatment of Chinese kale (*Brassica oleracea* var. alboglabra) by pulse light to removal of microbial load, pesticide residue and integrity of physicochemical quality and phytochemical constituent. *Journal of Pure and Applied Microbiology, 15*(4), 2252–2262. https://doi.org/10.22207/JPAM.15.4.47

Mohd Aziz, K., & Ding, P. (2018). Ozone application in fresh fruits and vegetables. *Pertanika Journal of Scholarly Research Review*, *4*(2), 29–35.

Moreira, M. R., Álvarez, M. V., Martín-Belloso, O., & Soliva-Fortuny, R. (2017). Effects of pulsed light treatments and pectin edible coatings on the quality of fresh-cut apples: A hurdle technology approach. *Journal of the Science of Food and Agriculture*, *97*(1), 261–268. https://doi.org/10.1002/jsfa.7723

Mortezaee, K., Najafi, M., Samadian, H., Barabadi, H., Azarnezhad, A., & Ahmadi, A. (2019). Redox interactions and genotoxicity of metal-based nanoparticles: A comprehensive review. *Chemico-Biological Interactions*, *312*, 108814. https://doi.org/10.1016/j.cbi.2019.108814

Murray, K., Moyer, P., Wu, F., Goyette, J. B., & Warriner, K. (2018). Inactivation of *Listeria monocytogenes* on and within apples destined for caramel apple production by using sequential forced air ozone gas followed by a continuous advanced oxidative process treatment. *Journal of Food Protection*, *81*(3), 357–364. https://doi:10.4315/0362-028X.JFP-17-306. https://doi.org/10.4315/0362-028X.JFP-17-306

Murray, K., Wu, F., Shi, J., Xue, J, X., & Warriner, K. (2017). Challenges in the microbiological food safety of fresh produce: Limitations of post-harvest washing and the need for alternative interventions. *Food Quality and Safety*, *1*(4), 289–301. https://doi.org/10.1093/fqsafe/fyx027

Mustapha, A. T., Zhou, C., Wahia, H., Amanor-Atiemoh, R., Otu, P., Qudus, A., Fakayode, O. A., & Ma, H. (2020). Sonozonation: Enhancing the antimicrobial efficiency of aqueous ozone washing techniques on cherry tomato. *Ultrasonics Sonochemistry*, *64*, 105059. https://doi.org/10.1016/j.ultsonch.2020.105059

Nie, M., Wu, C., Xiao, Y., Song, J., Zhang, Z., Li, D., & Liu, C. (2020). Efficacy of aqueous ozone combined with sodium metasilicate on microbial load reduction of fresh-cut cabbage. *International Journal of Food Properties*, *23*(1), 2065–2076. https://doi.org/10.1080/10942912.2020.1842446

Ochoa-Velasco, C. E., Ávila-Sosa, R., Hernández-Carranza, P., Ruíz-Espinosa, H., Ruiz-López, I. I., & Guerrero-Beltrán, J. Á. (2020). Mathematical modeling used to evaluate the effect of UV-C light treatment on microorganisms in liquid foods. *Food Engineering Reviews*, *12*(3), 290–308. https://doi.org/10.1007/s12393-020-09219-y

Öztekin, S. (2019). Application of ozone as a postharvest treatment. In K. Barma, S. Sharm a, & M. W. Siddiqui (Eds.), *Emerging postharvest treatment of fruits and vegetables* (64 pp.). Apple Academic Press. https://doi.org/10.1201/9781351046312

Palgan, I., Caminiti, I. M., Muñoz, A., Noci, F., Whyte, P., Morgan, D. J., Cronin, D. A., & Lyng, J. G. (2011). Effectiveness of high intensity light pulses (HILP) treatments for the control of *Escherichia coli* and *Listeria innocua* in apple juice, orange juice and milk. *Food Microbiology*, *28*(1), 14–20. https://doi.org/10.1016/j.fm.2010.07.023

Pankaj, S. K., & Keener, K. M. (2017). Cold plasma: Background, applications, and current trends. *Current Opinion in Food Science*, *16*, 49–52. https://doi.org/10.1016/j.cofs.2017.07.008

Pérez-Pérez, C., Regalado-González, C., Rodríguez-Rodríguez, C. A., Barbosa-Rodríguez, J. R., & Villaseñor-Ortega, F. (2016). Incorporation of antimicrobial agents in food packaging films and coatings. In R. G. Guevara-González & I. Torres-Pacheco (Eds.), *Advances in Agricultural and Food Biotechnology, 37*(2), 193–216. Research Signpost.

Perry, J. J., & Yousef, A. E. (2011). Decontamination of raw foods using ozone-based sanitization techniques. *Annual Review of Food Science and Technology, 2*, 281–298. https://doi.org/10.1146/annurev-food-022510-133637

Piechowiak, T., Antos, P., Kosowski, P., Skrobacz, K., Józefczyk, R., & Balawejder, M. (2019). Impact of ozonation process on the microbiological and antioxidant status of raspberry (*Rubus ideaeus* L.) fruit during storage at room temperature. *Agricultural and Food Science, 28*(1), 35–44. https://doi.org/10.23986/afsci.70291

Pignata, C., D'Angelo, D., Fea, E., & Gilli, G. (2017). A review on microbiological decontamination of fresh produce with nonthermal plasma. *Journal of Applied Microbiology, 122*(6), 1438–1455. https://doi.org/10.1111/jam.13412

Pillai, K. C., Kwon, T. O., Park, B. B., & Moon, I. S. (2009). Studies on process parameters for chlorine dioxide production using $IrO_2$ anode in an un-divided electrochemical cell. *Journal of Hazardous Materials, 164*(2–3), 812–819. https://doi.org/10.1016/j.jhazmat.2008.08.090

Pongener, A., Sharma, S., & Purbey, S. K. (2018). Heat treatment of fruits and vegetables. In M. W. Siddiqui (Ed.), *Postharvest disinfection of fruits and vegetables* (pp. 179–196). Academic Press. https://doi.org/10.1016/B978-0-12-812698-1.09988-X

Pou, K. R. J., & Raghavan, V. (2020). Recent advances in the application of high-pressure processing-based hurdle approach for enhancement of food safety and quality. *Journal of Biosystems Engineering, 45*(3), 175–187. https://doi.org/10.1007/s42853-020-00059-6

Pourbagher, R., Abbaspour-Fard, M. H., Sohbatzadeh, F., & Rohani, A. (2021). In vivo antibacterial effect of non-thermal atmospheric plasma on *Pseudomonas tolaasii*, a causative agent of *Agaricus bisporus* blotch disease. *Food Control, 130*(1), 108319. https://doi.org/10.1016/j.foodcont.2021.108319

Praeger, U., Herppich, W. B., & Hassenberg, K. (2018). Aqueous chlorine dioxide treatment of horticultural produce: Effects on microbial safety and produce quality-A review. *Critical Reviews in Food Science and Nutrition, 58*(2), 318–333. https://doi.org/10.1080/10408398.2016.1169157

Praeger, U., Scaar, H., Jedermann, R., Neuwald, D., König, M., & Geyer, M. (2016), Bonn Germany. Airflow conditions in apple bins and a commercial apple cold store. In *Proceedings of the 6th International Conference, Cold Chain Management* (pp. 99–104). ISBN-978-3-9812345-2-7.

Puligundla, P., Kim, J. W., & Mok, C. (2018). Broccoli sprout washing with electrolyzed water: Effects on microbiological and physicochemical characteristics. *LWT, 92*, 600–606. https://doi.org/10.1016/j.lwt.2017.09.044

Rikta, S. Y. (2019). Application of nanoparticles for disinfection and microbial control of water and wastewater. *Nanotechnology in Water and Wastewater Treatment, 1*, 159–176. https://doi.org/10.1016/B978-0-12-813902-8.00009-5

Ruffo Roberto, S., Youssef, K., Hashim, A. F., & Ippolito, A. (2019). Nanomaterials as alternative control means against postharvest diseases in fruit crops. *Nanomaterials, 9*(12), 1752. https://doi.org/10.3390/nano9121752

Sachadyn-Król, M., & Agriopoulou, S. (2020). Ozonation as a method of abiotic elicitation improving the health-promoting properties of plant products – A review. *Molecules, 25*(10), 2416. https://doi.org/10.3390/molecules25102416

Salehi, F. (2022). Application of pulsed light technology for fruits and vegetables disinfection: A review. *Journal of Applied Microbiology, 132*(4), 2521–2530. https://doi.org/10.1111/jam.15389

Shah, N. N. A. K., Supian, N. A. M., & Hussein, N. A. (2019). Disinfectant of pummelo (*Citrus grandis* L. Osbeck) fruit juice using gaseous ozone. *Journal of Food Science and Technology*, *56*(1), 262–272. https://doi.org/10.1007/s13197-018-3486-2

Shahi, S., Khorvash, R., Goli, M., Ranjbaran, S. M., Najarian, A., & Mohammadi Nafchi, A. M. (2021). Review of proposed different irradiation methods to inactivate food-processing viruses and microorganisms. *Food Science and Nutrition*, *9*(10), 5883–5896. https://doi.org/10.1002/fsn3.2539

Shehata, S. A., Abdelrahman, S. Z., Megahed, M. M. A., Abdeldaym, E. A., El-Mogy, M. M., & Abdelgawad, K. F. (2021). Extending shelf life and maintaining quality of tomato fruit by calcium chloride, hydrogen peroxide, chitosan, and ozonated water. *Horticulturae*, *7*(9), 309. https://doi.org/10.3390/horticulturae7090309

Shen, C., Rao, J., Wu, Q., Wu, D., & Chen, K. (2021). The effect of indirect plasma-processed air pretreatment on the microbial loads, decay, and metabolites of Chinese bayberries. *LWT*, *150*, 111998. https://doi.org/10.1016/j.lwt.2021.111998

Shinde, R., Rodov, V., Krishnakumar, S., & Subramanian, J. (2019). Active and intelligent packaging for reducing postharvest losses of fruits and vegetables. In G. Paliyath, J. Subramanian, L.-T. Lim, K. S. Subramanian, A. K. Handa, & A. K. Mattoo (Eds.), *Postharvest biology and nanotechnology* (pp. 171–189). John Wiley and Sons.

Shirai, A., & Yasutomo, Y. K. (2019). Bactericidal action of ferulic acid with ultraviolet-A light irradiation. *Journal of Photochemistry and Photobiology. B, Biology*, *191*(4), 52–58. https://doi.org/10.1016/j.jphotobiol.2018.12.003

Shirani, K., Shahidi, F., & Mortazavi, S. A. (2020). Investigation of decontamination effect of argon cold plasma on physicochemical and sensory properties of almond slices. *International Journal of Food Microbiology*, *335*, 108892. https://doi.org/10.1016/j.ijfoodmicro.2020.108892

Shu, X., Singh, M., Karampudi, N. B. R., Bridges, D. F., Kitazumi, A., Wu, V. C. H., & De los Reyes, B. G. (2021). Responses of *Escherichia coli* and *Listeria monocytogenes* to ozone treatment on non-host tomato: Efficacy of intervention and evidence of induced acclimation. *PLoS One*, *16*(10), e0256324. https://doi.org/10.1371/journal.pone.0256324

Siddique, Z., Malik, A. U., Asi, M. R., Anwar, R., & Inam Ur Raheem, M. I. U. (2021). Sonolytic-ozonation technology for sanitizing microbial contaminants and pesticide residues from spinach (*Spinacia oleracea* L.) leaves, at household level. *Environmental Science and Pollution Research International*, *28*(38), 52913–52924. https://doi.org/10.1007/s11356-021-14203-y

Smith, D. J., Ernst, W., & Herges, G. R. (2015). Chloroxyanion residues in cantaloupe and tomatoes after chlorine dioxide gas sanitation. *Journal of Agricultural and Food Chemistry*, *63*(43), 9640–9649. https://doi.org/10.1021/acs.jafc.5b04153

Song, H. J., Kim, M. H., & Ku, K.-. (2021). A double-edged sword of surfactant effect on hydrophobic surface broccoli leaf as a model plant: Promotion of pathogenic microbial contamination and improvement to disinfection efficiency of ozonated water. *Processes*, *9*(4), 679. https://doi.org/10.3390/pr9040679

Soni, A., Choi, J., & Brightwell, G. (2021). Plasma-activated water (PAW) as a disinfection technology for bacterial inactivation with a focus on fruit and vegetables. *Foods*, *10*(1), 166. https://doi.org/10.3390/foods10010166

Srivastava, S., & Mishra, H. N. (2021). Ecofriendly nonchemical/nonthermal methods for disinfestation and control of pest/fungal infestation during storage of major important cereal grains: A review. *Food Frontiers*, *2*(1), 93–105. https://doi.org/10.1002/fft2.69

Sun, J., Chen, H., Xie, H., Li, M., Chen, Y., Hung, Y. C., & Lin, H. (2022). Acidic electrolyzed water treatment retards softening and retains cell wall polysaccharides in pulp of postharvest fresh longans and its possible mechanism. *Food Chemistry: X*, *13*, 100265. https://doi.org/10.1016/j.fochx.2022.100265

Szeto, W., Yam, W. C., Huang, H., & Leung, D. Y. C. (2020). The efficacy of vacuum-ultraviolet light disinfection of some common environmental pathogens. *BMC Infectious Diseases*, *20*(1), 127. https://doi.org/10.1186/s12879-020-4847-9

Takaki, K., Takahashi, K., Hamanaka, D., Yoshida, R., & Uchino, T. (2021). Function of plasma and electrostatics for keeping quality of agricultural produce in post-harvest stage. *Japanese Journal of Applied Physics*, *60*(1), 010501. https://doi.org/10.35848/1347-4065/abcc13

Teixeira, P., Fernandes, B., Silva, A. M., Dias, N., & Azeredo, J. (2020). Evaluation by flow cytometry of *Escherichia coli* viability in lettuce after disinfection. *Antibiotics*, *9*(1), 14. https://doi:10.3390/antibiotics9010014

Thirumdas, R., Sarangapani, C., & Annapure, U. S. (2015). Cold plasma: A novel non-thermal technology for food processing. *Food Biophysics*, *10*(1), 1–11. https://doi.org/10.1007/s11483-014-9382-z

Tomás-Callejas, A., López-Gálvez, F., Sbodio, A., Artés, F., Artés-Hernández, F., & Suslow, T. V. (2012). Chlorine dioxide and chlorine effectiveness to prevent *Escherichia coli* O157:H7 and Salmonella cross-contamination on fresh-cut red chard. *Food Control*, *23*(2), 325–332. https://doi.org/10.1016/j.foodcont.2011.07.022

Tzortzakis, N., & Chrysargyris, A. (2017). Postharvest ozone application for the preservation of fruits and vegetables. *Food Reviews International*, *33*(3), 270–315. http://doi.org/10.1080/87559129.2016.1175015

Urban, L., Chabane Sari, D., Orsal, B., Lopes, M., Miranda, R., & Aarrouf, J. (2018). UV-C light and pulsed light as alternatives to chemical and biological elicitors for stimulating plant natural defenses against fungal diseases. *Scientia Horticulturae*, *235*, 452–459. https://doi.org/10.1016/j.scienta.2018.02.057

Varalakshmi, S. (2021). A review on the application and safety of nonthermal techniques on fresh produce and their products. *LWT*, *149*, 111849. https://doi.org/10.1016/j.lwt.2021.111849

Vivek, K., Suranjoy Singh, S. S., Ritesh, W., Soberly, M., Baby, Z., Baite, H., Mishra, S., & Pradhan, R. C. (2019). A review on postharvest management and advances in the minimal processing of fresh-cut fruits and vegetables. *Journal of Microbiology, Biotechnology and Food Sciences*, *8*(5), 1178–1187. https://doi.org/10.15414/jmbfs.2019.8.5.1178-1187

Vorobiev, E., & Lebovka, N. (2019). Pulsed electric field in green processing and preservation of food products. In F. Chemat & E. Vorobiev (Eds.), *Green food processing techniques preservation, transformation, and extraction* (pp. 403–430). Elsevier, Inc. https://doi.org/10.1016/B978-0-12-815353-6.00015-X

Wan, C., Kahramanoğlu, İ., & Okatan, V. (2021). Application of plant natural products for the management of postharvest diseases in fruits. *Folia Horticulturae*, *33*(1), 203–215. https://doi.org/10.2478/fhort-2021-0016

Wang, J., & Wu, Z. (2022). Combined use of ultrasound-assisted washing with in-package atmospheric cold plasma processing as a novel non-thermal hurdle technology for ready-to-eat blueberry disinfection. *Ultrasonics Sonochemistry*, *84*, 105960. https://doi.org/10.1016/j.ultsonch.2022.105960

Xu, H., Ma, R., Zhu, Y., Du, M., Zhang, H., & Jiao, Z. (2020). A systematic study of the antimicrobial mechanisms of cold atmospheric-pressure plasma for water disinfection. *Science of the Total Environment*, *703*(51), 134965. https://doi.org/10.1016/j.scitotenv.2019.134965

Xuan, X., & Ling, J. (2019). Generation of electrolysed water. In T. Ding, D.-H. Oh, & D. Liu (Eds.), *Electrolyzed Water in Food: Fundamentals and Applications* (pp. 1–16). Springer, Singapore. https://doi.org/10.1007/978-981-13-3807-6_1

Yan, P., Chelliah, R., Jo, K., & Oh, D. H. (2021). Research trends on the application of electrolyzed water in food preservation and sanitation. *Processes*, *9*(12), 2240. https://doi.org/10.3390/pr9122240

Yinxin, L., Can, Z., Menglu, H., Cui, S., Jinping, C., Jingyu, W., & Huang, L. (2022). Effect of cold atmospheric plasma on the gray mold rot of postharvest mulberry fruit. *Food Control, 137*, 108906. https://doi.org/10.1016/j.foodcont.2022.108906

Yousefi, M., Mohammadi, M. A., Khajavi, M. Z., Ehsani, A., & Scholtz, V. (2021). Application of novel non-thermal physical technologies to degrade mycotoxins. *Journal of Fungi, 7*(5), 395. https://doi.org/10.3390/jof7050395

Zainuri, J., Jayaputra, A., Sauqi, A., Sjah, T., & Desiana, R. Y. (2018). Combination of ozone and packaging treatments maintained the quality and improved the shelf life of tomato fruit. *IOP Conference Series: Earth and Environmental Science, 102*, 012027. https://doi.org/10.1088/1755-1315/102/1/012027

Zhang, W., & Jiang, W. (2019). UV treatment improved the quality of postharvest fruits and vegetables by inducing resistance. *Trends in Food Science & Technology, 92*, 71–80. https://doi.org/10.1016/j.tifs.2019.08.012

Zhang, X., Tang, N., Zhang, H., Chen, C., Li, L., Dong, C., & Cheng, Y. (2021). Comparative transcriptomic analysis of cantaloupe melon under cold storage with ozone treatment. *Food Research International, 140*, 109993. https://doi.org/10.1016/j.foodres.2020.109993

Zhao, J., Qian, J., Luo, J., Huang, M., Yan, W., & Zhang, J. (2021a). Morphophysiological changes in *Staphylococcus aureus* biofilms treated with plasma-activated hydrogen peroxide solution. *Applied Sciences, 11*(24), 11597. https://doi.org/10.3390/app112411597

Zhao, J., Qian, J., Zhuang, H., Luo, J., Huang, M., Yan, W., & Zhang, J. (2021b). Effect of plasma-activated solution treatment on cell biology of *Staphylococcus aureus* and quality of fresh lettuces. *Foods, 10*(12), 2976. https://doi.org/10.3390/foods10122976

Zhou, R., Zhou, R., Wang, P., Xian, Y., Mai-Prochnow, A., Lu, X., Cullen, P. J., Ostrikov, K., & Bazaka, K. (2020). Plasma activated water (PAW): Generation, origin of reactive species and biological applications. *Journal of Physics. Part D, 53*(30), 303001. https://doi.10.1088/1361-6463/ab81cf

# 6 Importance of Atmospheric Composition for Postharvest Handling of Fruits and Vegetables

*Hanifeh Seyed Hajizadeh*

## 6.1 INTRODUCTION

Horticultural crops, particularly fresh fruit and vegetable crops, are good sources of natural antioxidants including vitamins, bioactive molecules, various polyphenols and some micronutrients. They have promising abilities for being used as antiviral agents and leading to a reduction in diseases including cancer, satiety and high blood pressure, and recently their efficiency against COVID-19 has been proved also by most researchers (Hajizadeh, 2021a). One-fourth of all harvested fruits and vegetables are not consumed due to decay between harvesting and consumption by consumers (Kiaya, 2014), which leads to a big reduction in the nutritive value and general quality of the organoleptic features of fresh produce. This estimation is also greater in some countries due to poor systems for storage and processing. For example, postharvest losses are evaluated to be 5–25% in developed countries and 25–50% in developing countries (Buzby et al., 2014).

Indeed, fruits, vegetables and summer crops, as living organisms after harvesting, breathe and sustain damage. Also, the respiration level of fruits and vegetables, which is peculiar to each product, plays a vital role in their metabolism rate and senescence and in determining the final storage life. Before long, an increase in the respiration level reduces the storability of fresh horticultural products, and vice versa. Furthermore, as the respiration rate increases, the possibility of infection by microorganisms becomes greater, and horticultural crops spoilage increases.

Both the preharvest factors (mainly Ca and Fe fertilization during cell division and enlargement) and postharvest applications have significant roles in the storage quality of fruits and vegetables (Khakpour et al., 2022). In order to increase the value added, maintaining the quality of horticultural products seems to be essential. Also, meeting consumer demand for a variety of vegetables and fruits during the year is possible only through long-term storage of products. Fruits and vegetables are living organisms even after harvest, and they respire. During respiration, the oxidation of

DOI: 10.1201/9781003452355-8

carbohydrates takes place, whereby carbon dioxide, water and heat are produced. Storage of products at low temperature is the most important way of reducing the respiration rate and delaying senescence. Since oxygen is the main factor in respiration, modification of the atmospheric composition around the products is the second most important measure for preventing postharvest losses.

The atmospheric composition around harvested product affects not only the rate of respiration but also the rate of other common metabolic reactions. Oxygen ($O_2$) is the most important gaseous compound effective in product respiration. However, gases such as sulfur dioxide ($SO_2$), ozone ($O_3$) and propylene, which are considered pollutant gases if their concentration is too high, will have significant effects on producing respiration (Gheorghe & Ion, 2011).

The terms controlled atmosphere and modified atmosphere imply the addition or removal of gases from the ambient atmosphere (about 20% oxygen, 78% nitrogen, 0.9% argon and 0.03% carbon dioxide) of the storage area or package, respectively, to achieve a novel equilibrium of oxygen, carbon dioxide, nitrogen, ethylene, etc. that is different from the ambient. Today, this technique is commonly used in the storage of most horticultural crops including litchi fruit (Ali et al., 2016), pomegranate (Matityahu et al., 2016) and apple (Thewes et al., 2017). In line with this information, the present chapter aimed to discuss the importance of atmospheric composition surrounding fruits and vegetables during their handling.

## 6.2   CONTROLLED ATMOSPHERE (CA) STORAGE

Controlled atmospheric (CA) storages are among the most innovative techniques in bulk storage of fruits and vegetables. They help to regulate the composition of the gas in the storage according to the needs of products for reducing the metabolic activities and improving storage life. CA storage technology consists of reduced levels of $O_2$ and increased levels of $CO_2$ compared to normal air components with refrigeration. In CA storages, sometimes ethylene is removed and carbon monoxide is added to the storage atmosphere. The CA storage requires continuous monitoring and careful adjustment of these gases in the storage container to maintain a predetermined level (Thompson et al., 2018). The first studies on CA storages was done by Jacques Etienne Berard in the early 1800s in France (Dalrymple, 1969), who found that fruits did not ripen in the atmospheres without oxygen. At the same time, many studies were conducted on the effect of changing the atmosphere composition around the commodity on the ripening process and the quality of commodities, but the commercial use of CA storages was introduced by Kidd and West (1925) who investigated the effects of oxygen, carbon dioxide and ethylene on respiration rate and ripening process in pome fruits, especially apples and also berries, in the early 20th century. In fact, the controlled atmosphere storage has been successfully used for *Malus sylvestris* and *Pyrus communis* (Brecht et al., 2003). So, after the advancement of CA technology around the 1950s, its usage became more common throughout the world in the 1990s by means of enhanced technical innovations (Prange et al., 2005). So CA storages are probably one of the most effective technologies invented in the fruits and vegetables industry in the 20th century. It has been demonstrated to be effective in expanding the storability and qualitative parameters of most fresh crops.

In fact, the increase in market demand for fresh fruits and vegetables has led to the increased commercial application of CA storages. The concentration of $O_2$ in controlled atmosphere storage generally caused to decrease in respiration rate up to 0.5–2.5%, which is mostly dependent on the type of commodity and the variety. It is estimated that CA conditions normally increased the storability of fresh crops by a factor of 2 to 4 (Bodbodak & Moshfeghifar, 2016). However, as previously mentioned, the atmospheric composition is dependent on the crop species, although as a general rule the most common combinations are 2–5% oxygen and 3–10% carbon dioxide. The atmospheric composition is monitored daily for $CO_2$ and $O_2$ concentrations by sampling the CA room's atmosphere.

Increased carbon dioxide levels (up to 14%) are another component in CA storages, and their purpose is to reduce ethylene production, especially at 10% $CO_2$ while ethylene was not detectable in mangoes stored at 25% $CO_2$ treatment (Bender et al., 2021). Consequently, the ripening process was diminished, and storage life was prolonged. However, too high $CO_2$ concentrations can negatively affect the product's appearance, its taste, nutritional value and storage time. One of the main differences between MA packaging and CA storages is that in CA storages, gas composition should be continuously monitored and adjusted to keep the optimum concentration within a certain range. As CA storages are costly and expensive, they are more suitable for long-lasting fruits including apple, kiwi and pear. The main factor in using of CA storages individually for apple fruits is its unique benefits, such as the reduction of some disorders and diseases during CA storages, although other disorders may be exacerbated or initiated during CA storages. However, the general benefits of CA storages are predominant in apples (Prange & DeLong, 2006). However, various studies indicate the efficiency of controlled atmosphere conditions for commercial storage and recommends the combination of atmospheric gases for fruits, vegetables, fresh-cuts and also cut flowers and ornamentals (Brecht, 2006; Saltveit, 2003).

The **advantages** of controlled atmosphere storage can be summarized:

• Increased storability and quality of fresh produce
• Improved longevity of most horticultural crops
• Lack of chemical compounds in crops
• Inhibition of physiological disorders and decay caused by anaerobic microorganisms
• Retardation of senescence and associated biochemical and physiological changes
• Reduction of produce sensitivity to ethylene action at $O_2$ levels below 8% and/or $CO_2$ levels above 1%
• Useful tool for insect control in some commodities (Watkins, 2008)

In contrast, potential **harmful** effects of CA are that it:

• May cause irregular ripening after storage.
• May cause certain physiological disorders such as black heart in potatoes, brown stain of lettuce.

- May enhance anaerobic respiration and development of off-flavors and off-odors at very low $O_2$ concentrations.
- May cause susceptibility to decay.
- Requires costly technical know-how.

There are different types of controlled atmosphere storage depending mainly on the method or degree of control of the gases.

1. **Static controlled atmosphere storage**. The $O_2$ and $CO_2$ levels are set and generated for fruit and vegetable respiration, are commonly performed by product respiration and are stabilized by ventilation and scrubbing (Thompson et al., 2018).

2. **Flushed controlled atmosphere storage**. In this CA technology, the $O_2$ and $CO_2$ levels in the atmosphere are provided by the flowing gas stream, which purges the storage continuously by different gas scrubbers and adjusts the $O_2$ and $CO_2$ concentrations again in the storage by flushing the one that is under the set level (Thompson et al., 2018).

3. **Dynamic controlled atmosphere storage (DCA)**. Dynamic controlled atmosphere is a modern technology in the apple industry (Mditshwa et al., 2018) that Watkins (2008) called "A New Technology for The New York Storage Industry" and that allows a reduction in the oxygen content to a very low level (Watkins, 2008). It works by monitoring the stress level of the fruit in storage and adjusting the oxygen ($O_2$) and carbon dioxide ($CO_2$) levels by sensing the fruits responses via three sensors: including chlorophyll fluorescence (CF), respiration quotient (RQ) and ethanol (ET) (Thewes et al., 2018). The method provides the best conditions for suppressing ethylene synthesis and respiration rate during storage (Onursal & Koyuncu, 2021). It is very similar to ULO (ultra-low oxygen) storages Onursal and Koyuncu (2021) in that it obtained the same results in preserving firmness and titratable acidity of Scarlet Spur apple, but DCA technology gave better results in maintaining color.

## 6.3 MODIFIED ATMOSPHERE PACKAGING (MAP)

Packaging of fresh produce under CA conditions inside a gas-tight package, in fact, is a small-scale CA storage where the gaseous ambient with respect to $O_2$, $CO_2$, $N_2$, $H_2O_2$ and other trace gases has been modified as each of the gases have been monitored and adjusted to prolong the commodity storage life. According to this definition, there is no CA packaging system in commercial use. However, application of the combined package consists of oxygen and ethylene scrubbers, along with carbon dioxide release agents, at least at the beginning stages of storage-life, the packaged product could be classified as controlled atmosphere packaging. Such controlled atmosphere packages are called modified atmosphere packaging (MAP), which are discussed further in this section (Ben-Yehoshua et al., 2005).

Modified atmosphere packaging has been publicly accepted because the components are mostly $O_2$, $CO_2$ and $N_2$, it doesn't use any synthetic chemicals, and it leaves no toxic residue on the commodity; however, one of the most widely used inert gases for MAP is nitrogen. MAP at one time was used for the extension of apple shelf life in 1927 (Davies, 1995), and even older storage methods may have involved a modified atmosphere with elevated carbon dioxide and lowered oxygen levels for extending the longevity of horticultural crops (Dilley, 2006). Furthermore, in MA the packaging method, the gaseous combination is not continuously monitored; the atmospheric composition inside the packages is established by the respiration of the fruits and vegetables. The major aim of MA packaging is to stablish a constant gaseous combination inside the package that preserves the commodity without damage or losses (Zagory & Kader, 1988). Achieving an equilibrium gaseous combination inside a package of fresh produce relies on the following factors:

- Respiration rate of commodity
- Weight of commodity
- Storage temperature
- Relative humidity
- Package permeability
- Growing condition of the crop
- Preharvest applications
- Harvest quality
- Physiological stage of the product
- Storage duration

Most commodities may be packed under modified atmosphere conditions (Sandhya, 2010; Rai et al., 2011), and so the MA packaging method is performed in different ways, of which we mention single-unit packing systems, bags and trays. Along with modifications in atmosphere composition, lowering the temperature of the produce storage is the main factor for improvement of CA storage or MA packaging efficiency; postharvest losses are often faster in the absence of cold temperature as respiration releases vital heat and carbon dioxide. All efforts are in parallel with increasing the product sensory attributes related to quality. Hence, the crop metabolism should keep at a minimum level necessary to keep the commodity alive while preserving the organoleptic attributes during storage (Fonseca et al., 2002).

The MA technique is a suitable method as it needs a low level of equipment and technologies, and hence it is more suitable for small-scale producers for the storage of fresh fruits and vegetables (Flores et al., 2004). Furthermore, the method increases the storage life of the product by maintaining its quality and creates a physical barrier against insect attacks because all the packages are separate for each fruit (Conyers & Bell, 2007). Therefore, it is possible to storage various types of the commodities at the same place and time, of course, depending on their recommended storage temperature. MA packaging causes a reduction in the commodity's respiration level because fruit respiration and the gas permeability of the plastic film, combined, increase the concentration of carbon dioxide and reduce the level of oxygen inside the package. So the level of metabolic reactions will change and subsequently lead to a decrease in fruit ripening, spoilage and diseases (Caleb et al., 2012), weight loss (Sabir et al., 2011), enzymatic discoloration and internal breakdown (Guan & Dou, 2010). In fact, according to the fruit's respiratory activity and film's permeability, an increase in the concentration of $CO_2$ can cause anaerobic respiration, ethanol production (Ares et al., 2007) and physiological disorders in the commodity. Consequently, off-flavor and core breakdown will occur, which result in an undesirable fruit for consumption (Caleb et al., 2012). In this regard, attempts are continuing to introduce the best film for the longest possible fruit quality preservation.

MA packaging is commercially used for value-added fruits and vegetables, including broccoli florets and asparagus tips, since they are more susceptible to weight loss microbial decay and have the highest respiration rate and ethylene biosynthesis and action. At the same time, there are some limitations in using of MA packaging for fresh-cut commodities because access to a variety of packages with appropriate permeability to gases is limited and may result in fermentation in products. Also, the costs of packing and transporting some fruits, such as persimmon, over long distances are more justifiable and more useful than others, such as apple (Kader & Watkins, 2000).

### 6.3.1  DIFFERENT TYPES OF MA PACKAGING

#### 6.3.1.1  Active MA Packaging

With MA packaging, the gas composition in the rigid package can be adjusted by vacuum and then gas flush, called active MAP. In the active state, after placing the product inside the package, the air is evacuated and then the desired gas composition is injected into the package (Artés et al., 2006).

#### 6.3.1.2  Passive MA Packaging

However, in the passive state, the product is packaged in normal air. Then, due to product respiration and the special permeability of the film, the desired modified air is created after a few days up to the steady state (equilibrium state) (Artés et al., 2006).

However, the equilibrium state is done faster in active MAP, whereas in the passive method there is a lag phase. In fact, a dynamic equilibrium is created between the fruit-producing gases and the gases in the microatmosphere. In this equilibrium, the consumption of oxygen and the production of carbon dioxide are equal to the transfer of this gas through the package at a certain temperature. Packaging of fresh-cuts in polymeric films, which is a kind of passive MA, can also be a good supplement in managing the certain temperature for preserving quality (Fonseca et al., 2005).

In the modern active packaging approaches, essential oils (EOs) derived from plants are applied to extend the storage life of fruits and vegetables by preventing microbial activities and oxidative stress (Wyrwisz et al., 2022). Despite the synergistic effect of MA packaging and essential oils, the strong aroma is one hindrance in using EOs in packages of produce, limiting their application; this drawback can be solved by choosing the right essential oil according to the kind of food (Wen, 2016). In a static atmosphere for packaging agricultural products:

- The intensity of oxygen consumed by the crop should be equal to the total input oxygen flux from the plastic.
- The intensity of $CO_2$ production by the crop should be equal to the total $CO_2$ output flux from the plastic.

Furthermore, product respiration, film permeability, temperature and relative humidity are the main parameters that determine the equilibrium conditions inside the package.

## 6.3.2 DIFFERENT PLASTIC FILMS USED IN MA PACKAGING

Recently, the use of flexible plastic films has been increased due to the proliferation of new plastics and new application systems. The development of new forms of new packaging materials leads to the development of films with new characteristics. Much progress has been made in the form and fabrication of polymeric films with a different gas permeability. This has led to the expansion of $C_2H_4$, $O_2$ and $CO_2$ scrubbers (Yuvaraj et al., 2021). The most common materials applied in the MA packaging of fresh produce are PP, LDPE, PVC, ethylene vinyl alcohol copolymer and softened polystyrene. Some other important and new materials for packaging include synthetic polyolefins, linear low-density polyethylene (LLDPE), high-density polyethylene (HDPE), polyesters, polyethylene terephthalate (PET), polyvinylidene chloride (PVDC), ethylene-vinyl alcohol (EVOH), polyamide (nylon), polyvinyl alcohol (PVOH), ethylene vinyl acetate (EVA), cellulose-derived plastics such as cellophane and natural biodegradable polymers like polylactic acid (PLA). In Table 6.1, the gas permeabilities in most commercial polymers, along with their structure, are listed.

Recently, films have been used that change their permeability by changing the combination of the atmosphere within the package. If these smart films are used in the packaging of products with a long marketing process, the damage will be minimized even if during this time, the external temperature, relative humidity and amount of light changed. Despite the expansion of these new films, storage conditions

**TABLE 6.1**
**Permeabilities of Most Commercial Polymers (Zeman & Kubík, 2007)**

| Polymer | Nature of Polymer | Gas P (ml mm cm$^{-2}$ s cm Hg) | | |
|---|---|---|---|---|
| | | $O_2$ (30 °C) | $CO_2$ (30 °C) | Vapor (90% RH and 25 °C) |
| Polystyrene (PS) | Glassy | 11 | 88 | 12,000 |
| Cellulose acetate | Glassy | 7.8 | 68 | 75,000 |
| LDPE | Some crystallinity | 55 | 352 | 800 |
| HDPE | Crystalline | 10.6 | 35 | 130 |
| Polypropylene (PP) | Crystalline | 23 | 92 | 680 |
| Nylon 6 | Crystalline | 0.38 | 1.6 | 7000 |
| Poly ethylene terephthalate (PET) | Crystalline | 0.22 | 1.53 | 1300 |
| Poly vinylidene Dichloride (Saran) | Crystalline | 0.053 | 0.29 | 14 |

and the type of atmosphere produced by the product are also factors. Differences in the permeability of film used for MA packaging depend on various parameters such as the type of film, thickness, and internal and external conditions of the package.

The presence of possible ruptures and holes in the film will changes the film permeability. These factors may be due to mechanical damage during heat sealing of the film. One of the factors considered in packaged products is weight loss. The rate of weight loss in films with high vapor permeability is 8–9%, while in low vapor permeability films, it is 1–2%. Although the balanced atmosphere is directly related to the storage conditions and the interaction between the package and the product, the uniformity of the plastics used and good heat sealing mechanisms are also very important.

Already one of the basic components of air, nitrogen is a safe and trusted solution for creating a hypoxic air mix for modified atmosphere packaging. The main food products that benefit from the use of MA packaging are salads, fruits, vegetables, nuts, fresh chilled meats, cooked meats, poultry and fish (Church, 1994). By using MAP and respecting the other determining factors of food shelf life, such as storage temperature, quality of raw ingredients, processing method, hygiene grade and packaging materials, the packaged/stored food can reach a shelf time up to three to four times longer than standard. However, the effect of different kinds of packaging materials is not negligible. In a project comparing three kinds of polymeric films for packaging mango fresh-cuts, it was concluded that the mango fresh-cuts packed in PET could be stored 14 days at 3°C versus 11 days for the mango slices in the other polymeric films (Donadon et al., 2004). Also, in another project by Chonhenchob et al. (2007), it was found that the shelf life of mango slices increased by packaging in PET trays.

## 6.4   COMBINATION OF CA STORAGES/MA PACKAGING TO THE POSTHARVEST HANDLING OF FRESH-CUTS

Despite the advantages of the CA and MA techniques, a new challenge has arisen for using controlled or modified atmosphere on a commodity for which a 2- or 3-week-long transit time may indicate a very considerable part of their potential postharvest life (Brecht et al., 2003). Due to their short lifespan, fruit and vegetable slices need to be packaged under modified atmosphere in today's marketplace (Toivonen et al., 2009) in order to maintain their optimum quality attributes (Brecht et al., 2003).

Since fruits and vegetables are exposed to a wide range of temperatures during handling and transportation, and on the other hand, inasmuch as temperature has a direct effect on the crop's transpiration level and also on the amount of gas emissions from the packaging film, and given that the composition of the modified atmosphere inside the polymeric film must be kept at a constant level under different temperatures, the rate of perforation should also change according to the product transpiration rate (Talasila et al., 1994). Hence, due to the perishable nature of fresh produce slices (Hajizadeh & Kazemi, 2012) and their high sensitivity to temperature fluctuations, it is obligatory to maintain an optimal temperature of 0°C to keep it fresh until consumption because the changes in temperature will lead to decreased $O_2$ levels, below the optimum, creating fermentative and other related injuries. Nevertheless, many MA packaging products are minimally processed, handled and preserved at a range of temperature from 5 to 10°C (Verlinden & Nicolaı, 2000).

In recent years, manufacturers have claimed that smart packages have been commercially produced with the ability to adjust permeability according to the temperature fluctuation by changing the polymer structure (Clarke & De Moor, 1997). Also, combining the CA and MA techniques allows the transporter to send mixed loads of different products at the same time and in the same container at the least cost. In this way, each of the MA packages of fresh crops, which is individually optimized for any condition, is placed inside a chosen controlled atmosphere container so that the desired gaseous composition will develop under a given set of transport conditions (Brecht et al., 2003). In this way, the changes in respiration level and package permeability are adjusted in response especially to higher-temperature chambers or retail display conditions in postharvest handling.

The benefits of modified atmosphere packaging in temperature management and minimizing quality changes, including water loss, discoloration and softening, is greater than CA storage. For example, tomato stored in 4% $O_2$ + 2% $CO_2$ at 12°C experienced prolonged storage life (Nunes et al., 1996), whereas in CA storages under similar levels of oxygen and carbon dioxide at 6°C, tomato peel exhibited $CO_2$ injuries as well as chilling injury signs, including water soaking, pitting, discoloration and dullness.

Fluctuation in temperature causes the development of undesirable odors, softening and browning in MA-packed mushrooms, since the $O_2$ dropped below 2% and the $CO_2$ increased to more than 12% (Brecht et al., 2003). These findings show that the optimum concentrations of $O_2$ and $CO_2$ should be recommended based on the anticipated temperature during postharvest handling. According to the respiration rate of the fruits and vegetables in different temperatures, the best MA is recommended for each product. Table 6.2 presents some of recommended conditions for storing the fruits and vegetables.

## TABLE 6.2
## Respiration Rates of Some Fruits and Vegetables in Air and Recommended Atmosphere (Kader et al., 1989)

| Commodity | Temperature (°C) | Atmosphere Composition | Respiration (cm³ Kg⁻¹ h⁻¹) |
|---|---|---|---|
| Apple ('Granny Smith') | 0 | Air | 1 |
| | | 2% $O_2$ + 2% $CO_2$ | 0.1 |
| Green bean | 5 | Air | 17.5 |
| | | 3% $O_2$ + 5% $CO_2$ | 10.8 |
| Broccoli | 0 | Air | 10 |
| | | 1.5% $O_2$ + 10% $CO_2$ | 7 |
| Cabbage | 0 | Air | 1.5 |
| | | 3% $O_2$ | 1 |
| Chili pepper | 5 | Air | 3.4 |
| | | 2% $O_2$ | 2.7 |
| Mango | 10 | Air | 15 |
| | | 4% $O_2$ + 7% $CO_2$ | 8 |
| Tomato | 12.5 | Air | 9 |
| | | Air + 10% $CO_2$ | 6 |

## 6.5   EFFECT OF ATMOSPHERIC COMPOSITION ON ETHYLENE PRODUCTION AND SENSITIVITY

Oxygen is required to convert ACC to ethylene and to metabolize ethylene, which is produced in combination with the receptor. Low concentrations of $O_2$ can prevent the effect of EFE (ethylene forming enzyme) and increase the level of ACC. Carbon dioxide is a competitive inhibitor of ethylene action. In plant tissues, $CO_2$ accelerates the conversion of ethylene to ethylene oxide. $CO_2$ can prevent ethylene production by limiting the production of ACC and also prevent the ACC from converting to ethylene. In general, if fruit is exposed to a mixture of $O_2$ and $CO_2$ gases for a long time, the amount of volatile compounds is reduced.

In the oxygen concentrations of 2.5%, the ethylene production will be decreased by one-half compared to normal weather conditions. In general, in fresh produce exposed to oxygen, at oxygen levels less than 8%, the formation of ethylene is reduced, and the sensitivity of product to ethylene decreases as well. The reasons for that can be summarized in two factors: (1) Oxygen is necessary for converting ACC to ethylene, and (2) increasing the level of $CO_2$ has different effects on produce. It may be positive or negative, or it may have no effect on the produce, depending on the amount of the $CO_2$ and the produce. According to what was previously mentioned, the efficiency of modified and controlled atmospheres in maintaining the appearance and sensory attributes of products and extending storage life can be via the prevention of metabolic activity, decay, discoloration (Gunes & Lee, 1997) and especially ethylene biosynthesis and action (Kader et al., 1989).

## 6.6   STRUCTURE OF CA STORAGES AND METHODS FOR CONTROL GAS CONCENTRATION

Controlled atmosphere chambers are constructed as gas-tight walls and doors with structures capable of supporting the gases used. They are provided with an adequate system for generating the atmosphere around the produce and the best way of maintaining the atmospheric combination for a certain storability. The doors of CA chambers are commonly sealed by rubber to avoid leaks. Some facilities are applied for controlling the gas concentrations in the CA chambers including external burners, liquid or gaseous nitrogen, gas separator systems and hypobaric storage for controlling the $O_2$ concentration. Using of scrubbers based on NaOH, hydrated lime $(Ca(OH)_2)$, water, activated charcoal and molecular sieves for removing $CO_2$. Heated catalyst scrubbers, ethylene-absorbing beads and UV light are applied to control ethylene levels (Thompson et al., 2018).

## 6.7   AVAILABLE GASEOUS COMPONENTS OF CA STORAGES AND MA PACKAGING

### 6.7.1   OXYGEN AND THRESHOLD OF OXYGEN

Oxygen is an important factor in plant respiration, but its effectiveness depends entirely on the species and type of plant organ. Normal fluctuations in the atmospheric oxygen is are far less low than any that can affect the respiration of leaves and stems. The rate of oxygen emission inside the mentioned organs is very fast due to the high contact surface. In addition, the cytochrome oxidase enzyme tends to be so oxygenated that it can continue to function at very low $O_2$ concentrations (Pittman, 2011). In pome fruits, the surface-to-volume ratio is not large enough to allow oxygen to be emitted easily (Sapers et al., 1989). Therefore, due to the restriction of oxygen within the tissue, the rate of respiration may be reduced and cause to some atmospheric physiological disorders. Pasteur found that, when the concentration of $O_2$ around the yeasts and most of plant cells gradually decreased, the production of $CO_2$ derived from respiration also decreased until to the a certain point, and that, if the $O_2$ concentration decreases again, from this point onwards, $CO_2$ production increases rapidly. In fact, when the $O_2$ concentration is below the critical level, the carbohydrate is fermented and converted to ethanol and $CO_2$. At high concentrations of $O_2$, the Krebs cycle is re-stimulated, and $CO_2$ production is increased. This phenomenon is called the Pasteur effect (Yan et al., 2019), and in fact that is why yeast cells grow rapidly in the aerobic conditions, consume little carbohydrate, and produce little ethanol and $CO_2$. Under anaerobic conditions, these cells grow little but consume more carbohydrate and produce more ethanol and $CO_2$ (Erkan & Chien, 2006).

Low oxygen concentrations maintains desirable color, decrease respiration level, and slow ripening or senescence. High $O_2$ atmospheres prevent of certain bacterial growth and fungi, and anaerobic fermentation and preserve produce texture (Teixeira et al., 2016). In CA storages, $O_2$ level is reduced (down to 4% of $O_2$ and below depending on vegetables and fruits types) by supplementing with $N_2$ or/and $CO_2$. After reaching the desired oxygen level, the fruits and vegetables continue to reduce their internal oxygen concentration via respiration. Care should be taken not to lower the oxygen level below 1%, as this will cause irreversible chemical changes in the commodity and reduce its quality. The mechanism of $O_2$ can be summarized: as following items; Given the activity of lactic dehydrogenase in the lack of oxygen, fruits degrade glucose anaerobically by glycolysis to generate energy. In the glycolysis pathway, aldehydes, alcohols and lactates are produced, which lead to anaerobic respiration and fermentation, and subsequently off-flavor and off-odor can occur (Belay et al., 2019). In grapes, low concentrations of $O_2$ limits the enzymatic steps between oxaloacetate and citric acid, as well as the conversion of alpha-ketoglutaric acid to succinic acid. Even though the Pasteur effect has been reported in many products, it is not always common. For example, in avocados, anaerobic respiration will not increase even if the $O_2$ concentration is zero.

The Pasteur effect is so important in fruits and vegetables storage. The main purpose in the storage of climacteric fruits, such as apple, is the prevention of sugar diminishment and over-ripening of fruit. This goal can be obtaining by decreasing in the $O_2$ level to the point where the aerobic respiration rate is minimized. Of course, there are many exceptions in this regard. For example, in sweet potatoes, when the oxygen concentration drops to 5–7%, the product starts anaerobic respiration (Li et al., 2014). Oxygen depletion from the atmosphere depleted by altering the ambient composition with a modified atmospheric composition prevents of oxidation, diminishes respiration level, slows down fruit ripening and the degradation rate of flavor and color of the produce (Teixeira et al., 2016). Also, the super-atmospheric oxygen level could be able to prevent anaerobic respiration, and microbial growth and reduce the decay rate of fresh commodities (Belay et al., 2017). Elevated $O_2$ has been shown to be aeffective to in enhancinge the production of ROS, such as superoxide, hydroxyl and hydrogen peroxide, that cause to disruption in the cytoplasm and that prevent of different metabolic processes leading to reduced crop quality and deterioration (Belay et al., 2019).

Recently, super atmospheric $O_2$ (> 70 kPa) are using a good method to lower concentrations of oxygen which caused to preserve better quality in color and texture to lower bacterial count and higher to increase volatile compounds. For example, pomegranates stored at super-atmospheric oxygen (70 kPa $O_2$) considerably maintained with better color, the highest volatile compounds (quality and quantity) and the lowest aerobic mesophilic microorganisms at 5°C and ambient compared to other atmospheres (Belay et al., 2017). Treatment of apples slices with high levels of oxygen resulted in better texture, reduced browning and anaerobic volatile formation (Toivonen, 2006).

The suitable level of oxygen in a MA for fruits and vegetables for both safety and quality falls between 1– and 5% (Sandhya, 2010). In fact, when the pressure of oxygen is greatly reduced in controlled atmosphere conditions, hypobaric conditions occur, which, unlike controlled atmosphere and modified atmosphere storages, no gas other than air is required in this method, but the total pressure inside the hypobaric chambers is important as the level of oxygen is directly related to the $O_2$ pressure. This technique is commercially used in the floriculture industry for the storage and handling of cut flowers.

The use of superatmospheric $O_2$ concentrations is another way of gaseous altering around the commodity to preserve its quality and freshness, although the effectiveness of super-atmospheric $O_2$ concentrations depends on commodity developmental stage, the concentrations of $O_2$, $CO_2$ and $C_2H_4$, storage duration and temperature. In some commodities cyanide-resistant respiration is enhanced by high levels of oxygen in the headspace and cause to enhances ripening of mature-green fruits, in 30–80 kPa oxygen, but not at concentrations above 80 kPa (Kader & Ben-Yehoshua, 2000). Superatmospheric concentrations of $O_2$ cause to the production of ROS, which damages the cytoplasm, and inhibits different metabolic reactions and reduces fruit quality, but it seems that the damage rate depends on how rich the commodity is in phytochemicals, including vitamins C and E, to prevent damage to the cell membrane components.

Elevated oxygen concentrations diminish the negative effects of high levels of carbon dioxide and may control decay. On the other hand, high $O_2$ levels causes to less off-flavor due to the lower production of ethyl acetate (Kader & Ben-Yehoshua, 2000). There is less literatures about the efficiency of super-atmospheric oxygen concentration on phytochemicals in fruits and vegetables, as contradictory literatures have been presented regarding the increase or decrease of antioxidants in different commodities under superatmospheric concentrations of $O_2$. Obviously, many experiments must be performed to achieve accurate results.

### 6.7.2 CARBON DIOXIDE AND THRESHOLD OF CARBON DIOXIDE

Increasing the concentration of $CO_2$ in some harvested fruits and vegetables prevents the progression of the respiratory processes. Although this effect is not common, it has been reported in many intact or detached fruits and vegetables either in aerobic or anaerobic conditions. Increasing the concentration of $CO_2$, like decreasing the concentration of oxygen, prevents decarboxylation, which occurs in normal respiration and in the Krebs cycle. Also $CO_2$ is a competitive inhibitor of ethylene action, is able to decrease fruits' sensitivity to ethylene at concentrations above 1–2 kPa and can reduce the pH or inhibit the growth of some microorganisms. Elevated carbon dioxide by more than 20% reduces the rate of aerobic respiration and causes large amounts of ethanol and acetaldehyde to accumulate in the tissues, which in turn leads to atmospheric physiological disorders. The amount of these damages depends on the storage temperature, the level of oxygen and carbon dioxide and the duration of crop exposure to adverse conditions. The accurate mechanism of elevated $CO_2$ in decreasing of respiration is not understood completely, but it is clear that it can't be related to the injuries induced by the elevated $CO_2$. By returning to normal conditions, plant tissue respiration returns to its normal status, which is likely due to carbon dioxide fixation mechanisms and enzymes involved in the metabolism of phospho-enolpyruvate to malate (Lakso & Kliewer, 1978), as the first by-products of the fruits' exposure to the radioactivated $CO_2$ detected in lemon were malic acid, citric acid and aspartic acid. So high amounts of $CO_2$ facilitate the intermediate products of the Krebs cycle and stimulate respiration. On the other hand, an elevated $CO_2$ changes the cytoplasmic pH, and it seems that the alteration in cytoplasmic pH is the major reason for stimulation of respiration (Holcroft & Kader, 1999). What is certain is that elevated levels of $CO_2$ modulate at two points in the Krebs cycle – (1) succinate converting to malate and (2) malate converting to pyruvate – although high $CO_2$ concentrations mainly affect succinate dehydrogenase.

Elevated $CO_2$ concentration in the atmosphere around the produce can be summarized also in antimicrobial effects, which delay microbial spoilage and has a most pronounced effect against gram negative bacteria. Increased $CO_2$ levels reduce the product's acidity and thus prevent microbial growth. On the other hand, $CO_2$ is able to penetrate bacterial membranes, change the interior acidity of cells, and ultimately change cellular metabolic reactions. Prevention of substrate access by microorganisms and direct impact on enzymes by alteration of proteins are other aspects of the mode of action of elevated $CO_2$ in the headspace. Like oxygen concentration, the presence or the accumulation of high levels of carbon dioxide over the critical limit could also have a negative effect on produce quality by accelerating changes in color (h°) and texture and by increasing the solubilization of pectic compounds (Teixeira et al., 2016).

### 6.7.3 NITROGEN AND THRESHOLD OF NITROGEN

Oxygen is harmful to fresh fruits and vegetables in that it causes bacteria, as well as mold and mildew, to grow, making it spoiled and unsellable. Usually $N_2$ is applied to replace $O_2$ in MA packaging to prevent rancidity and to inhibit the growth of aerobic organisms. Nitrogen acts as a filler gas to render the oxygen inert and to prevent pack collapse and can be so useful for packaging the ready-to-use fresh-cuts through the year. Also, since the process doesn't involve any chemicals, there is no impact on the quality of the commodity. Commonly 87–95% $N_2$ is used in combination with 3–8% carbon dioxide and 2–5% oxygen used for fresh produce, while pasta and bakery products are packed in 100% $N_2$ (Church, 1993). The major factor for using of nitrogen as a filler gas is its low solubility in water and lipid compared with that of $CO_2$ (Farber, 1991).

### 6.7.4  Carbon Monoxide and Threshold of Carbon Monoxide (CO)

CO is odorless, colorless and tasteless, as well as very poisonous, and it has been demonstrated to be a very potent microbial inhibitor at low concentrations (<1 kPa). It can be combined with low concentrations of oxygen (2–5 kPa) to delay the browning of fresh produce. However, due to the toxic effect of CO and its explosive nature at 12.5–74.2 kPa in air, it must be handled with specific precautions and is applied less in MA packaging of fresh produce (Bodbodak & Moshfeghifar, 2016).

### 6.7.5  Sulfur Dioxide and Threshold of Sulfur Dioxide

Sulfur dioxide ($SO_2$) is very reactive in aquatic and acid solutions with a pH lower than 4 and has an inhibitory effect on bacteria, especially on grapes and dried fruits. However, due to the hypersensitivity of some people to sulfite compounds, $SO_2$ application has some limitations and cautions (Bodbodak & Moshfeghifar, 2016).

### 6.7.6  Other Gases such as Helium and Argon

The noble gases, including helium, argon, xenon, and neon, are not reactive and are used in certain kinds of food, such as potato-based snack products. Although other gases, including nitrous and nitric oxides, sulfur dioxide, ethylene, chlorine, as well as ozone and propylene oxide, have also been studied, they have not been used commercially due to the safety, regulatory and cost considerations (Sandhya, 2010). The most common gaseous environment consists of low levels of oxygen and high levels of carbon dioxide. Carbon monoxide is also sometimes applied for prevention of discoloration and microbial decay (Kader et al., 1989). Although using argon or helium instead of nitrogen causes more $O_2$, $CO_2$ and $C_2H_4$ to be emitted, it not only has no direct effect on the product texture but also is much more expensive than $N_2$ as a component of MAP.

## 6.8  DIFFERENT ATMOSPHERIC COMPOSITIONS AND FRUITS AND VEGETABLES QUALITY DURING HANDLING

The flavor of harvested produce depends on its taste and odor. Although taste and aroma are two major components of overall flavor, aroma has a dominant role in flavor. So it seems that both non-volatile and volatile constituents of aroma should be investigated in future research of flavor (Kader, 2008). One of the main factors affecting on volatiles compounds is the atmospheric composition around harvested produce. According to the results of Forney (2001), two challenges control volatiles and flavor alterations during marketing and storage, as the aim is to optimize fruit taste and odor upon delivery the consumer. It is not enough to harvest produce with good taste and aroma; taste and aroma must be preserved or increased during storage and marketing. Along with developments in storage technologies, more accurate control over the surrounding environment, including atmosphere composition, can be achieved and used for optimizing volatile composition and flavor.

In addition to the composition of the atmosphere during handling along with storage temperate and relative humidity, longer times between harvest and fruit consumption lead to lessened aroma and sometimes off-flavors in fruits (Kader et al., 1978). Similar results were found by my collaborators and me when an inappropriate combination of atmospheric gases was used during the active MAP of many fruits and vegetables such as mushroom (data not published), apple (Hajizadeh et al., 2006, 2008; Mostofi et al., 2006) and table grape (Mosayyebzadeh et al., 2010). Hence the statement of Kader (2008) that "Postharvest life of harvested produce should be determined on the basis of flavor rather than appearance". Low-oxygen-controlled storages can reduce or delay the production of volatile compounds, especially alcohols, aldehydes and esters, and can decrease the flavor of the produce, and this effect can be exacerbated by increasing storage time or decreasing oxygen levels in the atmosphere around the produce (Yahia, 2009). The mentioned inhibitory effect of low oxygen is likely due to the prevention of fruit ripening and senescence. Elevated $CO_2$ levels can also affect volatiles production and flavor.

Reduced levels of oxygen or increased levels of carbon dioxide lead to delay in fruit ripening. In a modified atmosphere, the diminishment in chlorophyll content and the increase in carotenoids and anthocyanin slow down, although high concentrations of $CO_2$ can reduce phenolic compounds, phenol oxidase activity and phenol oxidation. Otherwise, browning usually occurs when the fruits are placed in an undesired concentration of $O_2$ and $CO_2$. In addition, when the fruits are placed in the open air, high concentrations of $CO_2$ prevent the formation of natural color in them. Low concentrations of oxygen can have an insecticidal effect. Therefore, if the quality of the fruit is not affected, low oxygen concentrations may be used instead of insecticides for short storage times.

In non-climacteric fruits such as berries, treatment with low oxygen and high $CO_2$ decreased firmness and caused browning. However, it has no effect on acidity, pH, total soluble solids (TSS) and antioxidant activity (AA). For fruits and vegetables with minimal processing such as peeling or slicing, it is necessary to use packaging with a controlled atmosphere because in these products the amount of respiration is increased and the possibility of microbial contamination is higher. To extend the storage life of fresh produce on which minimum processing has been done, packages with appropriate concentrations of gases are used. Several factors must be considered in this case: type of product, film used, storage temperature and relative humidity inside the package.

One of the factors that is very effective on quality is the browning of the tissue, which occurs due to the oxidation of phenols by the enzymes polyphenol oxidase (PPO) and peroxidase (POD). The predominant microbial flora of fruits and vegetables are enterobacteria, lactic acid bacteria, yeasts and molds. The spread of fungi and molds depends on the relative humidity inside the package. Temperature also affects the spread of microbial flora. The low temperatures that vegetables can tolerate are not enough to inhibit the growth of some bacteria. Some bacteria that caused the decay of vegetables as they are also active at 10°C, and some fungi such as *Alternaria*, *Cladosporium* and *Botrytis* are active at zero degree.

Vegetables stored in display refrigerators are photosynthesized, which will change the gaseous combination inside the polymeric film. At equilibrium, the concentrations of $CO_2$ and $O_2$ are lower and higher than in the dark, respectively. Lettuce leaves were lighter in fluorescent light than those in the dark at the same time. Also, in the case of tomatoes, de-greening with light will reduce the quality, so it is better to use dark (dullness) packages.

Small fruits are identified according to their higher perishability, rapid deterioration, short life span, even if maintained under low temperature conditions compared to other fruits, resulting in value-added costs, problems in their marketability and selling, and postharvest losses in nutritional and commercial value (Ingrassia et al., 2016). The most efficient MA packaging for berries is characterized by extended storage life and preserved organoleptic attributes due to the prevention of fungi contamination.

Controlled atmosphere prevents polygalacturonase biosynthesis and causes a decrease in tissue softening. Increasing the $CO_2$ level in the atmosphere storage of asparagus and broccoli maintains them as tender spears and crispy heads. The toughening of asparagus is delayed considerably at a 12% concentration of $CO_2$ (Lipton, 1975) and 0–1°C. Eureka lemon fruits stored at 6% $O_2$ + 8% $CO_2$ had lower levels of alcohols and esters, while displaying high retention of terpenoids and aldehydes (Ma et al., 2019). Some recommended conditions for preserving produce according to their tolerated threshold to low levels of $O_2$ and elevated levels of $CO_2$ are categorized in Tables 6.3 and 6.4, respectively.

## TABLE 6.3

## Storage Recommendations for MAP of Fruits and Vegetables According to Their Minimum $O_2$ Concentration Tolerated (%) (Kader et al., 1989)

| Fruit | Minimum $O_2$ Concentration Tolerated (%) | Vegetable |
|---|---|---|
| Nuts, dried fruits | 0.5 | Dried vegetables |
| Some cultivars of apples and pears, most cut or sliced fruits and vegetables | 1 | broccoli, mushroom, garlic, onion |
| Most cultivars of apples and pears, kiwifruit, apricot, cherry, nectarine, peach, plum, strawberry, papaya, pineapple, olive, cantaloupe | 2 | Sweet corn, green bean, celery, lettuce, cabbage, cauliflower, Brussels sprouts |
| Avocado, persimmon | 3 | Tomato, pepper, cucumber, artichoke |
| Citrus fruits | 5 | Green pea, asparagus, potato, sweet potato |

## TABLE 6.4
## Storage Recommendation for MAP of Fruits and Vegetables According to Their Maximum $CO_2$ Concentration Tolerated (%) (Kader et al., 1989)

| Fruit | Maximum $CO_2$ Concentration Tolerated (%) | Vegetable |
|---|---|---|
| Apple (Golden Delicious), Asian pear, European pear, apricot, grape, olive | 2 | Tomato, pepper sweet, lettuce, endive, Chinese cabbage, celery, artichoke, sweet potato |
| Apple (most cultivars), peach, nectarine, plum, orange, avocado, banana, mango, papaya, kiwifruit, cranberry | 5 | Pea, pepper (chili), eggplant, cauliflower, cabbage, Brussels sprouts, radish, carrot |
| Grapefruit, lemon, lime, persimmon, pineapple | 10 | Cucumber, summer squash, snap bean, okra, asparagus, broccoli, parsley, leek, green onion, dry onion, garlic, potato |
| Strawberry, raspberry, blackberry, blueberry, cherry, fig, cantaloupe | 15 | sweet corn, mushroom, spinach, kale, Swiss chard |

Some of advantages of CA storage on fruits and vegetables quality include:

1. Changes in color by the prevention of chlorophyll loss and yellowing, biosynthesis of other pigments and biosynthesis and degradation of phenols.
2. Changes in texture by retarding polygalacturonase activity, solubilization of pectin, firmer preserved fruits and the prevention of lignification of leafy greens.
3. Changes in fruit taste and aroma by reducing loss of acidity, conversion of starch to sugar, interconversions of sugars and biosynthesis of volatile compounds.
4. Preservation of vitamin C and other antioxidants and subsequently potent nutritional quality of the fresh produce (Zagory & Kader, 1989).

Some fruits, such as mangoes, are usually harvested at full ripeness, and are subjected to deterioration in terms of more softening during handling. The efficiency of CA during the transport of mangoes to distant destinations started in the 1960s (Yahia, 2011). In this case, it was concluded that 25% $CO_2$ (Bender et al., 1995) at 12°C was the maximum tolerance threshold for mangoes to be stored safely for 3 weeks. Also, the lower limit for $O_2$ was estimated at 5% at the same temperature (Bender et al., 2000). The upper and lower levels of $CO_2$ and $O_2$ of 12% and 2%, respectively, caused the production of ethanol and off-flavor in mangoes. In another project, the optimum gas composition for packaging of fresh-cut kale was estimated as 1–2% $O_2$ + 15–20% $CO_2$ for 4 days at 20°C, while it was 1% $O_2$ + 10% $CO_2$ at 1°C (Fonseca et al., 2001). It can be concluded that the effect of fruit ripening stage at harvest time is the main parameter in determination of storage life although the efficiency of CA cannot be neglected.

## 6.9   MODE OF ACTION OF ATMOSPHERIC COMPOSITION IN ADDED QUALITY OF FRUITS AND VEGETABLES

Elevated $CO_2$ caused an increase in cellular pH, which is effective in reducing chlorophyll degradation to pheophytin (Zagory & Kader, 1989). Although the stability of carotenoids varies among chemical types, they are generally susceptible to oxidation because they are a kind of unsaturated pigments. Carotenoids are also sensitive to nonenzymatic breakdown during water loss and the presence of $O_2$. The low $O_2$, high relative humidity common in CA storages may improve some of these changes. Altering cellular pH and metal chelation both affect anthocyanin color. It has been demonstrated that the important reason in disruption of anthocyanin in berries such as blueberries, cherries, currants, grapes, raspberries and strawberries are $O_2$ and high temperature, which is prevented under CA storages. Also, anthocyanin ban be degraded by means of polyphenol oxidases in the presence of other phenolic compounds such as catechol.

Since polyphenol oxidase is evaluated to have a Michaelis constant (Km) higher than that of, $O_2$ may become insufficient for its activity at ambient levels as high as 2% $O_2$. As 2% $O_2$ concentration is common in controlled atmospheres, the maintenance of anthocyanin should be possible. Since the polyphenol oxidase has a Km many times higher than cytochrome oxidase, it seems that the activity of polyphenol oxidase is not possible at ambient concentrations as high as 2% $O_2$. So under CA conditions with the same concentration of $O_2$ more or less, anthocyanin degradation will be suppressed.

Some of enzymes catalyze the biosynthesis or oxidation of phenols, among them phenylalanine ammonia lyase, tyrosine ammonia lyase, cinnamic acid-4-hydroxylase, polyphenol oxidase and catechol oxidase. At the presence of $CO_2$, the activity of PPO is reduced, since $CO_2$ has been demonstrated to be a competitive inhibitor of PPO. $CO_2$ prevents the biosynthesis of phenolic substances, but the incidence of brown stain in lettuce, which is a symptom for $CO_2$ injury, did not have a positive correlation with total phenols.

Kader et al. (1989) suggested that the fruit RQ breakpoint is increased at elevated temperatures and is related to the greater difference between the oxygen concentration in the film and the atmospheric composition inside the package at elevated temperatures. Although the use of anti-browning chemicals, such as Nature Seal AS-1, is able to prevent the browning of fresh-cut apples along with increasing fruit quality slightly, choosing the suitable film which is match with apple slices plays a key role in the prevention of discoloration and other deteriorative processes. Otherwise, the level of $O_2$ can fall too much and cause an off-odor. In general, a range of 2–5% $O_2$ has been recommended for apples, while the desired $O_2$ level is 1.4–3.8 kPa in sliced salad savoy (Kim et al., 2004).

The best atmospheric combination for the inhibition of antioxidant compounds oxidation in sliced strawberry is 2.5 kPa $O_2$ + 7 kPa $CO_2$ (Odriozola-Serrano et al., 2010). Fresh-cut pineapples, stored in 8 kPa or lower $O_2$ concentrations, preserved a better yellow color of the pulp slices as demonstrated by higher values of chroma, while 10 kPa $CO_2$ concentrations caused a decrease in browning that reflected the higher values of L (Marrero & Kader, 2006). Our experiment on intact local Iranian apple, Golab kohanz and Shafi abadi (Hajizadeh et al., 2006; Mostofi et al., 2008), and mushroom (data not published), indicated that the storage of intact apples in PP films under active MA packaging was the most effective for an increase in the storage life of produce, so the same recommendation for storage of fresh-cuts of apple and mushroom can be used. However, the use of these storage conditions should be further investigated according to the different physiological mechanisms between sliced and intact fruit.

Dong et al. (2018) studied the effect of high $CO_2$ levels on the quality of lettuce, tomato and potato and concluded that the amount of fructose, glucose, TSS, antioxidants, phenol, flavonoid, vitamin C and calcium in the edible parts of vegetables increased by 14.2%, 13.2%, 17.5%, 59.0%, 8.9%, 45.5%, 9.5% and 8.2%, respectively, while the amount of protein, nitrate, magnesium, iron and zinc decreased by 9.5%, 18.0%, 9.2%, 16.0% and 9.4%. There were no changes in the content of TA, chlorophyll, carotenoid, lycopene, anthocyanin, phosphorus, potassium, sulfur, copper and manganese under elevated $CO_2$.

## 6.10 ATMOSPHERIC COMPOSITION AND PHYSIOLOGICAL DISORDERS OF FRUITS AND VEGETABLES

To understand physiological disorders and the mechanism involved in their incidence on fruits and vegetables, we have to first define it correctly. So a physiological disorder is introduced as a visible result of disruption in the normal metabolic reactions of a commodity that is not related to microbial infection such as fungi and pests. High $CO_2$ concentrations enable the control of microbial growth in fruits (Farber, 1991); however, off-flavor can be developed at $CO_2$ levels higher than 5 kPa in strawberry and raspberry (Van der Steen et al., 2002). The biological structure of individual fruit or vegetables determines the tolerance of plant organs to the emission of oxygen, carbon dioxide, ethylene and $H_2O$. Since tissue resistance to the gas emission affect the commodity's resistance to decreased concentrations of oxygen and elevated concentrations of carbon dioxide, this is related to the different parts of the produce, cultivar, maturity level but seems to be less influenced by temperature (Fallik & Aharoni, 2004).

Ethylene has a lot of roles from seed germination to fruit ripening and senescence, but the sensitivity of horticultural products to ethylene ($C_2H_4$), which is active in trace amounts (<0.1 ppm) in physiological processes, is different. The level of ethylene production in fresh horticultural commodities increases with the maturation stage at harvest, bruises, disease infection, high temperatures up to 30°C, water deficiency and decreases at low temperatures, low concentrations of oxygen, and high concentrations of carbon dioxide surrounded the produce. Very low levels of $O_2$ or very high levels of $CO_2$, along with the existence of high amounts of ethylene, may enhance the severity of physiological disorders related to storage conditions (Irtwange, 2006).

The atmospheric composition of the commodity can alleviate and/or aggravate disorders. In Table 6.5, some of physiological disorders related to the atmospheric composition around the commodity are listed. Chilling injury is one of the prevailing disorders that occurs in some tropical and warm season products during storage at temperatures below their recommended threshold. It is directly related also to the storage

**TABLE 6.5**

**Reduced Physiological Disorders under Lowering $O_2$ and Elevated $CO_2$ Concentrations**

| Physiological Disorder | Commodity | Effect of $O_2$ and $CO_2$ | Reference |
|---|---|---|---|
| Chilling injury | Zucchini | Reduced CI at 1–4 kPa $O_2$+5–10 kPa $CO_2$ | (Mencarelli et al., 1983) |
| Chilling injury | Plum | Reduced CI at 3 kPa $O_2$+2–8 kPa $CO_2$ | |
| Chilling injury | Olive | Accentuates at elevated $CO_2$ (2 or 5 kPa) alone or with reduced $O_2$ (2 kPa) | (Nanos et al., 2002) |
| superficial scald | Granny Smith apples | Induced on 2 or 4 days of anaerobiosis ($N_2$ induced) | (Prange & DeLong, 2006) |
| Russet spot | Crisphead lettuce | Prevented by reduced $O_2$ below 8 kPa or elevated $CO_2$ to 5 kPa or more | (Lougheed, 1987) |
| brown core | Apple | Reduced at 2 kPa $O_2$+ $CO_2$ as low as possible | (Sharples & Johnson, 1987) |
| brown core | Loquat | Severe internal browning in CA including 12 kPa $CO_2$, regardless of $O_2$ level | (Ding et al., 1999) |
| Bitter pit | Apple | Reduced at 1.5 kPa $O_2$ + 0.5 kPa $CO_2$ | (Mattheis et al., 2017) |
| Jonathan spot | Apple | Reduced at $CO_2$ concentrations as low as 0.7 kPa | (Meheriuk, 1994) |

period; if the commodity is stored for too long, the severity of the chilling injury will be greater. CI can be reduced in some commodities, such as grapefruit, zucchini, chili pepper, plum, melon, okra and avocado, in CA storages (Prange & DeLong, 2006). It is shown that the effect of high levels of carbon dioxide on reducing CI is more than reduced levels of $O_2$ (Mencarelli et al., 1983). They also inhibit the degradation of the ATP/ADP energy transport system during chilling (Vakis et al., 1970) and can be another reason for reducing CI in controlled atmosphere storages.

There is some evidence showing a positive correlation between higher proline levels, especially in fruit skins, and CI resistance. Hajizadeh and Safkhani (2017) showed that cherry tomatoes cv. Messina with more proline content and GPX activity had the least CI at 1°C. It is shown that the appearance and severity of bitter pit slow under CA conditions, although it is not completely understood which one has the greater importance, reducing in levels of oxygen or increasing in levels of carbon dioxide (Sharples & Johnson, 1987). Anthocyanin content is also affected by CA storages, and it can be converted from red to blue or purple in apple fruits (at low oxygen levels), in strawberry (at high carbon dioxide levels) and in sweet cherry, which occurs by either low oxygen or high carbon dioxide levels (Lidster et al., 1990). Reduced oxygen or high carbon dioxide concentrations can affect anthocyanin and lead to the accumulation of anthocyanin in the epidermal layer of fruits. Under CA storages, ATP and cytoplasmic pH decrease, and the activity of pyruvate dehydrogenase reduces. At the same time, the activity of some enzymes related to fermentation, such as pyruvate decarboxylase, alcohol dehydrogenase and lactate dehydrogenase, are enhanced (Prange & DeLong, 2006). One of the most important of CA storages of commodities is enzymatic browning, which causes the conversion of pigmented and non-pigmented phenolic compounds into brown pigmented polyphenols. Enzymatic browning occurs in two phases:

1. A change in phenol pigmentation that is not necrotic in nature and that is due to the pH of the cytoplasm and other metabolic activities that change cell pigment chemistry without killing the cells. This CA-induced discoloration is largely associated with anthocyanin-containing tissue: for example, epidermal dullness in apple, skin darkening in sweet cherry and dullness of the outer flesh of strawberry (Holcroft & Kader, 1999)
2. The second and more prevalent stage is cell necrosis caused by severe cytoplasmic acidosis due to insufficient ATP (due to anoxia), leading to a loss of membrane integrity and a wide ranges of disorders, in either the internal or external parts of the fruits. Commonly in commodities with high bulk density, such as beet, carrot, horseradish and potato, suffer from this form of enzymatic necrosis and will be the first to readily produce signs of disorder (Prange & DeLong, 2006). Some benefits of different atmospheric conditions in relation to reduced physiological disorders are shown in Table 6.5.

Low storage temperatures not only retard the growth of pathogens but also increase the inhibitory effects of MA packaging by increasing the solubility of $CO_2$ in the liquid phase surrounding the commodity. In some fruits such as McIntosh apples, when carbon dioxide is too high, browning disorders of the skin and flesh occur that make fruit unmarketable. Sometimes the high $CO_2$ (20%) detrimental effects are represented by flesh softening or the detection of off-flavors in peaches and nectarines (Nanos & Mitchell, 1991).

One of the other injuries induced by CA storages is fresh browning and cavities. Chiu et al. (2015) reported that 1-MCP sprayed Honeycrisp apple after 24 weeks of preservation under 2.5 kPa $O_2$ + 5 kPa $CO_2$ at 3°C shows this injury. In addition, low oxygen may cause injury on the peel and flesh of avocado (Yearsley & Lallu, 2001).

Although it is suggested that the primary reason for off-flavor seems to be the formation of acetaldehyde, ethyl acetate and ethanol via anaerobic respiration (Li & Kader, 1989), some literature has demonstrated that the main reason can be the direct effect of ethanol rather than other chemicals; in strawberries packed in materials at the lowest $CO_2$ concentration, acetaldehyde concentrations were highest (Almenar et al., 2007).

In the following study, some of injuries' symptoms related to low levels of oxygen during postharvest chain are summarized in Table 6.6. According to the Kishioka et al. (2004) results, the major factor in determining the threshold tolerance of commodities to oxygen level is the occurrence of off-flavor or off-odor. Due to the higher amounts of ethanol compared to acetaldehyde or other volatile compounds in all evaluated 20 commodities at day 7, they concluded that ethanol is the most important factor in developing off-flavor or off-odor. Ethanol production was not directly

**TABLE 6.6**

**Symptoms of Low-Oxygen-Induced Injuries in Some Fruits and Vegetables**

| Commodity | Low Oxygen Injury | Reference |
|---|---|---|
| Chinese chive | Off-odor, discoloration | (Kishioka et al., 2004) |
| Spinach | Off-odor, discoloration, water-soaked tissue | (Kishioka et al., 2004) |
| Potato, sweet potato, taro, onion, Carrot | Off-odor | (Kishioka et al., 2004) |
| Mini kiwi | Skin and flesh dulling and darkening | (Krupa & Tomala, 2021) |
| Cauliflower | Off-odor, discoloration, water-soaked tissue | (Kishioka et al., 2004) |
| Okra | Off-odor, skin pitting | (Kishioka et al., 2004) |
| Cucumber | Off-odor, skin pitting, discoloration | (Kishioka et al., 2004) |
| Eggplant | Off-odor, skin pitting, water-soaked tissue | (Kishioka et al., 2004) |
| Strawberry, banana | Off-flavor, discoloration | (Kishioka et al., 2004) |
| Pear | Off-flavor, discoloration, water-soaked tissue | (Kishioka et al., 2004) |
| Apple | Alcoholic fermentation | (Schlie et al., 2020) |
| Japanese pear, persimmon, apple, satsuma mandarin, blueberry, kiwifruit | Off-flavor, | (Kishioka et al., 2004) |
| Persimmon | Deastringency | (Orihuel-Iranzo et al., 2010; Zhu et al., 2018) |
| Persimmon | Softening | (Wu et al., 2020) |
| Pineapple | Internal browning | (Phonyiam et al., 2016) |
| Mango | Internal breakdown | (Lizada & Rumbaoa, 2017) |
| Tomato | Rot and decay | (Fallik et al., 2003) |

correlated with the activity of alcohol dehydrogenase (ADH) activity, but it was closely related to the level of TSS in the commodities. Therefore, commodities that have more TSS are more prone to development of off-odor or off-flavor, as demonstrated previously by Ke et al. (1991). Symptoms of low $O_2$-induced injuries in some fruits and vegetables are shown in Table 6.6.

## 6.11  ATMOSPHERIC COMPOSITION AND DISEASE OF FRUITS AND VEGETABLES

Postharvest losses could be due to the physical, physiological, biochemical and microbial reactions. However, microbial, physical and physiological factors cause most of the losses in fresh produce (Kader, 1992). It is demonstrated that controlled atmosphere storage has a major role in insect control and disinfection (Mitcham et al., 2003); for example, increasing the level of CO and decreasing the $O_2$, which is commonly used in MAP, is suitable for the growth and development of lactic acid bacteria, which can increase the decay of the lactic acid-sensitive products of bacteria such as lettuce, chicory and carrots (Nguyen & Carlin, 1994). Exposure of avocado to hypoxic conditions has been used as a good method instead of fumigation with chemicals; however, fruit flesh is sensitive to reduced levels of $O_2$. Optimum results in altering atmosphere composition during avocado handling were suggested at 3% $O_2$ and 97% $N_2$ for a period of 24 hours at 17°C before to storage at 2°C with 90% RH for 3 weeks, although treatment of the fruit with an insecticide of 1 and 0.25% $O_2$ for 1–3 days at 20 °C produced firmer fruits (Kassim et al., 2013).

Reducing respiration rate by limiting the presence of oxygen increases the storage life of fresh produce by the prevention of oxidative spoilage. On the other hand, concentrations less than 8% oxygen reduced the biosynthesis of ethylene in the product, which is a key component in fruit ripening. However, at very low oxygen concentrations (<1%), anaerobic respiration occurs, which can lead to tissue damage and development of off-flavor, as well as the growth of pathogenic microorganisms such as *Clostridium botulinum*. So the suggested amount of oxygen in the MA packaging for produce, either for safety or quality, is 1–5% (Sandhya, 2010) as mentioned. Among three gases applied in MA packaging, $CO_2$ is the only one that has antimicrobial properties. Many theories have been proposed for the antimicrobial effect of $CO_2$. In general, carbon dioxide in MA packaging prolongs the delay phase and the reproduction time becomes logarithmic. MA packaging is generally used to control four major types of microorganisms: aerobic microbes, anaerobic microbes, microaerophilic microbes and facultative anaerobic microbes. The theories that explain the antimicrobial properties of carbon dioxide were summarized by Farber (1991):

- Alterations in cell membrane function and nutrient absorption
- Direct inhibition of enzyme activity by increasing or decreasing enzymatic reaction
- Penetration into bacterial membranes and changes in intracellular pH
- Alterations in physiological and biochemical characteristics of proteins

As previously mentioned, $CO_2$ has different inhibitory effects on a microorganism's activity. Thus while moderate concentrations of $CO_2$ (10–20%) inhibit the growth of aerobic bacteria such as *Pseudomonas*, the growth of microorganisms such as lactic acid bacteria is stimulated by carbon dioxide (Amanatidou et al., 1999). In addition, pathogenic microorganisms, including *Clostridium botulinum*, *Clostridium perfringens* and *Loconostoc monocytogenes*, are partially influenced at concentrations less than 50% carbon dioxide, and there is a concern that, by inhibiting the activity of decay microorganisms, a food product still appears healthy while it contains a lot of pathogenic microorganisms, which, due to the lack of competition between pathogenic and decay microorganisms, the probability of multiplying the number of pathogenic microorganisms increases (Zagory, 1995).

However, further studies are needed on the interaction of the microbial flora of commodities and pathogenic microorganisms in different atmosphere compositions used for the products, as well as the effects of various gases on the growth and development of pathogens in fruits and vegetables, for either intact fruits or fresh-cuts, should be considered.

## 6.12 EFFECT OF CA STORAGE OR MA PACKAGING ON VISUAL AND ORGANOLEPTIC ANALYSIS OF FRUITS AND VEGETABLES ASSESSED BY PANEL TEST

The sensory characterization of fruits and vegetables has been usually obtained by recording sensory evaluations according to trained panel judges to assay certain parameters related to the produce quality during storage. Consumers have been considered to be capable only of making hedonic judgments; i.e., they try to give a qualitative exploratory response according to their preferences of internal sensory factors (peeling easily, seed number, sweetness, acidity, firmness, juiciness and taste) and external factors (appearance). For this reason, not only is it necessary to evaluate the efficiency of MAP and CA techniques based on physical and chemical parameters of fruits and vegetables, but it is also necessary to adapt it to consumers' sensory evaluation of the organoleptic characteristics of products. Therefore, some of the related literature in this field is mentioned here. The exposure of red currants to $SO_2$ gas under MAP conditions caused a good visual appearance and sensorial attributes along with extended shelf life over those packed in $CO_2$ gas. The sensorial quality of modified atmosphere packaging of $SO_2$-treated currants, including total aroma, herbaceous taste, astringent taste, acid taste, sweet taste and consistency after 30 days, were more similar to those of fresh berries at harvest time (Brondino et al., 2021).

## 6.13 COMBINED IMPACTS OF MAP AND BIOMATERIALS

Along with the increasing demand of consumers to the use of organic horticultural crops and not to use chemicals to preserve fruits and vegetables, many efforts have been made to use biomaterials as an edible coating in combination with packaging. Some examples follow. Hedonic analysis of MA-packaged table grape cv. Italia, plus treatment with chitosan, showed that the table grapes packed in active MA with elevated concentrations of carbon dioxide and lowered levels of oxygen in combination with chitosan did not show any symptoms of berry browning and rachis browning or of dehydration. Also, the highest score for all sensorial attributes was recorded for those packed under active MA than those packed under the passive method after 7 and 14 days shelf life (Liguori et al., 2021). In another work on the effects of black seed oil, propolis and fludioxonil on MA-packaged pomegranate, results showed treatment of fruit with both black seed oil and propolis plus MAP was more effective in the prevention of gray mold incidence and chilling injury. Also, the weight loss of the mentioned fruits was less than that in controls (Kahramanoğlu et al., 2018).

Strawberry, a perishable fruit, is a tasty berry with an individual flavor with a short shelf life of about 2–5 days at room temperature. Coating the strawberry in different concentrations of lemongrass oil, along with packaging in MA, extended the shelf life of strawberry up to 18 days, maintaining the organoleptic characteristics of the fruits. The effect of lemongrass oil in the prevention of microbial and chemical spoilage and off-odor of strawberry fruits was also evident (Kahramanoğlu, 2019). Coating cucumbers with eco-friendly biomaterials, including lemongrass oil, propolis alone or in combination with packaging in MA, extended the storage life of treated cucumbers to 20 days, when they were firmer and had lower chilling injury (Kahramanoğlu & Usanmaz, 2019).

The packaging of table grape under passive MA and treatment with %5 Arabic gum and %3.5 almond gum had a significant effect on the sensual attributes of berry and extended berry storage life. The coated berries had lower weight loss and better firmness than the uncoated samples. Also, coated fruits were well able to preserve vitamin C, total phenol, pigments and antioxidant activity of the skin and flesh of berry, and the amount of TSS/TA and total carbohydrates increased less than in the control samples (Abbasi & Seyed, 2020). Using nanoencapsulated essential oils in maintaining the quality of fruits and vegetables has a high potential as an alternative to artificial preservatives that are dangerous to health. MA packaged of apricot treated with nanoencapsulated rosemary had more phenolic compounds, vitamin C, total carbohydrate and antioxidant activity compared to controls (Hajizadeh, 2021b).

## 6.14  CONCLUSION

An important supplemented element to suitable temperature and RH% management is the use different atmospheric compositions as CA storages or MA packaging for harvested produce. The chapter encourages the commercial use of MA packaging by processors and retailers in addition to further research efforts for local commodities under local conditions. Atmospheric composition plays a key role in preserving product quality, delivering a healthy produce to the consumers and reducing produce losses. Increasing the storage life of the produce means increasing the chances of competing in the market and taking advantage of more opportunities to deliver the product to the customer.

$O_2$, $CO_2$ and $N_2$ are the most prevalent gases used in MA packaging. Among them, only carbon dioxide has antimicrobial properties, leading to a longer delay phase and reproduction time in the logarithmic phase. The suggested level of oxygen in MA packaging for fresh produce is between 0 and 1% in terms of both safety and quality, although in reality the oxygen level is less than 1% in modified atmosphere packed products. One of the benefits of change in the atmospheric composition around the product, especially in the MAP technique, is an alternative to storing products through freezing and reducing energy consumption without having negative effects on product quality during storage and sales. Producing a long-lasting product makes it possible to sell it in small and remote shops without sufficient cooling facilities. In addition, eliminating the need for refrigeration facilitates product handling and display in larger retails.

Ideally, a modified atmosphere package should maintain safe and effective partial pressures of oxygen and carbon dioxide over a range of temperatures because there is a distinct risk of temperature abuse during shipping, handling and marketing, as increased temperature during subsequent handling leads to anaerobic transpiration. Research has showed that slightly high levels of $O_2$ enhances some of the effects of ethylene on produce, including ripening, senescence and physiological disorders caused by ethylene, including bitterness in carrot and russet spotting in lettuce. Superatmospheric levels of $O_2$ prevent the growth of some pathogens, which is more efficient if matched with increased levels of carbon dioxide up to 15–20 kPa, referred to as a fungicide.

The natural microbial flora of products, which are mainly responsible for the spoilage of new products, can be completely different for each product and for each storage condition, and, due to the interaction of this microbial flora with pathogens in food safety, complete elimination is not realistic. Today, there are concerns about psychotropic food microorganisms such as *Lukonostok monocytogenes*, *Yersinia enterocolitica* and *Aeromonas hydrophila*, as well as non-proteolytic *Clostridium botulinum* and many other microorganisms, especially *Salmonella* and *Escherichia coli*, which, if they are present in atmospherically modified packaged fruits and vegetables, can be hazardous to consumer health.

Reduced physiological disorders of fruits and vegetables that are developing during storability increased the efficiency of the orchard and decreased in postharvest losses through the supply chain. In designing the optimum atmosphere for the transport, handling and retail display of fruits and vegetables, especially fresh-cut products with the least durability, the time and temperature conditions during storage, as well as maturity at harvest, are precisely evaluated. Also, it may be better to use a combination CA/MAP that can preserve recommended atmospheric composition during postharvest handling and transportation, especially the likelihood of mixed load conditions. Nowadays the use of a combination of biomaterials, along with the packaging of fruits and vegetables especially in fresh-cuts as an edible coating, has been expanded to preserve and improve the organoleptic quality and longevity of fruits and vegetables.

## REFERENCES LIST

Abbasi, P., & Seyed, H. H. (2020). Application of some edible coatings on biochemical properties and shelf life of grape (*Vitis vinifera* L.). *Journal of Plant Process and Function*, 153–168.

Ali, S., Khan, A. S., Malik, A. U., & Shahid, M. (2016). Effect of controlled atmosphere storage on pericarp browning, bioactive compounds and antioxidant enzymes of litchi fruits. *Food Chemistry*, *206*, 18–29. https://doi.org/10.1016/j.foodchem.2016.03.021

Almenar, E., Del-Valle, V., Hernández-Muñoz, P., Lagarón, J. M., Catalá, R., & Gavara, R. (2007). Equilibrium modified atmosphere packaging of wild strawberries. *Journal of the Science of Food and Agriculture*, *87*(10), 1931–1939. https://doi.org/10.1002/jsfa.2938

Amanatidou, A., Smid, E. J., & Gorris, L. G. M. (1999). Effect of elevated oxygen and carbon dioxide on the surface growth of vegetable-associated micro-organisms. *Journal of Applied Microbiology*, *86*(3), 429–438. https://doi.org/10.1046/j.1365-2672.1999.00682.x

Ares, G., Lareo, C., & Lema, P. (2007). Modified atmosphere packaging for postharvest storage of mushrooms. A review. *Fresh Produce*, *1*(1), 32–40.

Artés, F., Gómez, P. A., & Artés-Hernández, F. (2006). Modified atmosphere packaging of fruits and vegetables. *Stewart Postharvest Review*, *2*(5), 1–13. https://doi.org/10.2212/spr.2006.5.2

Belay, Z. A., Caleb, O. J., & Opara, U. L. (2017). Impacts of low and super-atmospheric oxygen concentrations on quality attributes, phytonutrient content and volatile compounds of minimally processed pomegranate arils (cv. wonderful). *Postharvest Biology and Technology*, *124*, 119–127. https://doi.org/10.1016/j.postharvbio.2016.10.007

Belay, Z. A., Caleb, O. J., & Opara, U. L. (2019). Influence of initial gas modification on physicochemical quality attributes and molecular changes in fresh and fresh-cut fruit during modified atmosphere packaging. *Food Packaging and Shelf Life*, *21*, 100359. https://doi.org/10.1016/j.fpsl.2019.100359

Bender, R. J., Brecht, J. K., Baldwin, E. A., & Malundo, T. M. M. (2000). Aroma volatiles of mature-green and tree-ripeTommy Atkins' mangoes after controlled atmosphere vs. air storage. *HortScience*, *35*(4), 684–686.

Bender, B. J., Brecht, J. K., & Campbell, C. A. (1995). Responses of 'Kent' and 'Tommy Atkins' mangoes to reduced $O_2$ and elevated $CO_2$. In *Florida State Horticultural Society. Meeting (USA)*.

Bender, R. J., Brecht, J. K., & Sargent, S. A. (2021). Reduced ethylene synthesis of mangoes under high CO2 atmosphere storage. *Acta Scientiarum. Agronomy*, *43*. https://doi.org/10.4025/actasciagron.v43i1.51540

Ben-Yehoshua, S., Beaudry, R. M., Fishman, S., Jayanty, S., & Mir, N. (2005). Modified atmosphere packaging and controlled atmosphere storage. In S. Ben-Yehoshua (Ed.), *Environmentally friendly technologies for agricultural produce quality* (pp. 51–73). Taylor & Francis Group LLC. https://doi.org/10.1201/9780203500361.ch4

Bodbodak, S., & Moshfeghifar, M. (2016). Advances in controlled atmosphere storage of fruits and vegetables. In M. W. Mohammed Wasim Siddiqui (Ed.), *Eco-friendly technology for postharvest produce quality* (pp. 39–76). Academic Press. https://doi.org/10.1016/B978-0-12-804313-4.00002-5

Brecht, J. K. (2006). Controlled atmosphere, modified atmosphere and modified atmosphere packaging for vegetables. *Stewart Postharvest Review*, *5*(2), 1–6. https://doi.org/10.2212/spr.2006.5.5

Brecht, J. K., Chau, K. V., Fonseca, S. C., Oliveira, F. A. R., Silva, F. M., Nunes, M. C. N., & Bender, R. J. (2003). Maintaining optimal atmosphere conditions for fruits and vegetables throughout the postharvest handling chain. *Postharvest Biology and Technology*, *27*(1), 87–101. https://doi.org/10.1016/S0925-5214(02)00185-0

Brondino, L., Cadario, D., & Giuggioli, N. R. (2021). Influence of a sulphur dioxide active storage system on the quality of *Ribes rubrum* L. Berries. *Polish Journal of Food and Nutrition Sciences*, 279–288. https://doi.org/10.31883/pjfns/139997

Buzby, J. C., Farah-Wells, H., & Hyman, J. (2014). The estimated amount, value, and calories of postharvest food losses at the retail and consumer levels in the United States. USDA-ERS Economic Information Bulletin. *SSRN Electronic Journal*, *121*. https://doi.org/10.2139/ssrn.2501659

Caleb, O. J., Opara, U. L., & Witthuhn, C. R. (2012). Modified atmosphere packaging of pomegranate fruit and arils: A review. *Food and Bioprocess Technology*, *5*(1), 15–30. https://doi.org/10.1007/s11947-011-0525-7

Chiu, G. Z., Shelp, B. J., Bowley, S. R., DeEll, J. R., & Bozzo, G. G. (2015). Controlled atmosphere-related injury in 'Honeycrisp' apples is associated with γ-aminobutyrate accumulation. *Canadian Journal of Plant Science*, *95*(5), 879–886. https://doi.org/10.4141/cjps-2015-061

Chonhenchob, V., Chantarasomboon, Y., & Singh, S. P. (2007). Quality changes of treated fresh-cut tropical fruits in rigid modified atmosphere packaging containers. *Packaging Technology and Science*, *20*(1), 27–37. https://doi.org/10.1002/pts.740

Church, P. N. (1993). Meat and meat products. In R. T. Parry (Ed.), *Principles and applications of modified atmosphere packaging of food* (pp. 170–187). Blackie.

Church, P. N. (1994). Developments in modified atmosphere packaging and related technologies. *Trends in Food Science and Technology*, *5*(11), 345–352. https://doi.org/10.1016/0924-2244(94)90211-9

Clarke, R., & De Moor, C. P. (1997). The future in film technology: A tunable packaging system for fresh produce. *Postharvest Horticulture Series – Department of Pomology, University of California Year*, *19*, 68–75.

Conyers, S. T., & Bell, C. H. (2007). A novel use of modified atmospheres: Storage insect population control. *Journal of Stored Products Research*, *43*(4), 367–374. https://doi. org/10.1016/j.jspr.2006.09.003

Dalrymple, D. G. (1969). The development of an agricultural technology: Controlled-atmosphere storage of fruit. *Technology and Culture*, *10*(1), 35–48. https://doi. org/10.1353/tech.1969.a892313

Davies, A. R. (1995). Advances in modified-atmosphere packaging. In G.W. Gould (Ed.), *New methods of food preservation* (pp. 304–320). Springer. https://doi.org/10.1007/ 978-1-4615-2105-1_14

Dilley, D.R. (2006). Advances in controlled atmosphere storage of fruits and vegetables. *Encyclopedia of Food Science and Nutrition*, *2*, 24–38.

Ding, C. K., Chachin, K., Ueda, Y., Imahori, Y., & Kurooka, H. (1999). Effects of high $CO_2$ concentration on browning injury and phenolic metabolism in loquat fruits. *Engei Gakkai Zasshi*, *68*(2), 275–282. https://doi.org/10.2503/jjshs.68.275

Donadon, J. R., Durigan, J. F., Teixeira, G. H. A., Lima, M. A., & Sarzi, B. (2004). Production and preservation of fresh-cut "Tommy Atkins" mango chunks. *Acta Horticulturae*, (645), 257–260. https://doi.org/10.17660/ActaHortic.2004.645.26

Dong, J., Gruda, N., Lam, S. K., Li, X., & Duan, Z. (2018). Effects of elevated $CO_2$ on nutritional quality of vegetables: A review. *Frontiers in Plant Science*, *9*, 924. https://doi. org/10.3389/fpls.2018.00924

Erkan, M., & Chien, Y.W. (2006). Modified and controlled atmosphere storage of subtropical crops. *Stewart Postharvest Review*, *5*(4), 1–8. https://doi.org/10.2212/spr.2006.5.4

Fallik, E., & Aharoni, Y. (2004). *Postharvest physiology, pathology and handling of fresh produce*. Lecture notes. International Research and Development course on Postharvest Biology and Technology. Volcani Center.

Fallik, E., Polevaya, Y., Tuvia-Alkalai, S., Shalom, Y., & Zuckermann, H. (2003). A 24-h anoxia treatment reduces decay development while maintaining tomato fruit quality. *Postharvest Biology and Technology*, *29*(2), 233–236. https://doi.org/10.1016/ S0925-5214(03)00109-1

Farber, J. M. (1991). Microbiological aspects of modified-atmosphere packaging technology-a review. *Journal of Food Protection*, *54*(1), 58–70. https://doi.org/10.4315/ 0362-028X-54.1.58

Flores, F. B., Martínez-Madrid, M. C., Ben Amor, M., Pech, J. C., Latché, A., & Romojaro, F. (2004). Modified atmosphere packaging confers additional chilling tolerance on ethylene-inhibited cantaloupe Charentais melon fruit. *European Food Research and Technology*, *219*(6), 614–619. https://doi.org/10.1007/s00217-004-0952-z

Fonseca, S. C., Oliveira, F. A. R., & Brecht, J. K. (2002). Modelling respiration rate of fresh fruits and vegetables for modified atmosphere packages: A review. *Journal of Food Engineering*, *52*(2), 99–119. https://doi.org/10.1016/S0260-8774(01)00106-6

Fonseca, S. C., Oliveira, F. A. R., Brecht, J. K., & Chau, K. V. (2001). *Evaluation of the physiological response of shredded Galega kale under low oxygen and high carbon dioxide concentrations*. International Controlled Atmosphere Research Conference, Vol. 600, pp. 389–391. https://doi.org/10.17660/ActaHortic.2003.600.56

Fonseca, S. C., Oliveira, F. A. R., Brecht, J. K., & Chau, K. V. (2005). Influence of low oxygen and high carbon dioxide on shredded Galega kale quality for development of modified atmosphere packages. *Postharvest Biology and Technology*, *35*(3), 279–292. https://doi. org/10.1016/j.postharvbio.2004.08.007

Forney, C. F. (2001). Horticultural and other factors affecting aroma volatile composition of small fruit. *HortTechnology*, *11*(4), 529–538. https://doi.org/10.21273/HORTTECH. 11.4.529

Gheorghe, I. F., & Ion, B. (2011). The effects of air pollutants on vegetation and the role of vegetation in reducing atmospheric pollution. In M. Khallaf (Ed.), *The impact of air pollution on health, economy, environment and agricultural sources* (pp. 241–280). Intech Open. https://doi.org/10.5772/17660

Guan, J., & Dou, S. (2010). The effect of MAP on quality and browning of cold-stored plum fruits. *Journal of Food, Agriculture and Environment, 8*(2), 113–116.

Gunes, G., & Lee, C. Y. (1997). Color of minimally processed potatoes as affected by modified atmosphere packaging and antibrowning agents. *Journal of Food Science, 62*(3), 572–575. https://doi.org/10.1111/j.1365-2621.1997.tb04433.x

Hajizadeh, H. S. (2021a). Potential mechanisms of antioxidants derived from horticultural crops against Covid-19. *International Journal of Agriculture Forestry and Life Sciences, 5*(1), 113–121.

Hajizadeh, H. S. (2021b). Effect of nano-encapsulation of rosemary in quality preserving and antioxidative activity of apricot (*Prunus armeniaca* L.) during storage life. *Journal of Food Science and Technology (Iran), 18*(117), 183–196.

Hajizadeh, H. S., & Kazemi, M. (2012). Investigation of approaches to preserve postharvest quality and safety in fresh-cut fruits and vegetables. *Research Journal of Environmental Sciences, 6*(3), 93–106. https://doi.org/10.3923/rjes.2012.93.106

Hajizadeh, H. S., Mostofi, Y., & Talaie, A. (2006). Modified atmosphere packaging (MAP) effects on quality maintenance and storage life extension of local Iranian apple "Golab Kohanz". *Acta Horticulturae, 768*, 274–274. https://doi.org/10.17660/ActaHortic.2008.768.12

Hajizadeh, H. S., & Safkhani, S. (2017). Effect of salicylic acid on prevention of chilling injury of cherry tomato (*Lycopersicun esculentum* cv. Messina). *Journal of Horticulture Science, 31*(3), 517–532.

Holcroft, D. M., & Kader, A. A. (1999). Controlled atmosphere-induced changes in pH and organic acid metabolism may affect color of stored strawberry fruit. *Postharvest Biology and Technology, 17*(1), 19–32. https://doi.org/10.1016/S0925-5214(99)00023-X

Ingrassia, M., Bacarella, S., Altamore, L., Sortino, G., & Chironi, S. (2016). *Consumer acceptance and primary drivers of liking for small fruits*. International Postharvest Symposium: Enhancing Supply Chain and Consumer Benefits-Ethical and Technological Issues 1194, pp. 1147–1154. https://doi.org/10.17660/ActaHortic.2018.1194.164

Irtwange, S. V. (2006, February). Application of modified atmosphere packaging and related technology in postharvest handling of fresh fruits and vegetables. *Agricultural Engineering International: The CIGR Ejournal*. Invited Overview No. 4, VIII.

Kader, A.A. (1992). *Postharvest technology of horticultural crops* (2nd ed., Vol. 3311, p. 296). University of California, Division of Agriculture and Natural Resources Publishing House.

Kader, A. A. (2008). Flavor quality of fruits and vegetables. *Journal of the Science of Food and Agriculture, 88*(11), 1863–1868. https://doi.org/10.1002/jsfa.3293

Kader, A. A., & Ben-Yehoshua, S. (2000). Effects of superatmospheric oxygen levels on postharvest physiology and quality of fresh fruits and vegetables. *Postharvest Biology and Technology, 20*(1), 1–13. https://doi.org/10.1016/S0925-5214(00)00122-8

Kader, A. A., Morris, L. L., Stevens, M. A., & Albright-Holton, M. (1978). Composition and flavor quality of fresh market tomatoes as influenced by some postharvest handling procedures. *Journal of the American Society for Horticultural Science, 103*(1), 6–13. https://doi.org/10.21273/JASHS.103.1.6

Kader, A. A., & Watkins, C. B. (2000). Modified atmosphere packaging – Toward 2000 and beyond. *HortTechnology, 10*(3), 483–486. https://doi.org/10.21273/HORTTECH.10.3.483

Kader, A. A., Zagory, D., Kerbel, E. L., & Wang, C. Y. (1989). Modified atmosphere packaging of fruits and vegetables. *Critical Reviews in Food Science and Nutrition, 28*(1), 1–30. https://doi.org/10.1080/10408398909527490

Kahramanoğlu, İ. (2019). Effects of lemongrass oil application and modified atmosphere packaging on the postharvest life and quality of strawberry fruits. *Scientia Horticulturae, 256*, 108527. https://doi.org/10.1016/j.scienta.2019.05.054

Kahramanoğlu, İ., Aktaş, M., & Gündüz, Ş. (2018). Effects of fludioxonil, propolis and black seed oil application on the postharvest quality of "Wonderful" pomegranate. *PLoS One, 13*(5), e0198411. https://doi.org/10.1371/journal.pone.0198411

Kahramanoğlu, İ., & Usanmaz, S. (2019). Improving postharvest storage quality of cucumber fruit by modified atmosphere packaging and biomaterials. *Hortscience, 54*(11), 2005–2014. https://doi.org/10.21273/HORTSCI14461-19

Kassim, A., Workneh, T. S., & Bezuidenhout, C. N. (2013). A review on postharvest handling of avocado fruit. *African Journal of Agricultural Research, 8*(21), 2385–2402. https://doi.org/10.5897/AJAR12.1248

Ke, D., Rodriguez-Sinobas, L., & Kader, A. A. (1991). Physiology and prediction of fruit tolerance to low-oxygen atmospheres. *Journal of the American Society for Horticultural Science, 116*(2), 253–260. https://doi.org/10.21273/JASHS.116.2.253

Khakpour, S., Seyed Hajizadeh, H. S., Hemati, A., Bayanati, M., Nobaharan, K., Mofidi Chelan, E., Asgari Lajayer, B., & Dell, B. (2022). The effect of preharvest treatment of calcium nitrate and iron chelate on post-harvest quality of apple (*Malus domestica* Borkh cv. Red Delicious). *Scientia Horticulturae, 304*, 111351. https://doi.org/10.1016/j.scienta.2022.111351

Kiaya, V. (2014). Postharvest losses and strategies to reduce them. Technical paper on postharvest losses. *Action Contre la Faim (ACF), 25*, 1–25.

Kidd, F., & West, C. (1925). A relation between the respiratory activity and the keeping quality of apples. *Report Food Investing Board, 26*, 37–41.

Kim, J. G., Luo, Y., & Gross, K. C. (2004). Effect of package film on the quality of fresh-cut salad savoy. *Postharvest Biology and Technology, 32*(1), 99–107. https://doi.org/10.1016/j.postharvbio.2003.10.006

Kishioka, I., Fujiwara, H., Tulio, Jr., A. Z., Ueda, Y., Chachin, K., Imahori, Y., & Uemura, K. (2004). Relationship between low-oxygen injury and ethanol metabolism in various fruits and vegetables. In *V. International Postharvest symposium 682* (pp. 1103–1108).

Krupa, T., & Tomala, K. (2021). Effect of oxygen and carbon dioxide concentration on the quality of Minikiwi fruits after storage. *Agronomy, 11*(11), 2251. https://doi.org/10.3390/agronomy11112251

Lakso, A. N., & Kliewer, W. M. (1978). The influence of temperature on malic acid metabolism in grape berries. II. Temperature responses of net dark $CO_2$ fixation and malic acid pools. *American Journal of Enology and Viticulture, 29*(3), 145–149. https://doi.org/10.5344/ajev.1978.29.3.145

Li, C., & Kader, A. A. (1989). Residual effects of controlled atmospheres on postharvest physiology and quality of strawberries. *Journal of the American Society for Horticultural Science, 114*(4), 629–634. https://doi.org/10.21273/JASHS.114.4.629

Li, X., Jiang, Y., Li, W., Tang, Y., & Yun, J. (2014). Effects of ascorbic acid and high oxygen modified atmosphere packaging during storage of fresh-cut eggplants. *Food Science and Technology International, 20*(2), 99–108. https://doi.org/10.1177/1082013212472351

Lidster, P. D., Blanpied, G. D., & Prange, R. K. (1990). *Controlled-atmosphere disorders of commercial fruits and vegetables.* Communications Branch, Agriculture Canada Publication 1847/E. Retrieved August 26, 2022, from https://publications.gc.ca/collections/collection_2012/agr/A53-1847-1990-eng.pdf

Liguori, G., Sortino, G., Gullo, G., & Inglese, P. (2021). Effects of modified atmosphere packaging and chitosan treatment on quality and sensorial parameters of minimally processed cv.'Italia' table grapes. *Agronomy, 11*(2), 328. https://doi.org/10.3390/agronomy11020328

Lipton, W. J. (1975). Controlled atmospheres for fresh vegetables and fruits – Why and when. In *Symposium: Postharvest biology and handling of fruits and vegetables.*

Lizada, M. C. C., & Rumbaoa, R. A. (2017). Ethylene and the adaptive response of mango to hypoxia. *Acta Horticulturae*, (1178), 143–146. https://doi.org/10.17660/ActaHortic.2017.1178.25

Lougheed, E. C. (1987). Interactions of oxygen, carbon dioxide, temperature, and ethylene that may induce injuries in vegetables. *HortScience*, 22(5), 791–794. https://doi.org/10.21273/HORTSCI.22.5.791

Ma, Y., Li, S., Yin, X., Xing, Y., Lin, H., Xu, Q., Bi, X., & Chen, C. (2019). Effects of controlled atmosphere on the storage quality and aroma compounds of lemon fruits using the designed automatic control apparatus. *BioMed Research International*, *2019*, 6917147. https://doi.org/10.1155/2019/6917147

Marrero, A., & Kader, A. A. (2006). Optimal temperature and modified atmosphere for keeping quality of fresh-cut pineapples. *Postharvest Biology and Technology*, 39(2), 163–168. https://doi.org/10.1016/j.postharvbio.2005.10.017

Matityahu, I., Marciano, P., Holland, D., Ben-Arie, R., & Amir, R. (2016). Differential effects of regular and controlled atmosphere storage on the quality of three cultivars of pomegranate (*Punica granatum* L.). *Postharvest Biology and Technology*, *115*, 132–141. https://doi.org/10.1016/j.postharvbio.2015.12.018

Mattheis, J. P., Rudell, D. R., & Hanrahan, I. (2017). Impacts of 1-methylcyclopropene and controlled atmosphere established during conditioning on development of bitter pit in "Honeycrisp" apples. *Hortscience*, 52(1), 132–137. https://doi.org/10.21273/HORTSCI11368-16

Mditshwa, A., Fawole, O. A., & Opara, U. L. (2018). Recent developments on dynamic controlled atmosphere storage of apples – A review. *Food Packaging and Shelf Life*, *16*, 59–68. https://doi.org/10.1016/j.fpsl.2018.01.011

Meheriuk, M. (1994). Postharvest disorders of apples and pears. *Agriculture and Agri-food Canada. Agriculture Canada publications, 1737/E*. Retrieved August 26, 2022, from https://ia801405.us.archive.org/26/items/postharvestdisor00mehe/postharvestdisor00mehe.pdf

Mencarelli, F., Lipton, W. J., & Peterson, S. J. (1983). Response of zucchini. *Journal of the American Society for Horticultural Science*, 108(6), 884–890. https://doi.org/10.21273/JASHS.108.6.884

Mitcham, E. J., Lee, T., Martin, A., Zhou, S., & Kader, A. A. (2003). Summary of CA for arthropod control on fresh horticultural perishables. *Acta Horticulturae*, (600), 741–745.

Mosayyebzadeh, A., Mostofi, Y., Jomeh, Z. E., Nikkhah, M. J., & Hajizadeh, H. S. (2010). Effect of modified atmosphere packaging (MAP) with increased levels of $O_2$ on postharvest quality of Iranian Shahroodi table grape. In *XXVIII international horticultural congress on science and horticulture for people (IHC2010): International symposium on* (Vol. 934, pp. 207–211).

Mostofi, Y., Hajizadeh, H. S., Reza Talaie, A., & Zadeh, M. M. E. (2006). The effect of modified atmosphere packaging (MAP) on some physiochemical characteristics and texture of Iranian apple "Shafi". *Acta Horticulturae*, *768*, 273–273. https://doi.org/10.17660/ActaHortic.2008.768.11

Nanos, G. D., Kiritsakis, A. K., & Sfakiotakis, E. M. (2002). Preprocessing storage conditions for green Conservolea and Chondrolia table olives. *Postharvest Biology and Technology*, 25(1), 109–115. https://doi.org/10.1016/S0925-5214(01)00164-8

Nanos, G. D., & Mitchell, F. G. (1991). Carbon dioxide injury and flesh softening following high-temperature conditioning in peaches. *Hortscience*, 26(5), 562–563. https://doi.org/10.21273/HORTSCI.26.5.562

Nguyen-the, C., & Carlin, F. (1994). The microbiology of minimally processed fresh fruits and vegetables. *Critical Reviews in Food Science and Nutrition*, 34(4), 371–401. https://doi.org/10.1080/10408399409527668

Nunes, M. C. N., Morais, A. M. M. B., Brecht, J. K., & Sargent, S. A. (1996). Quality of pink tomatoes (cv. buffalo) after storage under controlled atmosphere at chilling and nonchilling temperatures. *Journal of Food Quality*, *19*(5), 363–374. https://doi. org/10.1111/j.1745-4557.1996.tb00431.x

Odriozola-Serrano, I., Soliva-Fortuny, R., & Martín-Belloso, O. (2010). Changes in bioactive composition of fresh-cut strawberries stored under superatmospheric oxygen, low-oxygen or passive atmospheres. *Journal of Food Composition and Analysis*, *23*(1), 37–43. https://doi.org/10.1016/j.jfca.2009.07.007

Onursal, C. E., & Koyuncu, M. A. (2021). Role of controlled atmosphere, ultra low oxygen or dynamic controlled atmosphere conditions on quality characteristics of "scarlet spur" apple fruit. *Tarım Bilimleri Dergisi*, *27*(3), 267–275. https://doi.org/10.15832/ ankutbd.631956

Orihuel-Iranzo, B., Miranda, M., Zacarías, L., & Lafuente, M. T. (2010). Temperature and ultra low oxygen effects and involvement of ethylene in chilling injury of "Rojo Brillante" persimmon fruit. *Food Science and Technology International*, *16*(2), 159–167. https:// doi.org/10.1177/1082013209353221

Phonyiam, O., Kongsuwan, A., & Setha, S. (2016). Effect of short-term anoxic treatment on internal browning and antioxidant ability in pineapple cv. Phulae. *International Food Research Journal*, *23*(2), 521.

Pittman, R. N. (2011, April). Regulation of tissue oxygenation. In *Colloquium series on integrated systems physiology: From molecule to function* (Vol. 3, No. 3, pp. 1–100). Morgan & Claypool Life Sciences. https://doi.org/10.4199/C00029ED1V01Y201103ISP017

Prange, R. K., & DeLong, J. M. (2006). Controlled-atmosphere related disorders of fruits and vegetables. *Stewart Postharvest Review*, *5*(7), 1–10. https://doi.org/10.2212/spr.2006.5.7

Prange, R. K., DeLong, J. M., Daniels-Lake, B. J., & Harrison, P. A. (2005). Innovation in controlled atmosphere technology. *Stewart Postharvest Review*, *1*(3), 1–11.

Rai, D. R., Chadha, S., Kaur, M. P., Jaiswal, P., & Patil, R. T. (2011). Biochemical, microbiological and physiological changes in Jamun (*Syzyium cumini* L.) kept for long term storage under modified atmosphere packaging. *Journal of Food Science and Technology*, *48*(3), 357–365. https://doi.org/10.1007/s13197-011-0254-y

Sabır, A., Sabır, F. K., & Kara, Z. (2011). Effects of modified atmosphere packing and honey dip treatments on quality maintenance of minimally processed grape cv. Razaki (*V. vinifera* L.) during cold storage. *Journal of Food Science and Technology*, *48*(3), 312–318. https://doi.org/10.1007/s13197-011-0237-z

Saltveit, M. E. (2003). Is it possible to find an optimal controlled atmosphere? *Postharvest Biology and Technology*, *27*(1), 3–13. https://doi.org/10.1016/ S0925-5214(02)00184-9

Sandhya. (2010). Modified atmosphere packaging of fresh produce: Current status and future needs. *LWT – Food Science and Technology*, *43*(3), 381–392. https://doi.org/10.1016/j. lwt.2009.05.018

Sapers, G. M., Hicks, K. B., Phillips, J. G., Garzarella, L., Pondish, D. L., Matulaitis, R. M., McCormack, T. J., Sondey, S. M., Seib, P. A., & Ei-atawy, Y. S. (1989). Control of enzymatic browning in apple with ascorbic acid derivatives, polyphenol oxidase inhibitors, and complexing agents. *Journal of Food Science*, *54*(4), 997–1002. https://doi. org/10.1111/j.1365-2621.1989.tb07931.x

Schlie, T. P., Köpcke, D., Rath, T., & Dierend, W. (2020). *Importance of individual apple fruit characteristics for DCA-CF storage on the variety "Elstar, PCP"* (Vol. 14, No. 1, pp. 57–60). Springer.

Sharples, R. O., & Johnson, D. S. (1987). Influence of agronomic and climatic factors on the response of apple fruit to controlled atmosphere storage. *HortScience*, *22*(5), 763–766. https://doi.org/10.21273/HORTSCI.22.5.763

Talasila, P. C., Cameron, A. C., & Joles, D. W. (1994). Frequency distribution of steady-state oxygen partial pressures in modified-atmosphere packages of cut broccoli. *Journal of the American Society for Horticultural Science, 119*(3), 556–562. https://doi.org/10.21273/JASHS.119.3.556

Teixeira, G. H. A., Cunha Júnior, L. C. C., Ferraudo, A. S., & Durigan, J. F. (2016). Quality of guava (*Psidium guajava* L. cv. P. Sato) fruit stored in low-O2 controlled atmospheres is negatively affected by increasing levels of CO2. *Postharvest Biology and Technology, 111*, 62–68. https://doi.org/10.1016/j.postharvbio.2015.07.022

Thewes, F. R., Brackmann, A., de Oliveira Anese, R., Bronzatto, E. S., Schultz, E. E., & Wagner, R. (2017). Dynamic controlled atmosphere storage suppresses metabolism and enhances volatile concentrations of "Galaxy" apple harvested at three maturity stages. *Postharvest Biology and Technology, 127*, 1–13. https://doi.org/10.1016/j.postharvbio.2017.01.002

Thewes, F. R., Brackmann, A., de Oliveira Anese, R., Ludwig, V., Schultz, E. E., & Berghetti, M. R. P. (2018). 1-methylcyclopropene suppresses anaerobic metabolism in apples stored under dynamic controlled atmosphere monitored by respiratory quotient. *Scientia Horticulturae, 227*, 288–295. https://doi.org/10.1016/j.scienta.2017.09.028

Thompson, A. K., Prange, R. K., Bancroft, R., & Puttongsiri, T. (2018). *Controlled atmosphere storage of fruit and vegetables*. CABI Publishing. Retrieved September 29, 2022, from http://www.cabidigitallibrary.org/doi/book/10.1079/9781786393739.0000

Toivonen, P. M. A. (2006). Fresh-cut apples: Challenges and opportunities for multi-disciplinary research. *Canadian Journal of Plant Science, 86*, 1361–1368. https://doi.org/10.4141/P06-147

Toivonen, P. M. A., Brandenburg, J. S., & Luo, Y. (2009). Modified atmosphere packaging for fresh-cut produce. In E. M. Yahia (Ed.), *Modified and controlled atmospheres for the storage, transportation, and packaging of horticultural commodities* (pp. 463–489). CRC Press. https://doi.org/10.1201/9781420069587

Vakis, N., Grierson, W., & Soule, J. (1970). Chilling injury in tropical and subtropical fruits. III. The role of CO$_2$ in suppressing chilling injury of grapefruit and avocados. *Proceedings of the Tropical Region. American Society for Horticultural Science, 14*, 89–100.

Van der Steen, C., Jacxsens, L., Devlieghere, F., & Debevere, J. (2002). Combining high oxygen atmospheres with low oxygen modified atmosphere packaging to improve the keeping quality of strawberries and raspberries. *Postharvest Biology and Technology, 26*(1), 49–58. https://doi.org/10.1016/S0925-5214(02)00005-4

Verlinden, B. E., & Nicolaï, B. M. (2000). Fresh-cut fruits and vegetables. *Acta Horticulturae*, (518), 223–232. https://doi.org/10.17660/ActaHortic.2000.518.30

Watkins, C. B. (2008). Dynamic controlled atmosphere storage–a new technology for the New York storage industry. *New York Fruit Quarterly, 16*(1), 23–26.

Wen, P., Zhu, D. H., Wu, H., Zong, M. H., Jing, Y. R., & Han, S. Y. (2016). Encapsulation of cinnamon essential oil in electrospun nanofibrous film for active food packaging. *Food Control, 59*, 366–376. https://doi.org/10.1016/j.foodcont.2015.06.005

Wu, W., Wang, M. M., Gong, H., Liu, X. F., Guo, D. L., Sun, N. J., Huang, J. W., Zhu, Q. G., Chen, K. S., & Yin, X. R. (2020). High CO2/hypoxia-induced softening of persimmon fruit is modulated by DkERF8/16 and DkNAC9 complexes. *Journal of Experimental Botany, 71*(9), 2690–2700. https://doi.org/10.1093/jxb/eraa009

Wyrwisz, J., Karp, S., Kurek, M. A., & Moczkowska-Wyrwisz, M. (2022). Evaluation of modified atmosphere packaging in combination with active packaging to increase shelf life of high-in beta-glucan gluten free cake. *Foods, 11*(6), 872. https://doi.org/10.3390/foods11060872

Yahia, E. M. (Ed.). (2009). *Modified and controlled atmospheres for the storage, transportation, and packaging of horticultural commodities*. CRC Press. https://doi.org/10.1201/9781420069587

Yahia, E. M. (2011). Mango (Mangifera indica L.). In *Postharvest biology and technology of tropical and subtropical fruits* (pp. 492–567e). *Woodhead Publishing Series in Food Science, Technology and Nutrition*, 492–565, 566e–567e

Yan, Y., Chen, X., Wang, X., Zhao, Z., Hu, W., Zeng, S., Wei, J., Yang, X., Qian, L., Zhou, S., Sun, L., Gong, Z., & Xu, Z. (2019). The effects and the mechanisms of autophagy on the cancer-associated fibroblasts in cancer. *Journal of Experimental and Clinical Cancer Research, 38*(1), 171. https://doi.org/10.1186/s13046-019-1172-5

Yearsley, C., & Lallu, N. (2001). Symptoms of controlled atmosphere damage in avocados. *NZ Avocado Growers Association Annual Research Report, 1*, 26–32.

Yuvaraj, D., Iyyappan, J., Gnanasekaran, R., Ishwarya, G., Harshini, R. P., Dhithya, V., Chandran, M., Kanishka, V., & Gomathi, K. (2021). Advances in bio food packaging–An overview. *Heliyon, 7*(9), e07998. https://doi.org/10.1016/j.heliyon.2021.e07998

Zagory, D. (1995). Principles and practice of modified atmosphere packaging of horticultural commodities. In J.M. Farber & K. Dodds (Eds.), *Principles of modified-atmosphere and sous vide product packaging* (pp. 175–206). CRC Press. https://doi.org/10.1201/9780203742075

Zagory, D., & Kader, A.A. (1988). Modified atmosphere packaging of fresh produce. *Food Technology, 42*(9), 70–77.

Zagory, D., & Kader, A. A. (1989). *Quality maintenance in fresh fruits and vegetables by controlled atmospheres*. ACS Symposium Series, 405 (Chapter 14, pp. 174–188). Quality Factors of Fruits and Vegetables. https://doi.org/10.1021/bk-1989-0405.ch014, Chapter 14.

Zeman, S., & Kubík, L. (2007). Permeability of polymeric packaging materials. *Technical Sciences/University of Warmia and Mazury in Olsztyn, 10*, 26–34.

Zhu, Q. G., Gong, Z. Y., Wang, M. M., Li, X., Grierson, D., Yin, X. R., & Chen, K. S. (2018). A transcription factor network responsive to high CO2/hypoxia is involved in deastringency in persimmon fruit. *Journal of Experimental Botany, 69*(8), 2061–2070. https://doi.org/10.1093/jxb/ery028

# 7 Ethylene Control in Postharvest Handling of Fruits and Vegetables

*Nirmal Kumar Meena and Vinod B. R. Menaka M.*

## 7.1 INTRODUCTION

Ethylene ($C_2H_4$) is a biologically active natural substance that has manifold effects on the growth and developmental processes of plants. It plays an important role in postharvest physiology such as the ripening, pigment synthesis, seed maturation and senescence of fresh produce. Ethylene is a strong two-carbon gaseous plant hormone that can accelerate the physiological activities even at very low concentration, i.e., $1\ \mu l\ L^{-1}$ to $1\ nl\ L^{-1}$ (Saltveit, 1999). Upon the biosynthesis of ethylene, it starts as diffuse throughout the plant parts and stimulates the responses by binding ethylene receptors (Binder, 2020). The ethylene biosynthesis pathway includes certain complex metabolic processes, is sensitive to elevated $CO_2$ and requires oxygen (Saltveit, 1999).

The major response of ethylene, commonly known as triple response, is well studied (Guzman & Ecker, 1990; Binder, 2020). According to this response, reduced growth of the root and hypocotyl and an exaggerated apical hook and a thickening of the hypocotyl were observed in dark grown *Arabidopsis* seedlings. In climacteric produces, ethylene triggers the ripening process; however, the concentration may vary by cultivar, plant parts, stage of harvest and storage conditions, and it is not detected the during initial developmental stage of fruits (Meena & Asrey, 2018). On the basis of the presence of a climacteric peak (as a result of ethylene), fruits can be differentiated between climacteric and non-climacteric. Generally, non-climacteric fruits either do not produce ethylene or produce very little or a negligible amount of ethylene (citrus, grape), and no role of ethylene in ripening process has been reported. These fruits do not ripen off-tree and should be harvested at the right maturity (horticultural maturity) level for human consumption, whereas climacteric fruits can be harvested at physiological maturity and stored longer if the ethylene action is controlled. Then, the fruits can be ripened off-tree with the help of ethylene.

Ethylene is a key hormone for the ripening of fruits and vegetables. Apart from this, ethylene also helps in flowering, de-greening, seed maturation and sex inversion in some cucurbits. In climacteric fruits, $C_2H_4$ triggers color development and enhances softening and senescence, which leads to rapid deterioration and quality loss. An unripe climacteric fruit mostly has low levels of ethylene. As the fruit matures, ethylene is synthesized by fruits as a signal to induce ripening. A list of changes due to ethylene is given in Table 7.1. Ethylene is harmful for many vegetables since it causes

DOI: 10.1201/9781003452355-9

**TABLE 7.1**

**Some Major Impacts of Ethylene on Fresh Horticultural Produce**

| Name of Crop | Effects due to Ethylene |
|---|---|
| Most of climacteric fruits such as mango, papaya, guava, sapota, apple, pear, peach, plum, etc. | Loss of green color |
| | Reduction in firmness |
| | Increased the activity of softening enzymes such as pectin methyl esterase, cellulose and polygalacturonate |
| | Overripening and senescence |
| | Production of volatiles |
| Citrus fruits | De-greening |
| Pineapple | Increase flowering and synchronizing of flowering |
| Carrot | Bitterness due to isocoumarin |
| Broccoli and cucumber | Yellowing |
| Asparagus | Toughness |
| Rose | Bent neck, discoloration |
| Orchid, hibiscus and carnation | Wilting |

*Source:* Adapted from Fan & Mattheis (2000); Wills et al. (2007); Martínez-Romero et al. (2007) with slight modifications.

softening, disease susceptibility, storage disorders, undesirable changes and reduced shelf life (Martínez-Romero et al., 2007). To avoid the detrimental effects of ethylene in the supply chain, sustainable approaches need to be implemented. The agents that can suppress the ethylene biosynthesis, as well as check the action by absorbing surrounding ethylene, need to be addressed. In addition, altering the environmental conditions such as ventilation, storage structure, elevated $CO_2$ containing controlled storage and antisense silencing technologies can be effective ways to mitigate the excess ethylene in fresh produce. Some of the oxidizing agents, such as potassium permanganate ($KMnO_4$) and ozone, change the status of ethylene and thus could be an effective strategy (Scariot et al., 2014).

The application of ethylene biosynthesis inhibitors can check the level of endogenous ethylene. Such agents involve aminooxyacetic acid (AOA), aminoethoxyvinylglycine (AVG), methoxyvinylglycine (MVG) and cobalt ions; however, their use has some limitations. Nowadays, ethylene inhibitor 1-methylcyclopropene (1-MCP) is commercially used for the preservation of fresh horticultural produce. 1-MCP not only checks the ethylene action but also enhances the quality, aromatic volatiles, and it delays softening in many fruits and vegetables (Jia et al., 2018). Several climacteric fruits can be preserved by using elevated $CO_2$ either in controlled atmospheric storage or modified atmospheric packaging. Certain studies suggested that the use of two or more approaches in an integrated manner could be highly effective in mitigating some negative effects of ethylene. This chapter explains the various methods and approaches used for ethylene control in horticultural produce.

## 7.2 ETHYLENE BIOSYNTHESIS IN PLANTS

Ethylene biosynthesis is a complex metabolic pathway that requires oxygen (Figure 7.1). The physiology and biochemistry of the ethylene synthesis pathway has been extensively studied in different plant species. But most of the models were established on typical plants such as *Arabidopsis*, typical climacteric tomato and typical non-climacteric strawberry. Ethylene is highly sensitive to elevated $CO_2$. The extensive study revealed that in higher perennial plants, $C_2H_4$ is synthesized from an amino acid methionine that was converted to S-adensoylmethionine (SAM) (Martínez-Romero et al., 2007) and metabolic flux via the methionine/Yang cycle. After that, SAM is converted to 1-aminocyclopropane-1-carboylic acid (ACC) by the catalytic enzyme ACC synthase (ACS). ACC then is oxidized to ethylene by the action of ethylene, forming enzyme known as ACC oxidase (ACO). During this process, one by product i.e., 5-methylthioadenosine (MTA) is released (Yang & Hoffman, 1984). The comprehensive role of ACC in plant biology has been extensively reviewed by Poel and Straeten (2014). The previous studies on climacteric fruits suggested that once ethylene starts synthesizing, it steadily reaches higher levels and maintains a peak (climacteric peak) at a certain stage; this distinguishes the climacteric fruits with non-climacterics (Table 7.2). Both the enzymes ACS and ACO belong to multigene families and are associated with cytosol; however, some studies suggest that ACO may be associated with plasma membrane (Ramassamy et al., 1998; Hudgins et al., 2006). The regulation of ethylene biosynthesis is mainly dependent on these two key enzymes and their transcription, translation and protein stability (Pattyn et al., 2021).

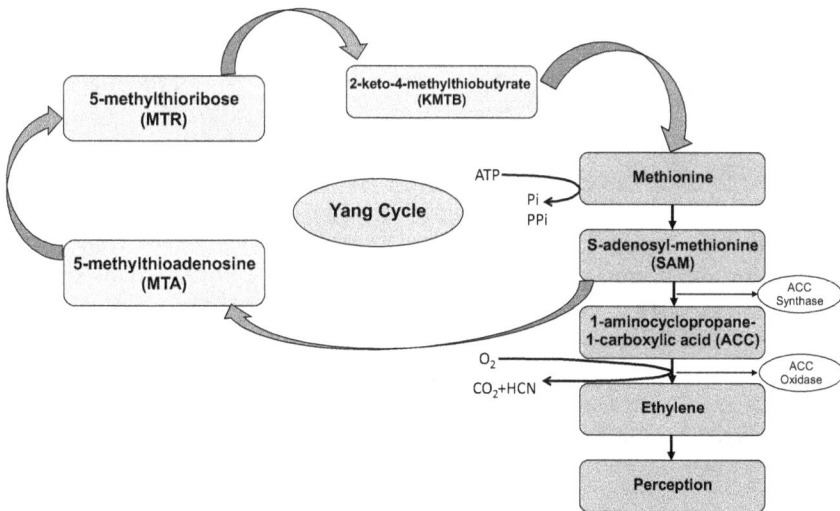

**FIGURE 7.1** Schematic diagram of ethylene biosynthesis pathway via methionine or Yang cycle (adapted from Arc et al. (2013) with slight modifications).

**TABLE 7.2**

**Classification of Fruits on the Basis of Climacteric Peak and Ethylene Evolution**

| Climacteric | Non-climacteric | Suppress Climacteric |
|---|---|---|
| Apple | Mandarin | Plum |
| Pear | Aonla | |
| Peach | Pomegranate | |
| Plum (except some cultivars) | Watermelon | |
| Papaya | Jujube | |
| Mango | Pomegranate | |
| Guava | Jamun | |
| Banana | Strawberry | |
| Sapota | Pineapple | |
| Cordia | Grape | |
| Tomato | Cherry | |
| | Sweet orange | |
| | Lime and lemon | |
| | Litchi | |
| | Cucumber | |

*Source:* Adapted from Wills et al. (2007); Martínez-Romero et al. (2007) with minor modifications.

## 7.3  ETHYLENE CONTROL

In order to overcome the detrimental effects of ethylene, it is necessary to manage the level of ethylene by suitable methods. The major aim of the ethylene management in postharvest is to enable the regular supply of sustainable and long-lived fresh produce. Several strategies and chemical-based technologies are used to inhibit biosynthesis: checking the action by binding their sites and removing the gas from the surrounding environment. Use of biosynthesis inhibitors is a costly technology and not very feasible with fruits and vegetables. However, use of action inhibitors such as 1-MCP has changed the scenario of the fruit industry and achieved great interest in the scientific community. Certain oxidizing agents, either of $KMnO_4$ or some nanoparticle-based agent like palladium – as well as $TiO_2$ – are also commercially used and have been found to be highly effective against exogenous ethylene. Ethylene scavengers are now available commercially in ready-to-use form. Some of the scavengers are given in Table 7.3.

**TABLE 7.3**

**Some Important Commercially Available Ethylene Scavengers**

| Sl. No. | Trade Name | Available Form | Oxidizing Agent |
|---|---|---|---|
| 1 | Purafil® | Sachet | KMnO$_4$ and activated alumina |
| 2 | Peakfresh™ | Films/sachet | Activated clays and zeolites |
| 3 | Air repair | Sachet | KMnO$_4$ beads |
| 4 | Mrs. Green extra life | cartridge | KMnO$_4$ and zeolite |
| 5 | Orega | Plastic film, filler | Pumice, activated clays, zeolites and metal oxides |
| 6 | Neupalon | Sachet | Activated carbon |
| 7 | BO films | Film | Crysburite ceramic |
| 8 | Green bags | Bag | Activated zeolites |
| 9 | Ethylene eliminator pack | Sachet | Activated zeolites |
| 10 | Bio-Kleen | Sachet | TiO$_2$ catalyst |
| 11 | Neupalon | Sachet | Activated carbon |
| 12 | Hatofresh | Paper/board | Activated carbon |
| 13 | Sendo-Mate | Plastic film | Palladium catalyst on activated carbon |
| 14 | Profresh | Sachet | Mineral |
| 15 | Bio-Fresh™ | Zipper bag | Activated clays and zeolites |
| 16 | DeltaTrack® | sachet/blankets | KMnO$_4$ |
| 17 | Evert-Fresh | LDPE Films | Oya-stone |
| 18 | Ever Fresh | Films | Oya-stone |
| 19 | BIO-KES | – | TiO$_2$ |
| 20 | Titan Aire | – | TiO$_2$ |

*Source:* Extracted from Sadeghi et al., 2021 with slight modifications.v

## 7.4 ETHYLENE ANTAGONISM

### 7.4.1 INHIBITION OF ACC SYNTHESIS

#### 7.4.1.1 Aminoethoxyvinylglycine (AVG)

ACS inhibitors are classified into two groups. The first is a class of vinylglycine analogues that function as potent inhibitors of pyridoxal-5'-phosphate (PLP)-dependent enzymes, including ACC synthases (Rando, 1974). Vinylglycine analogues, which include rhizobitoxine and vinylglycine, are structurally similar to SAM (Satoh & Yang, 1989). Aminoethoxyvinylglycine (AVG), however, was shown to be the most effective compound in both in vivo and in vitro assays (Amrhein & Wenker, 1979). The chemical proved essential in elucidating the ethylene biosynthesis mechanism (Adams & Yang, 1979; Yu et al., 1979). The covalent AVG-PLP ketimine adduct, which most likely prevents SAM from interacting with the catalytic site, was discovered after the isolation of recombinant apple ACS and further crystallization of the protein in association with AVG. Further biochemical analysis revealed a reduction in ACS activity in the presence of AVG, which was only countered by enhancing the amount of the PLP cofactor. Experiments using AVG verified the existence of a potential ACC transport pathway in waterlogged tomato (Depaepe & van der Straeten, 2020).

Researchers over the years have shown that AVG effectively delays ripening, enhances fruit quality and minimizes early fruit drop in climacteric fruit (Yuan & Carbaugh, 2007). Since 1997, AVG is developed and commercially accessible for horticultural uses as spray formulation ReTain that includes 15% AVG (Valent BioSciences, USA). It is used preharvest in the United States to slow down fruit maturity and ripening, enhance shelf life and avoid premature fruit drop in apples, pears, peaches and nut species (Byers et al., 2005). ReTain is sprayed on pears to increase fruit firmness. It is also used to decrease the abortion of female flowers in certain nuts. According to D'Aquino et al. (2010), the preharvest application of AVG to pears two weeks prior to harvest resulted in increased firmness, decreased respiration rate, ethylene production and suppressed internal browning under cold storage. Similarly, preharvest AVG application to Cavendish Banana on the 78th day of flowering reduced ACC, ACC oxidase, and ethylene production, resulting in a 42-day storage life (Van Toan & Thanh, 2011).

### 7.4.2　2-Aminooxyacetic Acid (AOA)

Hydroxylamine analogues, most prominently 2-aminooxyacetic acid (AOA), are a second class of compounds that compete with SAM to interact with ACS to decrease the reservoir of ACC. AOA. Its analogues suppressed ethylene biosynthesis in mung bean hypocotyls and apple, possibly due to decreased ACS activity (Amrhein & Wenker, 1979). AOA may potentially operate as a competitive inhibitor of ACS via irreversible interaction with the PLP cofactor (Yu et al., 1979).

AVG, AOA and their analogues all have inhibitory effects that are not confined to just ACS isoforms. These antagonists also inhibit other PLP-dependent enzymes, such as those involved in biosynthesis of auxin and the metabolism of nitrogen (Leblanc et al., 2008).

### 7.4.3　Inhibition of ACC Oxidases

Early research found that lipophilic and water-soluble metal chelating compounds such as 1,10-phenanthroline and ethylenediaminetetraacetic acid can block ethylene generation in apple and etiolated pea seedlings (Mattoo et al., 1979). This lent credence to the notion that a metalloenzyme is involved in the production of ethylene from ACC. ACOs are Fe(II)-containing enzymes that belong to the 2-oxoglutarate (2OxoGA)-dependent dioxygenase superfamily. Other transition metals that inhibit 2-OxoGA-dependent dioxygenases include cobalt ($Co^{2+}$), nickel ($Ni^{2+}$) and zinc ($Zn^{2+}$) (Sekirnik et al., 2010). Similarly, the presence of transition metals can prevent ACO-mediated conversion of ACC to ethylene.

### 7.4.3.1   Cobalt Ions (Co2+)

$Co^{2+}$ efficiently blocked methionine-to-ethylene conversion in mung bean and apple tissue, causing a reduction in ethylene levels in response to high auxin levels, which are reported to promote ethylene synthesis (Lau et al., 1976). More precisely, the magnitude of suppression of ethylene production in ACC-treated mung bean hypocotyl segments was comparable to that in segments treated with the indole-3-acetic acid (IAA), denoting that $Co^{2+}$ influenced the critical stage in the biosynthetic pathway, notably the conversion of ACC to ethylene (Yu et al., 1979). Although $Co^{2+}$ may be efficiently supplied as $CoCl_2$ in liquid solutions to delay fruit ripening, its use in agricultural techniques is restricted to minimizing heavy metal toxicity. Furthermore, $Co^{2+}$ has been shown to impact other processes like photosynthesis and to elicit oxidative stress (Depaepe & van der Straeten, 2020).

### 7.4.3.2   2,4-pyridinedicarboxylic Acid (PDCA)

A structural analogue of 2-oxoglutarate known as 2,4-pyridinedicarboxylic acid (PDCA) has been found to inhibit 2-OxoGA-dependent dioxygenases by interfering with 2-oxoglutarate and to inhibit ethylene biosynthesis in carnation flowers by interfering with ascorbate on ACO activity (Satoh et al., 2014). Vlad et al. (2010) reported that PDCA suppressed ethylene synthesis in carnation flowers and retarded flower senescence. Fragkostefanakis et al. (2013) demonstrated that PDCA decreased the in vitro activity of ACO produced from tomatoes as PDCA inhibits ACO by competing with ascorbate a co-substrate of the enzyme action. Satoh et al. (2014) established that PDCA suppressed ACO activity using a recombinant enzyme generated in *E. coli* from the carnation gene (*DcACO1 cDNA*).

### 7.4.3.3   α-aminoisobutyric Acid (AIB) and 2-aminooxyisobutyric Acid (AOIB)

The hunt is on for ACC structural analogues that function as alternative substrates and ACO competitive inhibitors. There are two categories of 2-alkyl analogues: those with precise stereochemistry that are transformed to 1-alkenes by ACS (Hoffman et al., 1982) and those that do not, such as α-aminoisobutyric acid (AIB) or 2-aminooxyisobutyric acid (AOIB) (Pirrung et al., 1998). Since 2-alkyl analogues are transformed to 1-alkenes, like propylene, which itself has little ethylene activity, AIB is the most often employed ACO inhibitor (Abeles et al., 1992; Mcmurchie et al., 1972). In vivo ethylene assays in cocklebur cotyledon tissue, however, revealed that $10 \times 10^{-3}$ m AIB decreased ethylene production to just 60%, showing that AIB is a weaker inhibitor of ethylene biosynthesis than those previously described (Satoh & Esashi, 1980). In a heterologous system, AOIB lowered the enzyme activity of both ACS and ACO, while reduced ethylene levels in carnations were thought to be mediated by its effects on ACS exclusively (Kosugi et al., 2014). This dual inhibitory activity is most likely owing to AOA and AIB structural similarities. So far, no ACO inhibitors have been approved for commercial usage.

### 7.4.4 ETHYLENE ACTION INHIBITORS

This is the first and most important point in postharvest handling and avoidance of ethylene action. The load of ethylene can be avoided by removal, oxidation and absorption of surrounding ethylene from the sensitive produce. It is well documented that by providing sufficient ventilation and space, the detrimental effects can be reduced. The use of activated carbon, membrane filtration, silver ions, ethylene scavengers, $KMnO_4$, etc. have been found to be effective. Under CA and MA storage, it is very difficult to provide space and ventilation as they are very precise, but the internal atmosphere can be modified and sensor-based technologies can be applied.

#### 7.4.4.1   Silver Ions (Ag+)

Plants are not protected from exogenously applied hormones when biosynthesis is blocked. As a result, techniques that affect the ethylene perception pathway or signal transduction have greater promise in research (Depaepe & van der Straeten, 2020). Silver ($Ag^+$) ions and strained alkenes are the two most common inhibitors identified and used to date. The capacity of $Ag^+$ to suppress ethylene responses as a pharmacological agent was identified in the 1970s. The early study showed that using silver nitrate ($AgNO_3$) or silver thiosulfate (STS) as a source of $Ag^+$ ions efficiently inhibited numerous ethylene actions, including the triple response in pea and arabidopsis, leaf and fruit abscission in cotton and blossom senescence in cattleya orchids (Beyer et al., 1976; Bleecker et al., 1988). Heterologous expression of ETR1 in yeast showed increased ethylene binding after $Ag^+$ treatment, showing that $Ag^+$ ions might replace the copper ion cofactor in the receptor (Rodríguez et al., 1999). However, $Ag^+$ had no effect on ethylene receptor affinity, which is stable with its suggested activity as a monopolistic inhibitor (McDaniel & Binder, 2012). The reduced ethylene response after $Ag^+$ treatment is suggested to be related to a decline in the accessibility of ethylene binding sites. Furthermore, the inhibitory action of $Ag^+$ on ethylene appears to be predominantly mediated by ETR1. Silver ions were also employed to inhibit ethylene synthesis and delay fruit ripening in apple, banana and tomato (Atta-Aly et al., 1987; Saltveit et al., 1978). But as a heavy metal, like cobalt, silver is barred from use in any fruit-and-vegetable-related applications (Sisler & Serek, 1997). Contrarily, Ag+ has been approved to prolong the vase life of cut flowers by delaying early abscission of petals induced by ethylene, provided that disposal is tightly regulated.

#### 7.4.4.2   2,5-norbornadiene (NBD)

Strained alkenes are structural analogues of ethylene that also behave as ethylene perception competitive inhibitors (Sisler et al., 1990). Ethylene functional analogues such as acetylene and propene can generate ethylene responses at much greater concentrations than ethylene. Several alkenes, on the other hand, have been found to have antiethylene actions at low concentrations while activating ethylene responses at extremely high dosages. 2,5-Norbornadiene (NBD) was the first alkene found in 1979 by Sisler with the capacity to completely prevent ethylene-induced reactions, such as the extension of carnation flower vase life and citrus leaf explants abscission (Sisler et al., 1985). Furthermore, NBD was shown to inhibit ethylene-stimulated responses to submergence in a deep-water rice variety (Bleecker et al., 1987). High quantities of NBD, on the other hand, stimulated gibberellic acid-dependent stem elongation following submergence, indicating that NBD displays weak ethylene action.

### 7.4.4.3  Trans-cyclooctene (TCO)

Sisler et al. (1990) reported that trans-cyclooctene (TCO), the simplest strained cyclic alkene, exhibited antiethylene effects in vivo in mung bean sprouts and tobacco in a competitive manner but is at least 50 times more powerful than NBD. Nonetheless, both NBD and TCO need continual exposure to block ethylene reactions. However, at standard temperature and pressure circumstances (1 atm and 20°C) both compounds are volatile, produce highly unpleasant odors and are suspected to have hazardous qualities and cancer-causing potential (Sisler & Serek, 1999). As a result, none of these compounds is often used in research, nor do they have any economic value.

### 7.4.4.4  Diazocyclopentadiene (DACP)

Sisler and Blankenship (1993) discovered diazocyclopentadiene (DACP) while looking for cyclic ligands with smaller ring sizes capable of binding ethylene receptors and efficiently blocking subsequent reactions. In mung bean and tobacco systems, DACP blocked ethylene binding at extremely low concentrations. It was also investigated to inhibit ethylene tissue sensitivity in apples and tomatoes (Blankenship & Sisler, 1993; Sisler & Lallu, 1994). Surprisingly, in darkness, ethylene binding to its receptors was regained after removing DACP; however, its light ethylene responsiveness was forever lost. When exposed to fluorescent light, a photolysis derivative of DACP is believed to bind irreversibly to the receptor, effectively preventing the ethylene response. Targeted plants acquire ethylene sensitivity only after de novo production of ethylene receptors. The specific photolysis derivative of DACP has yet to be identified, but they could contain cyclopropenes. It is a potential ethylene antagonist that, unlike NBD and TCO, can irreversibly block ethylene responses with a single white light pulse. However, at greater concentrations, DACP is unstable and highly explosive, making it similarly inappropriate for commercial use (Serek et al., 2006).

### 7.4.5  1-Methylcyclopropene (1-MCP)

This effort led to the discovery of a new class of highly strained alkenes with antiethylene properties, like cyclopropane (CP), 3,3-dimethylcyclopropene (3,3-DMCP), 1-methylcyclopropene (1-MCP) and their analogues during late 1980s by Sisler and Blankenship. Later, 1-MCP was patented by Sisler and Blankenship in 1996 (Blankenship & Dole, 2003). CP and 1-MCP are approximately a thousand times more effective than 3-DMCP. Furthermore, 1-MCP is much more stable than CP, hence most research on fruit preservation has been done using 1-MCP (Sisler & Serek, 1999). 1-MCP interferes with ethylene by permanently binding with receptors (ERR1, ETR2, EIN4, ERS1 and ETR3), which keeps CTR1 protein in its active (inhibiting) state and suppresses the expression of numerous transcription factors (ERF4, ERF6, ERF10 and ERF14) with a tenfold higher affinity, thereby inhibiting ethylene activity (Serek et al., 1994; Sisler & Serek, 1997). Adherence of 1-MCP to the receptor involves the removal of electrons from 1-MCP's orbital. Next, a copper ion ligand is rearranged in the receptor for high-affinity binding. The ligand in the trans region is either replaced or discharged, which is accompanied by further reconfiguration and binding (Sisler & Serek, 1997). As long as 1-MCP remains attached to the receptor, the plant will remain insensitive to ethylene (Sisler, 2006). Indeed, 1-MCP is so powerful at inhibiting climacteric fruit ripening that in some circumstances, such as banana, tomatoes and guava fruit, treatment with 1-MCP may be quite problematic, as it may irrevocably block the ripening process (Bassetto et al., 2005; Golding et al., 1998; Hurr et al., 2005). 1-MCP helps in extending shelf life and maintaining the postharvest quality of climacteric and non-climacteric fruits (Serek et al., 1995). Additionally, to the well-known and anticipated effects of 1-MCP in inhibiting the ripening of climacteric fruits, where ethylene plays a crucial role in regulating the ripening process, it was observed that treatment to 1-MCP can alter some ripening-related processes, such as color change, respiration, softening, and the development of physiological abnormalities and decay in non-climacteric fruits (Huber, 2008). Due to its low toxicity, high efficacy, relatively short exposure duration and long-term effects, 1-MCP has emerged as the antagonist of choice for studying ethylene reactions since its discovery (Blankenship & Dole, 2003). Due to its volatility at 1 atm pressure and 20°C temperature, 1-MCP is highly effective at extending the shelf life of horticultural produce in closed conditions, but the potential for additional uses seems to be limited. Daly and Kourelis (2001) developed a powdered α-cyclodextrin-based formulation to aid transportation, handling and discharge when 1-MCP comes in contact with water (Tze et al., 2007). 1-MCP is commercially registered and sold as SmartFresh® (AgroFresh) and AnsiP-G® to delay ripening in fruits and vegetables or as EthylBloc® (Floralife) to enhance ornamental flower vase life (Serek et al., 2006). As of now, 1-MCP has been registered for use on at least 18 different horticultural crops, which are all climacteric fruits. It has also been used commercially in dozens of nations throughout the world. By aggressively binding to ethylene receptors, two structural analogues of 1-MCP, 1-pentylcyclopropene (1-PentCP) and 1-octylcyclopropene (1-OCP) (Wang et al., 2015), can also effectively block ethylene action of respiratory climacteric fruits (Xu et al., 2019; Zhang et al., 2017).

### 7.4.5.1 Other Novel Chemicals

Recently many other compounds are identified as having ethylene antagonistic effects such as 3-cyclopropyl-1-enyl-propanoic acid sodium salt (CPAS), a novel water-soluble ethylene inhibitor reported by Goren et al. (2011). Because CPAS is a solid, water-soluble, non-phytotoxic and odorless ethylene inhibitor, it is a good choice for pre- and postharvest treatment in a variety of horticultural crops. Sun et al. in 2016 discovered a clinical drug, Pyrazinamide (PZA) and its derivatives (pyrazinecarboxylic acid) show anti-ethylene activity by suppressing ACO activity (Sun et al., 2017). Singh et al. (2018) discovered two stable compounds, namely 1H-cyclopropabenzene (BC) and 1H-cyclopropa[b]naphthalene (NC), which are distinct from 1-MCP but have the same ability and mode of action to inhibit ethylene action in plants. The advantage is that it can be applied to fruits both in aqueous form (dip) and via fumigation. Recently, Tokala et al. (2020, 2021) reported the efficacy of BC and NC as a fumigation treatment in inhibiting postharvest ethylene biosynthesis in Cripps Pink apple.

### 7.4.5.2   Use of Hormones

The regulation of ripening and senescence events by ethylene is the result of a relationship with a wide range of factors, one of which is plant growth regulators. The notion is that ethylene is alone responsible for ripening- and senescence-related action, but it occurs in the framework of a perfectly balanced interaction with other PGRs, which has been developed since ethylene's involvement in these operations was proposed. This hypothesis has started gaining biochemical, transcriptomic, proteomic and biotechnological assistance over time (Iqbal et al., 2017). In general, salicylic acid (SA), gibberellins (GAs), polyamines (PAs) and nitric oxide (NO) play an antagonistic role, whereas auxins and cytokinins (CKs) can have antagonistic or synergistic effects, depending on the developmental stage of the fruit or vegetable.

- **Auxin.** Auxins play an important role in fruit development, hampering ethylene synthesis and fruit ripening (Brady, 1987; Iqbal et al., 2017). Exogenous application of auxins (NAA or 2,4D) on immature fruits generally results in a delay in fruit ripening and senescence (e.g., in strawberries, grapes, banana, tomato, pear and peaches). Auxins, on the other hand, can enhance ripening in certain fruits when applied after initiation of ripening, mostly in climacteric fruits (peach, nectarine and apple), by stimulating the transcription of genes that encode ACS and, consequently, ethylene biosynthesis. For example, when applied at the preclimacteric stage, auxin reduced ethylene production in sliced apples while increasing its synthesis at the climacteric stage (Lieberman et al., 1977). Auxin response factor 2A (ARF2A), a transcription factor, has been identified as an auxin-signaling factor capable of controlling ripening (Breitel et al., 2016). Fruit ripening is closely linked to *SlIAA3*, *SlIAA4*, *SlIAA9*, *SlIAA15* and *SlIAA27* genes (Cruz et al., 2018). According to studies, ARF2, ARF4, ARF5, IAA3 and IAA27 are critical components in the interplay between auxin and ethylene. *SlARF2* downregulated the key ripening regulators like ripening inhibitor (RIN), colorless non-ripening (CNR), and non-ripening (NOR), as well as the expression of ETR and ERF genes and ethylene biosynthesis (Hao et al., 2015). Ethylene during tomato ripening induces the expression of *SlIAA3*, which positively regulates ERF gene expression (Li et al., 2017). Contradictorily, Bleecker and Kende (2000) discovered that auxins can prompt the biosynthesis of ethylene in climacteric fruit by inducing the expression of ACS (Abel & Theologis, 1996). Preharvest NAA and 1-MCP application delayed fruit ripening due to decreased ethylene biosynthesis (Acuna et al., 2010). Bananas treated with IAA produced less ethylene and ripened later. PpACS1a transcription can be induced by IAA treatment of pears (Yue et al., 2020). Tomatoes and other fleshy fruits also showed delayed ripening after auxin treatment (Kou et al., 2021).
- **Cytokinins (CK).** Cytokinin can also be said to play an ethylene-counteractive role, identical to that played by auxins. Exogenous applications of benzyl adenine or forchlorfenuron in both climacteric and non-climacteric fruits, as well as broccoli florets and leafy vegetables, inhibit ethylene-dependent chlorophyll degradation, resulting in a delay in postharvest de-greening,

loss of firmness and overall decay via downregulation of chlorophyllases. In terms of auxins, cytokinin application on immature fruits reduces ethylene biosynthesis and responsiveness, as well as the climacteric rise in respiration, whereas application on mature fruits may have the inverse result (Kou et al., 2021). For example, in kiwifruit, preharvest treatments of CK (CPPU) at the bloom stage increased starch accumulation, which was followed by rapid ripening. Broccoli florets treated with CKs have lower climacteric respiration, a delayed overall yellowing process, as well as an extended shelf life, even though there are increasing levels of ethylene, implying that some effects mediated by plant growth regulators may be elicited through different responses on specific ripening-associated events. According to recent research, CPPU reduced ethylene synthesis and the softening of the central placenta during fruit ripening (Ainalidou et al., 2016).

- **Gibberellins (GA)**. It was proposed that GAs would suppress ethylene biosynthesis and expression in order to extend the lifespan of horticultural produce. According to the signaling network between GAs and ethylene, the exogenous application of GAs will suppress transcription of ethylene biosynthesis and oxidases ACS and ACO (Kashyap & Banu, 2019). As a consequence, GAs reduce ethylene output in plants, slow the ripening process and increase shelf life. Furthermore, GAs can inhibit the transcription of ethylene receptors, particularly ETR1 and ERS1, which are both stimulators of CTR1 (the negative regulator of ethylene response). The upregulation of CTR1 and the downregulation of ETR1 and ERS1 will have a significant impact on ethylene perception. As a result, the use of GAs will inhibit ethylene biosynthesis and expression. GA has been shown to be a negative regulator of tomato ripening genes, ethylene synthesis genes and eventual ripening process. Gene expression for ethylene biosynthesis and production were preactivated in gibberellins-deficient fruits but reduced in wild-type fruits treated with exogenous GA3. GA hindered ripening by modulating the transcription of the ethylene synthesis genes, namely *SlACS2*, *SlACS4*, *SlACS6* and *SlACO3* in tomato (Chen et al., 2020). Multiple studies on the effects of GA3 treatment on climacteric and non-climacteric fruits have largely shown a delay in ripening. Some of the most commonly reported effects in treated fruits are increased firmness and delayed color changes (Khader, 1991; Lara, 2013).

### 7.4.5.3 Polyamines (PA)

PAs and ethylene are two metabolites that have antagonistic roles. Pas, namely spermine, spermidine, and putrescine, have an effect on fruit ripening-related events. Pre- and postharvest PA treatment have been shown to downregulate ethylene biosynthesis and responses in a variety of systems, improving fruit firmness by repressing cell wall enzymes. According to Sobolev et al. (2014) and Pandey et al. (2015), PAs and ethylene regulate each other's biosynthesis, either directly or indirectly by competing for SAM, a common precursor for their biosynthesis. PAs application reduces the expression of S-adenosylmethionine decarboxylase

(SAMDC), ACS, and ACO, resulting in delayed fruit ripening (Torrigiani et al., 2004; Khan et al., 2007). Exogenous application of PAs particularly restricts ethylene biosynthesis by suppressing ACS expression, according to Li et al. (1992). The proportion of ethylene inhibition is negatively related to ethylene production at the climacteric peak. Thus the deviation of SAMDC via PA biosynthesis could explain the significant decrease in ethylene synthesis observed in PUT-treated fruit, as a smaller pool of SAMDC would be available to synthesize ACC and thus ethylene. Polyamines also impeded auxin-induced ethylene synthesis and methionine and/or ACC conversion to ethylene (Li et al., 2004). Pre- and postharvest PAs application inhibits ethylene biosynthesis in tomato, avocado and pear (Kakkar & Rai, 1993; Saftner & Baldi, 1990). Exogenous application of 1 mM PUT increased endogenous PUT concentrations and Spd biosynthesis, while ethylene biosynthesis was impeded in apricots (Martinez-Romero et al., 2002), peaches (Martinez-Romero et al., 2000) and plum (Serrano et al., 2003). Putrescine effectively delayed plum fruit ripening by inhibiting ACS production and buildup to decrease endogenous ethylene (Davarynejad et al., 2015).

### 7.4.5.4 Nitric Oxide (NO)

Nitric oxide, as a plant intrinsic signal regulator, is linked to ethylene biosynthesis inhibition in activities that delay fruit ripening and senescence (Mukherjee, 2019). Interestingly, NO does not directly inhibit ethylene, and the most likely mechanism proposed is that ACO can integrate with NO to form a binary ACO-NO complex, which is later chelated by ACC to create a stable ternary ACC-ACO-NO complex. This directly resulted in a decrease in ethylene biosynthesis (Tierney et al., 2005; Manjunatha et al., 2010). Besides, NO can directly react with ACO via S-nitrosylation (Abat & Deswal, 2009), causing a drop in ACO activity and, as a consequence, a reduction in overall ethylene production (Zhu et al., 2006; Manjunatha et al., 2010). To support this, NO application to tomato fruit can reduce the expression of ACO genes such as *LeACO1*, *LeACOH2* and *LeACO4* (Eum et al., 2009). Furthermore, 1 mM SNP treatment significantly downregulated the expression of *LeACS2* and *LeACO1* (Lai et al., 2011). Similarly, Zhu et al. (2006) found that NO had no effect on ACO activity but significantly lowered ACS activity in peach fruit. This suggests that NO's suppression of ethylene biosynthesis is primarily related to ACS post-transcriptional regulation. Furthermore, methionine adenosyl transferase (MAT1) controlled the biosynthesis of SAM, and it was noticed in *Arabidopsis* that NO can decrease MAT1 activity via post-translational S-nitrosylation regulation, resulting in a reduction in ethylene biosynthesis (Lindermayr et al., 2006). To summarize, NO can either directly or indirectly inhibit ethylene biosynthesis, resulting in delayed postharvest fruit ripening and senescence. Zaharah and Singh (2011) reported that exogenous NO treatment (20 µL L$^{-1}$) decreased ACC content and ACO and ACS enzyme activity. Similarly, a preharvest spray of 50 µM sodium nitroprusside applied to apples 14 days before commercial harvest reduced the activities of ACS and ACO enzymes, inhibiting the biosynthesis of ethylene (Deng et al., 2013). Apple, peach, strawberry and tomato are examples of fruits where NO inhibits ethylene synthesis (Rudell & Mattheis, 2006; Zhu et al., 2006; Zhu & Zhou, 2007; Eum et al., 2009).

### 7.4.5.5   Salicylic Acid (SA)

Salicylic acid and ethylene were discovered to have an antagonistic relationship (Li et al., 2019). Leslie and Romani (1986) were the first to disclose that SA inhibited the transformation of ACC to ethylene under stress. Similarly, SA and its derivative acetyl salicylic acid (ASA) have been shown to inhibit ethylene biosynthesis in pear, carrot cell suspension, apple and pear discs, mung bean hypocotyls and a variety of other horticultural crops (Babalar et al., 2007; Romani et al., 1989), implying a role for SA as an ethylene antagonist. According to Srivastava and Dwivedi (2000), SA has delayed the ripening of banana fruit, most likely by inhibiting ethylene synthesis and/or action by reducing ACS2 and ACO1 production and activity. Zhang et al. (2003) reported that postharvest ASA treatment on kiwifruit resulted in reduced ACO and ACS activity as well as reduced ethylene biosynthesis during the initial stages of fruit ripening. Furthermore, *Arabidopsis* showed increased production of the ethylene response factor (ERF) repressor gene, implying that SA reduces ethylene production (Caarls et al., 2017). Similarly, Khan et al. (2013) reported that the activity of ACS was reduced after the application of SA. Surprisingly, enhanced ACS expression in pears is observed after SA treatment in a dose-dependent manner (Shi et al., 2013; Shi & Zhang, 2014) Specifically, at low concentrations, SA has been found to stimulate ethylene synthesis, whereas higher concentrations of SA (10−4 M) inhibit ethylene synthesis. As a result, ethylene biosynthesis dependent on SA is precisely regulated (Nissen, 1994).

### 7.4.5.6   Melatonin (MT)

Melatonin, a plant hormone, also acts as an ethylene antagonist. MT does not directly inhibit ethylene biosynthesis, but it may inhibit ethylene biosynthesis by increasing NO content. Pear fruit treated with MT (0.1 mM) resulted in reduced ethylene biosynthesis, maintained firmness and downregulation of the relative expression of *PcACS1*, *PcACS2*, *PcACO1* and *PcACO2* genes (Zhai et al., 2018). In banana, MT treatments (0.05 mM) reduced ethylene biosynthesis and related genes such as *MaACO1* and *MaACS1* (Hu et al., 2017). Similarly, postharvest melatonin application may regulate mango fruit ripening and softening by inhibiting ethylene and ABA biosynthesis (Liu et al., 2020). In tomato, however, MT treatment (0.05 mM) enhanced ethylene biosynthesis while having no effect on the related genes *ACS2* and *ACO1* but ramping up the relative expression of the *ACS4* gene (Sun et al., 2015). According to the findings, MT may well function not as a negative regulator by directly inhibiting the transcription of individual ethylene biosynthetic genes but rather as a regulator influencing the autocatalytic ethylene balance (Zhai et al., 2018). Although the mode of MT regulation on the ethylene synthesis of fruit is debatable, the beneficial impact of MT application on postharvest fruit ripening and senescence has been demonstrated in most studies.

### 7.4.6   Ethylene Scavengers

Ethylene scavengers are commonly used to retard physiological and biochemical changes of fresh produce during transportation and storage (Lee et al., 2015). An inexpensive method of removing ethylene from the air is using chemicals on a matrix

(Knee & Hatfield, 1981). This matrix increases the surface area available to chemicals and makes it as available as possible for ethylene removal. Ethylene scavengers are used to preserve fruits and vegetables by adsorption (zeolites, activated carbon) and catalytic ethylene oxidation ($KMnO_4$, ozone, $TiO_2$, etc.). Other types of ethylene absorbents include celite, alumina, vermiculite and $KMnO_4$ impregnated clay. Ethylene scavengers can be classified into three categories based on their mechanism of action.

1. **Catalysts**. These are often based on platinum/alumina and operate at high temperatures (>200 °C) to catalytically oxidize ethylene to carbon dioxide ($CO_2$) and water (Smith et al., 2009).
2. **Stoichiometric oxidants**. Based primarily on potassium permanganate ($KMnO_4$), these oxidize ethylene and are themselves reduced.
3. **Adsorbents**. These materials are used alone or with an oxidant. They are often based on high surface area of materials such as zeolites, activated carbons, clays, nanospheres, porous metal oxides, silica gels, halloysite nanotubes and montmorillonite (Smith et al., 2009; Prasad & Kochhar, 2014).

Ethylene scavenging substances are commonly used in the form of packaged sachets, as coatings on packaging materials or in the form of ethylene-scavenging active films (Tas et al., 2017). Ethylene is absorbed or adsorbed by various substances, including activated carbon, crystalline aluminosilicate molecular sieves, diatomaceous earth, bentonite, bleaching earth, brick dust, silica gel (Kays & Beaudry, 1987) and alumina (Goodburn & Halligan, 1988).

### 7.4.6.1   Zeolites

Zeolites are widely used for their unique structure and excellent performance (Wuttke et al., 2018). Among various porous materials, zeolites are most commonly used due to their large voids or cages for trapping or containing ethylene during adsorption (Ichiura et al., 2003). The superior performance of zeolite-based materials is attributed to their porous three-dimensional structure with cation exchange, adsorption and molecular sieve properties (Yildirim et al., 2018).

### 7.4.6.2   Activated or Brominated Charcoal

Charcoal air purifiers can absorb ethylene from the air, especially if they are brominated (Dhall, 2013). Their use is limited to laboratory conditions. A sachet containing activated charcoal, silica gel and a metal catalyst named Neupalon absorbs 500–1000 times its weight of water and 40 ml m$^{-2}$ of ethylene from the surface of the package (Brody et al., 2001; Coles et al., 2003).

### 7.4.6.3   Halloysite Nanotubes

Halloysite nanotubes (HNTs) are used as an alternative to traditional ethylene scavengers in applications where cost is a concern. HNTs are natural aluminosilicate nanoparticles characterized by hollow tubular nanostructures (Tas et al., 2017). The "green" material mined from natural deposits (Bodbodak & Rafiee, 2016) is environmentally friendly and recognized as safe by the U.S. Food and Drug Administration for food packaging (Lee et al., 2017). Their low price and high aspect ratio facilitate efficient dispersion within polymer matrices to form halloysite nanotube-polymer nanocomposites (Yuan et al., 2015; Gaikwad et al., 2018). The polymer nanocomposites thus formed adsorb ethylene released inside the food package.

### 7.4.6.4    Potassium Permanganate (KMnO₄)

Ethylene oxidation based on $KMnO_4$ is the most commonly used of all ethylene scavenging systems. The ability of $KMnO_4$ to reduce atmospheric ethylene concentrations around fresh crops was first demonstrated by Forsyth et al. (1967) in apples. It is a stable purple solid and a strong oxidant, readily oxidizing ethylene to carbon dioxide and water (Wills & Warton, 2004).

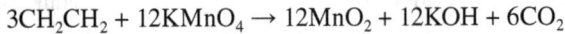

$$3CH_2CH_2 + 12KMnO_4 \rightarrow 12MnO_2 + 12KOH + 6CO_2$$

When $KMnO_4$ oxidizes ethylene, its color changes from purple to brown. The color change indicates residual ethylene absorption capacity. $KMnO_4$ is commonly embedded in various porous materials to improve cleaning ability. This includes pumice stone, brick, celite (Forsyth et al., 1967), vermiculite (Scott et al., 1970), alumina (Wills & Warton, 2004), clay (Picón et al., 1993) and zeolites (Oh et al., 1996).

Products containing $KMnO_4$ are toxic and should not come into direct contact with food (Prasad & Kochhar, 2014). Therefore, some commercially available $KMnO_4$ scavengers are found in packages or warehouse-located bags, filters, blankets and other special scavenger devices (Janjarasskul & Suppakul, 2018). Commercial products based on $KMnO_4$ and alumina beads are Purafil, Ethysorb, Circul-Aire and Bloomfresh. Abeles et al. (1992) stated that $KMnO_4$ content of these commercial products was typically 4–6 g 100 g⁻¹. One of the challenges associated with $KMnO_4$-based $C_2H_4$ scavengers is that they become less efficient after saturation over time, requiring frequent replacement, especially in high ethylene producing commodities (Pathak, 2019).

### 7.4.6.5    Ultraviolet Lamps (Ozone)

Currently available commercial devices use UV lamps in storage rooms. These lamps produce ozone, which is a strong oxidant and destroys ethylene gas (Kim et al., 2019). Ozone generators are very useful where ethylene-producing and ethylene-sensitive fruits and vegetables can be kept in the same room. The decomposition of ethylene by ozone is $2O_3 + C_2H_4 \rightarrow 2CO_2 + 2H_2O$.

### 7.4.6.6    Titanium Dioxide (TiO₂)

When exposed to UV radiation, the conventional semiconductor material $TiO_2$ exhibits antibacterial and ethylene photodegradation activity by producing reactive oxygen species (ROS) and hydroxyl radicals (OH•) on its surface, which then interact with organic molecules. $TiO_2$ can be used for the photocatalytic oxidation of ethylene at room temperature (Smith et al., 2009). This is mostly owing to its distinct photochemical reactivity and physical characteristics, which include ultrahigh brightness (caused by a high refractive index) and stain resistance (Weir et al., 2012).

According to Kim et al. (2019), the photocatalytic oxidation of ethylene scavenger coupled with zeolite adsorbents can be regarded as a beneficial approach for extending the shelf life of fruits. Low efficiency and the readiness of catalyst deactivation are the main drawbacks of photocatalytic devices. In an effort to increase photocatalytic efficiency, it is crucial to effectively utilize photons and minimize electron-hole recombination (Tytgat et al., 2012; Pathak et al., 2017).

### 7.4.6.7  Palladium

Materials based on palladium mostly function as adsorbers rather than as catalysts (Smith et al., 2009). Pd-based material has the capability to reduce ethylene under physiologically active levels at around 0.01 g $L^{-1}$. According to Terry et al. (2007), Pd-promoted powdered materials beat $KMnO_4$-based ethylene scavengers as they have an almost sixfold higher adsorption capacity of around 100% RH. Under low RH, this efficiency was raised to a level that was over 60 times greater than $KMnO_4$. To prevent the yellowing and loss of quality of broccoli, Cao et al. (2015) created a novel ethylene scavenger of the palladium chloride ($PdCl_2$) impregnated in acidified activated carbon powder. With the addition of copper sulfate, this scavenger's capacity to remove ethylene was considerably boosted ($CuSO_4$).

### 7.4.7  CATALYTIC OXIDIZERS

Ethylene is oxidized when it combines with oxygen at high temperature in the presence of a catalyst (such as platinum asbestos). Ethylene scrubbers that take advantage of this effect are now commercially available and overcome the difficulty of heating the incoming air by the clever use of a ceramic bed used as a heat sink and a reversible gas flow through the bed. These scrubbers are highly efficient, reducing the ethylene concentration in the air to 1% of the inlet concentration. It handles less air volume, making it ideal for tight spaces and long-term CA storage systems.

Kim et al. (2019) created nanocomposites with visible-light active platinum (Pt)-loaded monoclinic WO3 nanorods on ZSM-5 (ZSM-5/WO3-Pt), which have shown a synergistic impact for ethylene elimination. Jiang et al. (2013) reported a complete oxidation of ethylene using a 1% silica-based Pt catalyst with significant catalytic activity, in which trace amounts of ethylene were converted over 99.8% at 0°C. According to Bailén et al. (2006), activated carbon, enriched with 1% palladium (Pd), preserved tomato quality better than activated carbon used alone (Pathak et al., 2017; Pathak, 2019). Additionally, Pd-impregnated zeolite outperformed untreated zeolite in ethylene removal, prolonging shelf life while preserving color.

Other solid catalysts that have been extensively researched for the oxidation of ethylene include Ag/zeolites (Cisneros et al., 2019), Au/CO3O4 (Xue et al., 2011) and Ag/ZnO (Zhu et al., 2018). However, these ethylene capture materials are typically utilized in the form of sachets inserted within packaging, which subsequently seems to have limited popularity for end users due to health concerns and laws prohibiting their food contact uses. Scavengers being incorporated into packing films has proven to be a suitable solution to this issue.

### 7.4.8   Film-Based Packaging Containing Ethylene Scavenger

Scavengers included in packaging film may hold the key to solving sachet-related issues. The films can incorporate minerals such as pumice, zeolite, clays or nano-clays, or even Japanese oya stone. The minerals placed in the films can create pores within the bag by the two-dimensional surface adsorption of ethylene or the three-dimensional absorption of ethylene. This increases their gas permeability, which will lead to significantly faster ethylene diffusion.

According to Zhao et al. (2011), azeolite molecular sieve (ZMS) can be used as an ethylene-absorbing ingredient in packing films. Zeolites' crystalline porous, three-dimensional framework structure has been demonstrated in numerous papers to increase the gas permeability of packaging films when they are included in them (Yildirim et al., 2018; Zhao et al., 2011). Khosravi et al. (2013) looked at low-density polyethylene (LDPE) nanocomposites that were mixed with various concentrations of nano-$KMnO_4$ for usage in the fruit sector. The concentration of nano-$KMnO_4$ was enhanced to increase the nanocomposites' ability to absorb ethylene.

### 7.4.9   Packaging Material as Ethylene Adsorbents

Ethylene scavengers could be inserted into different layers of the container or implanted into a solid or disseminated in plastic (Ozdemir & Floros, 2004). Packaging materials, notably LDPE and HDPE films, are very effective at absorbing ethylene, ethanol, ammonia, hydrogen sulfide and ethyl acetate in the food industry. According to Scott et al. (1970), fruit storage in sealed PE packaging with an ethylene absorber at room temperature has been found to be just as effective as refrigeration. Various substances, such as different types of nanofillers (like nanoclays), have been introduced into the polymer matrix to overcome the low barrier characteristics of LDPE. As a result, there was notable reduction in permeability and the development of a tortuous channel for a permeant gas dispersing the nanocomposite (Cerisuelo et al., 2015). Due to a significant improvement in the barrier qualities of food packaging, LDPE/clay nanocomposites have attracted a lot of interest (Choudalakis & Gotsis, 2009; Siročić et al., 2014).

### 7.4.10   Genetic Strategies

The role of ethylene and its biosynthesis, mechanism and regulation in fruit physiology is well documented and established at the genetic level. It is believed that genetic control of ethylene perception and regulation is more effective (Scariot et al., 2014). However, it is more pronounced in flowers than in fruits and vegetables. It is reported by Savin et al. (1995) that Florigene succeeded in delaying carnation senescence by antisense downregulation of ACO enzyme. The activity of ACS and ACO is down-regulated, which leads to slower ripening, and shelf life can be extended. Recently, the CRISPR editing tool came into existence to modify the genetic constitution of the produce that alters the genes involved in ethylene biosynthesis. The example of FLAVR SAVR tomato is well-known in which the antisense gene was incorporated to downregulate the activity of polygalacturonase (a fruit softening enzyme).

## 7.5 CONCLUSION

A range of chemicals, hormones and oxidizing agents are currently in use to reduce the negative impacts of ethylene. However, none of single applications currently meets the need and requirement of the horticultural industry. the use of certain nanoparticles, 1-MCP and genetic modification could be effective technologies for the future. There is need to develop other safe formulations that can reduce the detrimental effects throughout the supply chain without impairing quality compositions. Certain eco-friendly approaches need to be harnessed for sustainable solutions. However, their efficacy along with cost-effectiveness need to be evaluated.

## REFERENCES LIST

Abat, J. K., & Deswal, R. (2009). Differential modulation of S-nitrosoproteome of Brassica juncea by low temperature: Change in S-nitrosylation of Rubisco is responsible for the inactivation of its carboxylase activity. *Proteomics*, *9*(18), 4368–4380. https://doi.org/10.1002/pmic.200800985

Abel, S., & Theologis, A. (1996). Early genes and auxin action. *Plant Physiology*, *111*(1), 9–17. https://doi.org/10.1104/pp.111.1.9

Abeles, F. B., Morgan, P. W., & Saltveit, M. E. (1992). *Ethylene in plant biology* (2nd ed). Academic Press. https://doi.org/10.1016/C2009-0-03226-7

Acuña, M. G. V., Biasi, W. V., Flores, S., Mitcham, E. J., Elkins, R. B., & Willits, N. H. (2010). Preharvest application of 1-methylcyclopropene influences fruit drop and storage potential of 'bartlett' pears. *HortScience*, *45*(4), 610–616. https://doi.org/10.21273/HORTSCI.45.4.610

Adams, D. O., & Yang, S. F. (1979). Ethylene biosynthesis: Identification of I-aminocyclopropane-1-carboxylic acid as an intermediate in the conversion of methionine to ethylene. *Proceedings of the National Academy of Sciences of the United States of America*, *76*(1), 170–174. https://doi.org/10.1073/pnas.76.1.170

Ainalidou, A., Tanou, G., Belghazi, M., Samiotaki, M., Diamantidis, G., Molassiotis, A., & Karamanoli, K. (2016). Integrated analysis of metabolites and proteins reveal aspects of the tissue-specific function of synthetic cytokinin in kiwifruit development and ripening. *Journal of Proteomics*, *143*, 318–333. https://doi.org/10.1016/j.jprot.2016.02.013

Amrhein, N., & Wenker, D. (1979). Novel inhibitors of ethylene production in higher plants. *Plant and Cell Physiology*, *20*(8), 1635–1642. https://doi.org/10.1093/oxfordjournals.pcp.a075966

Arc, E., Sechet, J., Corbineau, F., Rajjou, L., & Marion-Poll, A. (2013). ABA crosstalk with ethylene and nitric oxide in seed dormancy and germination. *Frontiers in Plant Science*, *4*, 63. https://doi.org/10.3389/fpls.2013.00063

Atta-Aly, M. A., Saltveit, M. E., & Hobson, G. E. (1987). Effect of silver ions on ethylene biosynthesis by tomato fruit tissue. *Plant Physiology*, *83*(1), 44–48. https://doi.org/10.1104/PP.83.1.44

Babalar, M., Asghari, M., Talaei, A., & Khosroshahi, A. (2007). Effect of pre- and post-harvest salicylic acid treatment on ethylene production, fungal decay and overall quality of Selva strawberry fruit. *Food Chemistry*, *105*(2), 449–453. https://doi.org/10.1016/j.foodchem.2007.03.021

Bailén, G., Guillén, F., Castillo, S., Serrano, M., Valero, D., & Martínez-Romero, D. (2006). Use of activated carbon inside modified atmosphere packages to maintain tomato fruit quality during cold storage. *Journal of Agricultural and Food Chemistry*, *54*(6), 2229–2235. https://doi.org/10.1021/JF0528761

Bassetto, E., Jacomino, A. P., Pinheiro, A. L., & Kluge, R. A. (2005). Delay of ripening of "Pedro Sato" guava with 1-methylcyclopropene. *Postharvest Biology and Technology*, *35*(3), 303–308. https://doi.org/10.1016/J.POSTHARVBIO.2004.08.003

Beyer, E. M., Johnson, A. L., & Sweetser, P. B. (1976). A new class of synthetic auxin transport inhibitors. *Plant Physiology, 57*(6), 839–841. https://doi.org/10.1104/PP.57.6.839

Binder, B. M. (2020). Ethylene signaling in plants. *Journal of Biological Chemistry, 295*(22), 7710–7725. https://doi.org/10.1074/jbc.REV120.010854

Blankenship, S. M., & Dole, J. M. (2003). 1-Methylcyclopropene: A review. *Postharvest Biology and Technology, 28*(1), 1–25. https://doi.org/10.1016/S0925-5214(02)00246-6

Blankenship, S. M., & Sisler, E. C. (1993). Response of apples to diazocyclopentadiene inhibition of ethylene binding. *Postharvest Biology and Technology, 3*(2), 95–101. https://doi.org/10.1016/0925-5214(93)90001-J

Bleecker, A. B., Estelle, M. A., Somerville, C., & Kende, H. (1988). Insensitivity to ethylene conferred by a dominant mutation in Arabidopsis thaliana. *Science, 241*(4869), 1086–1089. https://doi.org/10.1126/SCIENCE.241.4869.1086

Bleecker, A. B., & Kende, H. (2000). Ethylene: A gaseous signal molecule in plants. *Annual Review of Cell and Developmental Biology, 16*(1), 1–18. https://doi.org/10.1146/annurev.cellbio.16.1.1

Bleecker, A. B., Rose-John, S., & Kende, H. (1987). An evaluation of 2,5-norbornadiene as a reversible inhibitor of ethylene action in deepwater rice. *Plant Physiology, 84*(2), 395–398. https://doi.org/10.1104/PP.84.2.395

Bodbodak, S., & Rafiee, Z. (2016). Recent trends in active packaging in fruits and vegetables. In M. W. Siddiqui (Ed.), *Eco-friendly technology for postharvest produce quality* (pp. 77–125). Academic Press. https://doi.org/10.1016/B978-0-12-804313-4.00003-7

Brady, C. J. (1987). Fruit ripening. *Annual Review of Plant Physiology, 38*(1), 155–178. https://doi.org/10.1146/annurev.pp.38.060187.001103

Breitel, D. A., Chappell-Maor, L., Meir, S., Panizel, I., Puig, C. P., Hao, Y., Yifhar, T., Yasuor, H., Zouine, M., Bouzayen, M., Granell Richart, A., Rogachev, I., & Aharoni, A. (2016). AUXIN RESPONSE FACTOR 2 intersects hormonal signals in the regulation of tomato fruit ripening. *PLoS Genetics, 12*(3), e1005903. https://doi.org/10.1371/journal.pgen.1005903

Brody, A. L., Strupinsky, E. P., & Kline, L. R. (2001). *Active packaging for food applications.* CRC Press. https://doi.org/10.1201/9780367801311

Byers, R. E., Carbaugh, D. H., & Combs, L. D. (2005). Ethylene inhibitors delay fruit drop, maturity, and increase fruit size of "Arlet" apples. *HortScience, 40*(7), 2061–2065. https://doi.org/10.21273/HORTSCI.40.7.2061

Caarls, L., Van der Does, D., Hickman, R., Jansen, W., Verk, M. C. V., Proietti, S., Lorenzo, O., Solano, R., Pieterse, C. M., & Van Wees, S. C. (2017). Assessing the role of ethylene RESPONSE FACTOR transcriptional repressors in salicylic acid-mediated suppression of jasmonic acid-responsive genes. *Plant and Cell Physiology, 58*(2), 266–278. https://doi.org/10.1093/pcp/pcw187

Cao, J., Li, X., Wu, K., Jiang, W., & Qu, G. (2015). Preparation of a novel $PdC_{12}$–$CuSO_4$–based ethylene scavenger supported by acidified activated carbon powder and its effects on quality and ethylene metabolism of broccoli during shelf-life. *Postharvest Biology and Technology, 99*, 50–57. https://doi.org/10.1016/j.postharvbio.2014.07.017

Cerisuelo, J. P., Gavara, R., & Hernández-Muñoz, P. (2015). Diffusion modeling in polymer–clay nanocomposites for food packaging applications through finite element analysis of TEM images. *Journal of Membrane Science, 482*, 92–102. https://doi.org/10.1016/j.memsci.2015.02.031

Chen, S., Wang, X. J., Tan, G. F., Zhou, W. Q., & Wang, G. L. (2020). Gibberellin and the plant growth retardant paclobutrazol altered fruit shape and ripening in tomato. *Protoplasma, 257*(3), 853–861. https://doi.org/10.1007/s00709-019-01471-2

Choudalakis, G., & Gotsis, A. D. (2009). Permeability of polymer/clay nanocomposites: A review. *European Polymer Journal, 45*(4), 967–984. https://doi.org/10.1016/j.eurpolymj.2009.01.027

Cisneros, L., Gao, F., & Corma, A. (2019). Silver nanocluster in zeolites. Adsorption of ethylene traces for fruit preservation. *Microporous and Mesoporous Materials*, *283*, 25–30. https://doi.org/10.1016/j.micromeso.2019.03.032

Coles, R., McDowell, D., & Kirwan, M. J. (Eds.). (2003). *Food packaging technology*. https://doi.org/S138718111930174X

Cruz, A. B., Bianchetti, R. E., Alves, F. R. R., Purgatto, E., Peres, L. E. P., Rossi, M., & Freschi, L. (2018). Light, ethylene and auxin signaling interaction regulates carotenoid biosynthesis during tomato fruit ripening. *Frontiers in Plant Science*, *9*, 1370.

D'Aquino, S., Schirra, M., Molinu, M. G., Tedde, M., & Palma, A. (2010). Preharvest aminoethoxyvinylglycine treatments reduce internal browning and prolong the shelf-life of early ripening pears. *Scientia Horticulturae*, *125*(3), 353–360. https://doi.org/10.1016/J.SCIENTA.2010.04.020

Daly, J., & Kourelis, B. (2001). Synthesis methods, complexes and delivery methods for the safe and convenient storage, transport and application of compounds for inhibiting the ethylene response in plants. U.S. Patent 6313068.

Davarynejad, G. H., Zarei, M., Nasrabadi, M. E., & Ardakani, E. (2015). Effects of salicylic acid and putrescine on storability, quality attributes and antioxidant activity of plum cv. "Santa Rosa". *Journal of Food Science and Technology*, *52*(4), 2053–2062. https://doi.org/10.1007/s13197-013-1232-3

Deng, L., Pan, X., Chen, L., Shen, L., & Sheng, J. (2013). Effects of preharvest nitric oxide treatment on ethylene biosynthesis and soluble sugars metabolism in "Golden Delicious" apples. *Postharvest Biology and Technology*, *84*, 9–15. https://doi.org/10.1016/j.postharvbio.2013.03.017

Depaepe, T., & van der Straeten, D. (2020). Tools of the ethylene trade: A chemical kit to influence ethylene responses in plants and its use in agriculture. *Small Methods*, *4*(8), 1900267. https://doi.org/10.1002/SMTD.201900267

Dhall, R. K. (2013). Ethylene in postharvest quality management of horticultural crops: A review. *Journal of Crop Science and Technology*, *2*(2), 2319–2327. https://doi.org/292151893

Eum, H. L., Kim, H. B., Choi, S. B., & Lee, S. K. (2009). Regulation of ethylene biosynthesis by nitric oxide in tomato (*Solanum lycopersicum* L.) fruit harvested at different ripening stages. *European Food Research and Technology*, *228*(3), 331–338. https://doi.org/10.1007/s00217-008-0938-3

Fan, X., & Mattheis, J. P. (2000). Reduction of ethylene-induced physiological disorders of carrots and iceberg lettuce by 1-methylcyclopropene. *Hortscience*, *35*(7), 1312–1314. https://doi.org/10.21273/HORTSCI.35.7.1312

Forsyth, F. R., Eaves, C. A., & Lockhart, C. L. (1967). Controlling ethylene levels in the atmosphere of small containers of apples. *Canadian Journal of Plant Science*, *47*(6), 717–718. https://doi.org/10.4141/CJPS67-126

Fragkostefanakis, S., Kalaitzis, P., Siomos, A. S., & Gerasopoulos, D. (2013). Pyridine 2,4-dicarboxylate downregulates ethylene production in response to mechanical wounding in excised mature green tomato pericarp discs. *Journal of Plant Growth Regulation*, *32*(1), 140–147. https://doi.org/10.1007/s00344-012-9286-4

Gaikwad, K. K., Singh, S., & Lee, Y. S. (2018). High adsorption of ethylene by alkali-treated halloysite nanotubes for food-packaging applications. *Environmental Chemistry Letters*, *16*(3), 1055–1062. https://doi.org/10.1007/S10311-018-0718-7

Golding, J. B., Shearer, D., Wyllie, S. G., & McGlasson, W. B. (1998). Application of 1-MCP and propylene to identify ethylene-dependent ripening processes in mature banana fruit. *Postharvest Biology and Technology*, *14*(1), 87–98. https://doi.org/10.1016/S0925-5214(98)00032-5

Goodburn, K. E., & Halligan, A. C. (1988). *Modified-atmosphere packaging: A technology guide* (pp. 1–44). British Food Manufacturing Industries Research Association.

Goren, R., Huberman, M., Riov, J., Goldschmidt, E. E., Sisler, E. C., & Apelbaum, A. (2011). Effect of 3-cyclopropyl-1-enlyl-propanoic acid sodium salt, a novel water soluble antagonist of ethylene action, on plant responses to ethylene. *Plant Growth Regulation*, *65*(2), 327–334. https://doi.org/10.1007/s10725-011-9605-y

Guzman, P., & Ecker, J. R. (1990). Exploiting the triple response of Arabidopsis to identify ethylene-related mutants. *Plant Cell*, *2*(6), 513–523. https://doi.org/10.1105/tpc.2.6.513

Hao, Y., Hu, G., Breitel, D., Liu, M., Mila, I., Frasse, P., Fu, Y., Aharoni, A., Bouzayen, M., & Zouine, M. (2015). Auxin response factor SlARF2 is an essential component of the regulatory mechanism controlling fruit ripening in tomato. *PLoS Genetics*, *11*(12), e1005649. https://doi.org/10.1371/journal.pgen.1005649

Hoffman, N. E., Yang, S. F., Ichihara, A., & Sakamura, S. (1982). Stereospecific conversion of 1-aminocyclopropanecarboxylic acid to ethylene by plant tissues conversion of stereoisomers of 1-Amino-2-Ethylcyclopropanecarboxylic acid to 1-butene. *Plant Physiology*, *70*(1), 195–199. https://doi.org/10.1104/PP.70.1.195

Hu, W., Yang, H., Tie, W., Yan, Y., Ding, Z., Liu, Y., Wu, C., Wang, J., Reiter, R. J., Tan, D. X., Shi, H., Xu, B., & Jin, Z. (2017). Natural variation in banana varieties highlights the role of melatonin in postharvest ripening and quality. *Journal of Agricultural and Food Chemistry*, *65*(46), 9987–9994. https://doi.org/10.1021/acs.jafc.7b03354

Huber, D. J. (2008). Suppression of ethylene responses through application of 1-methylcyclopropene: A powerful tool for elucidating ripening and senescence mechanisms in climacteric and nonclimacteric fruits and vegetables. *Hortscience*, *43*(1), 106–111. https://doi.org/10.21273/HORTSCI.43.1.106

Hudgins, J. W., Ralph, S. G., Franceschi, V. R., & Bohlmann, J. (2006). Ethylene in induced conifer defense: cDNA cloning, protein expression, and cellular and subcellular localization of 1-aminocyclopropane-1-carboxylate oxidase in resin duct and phenolic parenchyma cells. *Planta*, *224*(4), 865–877. https://doi.org/10.1007/s00425-006-0274-4

Hurr, B. M., Huber, D. J., & Lee, J. H. (2005). Differential responses in color changes and softening of "Florida 47" tomato fruit treated at green and advanced ripening stages with the ethylene antagonist 1-methylcyclopropene. *HortTechnology*, *15*(3), 617–622. https://doi.org/10.21273/HORTTECH.15.3.0617

Ichiura, H., Kitaoka, T., & Tanaka, H. (2003). Removal of indoor pollutants under UV irradiation by a composite TiO2–zeolite sheet prepared using a papermaking technique. *Chemosphere*, *50*(1), 79–83. https://doi.org/10.1016/S0045-6535(02)00604-5

Iqbal, N., Khan, N. A., Ferrante, A., Trivellini, A., Francini, A., & Khan, M. I. R. (2017). Ethylene role in plant growth, development and senescence: Interaction with other phytohormones. *Frontiers in Plant Science*, *8*, 475. https://doi.org/10.3389/fpls.2017.00475

Janjarasskul, T., & Suppakul, P. (2018). Active and intelligent packaging: The indication of quality and safety. *Critical Reviews in Food Science and Nutrition*, *58*(5), 808–831. https://doi.org/10.1080/10408398.2016.1225278

Jia, X. H., Wang, W. H., Du, Y. M., Tong, W., Wang, Z. H., & Gul, H. (2018). Optimal storage temperature and 1-MCP treatment combinations for different marketing times of Korla Xiang pears. *Journal of Integrative Agriculture*, *17*(3), 693–703.

Jiang, C., Hara, K., & Fukuoka, A. (2013). Low-temperature oxidation of ethylene over platinum nanoparticles supported on mesoporous silica. *Angewandte Chemie*, *52*(24), 6265–6268. https://doi.org/10.1002/ANIE.201300496

Kakkar, R. K., & Rai, V. K. (1993). Plant polyamines in flowering and fruit ripening. *Phytochemistry*, *33*(6), 1281–1288. https://doi.org/10.1016/0031-9422(93)85076-4

Kashyap, K., & Banu, S. (2019). Characterizing ethylene pathway genes during the development, ripening, and postharvest response in Citrus reticulata Blanco fruit pulp. *Turkish Journal of Botany*, *43*(2), 173–184. https://doi.org/10.3906/bot-1711-45

Kays, S. J., & Beaudry, R. M. (1987). Techniques for inducing ethylene effects. *Acta Horticulturae*, *XXII*(201), 77–116. https://doi.org/10.17660/ActaHortic.1987.201.9

Khader, S. E. S. A. (1991). Effect of preharvest application of GA3 on postharvest behaviour of mango fruits. *Scientia Horticulturae*, *47*(3–4), 317–321. https://doi.org/10.1016/0304-4238(91)90014-P

Khan, A. S., Singh, Z., & Abbasi, N. A. (2007). Pre-storage putrescine application suppresses ethylene biosynthesis and retards fruit softening during low temperature storage in'Angelino'plum. *Postharvest Biology and Technology*, *46*(1), 36–46. https://doi.org/10.1016/j.postharvbio.2007.03.018

Khan, M. I. R., Iqbal, N., Masood, A., Per, T. S., & Khan, N. A. (2013). Salicylic acid alleviates adverse effects of heat stress on photosynthesis through changes in proline production and ethylene formation. *Plant Signaling and Behavior*, *8*(11), e26374. https://doi.org/10.4161/psb.26374

Khosravi, R., Hashemi, S. A., Sabet, S. A., & Rezadoust, A. M. (2013). Thermal, dynamic mechanical, and barrier studies of potassium permanganate-LDPE nanocomposites. *Polymer-Plastics Technology and Engineering*, *52*(2), 126–132. https://doi.org/10.1080/03602559.2012.719056

Kim, S., Jeong, G. H., & Kim, S. W. (2019). Ethylene gas decomposition using ZSM-5/WO3-Pt-nanorod composites for fruit freshness. *ACS Sustainable Chemistry and Engineering*, *7*(13), 11250–11257. https://doi.org/10.1021/ACSSUSCHEMENG.9B00584

Knee, M., & Hatfield, S. G. S. (1981). Benefits of ethylene removal during apple storage. *Annals of Applied Biology*, *98*(1), 157–165. https://doi.org/10.1111/j.1744-7348.1981.tb00433.x

Kosugi, Y., Matsuoka, A., Higashi, A., Toyohara, N., & Satoh, S. (2014). 2-Aminooxyisobutyric acid inhibits the in vitro activities of both 1-aminocyclopropane-1-carboxylate (ACC) synthase and ACC oxidase in ethylene biosynthetic pathway and prolongs vase life of cut carnation flowers. *Journal of Plant Biology*, *57*(4), 218–224. https://doi.org/10.1007/S12374-014-0180-4

Kou, X., Feng, Y., Yuan, S., Zhao, X., Wu, C., Wang, C., & Xue, Z. (2021). Different regulatory mechanisms of plant hormones in the ripening of climacteric and non-climacteric fruits: A review. *Plant Molecular Biology*, *107*(6), 477–497. https://doi.org/10.1007/s11103-021-01199-9

Lai, T., Wang, Y., Li, B., Qin, G., & Tian, S. (2011). Defense responses of tomato fruit to exogenous nitric oxide during postharvest storage. *Postharvest Biology and Technology*, *62*(2), 127–132. https://doi.org/10.1016/j.postharvbio.2011.05.011

Lara Ayala, I. (2013). Preharvest sprays and their effects on the postharvest quality of fruit. *Stewart Postharvest Review*, *9*(3), 1–12. https://doi.org/10.2212/spr.2013.3.5

Lau, O. L., Shang, A., & Yang, F. (1976). Inhibition of ethylene production by cobaltous ion. *Plant Physiology*, *58*(1), 114–117. https://doi.org/10.1104/PP.58.1.114

Leblanc, A., Renault, H., Lecourt, J., Etienne, P., Deleu, C., & le Deunff, E. (2008). Elongation changes of exploratory and root hair systems induced by aminocyclopropane carboxylic acid and aminoethoxyvinylglycine affect nitrate uptake and BnNrt2.1 and BnNrt1.1 transporter gene expression in oilseed rape. *Plant Physiology*, *146*(4), 1928–1940. https://doi.org/10.1104/PP.107.109363

Lee, M. H., Seo, H. S., & Park, H. J. (2017). Thyme oil encapsulated in halloysite nanotubes for antimicrobial packaging system. *Journal of Food Science*, *82*(4), 922–932. https://doi.org/10.1111/1750-3841.13675

Lee, S. Y., Lee, S. J., Choi, D. S., & Hur, S. J. (2015). Current topics in active and intelligent food packaging for preservation of fresh foods. *Journal of the Science of Food and Agriculture*, *95*(14), 2799–2810. https://doi.org/10.1002/JSFA.7218

Leslie, C. A., & Romani, R. J. (1986). Salicylic acid: A new inhibitor of ethylene biosynthesis. *Plant Cell Reports*, *5*(2), 144–146. https://doi.org/10.1007/BF00269255

Li, C. Z., Wei, X. P., Li, W., & Wang, G. X. (2004). Relationship between ethylene and spermidine in the leaves of Glycyrrhiza uralensis seedlings under root osmotic Stress1. *Russian Journal of Plant Physiology*, *51*(3), 372–378. https://doi.org/10.1023/B:RUPP.0000028683.70891.04

Li, J., Tao, X., Bu, J., Ying, T., Mao, L., & Luo, Z. (2017). Global transcriptome profiling analysis of ethylene-auxin interaction during tomato fruit ripening. *Postharvest Biology and Technology*, *130*, 28–38. https://doi.org/10.1016/j.postharvbio.2017.03.021

Li, N., Parsons, B. L., Liu, D. R., & Mattoo, A. K. (1992). Accumulation of wound-inducible ACC synthase transcript in tomato fruit is inhibited by salicylic acid and polyamines. *Plant Molecular Biology*, *18*(3), 477–487. https://doi.org/10.1007/BF00040664

Li, Y., Ma, Y., Zhang, T., Bi, Y., Wang, Y., & Prusky, D. (2019). Exogenous polyamines enhance resistance to *Alternaria alternata* by modulating redox homeostasis in apricot fruit. *Food Chemistry*, *301*, 125303. https://doi.org/10.1016/j.foodchem.2019.125303

Lieberman, M., Baker, J. E., & Sloger, M. (1977). Influence of plant hormones on ethylene production in apple, tomato, and avocado slices during maturation and senescence. *Plant Physiology*, *60*(2), 214–217. https://doi.org/10.1104/pp.60.2.214

Lindermayr, C., Saalbach, G., Bahnweg, G., & Durner, J. (2006). Differential inhibition of Arabidopsis methionine adenosyltransferases by protein S-nitrosylation. *Journal of Biological Chemistry*, *281*(7), 4285–4291. https://doi.org/10.1074/jbc.M511635200

Liu, S., Huang, H., Huber, D. J., Pan, Y., Shi, X., & Zhang, Z. (2020). Delay of ripening and softening in 'Guifei' mango fruit by postharvest application of melatonin. *Postharvest Biology and Technology*, *163*, 111136. https://doi.org/10.1016/j.postharvbio.2020.111136

Manjunatha, G., Lokesh, V., & Neelwarne, B. (2010). Nitric oxide in fruit ripening: Trends and opportunities. *Biotechnology Advances*, *28*(4), 489–499. https://doi.org/10.1016/j.biotechadv.2010.03.001

Martínez-Romero, D., Bailén, G., Serrano, M., Guillén, F., Valverde, J. M., Zapata, P., Castillo, S., & Valero, D. (2007). Tools to maintain postharvest fruit and vegetable quality through the inhibition of ethylene action: A review. *Critical Reviews in Food Science and Nutrition*, *47*(6), 543–560. https://doi.org/10.1080/10408390600846390

Martinez-Romero, D., Serrano, M., Carbonell, A., Burgos, L., Riquelme, F., & Valero, D. (2002). Effects of postharvest putrescine treatment on extending shelf life and reducing mechanical damage in apricot. *Journal of Food Science*, *67*(5), 1706–1712. https://doi.org/10.1111/j.1365-2621.2002.tb08710.x

Martínez-Romero, D., Valero, D., Serrano, M., Burló, F., Carbonell, A., Burgos, L., & Riquelme, F. (2000). Exogenous polyamines and gibberellic acid effects on peach (*Prunus persica* L.) storability improvement. *Journal of Food Science*, *65*(2), 288–294. https://doi.org/10.1111/j.1365-2621.2000.tb15995.x

Mattoo, A. K., Chalutz, E., & Lieberman, M. (1979). Effects of lipophilic and water-soluble membrane probes on ethylene synthesis in apple and *Penicillium digitatum*. *Plant and Cell Physiology*, *20*(6), 1097–1106. https://doi.org/10.1093/OXFORDJOURNALS.PCP.A075905

McDaniel, B. K., & Binder, B. M. (2012). Ethylene receptor 1 (ETR1) is sufficient and has the predominant role in mediating inhibition of ethylene responses by silver in Arabidopsis thaliana. *Journal of Biological Chemistry*, *287*(31), 26094–26103. https://doi.org/10.1074/jbc.M112.383034

Mcmurchie, E. J., Mcglasson, W. B., & Eaks, I. L. (1972). Treatment of Fruit with propylene gives Information about the Biogenesis of ethylene. *Nature*, *237*(5352), 235–236. https://doi.org/10.1038/237235a0

Meena, N. K., & Asrey, R. (2018). Tree age affects physicochemical, functional quality and storability of Amrapali mango (Mangifera indica L.) fruits. *Journal of the Science of Food and Agriculture*, *98*(9), 3255–3262. https://doi.org/10.1002/jsfa.8828

Mukherjee, S. (2019). Recent advancements in the mechanism of nitric oxide signaling associated with hydrogen sulfide and melatonin crosstalk during ethylene-induced fruit ripening in plants. *Nitric Oxide: Biology and Chemistry*, *82*, 25–34. https://doi.org/10.1016/j.niox.2018.11.003

Nissen, P. (1994). Stimulation of somatic embryogenesis in carrot by ethylene: Effects of modulators of ethylene biosynthesis and action. *Physiologia Plantarum, 92*(3), 397–403. https://doi.org/10.1111/j.1399-3054.1994.tb08827.x

Oh, S. Y., Shin, S. S., Kim, C. C., & Lim, Y. J. (1996). Effect of packaging films and freshness keeping agents on fruit quality of "Yumyung" peaches during MA storage. *Journal of the Korean Society for Horticultural Science* (Korea Republic), *37*(6), 781–786. https://doi.org/KR9700725

Ozdemir, M., & Floros, J. D. (2004). Active food packaging technologies. *Critical Reviews in Food Science and Nutrition, 44*(3), 185–193. https://doi.org/10.1080/10408690490441578

Pandey, R., Gupta, A., Chowdhary, A., Pal, R. K., & Rajam, M. V. (2015). Over-expression of mouse ornithine decarboxylase gene under the control of fruit-specific promoter enhances fruit quality in tomato. *Plant Molecular Biology, 87*(3), 249–260. https://doi.org/10.1007/s11103-014-0273-y

Pathak, N. (2019). *Photocatalysis and vacuum ultraviolet light photolysis as ethylene removal techniques for potential application in fruit storage* (Ph.D.). Technische Universität. http://doi.org/10.14279/depositonce-8313

Pathak, N., Caleb, O. J., Geyer, M., Herppich, W. B., Rauh, C., & Mahajan, P. V. (2017). Photocatalytic and photochemical oxidation of ethylene: Potential for storage of fresh produce – A review. *Food and Bioprocess Technology, 10*(6), 982–1001. https://doi.org/10.1007/S11947-017-1889-0

Pattyn, J., Vaughan-Hirsch, J., & Van de Poel, B. (2021). The regulation of ethylene biosynthesis: A complex multilevel control circuitry. *New Phytologist, 229*(2), 770–782. https://doi.org/10.1111/nph.16873

Picón, A., Martínez-Jávega, J. M., Cuquerella, J., Del Río, M. A., & Navarro, P. (1993). Effects of precooling, packaging film, modified atmosphere and ethylene absorber on the quality of refrigerated Chandler and Douglas strawberries. *Food Chemistry, 48*(2), 189–193. https://doi.org/10.1016/0308-8146(93)90056-L

Pirrung, M. C., Cao, J., & Chen, J. (1998). Ethylene biosynthesis: Processing of a substrate analog supports a radical mechanism for the ethylene-forming enzyme. *Chemistry and Biology, 5*(1), 49–57. https://doi.org/10.1016/S1074-5521(98)90086-2

Prasad, P., & Kochhar, A. (2014). Active packaging in food industry: A review. *IOSR Journal of Environmental Science, Toxicology and Food Technology, 8*(5), 1–7. https://doi.org/10.9790/2402-08530107

Ramassamy, S., Olmos, E., Bouzayen, M., Pech, J. C., & Latche, A. (1998). 1-aminocyclopropane-1-carboxylate oxidase of apple fruit is periplasmic. *Journal of Experimental Botany, 49*, 1909–1915. https://doi.org/10.1093/jxb/49.329.1909

Rando, R. R. (1974). Irreversible inhibition of aspartate aminotransferase by 2-amino-3-butenoic acid. *Biochemistry, 13*(19), 3859–3863. https://doi.org/10.1021/bi00716a006

Rodríguez, F. I., Esch, J. J., Hall, A. E., Binder, B. M., Schaller, G. E., & Bleecker, A. B. (1999). A copper cofactor for the ethylene receptor ETR1 from Arabidopsis. *Science, 283*(5404), 996–998. https://doi.org/10.1126/SCIENCE.283.5404.996

Romani, R. J., Hess, B. M., & Leslie, C. A. (1989). Salicylic acid inhibition of ethylene production by apple discs and other plant tissues. *Journal of Plant Growth Regulation, 8*(1), 63–69. https://doi.org/10.1007/BF02024927

Rudell, D. R., & Mattheis, J. P. (2006). Nitric oxide and nitrite treatments reduce ethylene evolution from apple fruit disks. *HortScience, 41*(6), 1462–1465. https://doi.org/10.21273/HORTSCI.41.6.1462

Sadeghi, K., Lee, Y., & Seo, J. (2021). Ethylene scavenging systems in packaging of fresh produce: A review. *Food Reviews International, 37*(2), 155–176. https://doi.org/10.1080/87559129.2019.1695836

Saftner, R. A., & Baldi, B. G. (1990). Polyamine levels and tomato fruit development: Possible interaction with ethylene. *Plant Physiology, 92*(2), 547–550. https://doi.org/10.1104/pp.92.2.547

Saltveit, M.E. (1999). Effect of ethylene on quality of fresh fruits and vegetables. *Postharvest Biology and Technology*, *15*(3), 279–292. https://doi.org/10.1016/S0925-5214(98)00091-X

Saltveit, M.E., Bradford, K.J., & Dilley, D.R. (1978). Silver ion inhibits ethylene synthesis and action in ripening Fruits1. *Journal of the American Society for Horticultural Science*, *103*(4), 472–475. https://doi.org/10.21273/JASHS.103.4.472

Satoh, S., & Esashi, Y. (1980). α-aminoisobutyric acid: A probable competitive inhibitor of conversion of 1-aminocyclopropane-1-carboxylic acid to ethylene. *Plant and Cell Physiology*, *21*(6), 939–949. https://doi.org/10.1093/oxfordjournals.pcp.a076082

Satoh, S., Kosugi, Y., Sugiyama, S., & Ohira, I. (2014). 2,4-Pyridinedicarboxylic acid prolongs the vase life of cut flowers of spray carnations. *Journal of the Japanese Society for Horticultural Science*, *83*(1), 72–80. https://doi.org/10.2503/jjshs1.CH-082

Satoh, S., & Yang, S.F. (1989). Inactivation of 1-aminocyclopropane-1-carboxylate synthase by l-Vinylglycine as related to the mechanism-based inactivation of the enzyme by S-adenosyl-l-methionine. *Plant Physiology*, *91*(3), 1036–1039. https://doi.org/10.1104/pp.91.3.1036

Savin, K.W., Baudinette, S.C., Graham, M.W., Michael, M.Z., Nugent, G.D., Lu, C.-Y., Chandler, S.F., & Cornish, E.C. (1995). Antisense ACC oxidase RNA delays carnation petal senescence. *HortScience*, *30*(5), 970–972. https://doi.org/10.21273/HORTSCI.30.5.970

Scariot, V., Paradiso, R., Rogers, H., & De Pascale, S. (2014). Ethylene control in cut flowers: Classical and innovative approaches. *Postharvest Biology and Technology*, *97*, 83–92. https://doi.org/10.1016/j.postharvbio.2014.06.010

Scott, K.J., McGlasson, W.B., & Roberts, E.A. (1970). Potassium permanganate as an ethylene absorbent in polyethylene bags to delay ripening of bananas during storage. *Australian Journal of Experimental Agriculture*, *10*(43), 237–240. https://doi.org/10.1071/EA9700237

Sekirnik, R., Rose, N.R., Mecinović, J., & Schofield, C.J. (2010). 2-oxoglutarate oxygenases are inhibited by a range of transition metals. *Metallomics*, *2*(6), 397–399. https://doi.org/10.1039/C004952B

Serek, M., Sisler, E.C., & Reid, M.S. (1994). Novel gaseous ethylene binding inhibitor prevents ethylene effects in potted flowering plants. *Journal of the American Society for Horticultural Science*, *119*(6), 1230–1233. https://doi.org/10.21273/JASHS.119.6.1230

Serek, M., Sisler, E.C., & Reid, M.S. (1995). Effects of 1-MCP on the vase life and ethylene response of cut flowers. *Plant Growth Regulation*, *16*(1), 93–97. https://doi.org/10.1007/BF00040512

Serek, M., Woltering, E.J., Sisler, E.C., Frello, S., & Sriskandarajah, S. (2006). Controlling ethylene responses in flowers at the receptor level. *Biotechnology Advances*, *24*(4), 368–381. https://doi.org/10.1016/J.BIOTECHADV.2006.01.007

Serrano, M., Martinez-Romero, D., Guillén, F., & Valero, D. (2003). Effects of exogenous putrescine on improving shelf life of four plum cultivars. *Postharvest Biology and Technology*, *30*(3), 259–271. https://doi.org/10.1016/S0925-5214(03)00113-3

Shi, H.Y., Wang, Y., Qi, A., Zhang, Y., Xu, J., Wang, A., & Zhang, Y. (2013). PpACS1b, a pear gene encoding ACC synthase, is regulated during fruit late development and involved in response to salicylic acid. *Scientia Horticulturae*, *164*, 602–609. https://doi.org/10.1016/j.scienta.2013.09.055

Shi, H.Y., & Zhang, Y.X. (2014). Expression and regulation of pear 1-aminocyclopropane-1 carboxylic acid synthase gene (PpACS1a) during fruit ripening, under salicylic acid and indole-3-acetic acid treatment, and in diseased fruit. *Molecular Biology Reports*, *41*(6), 4147–4154. https://doi.org/10.1007/s11033-014-3286-3

Singh, Z., Payne, A.D., Khan, S.A.K.U., & Musa, M.M.A. (2018). Method of retarding an ethylene response. U.S. Patent application No, 15/772, 324, filed.

Siročić, A. P., Rešček, A., Ščetar, M., Krehula, L. K., & Hrnjak-Murgić, Z. (2014). Development of low density polyethylene nanocomposites films for packaging. *Polymer Bulletin*, *71*(3), 705–717. https://doi.org/10.1007/s00289-013-1087-9

Sisler, E. C. (2006). The discovery and development of compounds counteracting ethylene at the receptor level. *Biotechnology Advances*, *24*(4), 357–367. https://doi.org/10.1016/J.BIOTECHADV.2006.01.002

Sisler, E. C., & Blankenship, S. M. (1993). Diazocyclopentadiene (DACP), a light sensitive reagent for the ethylene receptor in plants. *Plant Growth Regulation*, *12*(1–2), 125–132. https://doi.org/10.1007/BF00144593

Sisler, E. C., Blankenship, S. M., & Guest, M. (1990). Competition of cyclooctenes and cyclooctadienes for ethylene binding and activity in plants. *Plant Growth Regulation*, *9*(2), 157–164. https://doi.org/10.1007/BF00027443

Sisler, E. C., Goren, R., & Huberman, M. (1985). Effect of 2,5-norbornadiene on abscission and ethylene production in citrus leaf explants. *Physiologia Plantarum*, *63*(1), 114–120. https://doi.org/10.1111/J.1399-3054.1985.TB02828.X

Sisler, E. C., & Lallu, N. (1994). Effect of diazocyclopentadiene (DACP) on tomato fruits harvested at different ripening stages. *Postharvest Biology and Technology*, *4*(3), 245–254. https://doi.org/10.1016/0925-5214(94)90034-5

Sisler, E. C., & Serek, M. (1997). Inhibitors of ethylene responses in plants at the receptor level: Recent developments. *Physiologia Plantarum*, *100*(3), 577–582. https://doi.org/10.1111/J.1399-3054.1997.TB03063.X

Sisler, E. C., & Serek, M. (1999). Compounds controlling the ethylene receptor. *Botanical Bulletin of Academia Sinica*, *40*, 1–7.

Smith, A. W. J., Poulston, S., Rowsell, L., Terry, L. A., & Anderson, J. A. (2009). A new palladium-based ethylene scavenger to control ethylene-induced ripening of climacteric fruit. *Platinum Metals Review*, *53*(3), 112–122. https://doi.org/10.1595/147106709X462742

Sobolev, A. P., Neelam, A., Fatima, T., Shukla, V., Handa, A. K., & Mattoo, A. K. (2014). Genetic introgression of ethylene-suppressed transgenic tomatoes with higher-polyamines trait overcomes many unintended effects due to reduced ethylene on the primary metabolome. *Frontiers in Plant Science*, *5*, 632. https://doi.org/10.3389/fpls.2014.00632

Srivastava, M. K., & Dwivedi, U. N. (2000). Delayed ripening of banana fruit by salicylic acid. *Plant Science*, *158*(1–2), 87–96. https://doi.org/10.1016/s0168-9452(00)00304-6

Sun, Q., Zhang, N., Wang, J., Zhang, H., Li, D., Shi, J., Li, R., Weeda, S., Zhao, B., Ren, S., & Guo, Y. D. (2015). Melatonin promotes ripening and improves quality of tomato fruit during postharvest life. *Journal of Experimental Botany*, *66*(3), 657–668. https://doi.org/10.1093/jxb/eru332

Sun, X., Li, Y., He, W., Ji, C., Xia, P., Wang, Y., Du, S., Li, H., Raikhel, N., Xiao, J., & Guo, H. (2017). Pyrazinamide and derivatives block ethylene biosynthesis by inhibiting ACC oxidase. *Nature Communications*, *8*(1), 1–14, 15758. https://doi.org/10.1038/ncomms15758

Tas, C. E., Hendessi, S., Baysal, M., Unal, S., Cebeci, F. C., Menceloglu, Y. Z., & Unal, H. (2017). Halloysite nanotubes/polyethylene nanocomposites for active food packaging materials with ethylene scavenging and gas barrier properties. *Food and Bioprocess Technology*, *10*(4), 789–798. https://doi.org/10.1007/S11947-017-1860-0

Terry, L. A., Ilkenhans, T., Poulston, S., Rowsell, L., & Smith, A. W. J. (2007). Development of new palladium-promoted ethylene scavenger. *Postharvest Biology and Technology*, *45*(2), 214–220. https://doi.org/10.1016/j.postharvbio.2006.11.020

Tierney, D. L., Rocklin, A. M., Lipscomb, J. D., Que, L., & Hoffman, B. M. (2005). ENDOR studies of the ligation and structure of the non-heme iron site in ACC oxidase. *Journal of the American Chemical Society*, *127*(19), 7005–7013. https://doi.org/10.1021/ja0500862

Tokala, V.Y., Singh, Z., & Kyaw, P.N. (2020). Fumigation and dip treatments with 1H-cyclopropabenzene and 1H-cyclopropa[b]naphthalene suppress ethylene production and maintain fruit quality of cold-stored "Cripps Pink" apple. *Scientia Horticulturae, 272*, 109597. https://doi.org/10.1016/J.SCIENTA.2020.109597

Tokala, V.Y., Singh, Z., & Kyaw, P. N. (2021). 1H-cyclopropabenzene and 1H-cyclopropa[b] naphthalene fumigation downregulates ethylene production and maintains fruit quality of controlled atmosphere stored "Granny Smith" apple. *Postharvest Biology and Technology, 176*, 111499. https://doi.org/10.1016/J.POSTHARVBIO.2021.111499

Torrigiani, P., Bregoli, A. M., Ziosi, V., Scaramagli, S., Ciriaci, T., Rasori, A., Biondi, S., & Costa, G. (2004). Pre-harvest polyamine and aminoethoxyvinylglycine (AVG) applications modulate fruit ripening in Stark Red Gold nectarines (*Prunus persica* L. Batsch). *Postharvest Biology and Technology, 33*(3), 293–308. https://doi.org/10.1016/j.postharvbio.2004.03.008

Tytgat, T., Hauchecorne, B., Abakumov, A.M., Smits, M., Verbruggen, S.W., & Lenaerts, S. (2012). Photocatalytic process optimisation for ethylene oxidation. *Chemical Engineering Journal, 209*, 494–500. https://doi.org/10.1016/j.cej.2012.08.032

Tze, L. N., Yamauchi, K., Yoshii, H., & Furuta, T. (2007). Kinetics of molecular encapsulation of 1-methylcyclopropene into α-cyclodextrin. *Journal of Agricultural and Food Chemistry, 55*(26), 11020–11026. https://doi.org/10.1021/JF072357T

Van de Poel, B., & Van Der Straeten, D. (2014). 1-aminocyclopropane-1-carboxylic acid (ACC) in plants: More than just the precursor of ethylene! *Frontiers in Plant Science, 5*, 640. https://doi.org/10.3389/fpls.2014.00640

Van Toan, N., & Thanh, C. D. (2011). Effects of aminoethoxyvinylglycine (AVG) spraying time at preharvest stage to ethylene biosynthesis of Cavendish banana (Musa AAA). *Journal of Agricultural Science, 3*(1), 206. https://doi.org/10.5539/jas.v3n1p206

Vlad, F., Tiainen, P., Owen, C., Spano, T., Daher, F. B., Oualid, F., Senol, N. O., Vlad, D., Myllyharju, J., & Kalaitzis, P. (2010). Characterization of two carnation petal prolyl 4 hydroxylases. *Physiologia Plantarum, 140*(2), 199–207. https://doi.org/10.1111/J.1399-3054.2010.01390.X

Wang, Y., Xu, F., Feng, X., & MacArthur, R. L. (2015). Modulation of Actinidia arguta fruit ripening by three ethylene biosynthesis inhibitors. *Food Chemistry, 173*, 405–413. https://doi.org/10.1016/J.FOODCHEM.2014.10.044

Weir, A., Westerhoff, P., Fabricius, L., Hristovski, K., & Von Goetz, N. (2012). Titanium dioxide nanoparticles in food and personal care products. *Environmental Science and Technology, 46*(4), 2242–2250. https://doi.org/10.1021/es204168d

Wills, R.B.H., McGlasson, W.B., Graham, D., & Joyce, D.C. (Eds.). (2007). Physiology and biochemistry. In *Postharvest: An introduction to the physiology and handling of fruit, vegetables and ornamentals* (4th ed., pp. 28–51). Oxford University Press.

Wills, R. B.H., & Warton, M.A. (2004). Efficacy of potassium permanganate impregnated into alumina beads to reduce atmospheric ethylene. *Journal of the American Society for Horticultural Science, 129*(3), 433–438. https://doi.org/10.21273/JASHS.129.3.0433

Wuttke, S., Medina, D.D., Rotter, J.M., Begum, S., Stassin, T., Ameloot, R., Oschatz, M., & Tsotsalas, M. (2018). Bringing porous organic and carbon-based materials toward thin-film applications. *Advanced Functional Materials, 28*(44), 1801545. https://doi.org/10.1002/adfm.201801545

Xu, F., Liu, S., Liu, Y., Xu, J., Liu, T., & Dong, S. (2019). Effectiveness of lysozyme coatings and 1-MCP treatments on storage and preservation of kiwifruit. *Food Chemistry, 288*, 201–207. https://doi.org/10.1016/J.FOODCHEM.2019.03.024

Xue, W. J., Wang, Y. F., Li, P., Liu, Z. T., Hao, Z. P., & Ma, C. Y. (2011). Morphology effects of $Co_3O_4$ on the catalytic activity of Au/Co3O4 catalysts for complete oxidation of trace ethylene. *Catalysis Communications, 12*(13), 1265–1268. https://doi.org/10.1016/j.catcom.2011.04.003

Yang, S. F., & Hoffman, N. E. (1984). Ethylene biosynthesis and its regulation in higher plants. *Annual Review of Plant Physiology*, *35*(1), 155–189. https://doi.org/10.1146/annurev. pp. 35.060184.001103

Yildirim, S., Röcker, B., Pettersen, M. K., Nilsen-Nygaard, J., Ayhan, Z., Rutkaite, R., Radusin, T., Suminska, P., Marcos, B., & Coma, V. (2018). Active packaging applications for food. *Comprehensive Reviews in Food Science and Food Safety*, *17*(1), 165–199. https://doi.org/10.1111/1541-4337.12322

Yu, Y. B., Shang, A., & Yang, F. A. (1979). Auxin-induced ethylene production and its inhibition by Aminoethoxyvinylglycine and cobalt Ion. *Plant Physiology*, *64*(6), 1074–1077. https://doi.org/10.1104/PP.64.6.1074

Yuan, P., Tan, D., & Annabi-Bergaya, F. (2015). Properties and applications of halloysite nanotubes: Recent research advances and future prospects. *Applied Clay Science*, *112–113*, 75–93. https://doi.org/10.1016/J.CLAY.2015.05.001

Yuan, R., & Carbaugh, D. H. (2007). Effects of NAA, AVG, and 1-MCP on ethylene biosynthesis, preharvest fruit drop, fruit maturity, and quality of "golden supreme" and "golden delicious" apples. *Hortscience*, *42*(1), 101–105. https://doi.org/10.21273/HORTSCI.42.1.101

Yue, P., Lu, Q., Liu, Z., Lv, T., Li, X., Bu, H., Liu, W., Xu, Y., Yuan, H., & Wang, A. (2020). Auxin-activated MdARF5 induces the expression of ethylene biosynthetic genes to initiate apple fruit ripening. *New Phytologist*, *226*(6), 1781–1795. https://doi.org/10.1111/nph.16500

Zaharah, S. S., & Singh, Z. (2011). Postharvest nitric oxide fumigation alleviates chilling injury, delays fruit ripening and maintains quality in cold-stored "Kensington Pride" mango. *Postharvest Biology and Technology*, *60*(3), 202–210. https://doi.org/10.1016/j.postharvbio.2011.01.011

Zhai, R., Liu, J., Liu, F., Zhao, Y., Liu, L., Fang, C., Wang, H., Li, X., Wang, Z., Ma, F., & Xu, L. (2018). Melatonin limited ethylene production, softening and reduced physiology disorder in pear (Pyrus communis L.) fruit during senescence. *Postharvest Biology and Technology*, *139*, 38–46. https://doi.org/10.1016/j.postharvbio.2018.01.017

Zhang, J., Cheng, D., Wang, B., Khan, I., & Ni, Y. (2017). Ethylene control technologies in extending postharvest shelf life of climacteric fruit. *Journal of Agricultural and Food Chemistry*, *65*(34), 7308–7319. https://doi.org/10.1021/acs.jafc.7b02616

Zhang, Y., Chen, K., Zhang, S., & Ferguson, I. (2003). The role of salicylic acid in postharvest ripening of kiwifruit. *Postharvest Biology and Technology*, *28*(1), 67–74. https://doi.org/10.1016/S0925-5214(02)00172-2

Zhao, Z. N., Nie, X. X., & Wang, R. (2011). Study on the properties of LDPE/POE/Zeolite molecular sieves composite film. In *Advanced materials research*. Trans Tech. Publications Ltd. https://doi.org/10.4028/AMR.194-196.2347

Zhu, S.-H., Liu, M., & Zhou, J. (2006). Inhibition by nitric oxide of ethylene biosynthesis and lipoxygenase activity in peach fruit during storage. *Postharvest Biology and Technology*, *42*(1), 41–48. https://doi.org/10.1016/j.postharvbio.2006.05.004

Zhu, S.-H., & Zhou, J. (2007). Effect of nitric oxide on ethylene production in strawberry fruit during storage. *Food Chemistry*, *100*(4), 1517–1522. https://doi.org/10.1016/j.foodchem.2005.12.022

Zhu, X., Liang, X., Wang, P., Dai, Y., & Huang, B. (2018). Porous Ag-ZnO microspheres as efficient photocatalyst for methane and ethylene oxidation: Insight into the role of Ag particles. *Applied Surface Science*, *456*, 493–500. https://doi.org/10.1016/j.apsusc.2018.06.127

# 8 Eco-friendly, Nanoenabled Applications in Postharvest Handling of Fruits and Vegetables
## Quality and Safety of Produce

*Neela Badrie*

## LIST OF DEFINITIONS

**Nanobiopolymers**: Polymer or copolymer materials of different structures, shapes and functional forms containing dispersed nanoparticles

**Nanocapsules**: Nanometric colloidal carriers

**Nanocelluloses**: Cellulosic-based materials having a dimension of 100 nm or less as nanofillers

**Nanochitosan**: Chitosan nanoparticles

**Nanoclays**: Nanoparticles containing layered mineral silicates with potential uses in polymer nanocomposites

**Nanocoatings**: Nanoscale engineering of surfaces and layers for improved functionalities and physical effects

**Nanocomposites**: Multiphasic materials, with at least one phase of nanomaterial, of less than 100 nm to improve a particular property of the material

**Nanoemulsions**: Nanosized emulsions that serve as delivery channels for additives, which are typically oil-based

**Nanofibrils**: Nanofibrils, such as cellulose nanofibrils, having strong mechanical properties for composites

**Nanofilms**: Thin layers of barrier materials that influence behavior, such as by having antimicrobial prevention properties to counter microbial spoilage and oxygen absorption

**Nanofungicides**: Fungicides that include entities in the nanometer size range, such as nanoparticles, to provide anti-fungi toxicity efficacy

DOI: 10.1201/9781003452355-10

**Nanolaminates**: Laminates fabricated from nanoscopically thin laminae as nanocomposite materials for their mechanical and physical properties

**Nanomaterials**: Any materials that have one or more dimensions in the nanoscale, sized between 1 and 100 nm

**Nanomicrobicides**: Microbicide encapsulated in nanoparticles

**Nanopackaging**: Packaging of modules and systems with nanomaterials to form constituents with improved performance, functionality miniaturization and reliability

**Nanoparticles**: Nanoscale particles ranging in dimension from 1 to 100 nm, having unique properties compared to bulk equivalents, and serving as connecting links between molecular structures and macromolecular/bulk materials

**Nanosafety**: Assessment of risks to human health and the environment and of the ecological risks from the use of designed nanomaterials

**Nanosensors**: Sensors that evaluate nanosized biochemical and physiological changes at the macro level

**Nanotechnology**: Branch of technology that deals with dimensions and tolerances at the nanoscale, about 1 to 100 nm

## 8.1 PERISHABLE FRUITS AND VEGETABLES

It is estimated that globally about 45% of fruits and vegetables are wasted and about one-third of the fresh produce is subjected to postharvest loss during handling and storage (Pang et al., 2021; Gustavsson et al., 2013). Fruits and vegetables are perishable and subject to microbial decay, physical damage, moisture loss, biochemical changes and reduction in quality attributes during storage (Basumatary et al., 2022; Kumar et al., 2020a). Nanotechnology has advanced food packaging systems such as active and intelligent forms for better retention of the quality of fruits and vegetables (Ashfaq et al., 2022; Kim et al., 2022). The NPs can inhibit ethylene production, impede ripening in postharvest produce (Bhuyan et al., 2019) and control plant diseases (Alghuthaymi et al., 2021). Some nanoapplications for fruits and vegetables are in the form of nanocoatings/films, nanoemulsions, nonfungicides, nanopesticides and nanosensors (Neme et al., 2021; Vimala Devi et al., 2019).

## 8.2 WHAT IS NANOTECHNOLOGY?

Nanotechnology involves the control of matter at the nanoscale in the synthesis and manipulation of a particle's structure to develop functional materials and structures (Satalkar et al., 2016). It deals with the synthesis of nanoscale materials, which are referred to as nanoparticles (NPs). The NPs are defined as particles with sizes 1–100 nanometers (nm) that can be manufactured or natural (Rivero-Montejo et al., 2021) and have different qualities than bulk materials due to their high surface:volume ratio (Satalkar et al., 2016; Singh & Prasad, 2017). The synthesized nanomaterials (NMs) have structures of less than one 100 nm (Javed et al., 2022; Bhusare & Kadam, 2021; Vimala Devi et al., 2019).

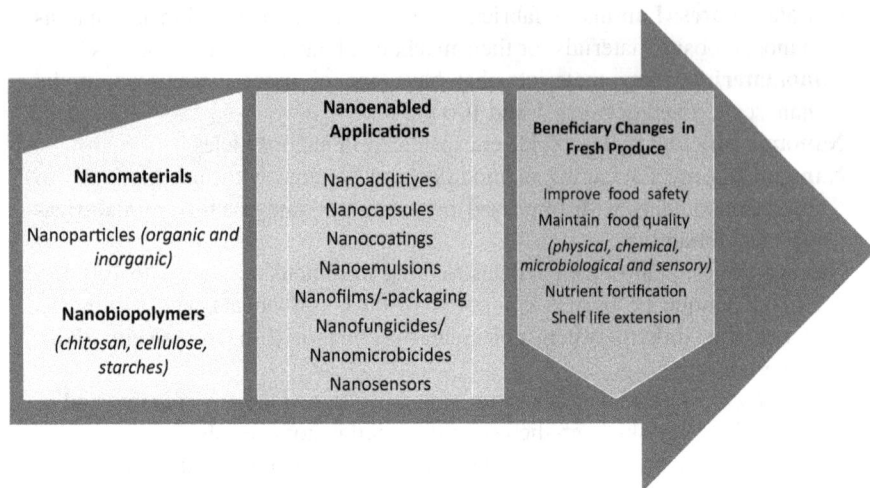

**FIGURE 8.1** Nanoenabled applications to improve the postharvest handling of fruits and vegetables.

At the nanoscale, the physical, chemical and biological properties of NMs can vary in fundamental and useful ways from the properties of bulk matter, thereby creating new applications for existing materials. Numerous novel NMs have beneficial distinctive properties such as larger surface:volume ratio, broader-spectrum antimicrobial properties and effective barrier properties, which could surmount the disadvantages of traditional preservation technologies (Liu et al., 2020a). The synthesis of individual nanostructures with the incorporation of NPs could result in hybrid nanocomposites with synergistic and promising functionalities (Winter et al., 2020). Figure 8.1 shows some nanoenabled applications for the postharvest handling of fruits and vegetables and the beneficial effects on fresh produce.

## 8.3 NANOMATERIALS

Nanomaterials (NMs) are any materials with at least one dimension equivalent to 100 nm thickness or less (Kaphle et al., 2018). They can also be categorized as films or coating attached to a substrate or as nanoscale pores and membranes on a substrate and porous films. The NMs are classified from the material bulk such as metal or carbon based, metal salts and nanosized polymers (Rivero-Montejo et al., 2021). In this chapter, various NMs of some metal oxide NPs, the edible nanobiopolymers (chitosan, cellulose and various starches), the green synthesis of NPs and the various nanoapplications to fruits and vegetables are included and discussed.

### 8.3.1 NANOPARTICLES

Nanoparticles (NPs) can exist as nanostructures or as composites (Mageswari et al., 2016) and can be classified into organic, inorganic and carbon-based materials

(Mokhena et al., 2020). NPs have been applied as nanopesticides, nanofertilizers, nanocoatings, nanodippings, nanofilms/-laminates, nanoactives, nanosmart and nanointelligent packaging (Kamatyanatti et al., 2019; Qadri et al., 2019), nanosensors and nanodisinfecting agents (He et al., 2019). Organic NPs are nontoxic and not biopersistent (Divya & Jisha, 2018). They may vary in toxicity subject to concentration level, type, exposure time and the sensitivity of the individual (Dimitrijevic et al., 2015). Research methodologies are being developed to detect nanoscale materials in samples (Patri, 2022).

### 8.3.1.1 Application of Silver Nanoparticles to Fruits and Vegetables

Nanoparticles (NPs) are effective antimicrobial NMs due to their particle sizes of 1–10 nm, which promotes better contact with microorganisms (Khan et al., 2018). Microbial cell damage or cell death is related to the inhibition of DNA replication and ATP formation, in turn related to the generation of reactive oxygen species by NPs (Ashfaq et al., 2022). Silver NPs have a natural antimicrobial effect against many pathogens due to their easy infiltration into the microbial cell through the cell membranes (Ali et al., 2020; Bora & Mishra, 2019, Hussain et al., 2019). The main findings of several studies on the impacts of nanoenabled coating/emulsion applications on different fruits and vegetables are summed in Table 8.1. It is clear from Table 8.1 that Ag-NPs-based coatings maintained the postharvest quality of loquat fruit during storage at 4°C and 8°C (Ali et al., 2020). The Ag-NPs formulations increased bioactive compounds in postharvest carrots. The exogenous application of Ag-NPs controlled abiotic stress and produced high-value secondary bioactive compounds such as phenolic compounds in carrot (Santoscoy-Berber et al., 2021). The effects were noted to be significantly related with concentrations and exposure time. Synthesized Ag-NPs have been applied for the detection of sodium sulfite and as a sensor for volatile organosulfides in garlic spoilage during postharvest loss (Motol et al., 2020). In another nanotechnology application, the Ag-NPs were combined with casein-NPs to form sodium alginate and casein bilayer film. This film controlled resistance against oxidation and inhibited the growth of *Escherichia coli* of stored almond oil for 60 days (Bora & Mishra, 2019).

### 8.3.1.2 Application of Zinc Oxide Nanoparticles to Fruits and Vegetables

Zinc oxide nanoparticles (ZnO-NPs) are eco-friendly, are easy to prepare, adapt to the environment and are recognized as safe materials (Zare et al., 2022; Tamimi et al., 2021; Jamdagni et al., 2018; FDA, 2019). Postharvest spraying of ZnO-NPs has been reported to enhance tomato quality, increase zinc concentration, facilitate postharvest ripening rate, promote antimicrobial protection and, as a consequence of all that, helped to prolong the shelf life of tomato fruits (Sharifan et al., 2021). The ZnO-NPs were also reported to be able to limit the presence of oxygen in the headspace for better antioxidant activity of the packaging materials (Zare et al., 2022). The migration of ZnO-NPs did not pose a risk to foods (Anugrah et al., 2020). The ZnO-NPs-containing nanocomposites based on gelatin and tragacanth had antimicrobial activity, better mechanical stress, improved barrier and thermal characteristics and reduced water vapor permeability of films (Shahvalizadeh et al., 2021).

**TABLE 8.1**

**Nanoenabled Coating/Emulsion Applications for Fruits and Vegetables**

| Treatment | Fruits/Vegetables | Treatment | Main Findings | Reference |
|---|---|---|---|---|
| Nanoemulsion with cinnamon oil | Tomato | Effect of edible coating with cinnamon oil nanoemulsion on storage at 27°C on quality of tomato fruits | Coating preserved physicochemical and microbiological quality better than uncoated tomatoes and had better quality for 15 days. | (Aisyah et al., 2022) |
| Nanoemulsion with whey proteins and thyme essential oil (EO) | Zucchini | Efficacy of nanoemulsion with thyme EO and whey proteins and comparison with guar and Arabic gum mix stabilizer (S) to Tween 20 (T) as coatings for zucchinis to prolong shelf life | S-coated had better integration of NPs than T-coated. T-coated had better firmness than S-coated zucchini at the end of storage at 10°C and were superior to the control T-coated increase shelf life of zucchini between 21–42 days. | (Bleoanca et al., 2022) |
| Coatings of zein NPs (ZNPs) and ε-polylysine (ZPLs) | Avocado | Effect of ZNPs, ZPLs, zein solution (ZS) and ε-polylysine solution (PL) coatings on fungal decay and quality on storage at 15°C | ZPL treated avocado lost least weight, compared to control. Pulp color of ZPL treatment avocado had least color difference in comparison to total color change. By day 15 of storage at 25°C control avocados were scarred, had major quality changes and showed fungal decay. | Garcia et al., 2022) |
| Zinc oxide NPs (ZnO-NPs) | Tomato | Investigate role of external application of ZnO-NPs to increase Zn content and on shelf life postharvest tomato for 4 weeks at different temperatures | ZnO-NPs mediated ripening rate and enhanced total lycopene content by 6% and Zn content by 17% and inhibited microbial growth by 47%. | (Sharifan et al., 2021) |
| Starch-silver NPs (starch-Ag-NPs) | Strawberry | Effect of starch-AgNPs coating on quality of strawberries | Starch-Ag-NPs had good microbiological quality for 6 days at 25 ± 3°C and for 16 days at 6 ± 2°C. NPs were spherical with average size of 12.7 nm. | (Taha et al., 2022) |

| Nanomaterial | Produce | Treatment | Results | Reference |
|---|---|---|---|---|
| Cellulose nanocrystals (CNC)/gellan gum-based coating | Mushroom | Spray CNCs/gellan gum coating on mushroom surface and record changes in respiration rate and color during storage of mushroom at 4°C | Higher levels of CNCs in coating showed lower respiration rate and extended shelf life. Coating decreased color changes compared to control. | (Criado et al., 2021) |
| Carboxymethyl cellulose (CMC)/guar gum-based silver nanoparticles (Ag-NPs) | Mango | Coat with the CMC-Ag-NPs and guar gum-Ag-NPs, and store at 13°C for 4 weeks | Coatings delayed fruit ripening by reducing respiration and maintained fruit quality. No trace of silver was detected in fruit pulp. | (Hmmam et al., 2021) |
| Nanoemulsion with fenugreek and flaxseed polysaccharide | Apple | Apply nanoemulsion treatment formulation to apple stored at 20°C for 14 days and investigate effects on antioxidant capacity activity, phenolic content and microbial quality | Response surface methodology indicated that a blend of 1.5 g fenugreek and 1.0 g flaxseed resulted in lowest weight loss, and microbial decay, kept firmness, total soluble solids and phenols and antioxidant capacity. | (Rashid et al., 2020) |
| Edible nanoemulsion coatings with sweet orange essential oil (EO) and sodium alginate | Tomato | Prepare and characterize nanoemulsion coating and examine coating effect on quality attributes of tomatoes at 22 ± 2°C for 15 day storage period | Stable nanoemulsion with NPs of 43.23 nm in size. Edible coating improved firmness, reduced weight loss and lowered total mesophilic bacteria. | (Das et al., 2020) |
| Titanium oxide NPs (TiO$_2$-NPs) with alginate and aloe vera | Tomato | Incorporation of (TiO$_2$-NPs) in alginate and aloe vera coating | Alginate/aloe vera film with 5 wt.% TiO$_2$-NPs resisted mass loss and spoilage. TiO$_2$-NPs improved mechanical and antimicrobial quality of tomatoes. | (Salama & Aziz, 2020) |
| Eugenol-entrapped casein NPs (EC-NPs) | pear | Optimize EC-NPs for improve antifungal effects of eugenol | EC-NPs were optimized with a mean size of 307.4 nm with efficacy of entrapment being 86.3%. EC-NPs had higher antifungal activity (>95.7%) against fungal spore. germination of fungus than for eugenol. | (Xue et al., 2019) |

(Continued)

**TABLE 8.1 (Continued)**
**Nanoenabled Coating/Emulsion Applications for Fruits and Vegetables**

| Treatment | Fruits/Vegetables | Treatment | Main Findings | Reference |
|---|---|---|---|---|
| Silver NPs (Ag-NPs) from pomegranate and orange peel | Tomato | Ag-NPs were synthesized from pomegranate and orange peel extract and characterized by UV spectroscopy. Ag-NPs assessed for antifungal activity on blight (*Alternaria solani*) of tomato. | Ag-NPs prevented the growth of *A. solani*, and average wavelengths for Ag-NPs for pomegranate and orange peels were, respectively, 437 nm and 450 nm. | (Mostafa et al., 2021) |
| Silver nanoparticles (Ag-NPs) | Carrot | Immerse carrot in various Ag-NPs concentration for 3 min and analyze for phenolic compounds | Ag-NPs increase the antioxidant capacity and phenolic content. | (Santoscoy-Berber et al., 2021) |
| Carnauba wax nanoemulsion (CWN) coating | Papaya | Apply CWN coating to fruits with various levels and compare to untreated and store fruits for 12–20 days between 16°C and 18°C. | High CWN concentration of 13.5% and 18.0% reduced weight loss, delayed ripening and decreased ethylene production compared to the control and coatings on the market. | (Zucchini et al., 2021) |
| Carboxymethyl cellulose (CG)/guar gum-based (GG)/Ag-NPs coating | Loquat | Effect of coating on postharvest storage of loquat at 4°C and 8°C for 4 weeks | During 1 month of storage at the two storage temperatures, coated fruits had less weight loss and total sugars, and increased ascorbic acid and total phenolic compounds to reduce postharvest losses. | (Ali et al., 2020) |
| Green silver nanoparticles (Ag-NPs) | Orange peel | Use 2, 4 and 8% w/v orange peel extracts to synthesize Ag-NPs as capping agent | Higher peel extract increased the formation of Ag-NPs. | (Bratovcic, 2020) |
| Nanocarrier of salicylate | Blueberry | Prepare polyethylene packaging and incorporate with a nanocarrier of salicylate and enclose blueberries for 13 days at 8°C | Blueberries had extended shelf life for 13 days at 8°C. Firmness was retained and sensory traits unaltered on release of release salicylate on blueberries. | (Bugatti et al., 2020) |

| Beeswax solid lipid NPs (BSL-NPs) | Strawberry | Prepare BSL-NPs by homogenization and apply nanocoating | Nanocoatings with 10 g L$^{-1}$ of BSLNPs increased shelf life of fruit stored in refrigeration. Higher level of 30 g L$^{-1}$ of BSL-NPs showed physiological damage due to limitation on respiration. Edible nanocoated strawberry maintained quality for 21 days at 4°C. | (Zambrano-Zaragoza et al., 2020) |
| --- | --- | --- | --- | --- |
| Nanofibrillated cellulose (NFC) | Tomato | Develop NFC from fibrillated cellulose of banana pseudostem fiber and combine with polyvinyl alcohol and polyacrylic acid by solvent casting method | Nanofilm wrapped tomatoes maintained fruit freshness for 15 days, conventional films for 8 days and unwrapped fruits for 6 days. NFC displayed UV protection, had high tensile strength and was thermally stable in comparison to conventional film. | (Ponni et al., 2020) |

### 8.3.1.3 Green Synthesis of Metal Nanoparticles

Metal-based NPs such as Ag, Au, Zn, $TiO_2$, $SiO_2$, ZnO and MgO, which are applied in the agri-food industry, may cause health risks due to their easy migration to human and animal organs and plant cells (Neme et al., 2021). Such risks could be minimized by using the "green" synthesis of NPs. The development of plant-based natural NPs is safer, environmentally friendly in nature, effective, clean and nontoxic (Bahrulolum et al., 2021; Bawazeer et al., 2022). The green synthesis of silver from rowanberries (*Sorbus aucuparia*) (Singh & Mijakovic, 2022), green tea leaf extract (Widatalla et al., 2022), the fibrous structure of pumpkin, *Cucurbita maxima* (Aktepe & Baran, 2022), ethanol extract of *Cucurbita pepo* leaves (Kumar, Mukherjee et al., 2020), fruit peel aqueous extract of eggplant, *Solanum melongena* L. (Das & Bhuyan, 2019), leaf extract of common grape (*Vitis vinifera*) (Acay et al., 2019) and maize leaves (*Zea mays* L.) (Eren & Baran, 2019) have served as capping and reducing agents. Ag-NP application was reported to control the early blight of tomato (*Solanum lycopersicum*), Ag-NPs also had antifungal activity against *Alternaria solani* and thus could be applied as alternatives to chemical fungicides (Mostafa et al., 2021). The green Ag-NPs have been synthesized from 2, 4 and 8% w/v orange peel extracts, which served as capping agents for the stabilization of silver NPs in biosynthesis (Bratovcic, 2020).

Starch has functioned as an environmentally safe capping agent for the synthesis of Ag-NPs with an average size range of 12.7 nm. The application of starch Ag-NPs as coating to strawberry fruits resulted in good microbiological quality for the fruits stored at 6 days at $25 \pm 3°C$ and 16 days at $6 \pm 2°C$ (Taha et al., 2022). The silver concentration was reduced in strawberries by washing. In other research, the Ag-NPs were synthesized by using *Moringa oleifera* leaves as a main reducing and stabilizing agent. A concentration of 30 ppm Ag-NPs was found to be the most suitable to create resistance against canker disease in kinnow (*Citrus reticulata*) (Hussain et al., 2019). The synthesized Ag-NPs developed resistance against *Xanthomonas axonopodis* pv. citri and altered the biochemical profiling in *C. reticulata*. For the "green" synthesis of ZnO-NPs, the flower extract of *Nyctanthes arbor-tristis* was employed with the NPs having antifungal potential (Jamdagni et al., 2018).

### 8.3.2 NANOBIOPOLYMERS

Nanopolymers are broad polymer/co-polymer-based nanotechnologies of various structures, shapes and functional forms based on the number of dimensions in the nanometer range (Larena et al., 2008). Biopolymer-based nanocomposite films and coatings are environmentally friendly alternatives to increase the shelf life and/or storage duration of fruits and vegetables during postharvest handling (Ashfaq et al., 2022). Some common natural polymers are gelatin, starch, chitosan, cellulose and polylactic acid.

The NMs could bolster physicochemical, mechanical, barrier properties and add functionalities to the biopolymers (Basumatary et al., 2022). Hybrid NPs produced from two or more biopolymers deliver combined multiple biological efficacies (Luo et al., 2020). The incorporation of NPs and nanoencapsulated active compounds into blended biopolymer-based films could improve the functional performance of packaging materials (Khezerlou et al., 2021; Chisenga et al., 2020).

### 8.3.2.1 Nanochitosan

Chitosan (CS) is a linear polysaccharide that comes from the outer skeleton of shellfish, including shrimp, crab and lobster. The CS treatments can the extend shelf life of fresh produce and are suitable alternatives for synthetic chemicals. The CS is recognized as safe, biodegradable and antimicrobial (Romanazzi et al., 2017; Yuan et al., 2016). The CS is used in various chitosan formulations such as nanofibers, nanocomposites or nanocapsules, as carriers for the immobilization or encapsulation of bioactive compounds for the extension of the shelf life of perishables (Meena et al., 2020). The NPs added to blended films enhance the beneficial functionalities of packaging materials (Chisenga et al., 2020). Chitosan nanoemulsions provide a protective coating for fresh produce and deliver antimicrobial agents and functional compounds (Chaudhary et al., 2020). They have good mechanical strength, transparency, good dispersibility, physical barrier properties and thermodynamic stability, which are highly beneficial for the improvement of fresh produce's storability.

The main effects of nanoenabled CS applications on fruits and vegetables are summarized in Table 8.2. For example, cucumbers are perishable and have a short storage life of about 10–14 days due to their high moisture content. Chitosan/nano-titanium oxide crystals/sodium tripolyphosphate films have extended the shelf life of cucumbers by maintaining quality attributes such as color, skin toughness, crispness index, reduced sugars and total sugars and antimicrobial contamination against *Salmonella* spp. for 21 days at 10°C (Helal et al., 2022). In another study, cucumbers coated with blended chitosan/sodium tripolyphosphate/titanium dioxide NPs and a blend of chitosan with titanium dioxide NPs remained moist and green at 10°C for 21 days (Khojah et al., 2021). The sodium tripolyphosphate served as cross-linker for stabilizing the nanoparticle polymers (Helal et al., 2022; Khojah et al., 2021).

In another study, it was reported that the chitosan solutions that were supplemented with 2% chitosan-montmorillonite coating reduced ethylene production and respiration rate of banana fruits (Wantat et al., 2022). Cu-chitosan-NPs were able to inhibit microbial decay (0–5%) in stored tomatoes up to 21 days at $27 \pm 2°C$. The Cu-chitosan NPs formed an invisible and intangible nanonet over the tomato surface as a potential hurdle for all openings (Meena et al., 2020).

The antioxidant activities and postharvest quality of nanocoated mushroom with titanium and silica or in combination were evaluated on storage at 4°C (Sami et al., 2021b). The silica-CS film delayed the respiratory spike of mushrooms, and the titanium-CS film showed low oxygen production rate and the production of thiobarbituric acid-reactive substances. The combined nanocoating films of either titanium-CS or silica-CS can reduce cell degradation and the oxidation processes of crops.

Postharvest guava fruits dipped in a solution of CS (44.5 kDa, 1%) in combination with 0.04% nano-SiO2 for 1 min exhibited the best antimicrobial inhibition against the common microorganisms of guava fruits (To et al., 2022). Chitosan film coatings with silicon dioxide and nisin were applied for packaging of fresh blueberry during postharvest storage. This coating limited the shrinking, decay rates and texture changes of blueberries on storage for 8 days at 28°C (Eldib et al., 2020).

**TABLE 8.2**

**Nanoenabled Chitosan Applications for Fruits and Vegetables**

| Nanotechnology Applications | Fruits and Vegetables | Treatments | Main Effects | References |
|---|---|---|---|---|
| Chitosan/nano titanium oxide crystals/sodium tripolyphosphate (CS-TiO$_2$-ST) | Cucumber | Effect of CS-TiO$_2$-ST to enhance cucumber quality during storage | The CS-TiO$_2$-ST retains maximum greenness and delayed the microbial contamination of *Salmonella* compared to CS-nanosamples. The highest crispness index was detected for CS-nano. | (Helal et al., 2022) |
| Chitosan (CS) supplemented with chitosan-montmorillonite (CS-MMT) nanocomposites | Banana | CS and CS supplemented with CS-MMT nanocomposite solutions as banana fruit coating by the dipping technique | CS supplemented with 2% CS-MMT nanocomposites maintain postharvest quality of banana fruit. Retarded peel color change and reduce electrolyte leakage, | (Wantat et al., 2022) |
| Chitosan (CS) with SiO$_2$-NPs | Guava | Evaluate the effectiveness of CS-SiO$_2$-NPs for common antibacterial and antifungal | Mixture of 0.04% nano-SiO$_2$ and 1% low-molecular-weight CS 44.5 kDa inhibited bacteria and fungi. | (To et al., 2022) |
| Chitosan nanoemulsion (CS) with CS-NPs | Raspberries | Coat fruits in CS emulsion with CS-NPs of and store for 9 days at 4°C, 85–95% RH | Treated fruits with CS-NPs of 5 g L$^{-1}$ had highest phenolic compounds, enzyme, and antioxidant activity after storage. | (Ishkeh et al., 2021) |
| Chitosan/sodium tripolyphosphate/TiO$_2$-NPs CS-TiO$_2$-NPs-ST) and chitosan/TiO$_2$-NPs | Cucumber | Evaluate physicochemical properties and fungal load of cucumber treated with CS-TiO$_2$-NPs-ST) and CS-TiO$_2$-NPs during storage at 10°C | Cucumbers with CS-TiO$_2$-NPs-ST) were hydrated and green-colored up to day 21 of storage with excellent quality for consumption. Shelf life was 14 days for cucumbers stored with CS coating. CS-TiO$_2$-NPs-ST treatment delayed chilling injury, reduced loss of ascorbic acid with highest chlorophyll content. | (Khojah et al., 2021) |

| Material | Commodity | Objective | Results | Reference |
|---|---|---|---|---|
| CS nanomaterial films with silicon (CS-SiO$_2$-NPs) and titanium (CS-TiO$_2$-NPs) | Blueberry | Develop CS coating with nanomaterial films and detect physical, mechanical and microbiological effects on fruits at commercial storage temperature | CS-TiO$_2$-NPs showed steady increase in enzyme activities of polyphenoloxidase and peroxidase. CS-SiO$_2$-NPs showed minor variation in acidity, anthocyanin and restricted microbial growth. | (Li et al., 2021) |
| Silica-CS film and Titanium-CS film | Mushroom | Determine postharvest quality of CS-nanocoated mushrooms at 4°C for 12 days | Silica-CS film was most effective for polyphenol constituent and antioxidant activity and highest oxidative enzyme activities. Titanium-CS film show lowest O$_2$ production and thiobarituric acid reactive components. | (Sami et al., 2021b) |
| Nano-chitosan silicon (CS-SiO$_2$) and nisin | Cantaloupes | Coating effects of chitosan and CS-SiO$_2$ with nisin on microbial stability, physical, chemical and sensory properties on storage at 4°C for 8 days for fresh-cut cantaloupes | Both CS-SiO$_2$ and CS-SiO$_2$-nisin extended the shelf life by retaining color, vitamin C and peroxidase activity up to 8 days. | (Sami et al., 2021a) |
| Chitosan-thyme essential oil (EO) nanocoating (CS-TEO) chitosan nanoparticles (CS-NPs) | Green bell pepper | Apply different applications based on CS-NPs and CS-TEO-NPs (15, 30 and 45%) for green bell pepper | Nanocoating formulation of 15% CS-NPs showed the lowest microbial colony-forming units and disease incidence. | (Correa-Pacheco et al., 2021) |
| Clove essential oil (EO) encapsulated in CS-NPs (CEO-CS-NPs or encapsulated oil) | Pomegranate | Effects of coatings and encapsulated oil on fungal growth and shelf life of pomegranate arils on storage | CEO-CS-NPs extended the pomegranate aril shelf life for 54 days at 5°C. Fungal decay was detected on day 18 at 5°C for uncoated arils. | (Hasheminejad & Khodaiyan, 2020) |

*(Continued)*

**TABLE 8.2 (*Continued*)**
**Nanoenabled Chitosan Applications for Fruits and Vegetables**

| Nanotechnology Applications | Fruits and Vegetables | Treatments | Main Effects | References |
|---|---|---|---|---|
| CS-NPs with α-pinene (P-CS-NPs) and a nanostructured edible coating (EOC-P-CS-NPs) | Bell peppers | Apply P-CS-NPs and the EOC-P-CS-NPs to bell peppers to assess postharvest quality of bell peppers Inoculate with *Alternaria alternata* under cold storage for 21 days at 12°C and store for 5 days at 20°C | Major differences in weight loss were obtained for P-CS-NPs and EOC-P-CS-NPs at 3% and 6% compared to the control but no difference in quality. Higher carotene for peppers without *A. alternata*. | (Hernández-López et al., 2020) |
| Cu-chitosan nanoparticles (Cu-CS-NPs) | Tomato | Extend shelf life of tomato nanonet effect Cu-CS-NPs on shelf life of stored tomato | Keeping quality of tomato up to 21 days at 27 ± 2°C, 55 ± 2% RH Cu-CS-NPs preserve color loss, limit microbial decay, physiological loss, respiration rate and maintained fruit firmness during storage. | (Meena et al., 2020) |
| Gelatin with propolis (PEE) within zein nanocapsules | Raspberries | Incorporate PEE into gelatin-based films and encapsulate within zein nanocapsules Evaluate antifungal activity of PEE during storage at 5°C | PEE enhanced antifungal activity. Showed greater inhibition on *Penicillium digitatum* and *Botrytis cinerea*. | (Moreno et al., 2020) |
| Chitosan film with ZnO and Ag-NPs loaded with citronella oil (CS-ZnO-Ag-NPs/CEO) | Grapes | Fabricate CS films with ZnO and Ag-NPs by casting method as filler/antimicrobial agents | CEO-NPs enhanced the antimicrobial activity of film. Nanocomposite films had lower water vapor permeability compared to CS control film. | (Motelica et al., 2020) |

| Material | Target | Objective | Results | Reference |
|---|---|---|---|---|
| Chitosan-titanium dioxide nanocomposite (CS-TiO$_2$-NPs) | Mangoes | Effects of CS and CS/TiO$_2$-NPs composite coating on physiology and storage of mangoes | Decay index was 14.49% lower for nanocomposite CS-TiO$_2$-NPs coated fruits than for the control mangoes. Nanocomposite film enhanced firmness of fruits. | (Xing et al., 2020b) |
| Chitosan-nanosilicon dioxide (CS-SiO$_2$) and chitosan-nanosilicon oxide with nisin (N) (CS-SiO$_2$-N) | Blueberries | Add of 1% CS-SiO$_2$-NPs and 1% N to CS solution for coating of blueberries | CS-SiO$_2$ and CS-SiO$_2$-N controlled shrinking and decay rates of blueberries stored under ambient temperature for over 8 days. | (Eldib et al., 2020) |
| Edible fungal CS (gel, NPs and gel-NPs) | strawberries | NPs were prepared by ionic gelation method. Evaluate effects of CS coatings (gel, NPs, and gel-NPs) on physicochemical, sensorial, and microbiological characteristics of strawberries | Nanocomposite showed changes to fungal morphology in artificially infected strawberries. CS-Gel, CS-NPs and gel-NPS showed minimum inhibitory concentration values. | (Melo et al., 2020) |
| Clay/chitosan nanocomposite (C-CS) | Orange | Test in vitro and in vivo fungicidal activity of C-CS against *Penicillium digitatum* | *P. digitatum* was inhibited at 20 µg mL$^{-1}$ with C-CS. | (Youssef & Hashim, 2020) |
| Chitosan nanoparticles (CS-NPs) from chitosan (CS) | Tomato | Investigate antifungal effects of CS and CS-NPs against phytopathogenic fungi. | CS-NPs show maximum inhibition on *Fusarium oxysporum* and next by *Phytophthora capsici*. Both CS and CS-NPs inhibited the growth *Xanthomonas campestris* and *Erwinia curatoria* | (Oh et al., 2019) |

*(Continued)*

**TABLE 8.2 (Continued)**

**Nanoenabled Chitosan Applications for Fruits and Vegetables**

| Nanotechnology Applications | Fruits and Vegetables | Treatments | Main Effects | References |
|---|---|---|---|---|
| Silica (S) chitosan (CS) and copper (Cu) nanoparticles (NPs) | Table grapes | Investigate in vitro and in vivo fungicidal activity of S, CS and CU-NPs and combination against gray mold *Botrytis cinerea* on table grapes | CS and S-NPs inhibited hyphal growth and/or altered hyphal morphology as shown by scanning electron microscopy. NPs interact with fungal DNA. Highest concentration of S and CS–NPs degraded DNA by affecting its integrity. | (Hashim et al., 2019) |
| Chitosan-nanosilica-sodium alginate film (CS-S-A) and abscisic acid (ABA) | Jujube | Effect of ABA and a CS-S-A composite film on color change and quality in cold storage | ABA induced ripening. Composite films prolonged the shelf life for about 1 month. | (Kou et al., 2019) |
| Chitosan-nano-titanium dioxide (CS-$TiO_2$). Chitosan with titanium, thymol, and tween (CS/$TiO_2$/TT) | Cantaloupe | Effect of CS-$TiO_2$ coating with antimicrobial agents on chilled ready-to-eat cantaloupe fruit | CS-$TiO_2$-TT film controlled mold and yeast growth and maintained quality such as ascorbic acid compared to the uncoated fruit. | (Qiao et al., 2019) |

In another study, nisin was also added to the chitosan/nanosilica coating treatment. This treatment was most effective by forming semi-films against aerobic microorganisms, enzyme activities, keeping color and microbiological quality on the storage of cantaloupes at 4°C for 8 days (Sami et al., 2021a). Nanosilicon dioxide and nisin are approved food safety additives (Sami et al., 2021a; Kou et al., 2019).

In this research, a CS composite film incorporating nanosilica-sodium alginate was noted to prolong the storage quality of winter jujube fruit for about 1 month, while the abscisic acid induced ripening and coloring (Kou et al., 2019). The CS edible films and coatings are carriers for essential oils (EO) such as thyme. clove, α-pinene and citronella. These EOs augment the antioxidant, antibacterial and antifungal efficacy of chitosan film to maintain fruit and vegetable quality and to control postharvest decay (Yuan et al., 2016).

Various combinations of chitosan nanoparticles (CS-NPs) with chitosan-thyme essential oil (EO) NPs were used to provide resistance against pathogenic bacteria for bell peppers (Correa-Pacheco et al., 2021). A nanocoating of 15% CS-NPs demonstrated the lowest disease incidence and protected the bell peppers against *Pectobacterium carotovorum*. In this study, chitosan NPs and nanostructured edible coating with CS-NPs and α-pinene inhibited the postharvest diseases due to *Alternaria alternata* and preserved the physicochemical quality of bell pepper during the cold storage period. The pepper disease occurred from day 22 on storage at 20°C. (Hernández-López et al., 2020). In another study the CS-NPs with encapsulated clove EO coating preserved the quality of pomegranate arils compared to the uncoated ones and controlled undesirable quality changes such as the microbial physicochemical and sensory properties (Hasheminejad & Khodaiyan, 2020). A CS film modified with ZnO-NPs and Ag-NPs with loaded citronella EO improved the antimicrobial performance of the nanocomposite film (Motelica et al., 2020).

A CS coating with nanomaterial films such as silicon and titanium retained nutrients, had minimal changes in acidity and anthocyanins and controlled microbial growth to lengthen the shelf life of fresh blueberry fruits (Li et al., 2021). Raspberries treated with emulsion containing CS-NPs at 5 g L$^{-1}$ had the highest phenolic compound, antioxidant and enzymatic activity on storage at 4°C for 9 days (Ishkeh et al., 2021). Chitosan-based coating films, which were an incorporated TiO$_2$-NPs coating, retained the nutrient composition of mangoes and preserved fruit quality at 13°C (Xing et al., 2020b). There was a lower decay index of 14.49% for nanocomposite-coated fruits compared to the control group.

In a different study, several combinations of edible CS coatings (gel, NPs and gel-NPs) were tested and were noted to have high scavenging activity on strawberries (Melo et al., 2020). The application of CS-NPs alone in the edible coating resulted in better fruit preservation than the use of CS gel edible coating or coating composed with gel enriched with CS-NPs. The gel and NPs decreased moisture and weight loss, microbiological growth and maintained the anthocyanin content and sensory quality of the strawberries.

In the control of green mold, *Penicillium digitatum* of citrus, the clay/CS nanocomposite aimed to have no fungicide residues. Clay/CS nanocomposite was prepared by anion exchange reaction (Youssef & Hashim, 2020). This nanocomposite inhibited the green mold of citrus in vitro and in vivo, and its effects on pathogens were shown by the genotoxicity of *P. digitatum* by DNA, malformation and irregular branching of the hyphae. The NMs compounds such as Si-NPs and CS-NPs have been effective antifungal agents for the control of *Botrytis cinerea*, the gray mold of table grapes. The CS-NPs either prevented hyphal growth or/and varied the hyphal morphology through cell wall disruption and withering and showed too much septation as revealed by scanning electron microscopy (Hashim et al., 2019). Edible CS-NPs coatings have delayed the ripening of grapes, decreased weight loss and soluble solids, reduced sugar, increased moisture retention and maintained sensory quality (Melo et al., 2018). According to the studies of Oh et al. (2019), the CS-NPs demonstrated inhibition against a broad range of phytopathogens of tomato. The chitosan-TiO$_2$ nanocomposite film delayed ripening and quality changes of cherry tomato and exhibited ethylene photodegradation capability (Kaewklin et al., 2018) in comparison to CS film only and control (without film). A chitosan/nano-titanium dioxide coating incorporating antimicrobial agents of thymol and tween on ready-to-eat cantaloupes inhibited mold and yeasts and enhanced shelf life (Qiao et al., 2019). This nanocoating also maintained the ascorbic acid content and had less loss of polyphenol oxidase activity in comparison to the uncoated fruits.

### 8.3.2.2  Nanocellulose

Cellulose nanocrystals have major physical and mechanical properties that facilitate uses in bio-based packaging films for crops (Metzger et al., 2018). Nanocellulose includes cellulose nanofibers (CNF), cellulose nanocrystals (CNCs) and cellulose nanofibrils (CNFs), which are applied as fruit coatings for postharvest physiological requisites and to enhance the storage of fruits (Jung et al., 2019). The cellulose nanostructures serve as reinforcement for food packaging (Azeredo et al., 2017). A nanofilm from nanofibrillated cellulose of banana pseudostem fiber with the inclusion of polyvinyl alcohol and polyacrylic acid was produced by the solvent casting method and tested on tomatoes. The shelf life of the wrapped tomatoes in the nanofilms had a fresh storage life of 15 days in comparison to those on the conventional film for 8 days and the control (without wrapping) for 6 days (Ponni et al., 2020).

In a different study, the surface of mushroom was sprayed with cellulose nanocrystals gellan gum-based coating (Criado et al., 2021). The coating slowed the respiration rate and showed less color difference in mushrooms compared to the untreated mushrooms on storage at 4°C (Table 8.1). In a different study, it was recommended that the inclusion of 5% cellulose nanocrystals in chitosan coating delayed green chlorophyll degradation of green D'Anjou pear peels, inhibited internal browning, retained fruit firmness delayed senescence scalding during 3 weeks of storage at 20°C for 3 weeks compared to commercial coating (Deng et al., 2017). Similarly, the carboxymethyl cellulose (CMC) and guar gum-based Ag-NPs delayed mango fruit ripening and retained the fruit quality during cold storage (Hmmam et al., 2021). The CMC and guar gum improved the stability and mobility of Ag-NPs, and the guar gum was the stabilizer and capping agent.

### 8.3.2.3  Starches and Other Nanobiopolymers

Starch films as packaging materials have limitations due to poor mechanical and barrier properties. Cellulose nanofibers (CNFs) and thymol were included in starch to reinforce the properties of nanocomposite films (Othman, 2021). The thymol improved the vapor barrier properties without affecting the color and opacity of the film. The effects of inclusion of zinc oxide nanoparticles (ZnO-NPs) and fennel EO in potato-starch-based nanocomposite film improved its tensile strength, lowered water vapor and oxygen permeability and increased antimicrobial activities (Babapour et al., 2021). In other research, the nanosphere form of ZnO-NPs was most influential on the mechanical strength and barrier properties of the tapioca starch bionanocomposite film (Tamimi et al., 2021). The starch nanocomposite films had good barrier resistance against UV light. The incorporation of ZnO-NPs and stearic acid on the characteristics of edible cassava-starch-based bionanocomposite films increased the thickness, color difference and tensile strength and decreased water vapor transmission rate (Wardana et al., 2018). Also, the improvement in mechanical and physical properties was achieved with cassava starch + 2% ZnO-NP and cassava starch + 2% ZnO-NP + 30% stearic acid, respectively. Avocadoes were coated with biopolymeric coatings made of zein NPs and ε-poly-L-lysine NPs. By day 36 at 25°C storage, both the zein NPs and ε-poly-L-lysine NPs-treated avocados retained their original physical appearance and texture, unlike the untreated avocadoes (Garcia et al., 2022). The addition of ε-poly-L-lysine NPs diminished the severity of postharvest fungal disease.

## 8.4 NANOCOATINGS

Nanocoatings incorporate bioactive or functional compounds, which allow for the controlled release of these compounds (Poonia & Mishra, 2022). Edible coatings are biodegradable alternatives to resolve postharvest issues during the storage of fresh produce (Abhirami et al., 2020; Rashid et al., 2020) and have been developed at the nanoscale with improved gas barrier properties and mechanical strength (Poonia & Mishra, 2022). These coatings are safe due to either the natural biocide activity or inclusion of antimicrobial compounds, and their being residue free and a sustainable substitute for common packaging materials (Duguma, 2022; Mohamed et al., 2020; Ghadermazi et al., 2019). The edible coating on fruits was also reported for preventing viral transmission for COVID-19 patient recovery (Baranwal et al., 2022).

The advent of nanotechnology has made it possible to engineer nanostructures in edible coatings to achieve desirable functional properties. The efficacy of nanocoatings can be further enhanced through electrospraying rather than the traditional spraying technique (Marboh & Gupta, 2020). In this section some edible nanocoatings from methylcellulose, cellulose hydrogel, carrageenan, alginate and aloe vera gel, beeswax solid lipid NPs and propolis extract, as applied to some fruits such as strawberry, tomato, raspberries, are reviewed.

### 8.4.1 NANOCOATING MATERIALS AND APPLICATIONS TO FRUITS AND VEGETABLES

Edible coating materials such as methylcellulose, carrageenan, alginate, proteins and lipids are thin layers prepared from edible materials to reduce the microbial load on fruit and vegetable surface after harvesting (Kumar et al., 2020b; Salama et al.,

2019). The incorporation of antimicrobial plant extracts in the medium of edible coatings and/or films by nanotechnology could offer postharvest safety of produce (Nxumalo et al., 2021).

The synthesis of individual nanostructures and their integration into hybrid nanocomposites generate structures with synergistic and new functionalities due to the joining of the properties of the individual NPs (Winter et al., 2020). They have improved the antimicrobial and antifungal coatings efficacy compared to use of only NPs (Liu et al., 2020b).

A biocompatible, antibacterial hybrid material system was based on Ag nano-clusters fixed in cellulose hydrogel. (Liu et al., 2020c). In comparison to the silver nanoclusters, the Ag-nanocluster-based fabricated hydrogel demonstrated extensive antimicrobial efficacy against bacteria due to the regulated release of silver particles. The properties of chitosan can be improved when employed in the form of NPs (Melo et al., 2018). Chitosan as an edible film could replace waxy coatings of fruits and confer antimicrobial activity and longer postharvest shelf life (Adiletta et al., 2021; Motelica et al., 2020; Xing et al., 2016, 2020a). Edible nanolaminate coatings with alginate, chitosan, and extracts of *Flourensia cernua* were developed for tomato and inhibited microbial growth, decreased the weight loss, and lengthened shelf life (de Jesús Salas-Méndez et al., 2018).

Table 8.1 shows that nanocoatings with 10 g $L^{-1}$ of beeswax solid lipid nanopar-ticles (BSL-NPs) increased the shelf life of stored strawberry at 4°C (Zambrano-Zaragoza et al., 2020). A treatment with 10 g $L^{-1}$ of BSL-NPs resulted in the lowest weight loss and decay index and loss of firmness, and a color change upon appli-cation of 30 g $L^{-1}$ of BSL-NPs resulted in physiological damage to the strawber-ries. Proteins and propolis extract (PEE) collected from hives were enveloped in active edible coatings and applied to raspberries (Moreno et al., 2020). The PEE was included in the gelatin edible films by mixing PEE directly in the protein matrix and by forming nanocapsules by encapsulation of the PPE in the zein. Table 8.2 shows that the encapsulation of PEE into zein capsules had antifungal activities and reduced the infection in raspberries stored at 4°C for a longer period.

Alginate-based edible coatings augmented with aloe vera gel were reinforced with TiO$_2$-NPs of a size range of 201.38–28.81 nm to extend the storage of tomatoes. The inclusion of TiO$_2$-NPs in various percentages improved both mechanical and antimicrobial properties. The multifunctional alginate/aloe vera film containing 5 wt.% of TiO$_2$-NPs resisted weight loss and reduced spoilage of tomatoes (Table 8.1, Salama & Aziz, 2020).

## 8.5   NANOEMULSIONS

As kinetically stable systems, nanoemulsions offer several advantages over con-ventional emulsions such as small droplet size, improved homogeneity, increased transparency, and they display excellent physical resistance against gravitational par-tition and droplet accumulation (Rashid et al., 2020; Prakash et al., 2018; Hernández-Fuentes et al., 2017). In addition, nanoemulsions offer a gradual and controlled release of compounds that increase bioactivity, extend shelf life and improve the nutritional quality of fruits (de Oliveira Filho et al., 2021). Edible oil nanoemulsions have synergistic benefits when combined with other hurdles such as either as a washing disinfectant or applied to edible coatings to provide microbiological safety of fresh and minimally processed produce (Prakash et al., 2018). Essential oils from aromatic plant species possess antimicrobial and antioxidant activities that aid in postharvest preservation (Ziv & Fallik, 2021; Prakash et al., 2018). The toxic effects of nanoemulsions are related their nanodimension, which impacts absorption, distribution, metabolism and excretion in human health (Pradhan et al., 2015). In this section, some nanoemulsion-loaded oils from thyme, cinnamon, orange and corn and Arabic gum, Tween 20, fenugreek, flaxseed polysaccharide, sodium alginate and carnaubax wax are reviewed as applied to some fruits and vegetables.

## 8.5.1 Nanoemulsion Materials and Applications to Fruits and Vegetables

The nanoemulsions based on thyme essential oil (EO) were synthesized by the ultrahigh agitation method. The nanoemulsions were used in the formation of chitosan nanocapsules (González-Reza et al., 2021). The nanocapsules of thyme EO showed good antioxidant stability on storage at 4–25°C after 5 weeks of storage, and its capacity was related to the residual oil concentration. In another related study, nanoemulsion formulae containing thyme EO and whey proteins were prepared in either Arabic gum and Arabic mix or Tween 20 and applied as coatings to zucchini. Table 8.1 shows that the nanoemulsions with guar and Arabic gum mix stabilizer showed a better rheological restructuring quality than Tween 20 (Bleoanca et al., 2022). The Tween-nanoemulsion-coated zucchinis were firmer than both coated and the control at the end of storage at 10°C.

In a different study, Rashid et al. (2020) reported that nanoemulsion treatment with 1.5% fenugreek and 1.0% flaxseed polysaccharide using corn oil provided the best overall quality attributes of apples compared to other treatments. Polysaccharide-based coating could serve as an eco-friendly natural approach to enhance postharvest quality of organic fruits. An edible coating containing cinnamon EO was applied as a nanoemulsion coating to tomato fruits at 27°C for 15 days. The coating delayed quality changes and maintained the physicochemical properties of tomato better than the control throughout storage (Aisyah et al., 2022). It was reported that 1% (v/v) cinnamon EO nanoemulsion coating had inhibited the growth of *Staphylococcus aureus* and *Escherichia coli* (Aisyah et al., 2018). The nanoemulsion showed better antibacterial activity compared to coarse emulsion and the cinnamon EO emulsion.

Sodium-alginate-based edible coating incorporating sweet orange EO delayed the ripening and microbial spoilage of tomatoes (Table 8.1, Das et al., 2020). This coating was effective against *Salmonella* and *Listeria* either individually or as combined cultures. The higher whiteness index of the nanoemulsion coatings was desirable for fruit quality and product marketing perspectives. Treatments with 13.5% and 18.0% carnauba wax nanoemulsion were effective in reducing weight loss and reducing the ripening of papaya fruits in comparison to the untreated, commercial coating and low carnauba wax nanoemulsion (Zucchini et al., 2021). Also, there was reduced disease severity of fruit at high carnauba concentration upon nanoemulsion application.

## 8.6   NANOPACKAGING/FILMS

There is an increasing demand for green and safe food packaging such as biopolymer-based films and coatings on produce surfaces as alternatives to synthetic packaging (Kumar et al., 2019; Shankar & Rhim, 2018). Natural polymers, due to their abundant and renewable sources, have been used as NMs for the fabrication of biodegradable active cling films (Costa et al., 2021; Valdés et al., 2021; Malhotra et al., 2015) for the preservation of food quality. This growing interest in biodegradable packaging stems from the various policy and legislative changes moving away from the use of synthetic plastic. The need for post-consumer waste education and management in the use of NMs in food packaging is warranted for the positive environmental and economical impact (Ahmad et al., 2023). Biodegradable packaging materials are brittle and have low transparency (Chisenga et al., 2020) and thus, with the incorporation of NPs, could improve these functionalities (Ediyilyam et al., 2021).

A nanocomposite is a multilevel solid material where one of the levels has one, two or three dimensions of less than 100 nm, and at least one nanomaterial has different physical and chemical properties (Kumar & Krishnamoorti, 2010). It is comprised of polymers and in combination with NPs provide increase functional properties (Pradhan et al., 2015). Nanofilms carry metal NPs as antimicrobial agents in the form of protective barriers that diminish respiration rate, control decomposition and color changes, and balance and lengthen the shelf life of fruits and vegetables (Rana et al., 2021). Organic NPs are increasingly being employed to develop active intelligent packaging to extend the shelf life of produce and to aid in the communication system throughout the postharvest chain (Ashfaq et al., 2022). Nanoclays and nanolaminates have been used in nanopackaging material for fresh produce due to their mechanical, water and oxygen barrier and antimicrobial functionalities (Ashfaq et al., 2022; Sharma et al., 2017; Echegoyen et al., 2016). Nanotechnology has allowed for the controlled release of preservatives/antimicrobials within the package (Casalini et al., 2022; Sharma et al., 2017).

In this section, some novel nanofunctional films from pullulan and carboxylated cellulose nanocrystal, carboxymethyl chitosan, sago starch-based, corn starch/polyvinyl alcohol (PVA) blends, cellulose nanofibrils from bleached bagasse pulp with inclusion of metal oxide NPs and their applications to fruits and vegetables are included.

### 8.6.1   BIOPOLYMER-BASED ACTIVE NANOFILM/PACKAGING MATERIAL

Active packaging is defined as the incorporation of active components into packaging film or a container to preserve the produce (Kumar et al., 2019). Nanofilm/packaging materials could have high barrier properties, incorporate active compounds, be heat tolerant and biodegradable and possess nanosized antimicrobials, nanopreservatives and nanosensors that provide information on the freshness of produce (Pascuta & Vodnar, 2022; Bhuyan et al., 2019). Nanoenabled sensors help to reduce foodborne illness in produce safety (Bhusare & Kadam, 2021) by detection of pathogenic bacteria, food-contaminating toxins and adulterants (Hossain et al., 2021). Nanocomponents have been integrated in ultra-thin polymer substrates for radio frequency identification reader chips with biosensors to detect foodborne pathogens and to sense other parameters such as moisture, temperature and odor of food products (Majid et al., 2018).

## 8.6.2 Novel Nanofunctional Films

Pullulan and carboxylated cellulose nanocrystals were combined with tea polyphenol (TP) by the solution casting method to form active food films. Scanning electron microscopy revealed that the TP was distributed within the bionanocomposite matrix (Chen & Chi, 2021). The addition of TP enhanced the ultraviolet barrier properties, antioxidant activity and antimicrobial activity of the bionanocomposite films with reduced transmittance.

Functional films of carboxymethyl chitosan were fabricated by incorporating gliadin/phlorotannin nanoparticles (GP-NPs) using a solution casting method (Zhao et al., 2022). The added GP-NPs improved the physical, antimicrobial and antioxidant characteristics of the film (Zhao et al., 2022).

Essential oils have been included in nanocomposites to improve and/or convey functional properties to packaged food products (Nath et al., 2022). Various levels of nano-titanium dioxide ($TiO_2$-N) and cinnamon EO were incorporated into sago starch film (Arezoo et al., 2020). The combination of EOs and NPs had synergistic effects that improved the functional properties and antibacterial activity of the sago starch films. Innovative composite films from corn starch/polyvinyl alcohol (PVA) blends were developed. The chitosan nanoparticles (CS-NPs) of diamers (100 nm) served as reinforcement agents to improve the starch-PVA matrix for enhancement of the physical and mechanical properties suitable for packaging (Garavand et al., 2022). The films demonstrated higher inhibitory effect of CS-NPs against gram positive than gram negative bacteria.

Cellulose nanofibrils (CNFs) from bleached bagasse pulp could serve as green functional material for chitosan/oregano essential oil (CS-OEO) biocomposite packaging film. Films with 2% (w/w) OEO were effective against *Escherichia coli* and *Listeria monocytogenes* (Chen et al., 2020). The added CNFs to CN-OEO film improved the tensile strength but did not improve barrier properties with the increase in CNFs contents.

## 8.6.3 Incorporation of Metal Nanoparticles (NPs) in Packaging Materials

Active packaging has engaged metal and metal oxide NPs (Pathakoti et al., 2017). In this chapter, the $TiO_2$-NPs and ZnO-NPs are reviewed. The $TiO_2$-NPs have served as functional nanofiller in the fabrication of eco-friendly smart/active packaging films (Sani et al., 2022). They also show significant UV-blocking, gas scavenging activity and have wide antimicrobial activity and negligible migration rates. The $TiO_2$-NPs and Miswak (*Salvadora persica* L.) extract were used in the fabrication carboxymethyl cellulose-based bionanocomposites. The $TiO_2$-NPs improved the protection from UV light, while the Miswak-extract-containing nanocomposites completely blocked it (Ahmadi et al., 2019).

The addition of zinc-oxide-reduced graphene oxide improved the thermal-mechanical properties of polybutylene, adipate-co-terephthalate nanocomposite films (Charoensri et al., 2021). This nanocomposite film showed antibacterial activity against *Escherichia coli* and *Staphylococcus aureus*, and the migration of the Zn ions from the film was found to be safe. The antibacterial properties of the films could be associated with the interaction of positive charges of the bio-nanocomposite film and with the associated negatively charged bacterial cell walls. In another related research, the zinc migration on the packaging functionality of polyethylene-ZnO low-density nanocomposite films into food stimulants was investigated (Bumbudsanpharoke et al., 2019). The loss of ZnO-NPs due to the dissolution on the film surface lowered the antimicrobial activity and UV-blocking functionality.

Nanocomposite films using regenerated cellulose and ZnO-NPs resulted in acceptable transparency for food packaging (Saedi et al., 2021). The addition of 7 wt.% of ZnO-NP to the regenerated cellulose film improved the thermal stability and the UV and oxygen barrier properties of the nanocomposite films and inhibited foodborne bacteria.

### 8.6.4 APPLICATIONS OF NANOPACKAGING TO FRUITS AND VEGETABLES

Several different applications of nanopackaging to fruits and vegetables are listed and summed in Table 8.1. For example, active packaging from polyethylene coated with acrylic resin (food grade) and filled with a nanocarrier of antimicrobial salicylate prolonged the shelf life of blueberries by 50%, which had been stored for 13 days at 8°C (Bugatti et al., 2020). Biodegradable NPs with incorporated EOs reduced the water loss on coated fruit and damaged the spore membrane integrity in decay control (Chávez-Magdaleno et al., 2018). After 10 days of storage, both chitosan nanoparticles (CS-NPs) and CS biocomposites with incorporation with pepper tree EO (*Schinus molle*) were effective against anthracnose *Colletotrichum gloeosporioides* and showed no internal damage in avocado (Chávez-Magdaleno et al., 2018). In another study, chitosan films with varying amounts of ZnO-NPs loaded with gallic acid were synthesized, and these biocomposite films possessed strong antioxidant behavior and had better antibacterial potential compared to CS film (Yadav et al., 2021). A composite film of whey protein isolate chitosan incorporated with $TiO_2$-NPs and the essential oil of *Zataria multiflora* (ZEO) was developed. This composite film was suitable for active packaging materials due to its antimicrobial and physicochemical properties (Gohargani et al., 2020).

### 8.7 NANOMICROBICIDES/FUNGICIDES

An alternative to reduce the usage of synthetic fungicides in the control of postharvest diseases for fruit and vegetables could be the synthesis of NMs (Roberto et al., 2019), which are nontoxic for consumers and the environment (Nxumalo et al., 2021; Zhang et al., 2011). Nanofungicides could offer additional advantages such as high efficacy, durability and delivery of small and specific doses of active ingredients (Rai et al., 2018). In this section, the focus has been on the synthesis and application of the biological metal NPs as antifungal agents.

The Ag-NPs and ZnO-NPs have been synthesized into antifungal agents for a broad range of common fungi (Hassan Basri et al., 2020; Bratovcic, 2019). Antifungal management requires the size control of nanocomposites and nanosensors for the detection and quantification of fungal pathogens in the postharvest handling of produce (Alghuthaymi et al., 2021). Biological Ag-NPs were synthesized by the fungus *Trichoderma longibrachiatum*. The extracellular fungal cell filtrate functioned as a reducing and stabilizing agent in the production of the NPs. The Ag-NPs reduced many plant-pathogenic fungi and were stable for up to 2 months (Elamawi et al., 2018).

Zinc oxide, due to its morphology, exhibits antibacterial/antifungal activity toward various microorganisms (Hassan Basri et al., 2020). The fabricated nanocomposite-based reduced graphene oxide and metal oxides showed antimicrobial inhibitions against *Staphylococcus aureus*, *Candida albicans* and *Escherichia coli* (Elbasuney et al., 2022). The antibacterial effect was related to the leakage of protein from the cytoplasm of *S. aureus*. The NMs, Carbon 60 ($C_{60}$), CuO and $TiO_2$ were engineered to control soft rot, *Rhizopus stolonifer*, in sweet potatoes during their storage and transportation. The CuO nanomaterial at 50 mg $L^{-1}$ exhibited the best antifungal activity (Pang et al., 2021). Optimized eugenol-casein NPs of size 307.4 nm and an entrapment efficiency of 86.3% improved the antifungal efficacy against anthracnose (Xue et al., 2019). The eugenol-casein NPs showed greater (>95.7%) antifungal activity against fungal spore germination, compared to eugenol.

## 8.8 NANOSAFETY

Despite the beneficial applications of nanotechnology, the insoluble and bio-persistent NPs must be evaluated in foods to ensure consumer health and environmental safety (Hossain et al., 2021). Investigations are required to determine the toxicity of NPs, as not all NPs are more toxic than fine-size particles of the same chemical composition (Upadhyay et al., 2022). The toxic effects of NPs vary due to the type, concentration, length of exposure and response of the individual (Dimitrijevic et al., 2015). The greater surface:volume ratio of NPs results in changes to original forms that facilitate easy migration into foods and into the human body (Mortezaee et al., 2019). The genotoxicity of most metal NPs is associated with the higher production of the reactive oxygen species (ROS). Lebre et al. (2022) have presented an overview of the possible interactions, routes of exposure and adverse effects that can be triggered by exposure of humans and the environment to NMs. The health hazards associated with many nanocomposite materials are produced from exposure to toxic NMs associated with the risk of carcinogenesis (Alghuthaymi et al., 2021). Toxicological studies in higher animals are critical, as the research can guide legislative regulations on the "safe" use of nanotechnology in the food chain and the environment (Rajwade et al., 2020).

The toxic effects of nanoemulsions are related to the size of nanodimension that affect the physiological bodily functions of humans (Pradhan et al., 2015). Safety concerns are related to the incorporation of NPs in active packaging due to the diffusion of NPs from packaging into food with potential human toxicological consequence (Ndwandwe et al., 2021). There is a need for research to investigate the impact of the migration of NMs into foodstuffs from packages and the associated toxicity risk and the permissible allowance for NPs in food packaging polymers (Ashfaq et al., 2022).

There is the possibility of the minimal migration into produce if the NPs have been embedded into the polymer matrix, but there could be other external influences. The nanotechnology-related intelligent labeling system has been targeted for future research (Liu et al., 2020a). The indiscriminate use of NMs may have deleterious impacts on the environment just like other synthetic agrochemicals; hence the need to foster useful nanoenabled applications that would protect the environment and sustain crop production systems (Hossain et al., 2020).

There are hurdles in the marketing of produce from nanotechnology due to the uncertainty of the technical benefits, safety, public opinion and legislation. Also, research is warranted as to the delivery of bioactive compounds, due to the variations in physical and chemical properties of molecular interactions with NMs (Luo et al., 2020). Risk and benefit analysis of nanotechnology can identify potential unforeseen hazards associated with the postharvest handling of fruits and vegetables, which can pose health risks to the consumers (Qadri et al., 2019).

## 8.9   CONCLUSION AND FUTURE RESEARCH

There have been many nanoenabled applications in the postharvest handling of fruits and vegetables. Some covered in this chapter were nanocoatings, nanoemulsions, nanopackaging/films and nanoantimicrobials, which could improve food safety and the quality of produce. Many reactive metal-based NPs have useful applications such as for the delivery of antimicrobials, antioxidants and barrier properties and add technical, structural and durability properties to biopolymers. The migration of NPs into food or the breakdown of nanocomposite polymers and the subsequent transfer of NPs into food may expose consumers to hazards. The small size of NMs even at trace levels could also create environmental and health risks.

The green method of NPs synthesis using biological materials such as plant materials and fruit peels is eco-friendly, safe, feasible with many functionalities for the fruit and vegetable industry. The focus of the nanoenabled applications should be on the minimal quantity usage of NMs with higher efficiencies. For instance, Ag-NPs have been synthesized from fruit peels to control fungal plant pathogens and may be applied safely as an alternative to chemical fungicides. Synthesized Ag-NPs have served as nanosensors for detecting the volatile organosulfides in garlic spoilage during postharvest. Novel nanosensors could be developed to detect, quantify and analyze the fungal pathogens during the postharvest period. Research is required to investigate the interface of science between nanotechnologies and produce physiology to ascertain the specific mechanism of how NPs inhibit ethylene production and delay ripening to lengthen the postharvest shelf life of perishable crops.

Researchers have encapsulated natural organic ingredients derived from plants (inorganic NPs, edibles, and EOs such as thyme, cinnamon, clove, citronella) in the development of edible nanocoatings and nanoemulsions to maintain produce safety and quality. However, there has been limited research on the nanofabrication of fruit coatings with antiviral activity such as from the safety perspective of the COVID 19 and other viral pathogenic crop diseases.

The use of eco-friendly materials is a growing trend for the fabrication of nano-composite film/packaging. Recent research studies indicate that chitosan is a versatile biodegradable polymer suitable in many nanoenabled postharvest handling applications, such as nanocoatings, nanoemulsions, nanofilms/packaging and nano-antimicrobials, to a wide range of fruits and vegetables. Chitosan serves as a matrix for encapsulating various bioactive compounds and supports various functionalities. Other biopolymer NMs such as celluloses, pullulan, bargasse and starches have been reinforced with NMs to improve their physicochemical, mechanical and barrier, anti-microbial, and antioxidant properties. Various NPs have been employed for active and intelligent packaging, to lengthen the storage life of fruits and vegetables and to communicate with the various contributors along the postharvest handling chain.

The application of nanotechnology in CS-based coatings and films/packaging has effectively extended the quality and shelf life of fresh produce due to their excellent mechanical, water barrier, antioxidant, antibacterial and antifungal functionalities. Many studies have shown the role of NPs in improving the properties of biopoly-meric packaging/nanoedible coatings. Chitosan NPs have inhibited fungal hyphal growth and/or altered the hyphal morphology. The CS-film coatings with various incorporations such as nisin have maintained the fruit texture, nutrients and micro-bial quality of fresh blueberry during storage for over 8 days.

There have been advances in the synthesis of individual nanostructures with the focus on their integration into nanocomposites. Chitosan nanobiocomposites have been popular as suitable alternatives to synthetic polymers in nanofilm/packaging. Cellulose nanostructures have been researched as components for a variety of nanoe-nabled applications such as for food packaging. Novel eco-friendly nanofilms have been developed from nanofibrillated cellulose using banana pseudostem fiber, pullu-lan and carboxylated cellulose nanocrystals for applications to fruits and vegetables. The fabrication of nanocomposites hybrid films has resulted in the wide application such as antimicrobial activity and antioxidant properties. Hybrid nanocomposites unite the properties of individual NPs and those of other materials resulting in syn-ergistically beneficial applications.

## REFERENCES LIST

Abhirami, P., Modupalli, N., & Natarajan, V. (2020). Novel postharvest intervention using rice bran wax edible coating for shelf-life enhancement of *Solanum lycopersicum* fruit. *Journal of Food Processing and Preservation*, *44*(12), e14989. https://doi.org/10.1111/jfpp.14989

Acay, H. I., Baran, M. F., & Eren, A. (2019). Investigating antimicrobial activity of silver nanoparticles produced through green synthesis using leaf extract of common grape (Vitis vinifera). *Applied Ecology and Environmental Research*, *17*(2), 4539–4546. https://doi.org/10.15666/aeer/1702_45394546

Adiletta, G., Di Matteo, M.D., & Petriccione, M. (2021). Multifunctional role of chitosan edible coatings on antioxidant systems in fruit crops: A review. *International Journal of Molecular Sciences*, *22*(5), 2633. https://doi.org/10.3390/ijms22052633

Ahmad, A., Qurashi, A., & Sheehan, D. (2023). Nano packaging – Progress and future per-spectives for food safety, and sustainability. *Food Packaging and Shelf Life*, *35*, 100997. https://doi.org/10.1016/j.fpsl.2022.100997

Ahmadi, R., Tanomand, A., Kazeminava, F., Kamounah, F. S., Ayaseh, A., Ganbarov, K., Yousefi, M., Katourani, A., Yousefi, B., & Kafil, H. S. (2019). Fabrication and characterization of a titanium dioxide ($TiO_2$) nanoparticles reinforced bio-nanocomposite containing Miswak (*Salvadora persica* L.) extract – the antimicrobial, thermo-physical and barrier properties. *International Journal of Nanomedicine*, *14*, 3439–3454. https://doi.org/10.2147/IJN.S201626

Aisyah, Y., Haryani, S., Safriani, N., & El Husna, N. (2018). Optimization of emulsification process parameters of cinnamon oil nanoemulsion. *International Journal on Advance Science Engineering Information Technology*, *8*(5), 2092–2098.

Aisyah, Y., Murlida, E., & Maulizar, T. A. (2022). Effect of the edible coating containing cinnamon oil nanoemulsion on storage life and quality of tomato (*Lycopersicum esculentum* Mill) fruits. *IOP Conference Series: Earth and Environmental Science*, *951*(1), 012048. https://doi:10.1088/1755-1315/951/1/012048

Aktepe, N., & Baran, A. (2022). Green synthesis and antimicrobial effects of silver nanoparticles by pumpkin *Cucurbita maxima* fruit fiber. *Medicine Science*, *11*(2), 794–799. https://doi.org/10.5455/medscience.2022.02.036

Alghuthaymi, M. A., Rajkuberan, C., Rajiv, P., Kalia, A., Bhardwaj, K., Bhardwaj, P., Abd-Elsalam, K. A., Valis, M., & Kuca, K. (2021). Nanohybrid antifungals for control of plant diseases: Current status and future perspectives. *Journal of Fungi*, *7*(1), 48. https://doi.org/10.3390/jof7010048

Ali, M., Ahmed, A., Shah, S. W. A., Mehmood, T., & Abbasi, K. S. (2020). Effect of silver nanoparticle coatings on physicochemical and nutraceutical properties of loquat during postharvest storage. *Journal of Food Processing and Preservation*, *44*(10), 14808. https://doi.org/10.1111/jfpp.14808

Anugrah, D. S. B., Alexander, H., Pramitasari, R., Hudiyanti, D., & Sagita, C. P. (2020). A review of polysaccharide-zinc oxide nanocomposites as safe coating for fruits preservation. *Coatings*, *10*(10), 988. https://doi.org/10.3390/coatings10100988

Arezoo, E., Mohammadreza, E., Maryam, M., & Abdorreza, M. N. (2020). The synergistic effects of cinnamon essential oil and nano TiO2 on antimicrobial and functional properties of sago starch films. *International Journal of Biological Macromolecules*, *157*, 743–751. https://doi.org/10.1016/j.ijbiomac.2019.11.244

Ashfaq, A., Khursheed, N., Fatima, S., Anjum, Z., & Younis, K. (2022). Application of nanotechnology in food packaging: Pros and cons. *Journal of Agriculture and Food Research*, *7*, 100270. https://doi.org/10.1016/j.jafr.2022.100270

Azeredo, H. M. C., Rosa, M. F., & Mattoso, L. H. C. (2017). Nanocellulose in bio-based food packaging applications. *Industrial Crops and Products*, *97*, 664–671. https://doi.org/10.1016/j.indcrop.2016.03.013

Babapour, H., Jalali, H., & Mohammadi Nafchi, A. M. (2021). The synergistic effects of zinc oxide nanoparticles and fennel essential oil on physicochemical, mechanical, and antibacterial properties of potato starch films. *Food Science and Nutrition*, *9*(7), 3893–3905. https://doi.org/10.1002/fsn3.2371

Bahrulolum, H., Nooraei, S., Javanshir, N., Tarrahimofrad, H., Mirbagheri, V. S., Easton, A. J., & Ahmadian, G. (2021). Green synthesis of metal nanoparticles using microorganisms and their application in the agrifood sector. *Journal of Nanobiotechnology*, *19*(1), 86. https://doi.org/10.1186/s12951-021-00834-3

Baranwal, J., Barse, B., Fais, A., Delogu, G. L., & Kumar, A. (2022). Biopolymer: A sustainable material for food and medical applications. *Polymers*, *14*(5), 983. https://doi.org/10.3390/polym14050983

Basumatary, I. B., Mukherjee, A., Katiyar, V., & Kumar, S. (2022). Biopolymer-based nanocomposite films and coatings: Recent advances in shelf-life improvement of fruits and vegetables. *Critical Reviews in Food Science and Nutrition*, *62*(7), 1912–1935. https://doi.org/10.1080/10408398.2020.1848789

Bawazeer, S., Khan, I., Rauf, A., Aljohani, A. S. M., Alhumaydhi, F. A., Khalil, A. A., Qureshi, M. N., Ahmad, L., & Khan, S. A. (2022). Black pepper (*Piper nigrum*) fruit-based gold nanoparticles (BP-AuNPs): Synthesis, characterization, biological activities, and catalytic applications – A green approach. *Green Processing and Synthesis, 11*(1), 11–28. https://doi.org/10.1515/gps-2022-0002

Bhusare, M. N., & Kadam, S. V. (2021). Applications of nanotechnology in fruits and vegetables. *Food and Agriculture Spectrum Journal, 2*(1), 90–95. https://fasj.org/index.php/fasj/article/view/51

Bhuyan, D., Greene, G. W., & Das, R. K. (2019). Prospects and application of nanobiotechnology in food preservation: Molecular perspectives. *Critical Reviews in Biotechnology, 39*(6), 759–778. https://doi.org/10.1080/07388551.2019.1616668

Bleoanca, I., Lanciu, A., Patraşcu, L., Ceoromila, A., & Borda, D. (2022). Efficacy of two stabilizers in nano-emulsions with whey proteins and thyme essential oil as edible coatings for zucchini. *Membranes, 12*(3), 326. https://doi.org/10.3390/membranes12030326

Bora, A., & Mishra, P. (2019). Casein and Ag nanoparticles: Synthesis, characterization, and their application in biopolymer-based bilayer film. *Journal of Food Processing and Preservation, 43*(9), e14062. https://doi.org/10.1111/jfpp.14062

Bratovcic, A. (2019). Different applications of nanomaterials and their impact on the environment. *International Journal of Material Science and Engineering, 5*(1), 1–7. https://doi.org/10.14445/23948884/IJMSE-V5I1P101

Bratovcic, A. (2020). Biosynthesis of green silver nanoparticles and its UV–vis characterization. *International Journal of Innovative Science Engineering and Technology, 7*(7), 170–176.

Bugatti, V., Cefola, M., Montemurro, N., Palumbo, M., Quintieri, L., Pace, B., & Gorrasi, G. (2020). Combined effect of active packaging of polyethylene filled with a nano-carrier of salicylate and modified atmosphere to improve the shelf life of fresh blueberries. *Nanomaterials, 10*(12), 2513. https://doi.org/10.3390/nano10122513

Bumbudsanpharoke, N., Choi, J., Park, H. J., & Ko, S. (2019). Zinc migration and its effect on the functionality of a low-density polyethylene-ZnO nanocomposite film. *Food Packaging and Shelf Life, 20*, 100301. https://doi.org/10.1016/j.fpsl.2019.100301

Casalini, S., & Baschetti, M. G. (2022). The use of essential oils in chitosan or cellulose-based materials to produce active food packaging solutions: A review. *Journal of the Science of Food and Agriculture.* https://doi.org/10.1002/jsfa.11918

Charoensri, K., Rodwihok, C., Ko, S. H., Wongratanaphisan, D., & Park, H. J. (2021). Enhanced antimicrobial and physical properties of poly (butylene adipate-co-terephthalate)/zinc oxide/reduced graphene oxide ternary nanocomposite films. *Materials Today Communications, 28*, 102586. https://doi.org/10.1016/j.mtcomm.2021.102586

Chaudhary, S., Kumar, S., Kumar, V., & Sharma, R. (2020). Chitosan nano-emulsions as advanced edible coatings for fruits and vegetables: Composition, fabrication, and developments in last decade. *International Journal of Biological Macromolecules, 152*, 154–170. https://doi.org/10.1016/j.ijbiomac.2020.02.276

Chávez-Magdaleno, M. E., González-Estrada, R. R., Ramos-Guerrero, A., Plascencia-Jatomea, M., & Gutiérrez-Martínez, P. (2018). Effect of pepper tree (*Schinus molle*) essential oil-loaded chitosan bio-nanocomposites on postharvest control of *Colletotrichum gloeosporioides* and quality evaluations in avocado (*Persea americana*) cv. Hass. *Food Science and Biotechnology, 27*(6), 1871–1875. https://doi.org/10.1007/s10068-018-0410-5

Chen, F., & Chi, C. (2021). Development of pullulan/carboxylated cellulose nanocrystal/tea polyphenol bio-nanocomposite films for active food packaging. *International Journal of Biological Macromolecules, 186*, 405–413. https://doi.org/10.1016/j.ijbiomac.2021.07.025

Chen, S., Wu, M., Wang, C., Yan, S., Lu, P., & Wang, S. (2020). Developed chitosan/oregano essential oil bio-composite packaging film enhanced by cellulose nanofibril. *Polymers, 12*(8), 780. https://doi.org/10.3390/polym12081780

Chisenga, S.M., Tolesa, G.N., & Workneh, T.S. (2020). Biodegradable food packaging materials and prospects of the fourth Industrial Revolution for tomato fruit and product handling. *International Journal of Food Science, 2020,* 8879101. https://doi.org/10.1155/2020/8879101

Correa-Pacheco, Z.N., Corona-Rangel, M.L., Bautista-Baños, S., & Ventura-Aguilar, R.I. (2021). Application of natural-based nano-coatings for extending the shelf life of green bell pepper fruit. *Journal of Food Science, 86*(1), 95–102. https://doi.org/10.1111/1750-3841.15542

Costa, S. M., Ferreira, D. P., Teixeira, P., Ballesteros, L. F., Teixeira, J. A., & Fangueiro, R. (2021). Active natural-based films for food packaging applications: The combined effect of chitosan and nanocellulose. *International Journal of Biological Macromolecules, 177,* 241–251. https://doi.org/10.1016/j.ijbiomac.2021.02.105

Criado, P., Fraschini, C., Shankar, S., Salmieri, S., & Lacroix, M. (2021). Influence of cellulose nanocrystals gellan gum-based coating on color and respiration rate of Agaricus bisporus mushrooms. *Journal of Food Science, 86*(2), 420–425. https://doi.org/10.1111/1750-3841.15580

Das, R.K., & Bhuyan, D. (2019). Microwave-mediated green synthesis of gold and silver nanoparticles from fruit peel aqueous extract of *Solanum melongena* L. and study of antimicrobial property of silver nanoparticles. *Nanotechnology for Environmental Engineering, 4*(5), 1–6. https://doi:10.1007/s41204-018-0052-0

Das, S., Vishakha, K., Banerjee, S., Mondal, S., & Ganguli, A. (2020). Sodium alginate-based edible coating containing nano-emulsion of *Citrus sinensis* essential oil eradicates planktonic and sessile cells of food-borne pathogens and increased quality attributes of tomatoes. *International Journal of Biological Macromolecules, 162,* 1770–1779. https://doi.org/10.1016/j.ijbiomac.2020.08.086

de Jesús Salas-Méndez, E., Vicente, A., Pinheiro, A.C., Ballesteros, L.F., Silva, P., Rodríguez-García, R., Hernández-Castillo, F.D., Díaz-Jiménez, M.L.V., Flores-López, M.L., Villarreal-Quintanilla, J.A., Peña-Ramos, F.M., Carrillo-Lomelí, D.A., & de Rodríguez, D. J. (2018). Application of edible nanolaminate coatings with antimicrobial extract of Flourensia cernua to extend the shelf life of tomato (*Solanum lycopersicum* L.) fruit. *Postharvest Biology and Technology, 150,* 19–27. https://doi.org/10.1016/j.postharvbio.2018.12.008

Deng, Z., Jung, J., Simonsen, J., Wang, Y., & Zhao, Y. (2017). Cellulose nanocrystal reinforced chitosan coatings for improving the storability of postharvest pears under both ambient and cold storages. *Journal of Food Science, 82*(2), 453–462. https://doi.org/10.1111/1750-3841.13601

de Oliveira Filho, J. G., Miranda, M., Ferreira, M. D., & Plotto, A. (2021). Nanoemulsions as edible coatings: A potential strategy for fresh fruits and vegetables preservation. *Foods, 10,* 2438. https://doi.org/10.3390/foods10102438

Dimitrijevic, M., Karabasil, N., Boskovic, M., Teodorovic, V., Vasilev, D., Djordjevic, V., Kilibarda, N., & Cobanovic, N. (2015). Safety aspects of nanotechnology applications in food packaging. *Procedia Food Science, 5,* 57–60. https://doi.org/10.1016/j.profoo.2015.09.015

Divya, K., & Jisha, M.S. (2018). Chitosan nanoparticles preparation and applications. *Environmental Chemistry Letters, 16*(1), 101–112. https://doi.org/10.1007/s10311-017-0670-y

Duguma, H.T. (2022). Potential applications and limitations of edible coatings for maintaining tomato quality and shelf life. *International Journal of Food Science and Technology, 57*(3), 1353–1366. https://doi.org/10.1111/ijfs.15407

Echegoyen, Y., Rodríguez, S., & Nerín, C. (2016). Nano-clay migration from food packaging materials. *Food Additives and Contaminants. Part A, Chemistry, Analysis, Control, Exposure and Risk Assessment, 33*(3), 530–539. https://doi.org/10.1080/19440049.2015.1136844

Ediyilyam, S., George, B., Shankar, S. S., Dennis, T. T., Wacławek, S., Černík, M., & Padil, V. V. T. (2021). Chitosan/gelatin/silver nanoparticles composites films for biodegradable food packaging applications. *Polymers*, *13*(11), 1680. https://doi.org/10.3390/polym13111680

Elamawi, R. M., Al-Harbi, R. E., & Hendi, A. A. (2018). Biosynthesis and characterization of silver nanoparticles using *Trichoderma longibrachiatum* and their effect on phytopathogenic fungi. *Egyptian Journal of Biological Pest Control*, *28*(1). https://doi.org/10.1186/s41938-018-0028-1

Elbasuney, S., Yehia, M., Ismael, S., Al-Hazmi, N. E., El-Sayyad, G. S., & Tantawy, H. (2022). Potential impact of reduced graphene oxide incorporated metal oxide nanocomposites as antimicrobial, and antibiofilm agents against pathogenic microbes: Bacterial protein leakage reaction mechanism. *Journal of Cluster Science*, *34*(2), 823–840. https://doi.org/10.1007/s10876-022-02255-0

Eldib, R., Khojah, E., Elhakem, A., Benajiba, N., & Helal, M. (2020). Chitosan, nisin, silicon dioxide nanoparticles coating films effects on blueberry (*Vaccinium myrtillus*) quality. *Coatings*, *10*(10). https://doi.org/10.3390/coatings10100962

Eren, A. I., & Baran, M. F. (2019). Green synthesis, characterization, and antimicrobial activity of silver nanoparticles (AgNPs) from maize (*Zea mays* L.). *Applied Ecology and Environmental Research*, *17*(2), 4097–4105. http://doi.org/10.15666/aeer/1702_40974105

FDA. (2019). *U.S. Food and Drug Administration (FDA) GRAS notice*. Retrieved July 22, 2022, from http://www.accessdata.fda.gov/scripts/cdrh/cfdocs/cfCFR/CFRSearch.cfm?fr=182.8991

Garavand, Y., Taheri-Garavand, A., Garavand, F., Shahbazi, F., Khodaei, D., & Cacciotti, I. (2022). Starch-polyvinyl alcohol-based films reinforced with chitosan nanoparticles: Physical, mechanical, structural, thermal, and antimicrobial properties. *Applied Sciences*, *12*(3), 1111. https://doi.org/10.3390/app12031111

Garcia, F., Lin, W.-J., Mellano, V., & Davidov-Pardo, G. (2022). Effect of biopolymer coatings made of zein nanoparticles and ε-polylysine as postharvest treatments on the shelf-life of avocados (*Persea americana* Mill. Cv. Hass). *Journal of Agriculture and Food Research*, *7*(1), 100260. https://doi.org/10.1016/j.jafr.2021.100260

Ghadermazi, R., Hamdipour, S., Sadeghi, K., Ghadermazi, R., & Asl, A. K. (2019). Effect of various additives on the properties of the films and coatings derived from hydroxypropyl methylcellulose – a review. *Food Science and Nutrition*, *7*(11), 3363–3377. https://doi.org/10.1002/fsn3.1206

Gohargani, M., Lashkari, H., & Shirazinejad, A. (2020). Study on biodegradable chitosan-whey protein-based film containing bio-nanocomposite $TiO_2$ and *Zataria multiflora* essential oil. *Journal of Food Quality*, *2020*, 1–11. https://doi.org/10.1155/2020/8844167

González-Reza, R. M., Hernández-Sánchez, H., Quintanar-Guerrero, D., Alamilla-Beltrán, L., Cruz-Narváez, Y., & Zambrano-Zaragoza, M. L. (2021). Synthesis, controlled release, and stability on storage of chitosan-thyme essential oil nanocapsules for food applications. *Gels*, *7*(4), 212. https://doi.org/10.3390/gels7040212

Gustavsson, J., Cederberg, C., & Sonesson, U. (2013). The methodology of the FAO study: Global food losses and food waste – extent, causes and prevention – FAO, 2011, 70 pp, *Save Food Congress, Düsseldorf*, 16 May 2011. The Swedish Institute for Food and Biotechnology, Food and Agriculture Organization of the United Nations (FAO), Rome, Italy.

Hasheminejad, N., & Khodaiyan, F. (2020). The effect of clove essential oil loaded with chitosan nanoparticles on the shelf life and quality of pomegranate arils. *Food Chemistry*, *309*, 125520. https://doi.org/10.1016/j.foodchem.2019.125520

Hashim, A. F., Youssef, K., & Abd-Elsalam, K. A. (2019). Ecofriendly nanomaterials for controlling gray mold of table grapes and maintaining postharvest quality. *European Journal of Plant Pathology*, *154*(2), 377–388. https://doi.org/10.1007/s10658-018-01662-2

Hassan Basri, H., Talib, R. A., Sukor, R., Othman, S. H., & Ariffin, H. (2020). Effect of synthesis temperature on the size of ZnO nanoparticles derived from pineapple peel extract and antibacterial activity of ZnO-starch nanocomposite films. *Nanomaterials, 10*(6), 1061. https://doi.org/10.3390/nano10061061

He, X., Deng, H., & Hwang, H. M. (2019). The current application of nanotechnology in food and agriculture. *Journal of Food and Drug Analysis, 27*(1), 1–21. https://doi.org/10.1016/j.jfda.2018.12.002

Helal, M., Sami, R., Algarni, E., Alshehry, G., Aljumayi, H., Al-Mushhin, A. A. M., Benajiba, N., Chavali, M., Kumar, N., Iqbal, A., Aloufi, S., Alyamani, A., Madkhali, N., & Almasoudi, A. (2022). Active bionanocomposite coating quality assessments of some cucumber properties with some diverse applications during storage condition by chitosan, nano titanium oxide crystals, and sodium tripolyphosphate. *Crystals, 12*(2), 131. https://doi.org/10.3390/cryst12020131

Hernández-Fuentes, A. D., López-Vargas, E., Pinedo-Espinoza, J. M., Campos-Montiel, R. G., Valdés-Reyna, J., & Juárez-Maldonado, A. (2017). Postharvest behavior of bioactive compounds in tomato fruits treated with Cu nanoparticles and NaCl stress. *Applied Sciences, 7*(10), 980. https://doi.org/10.3390/app7100980

Hernández-López, G., Ventura-Aguilar, R. I., Correa-Pacheco, Z. N., Bautista-Baños, S., & Barrera, L. L. (2020). Nanostructured chitosan edible coating loaded with α-pinene for the preservation of the postharvest quality of *Capsicum annuum* L. and *Alternaria alternata* control. *International Journal of Biological Macromolecules, 165*(Pt B), 1881–1888. https://doi.org/10.1016/j.ijbiomac.2020.10.094

Hmmam, I., Zaid, N., Mamdouh, B., Abdallatif, A., Abd-Elfattah, M., & Ali, M. (2021). Storage behavior of "Seddik" mango fruit coated with CMC and guar gum-based silver nanoparticles. *Horticulturae, 7*(3), 44. https://doi.org/10.3390/horticulturae7030044

Hossain, A., Kerry, R. G., Farooq, M., Abdullah, N., & Islam, M. T. (2020). Application of nanotechnology for sustainable crop production systems. In D. Thangadurai, J. Sangeetha, & R. Prasad (Eds.), *Nanotechnology for food, agriculture, and environment, nanotechnology in the life sciences* (Chapter 7, pp. 135–159). Springer Nature. https://doi.org/10.1007/978-3-030-31938-0_7

Hossain, A., Skalicky, M., Brestic, M., Mahari, S., Kerry, R. G., Maitra, S., Sarkar, S., Saha, S., Bhadra, P., Popov, M., Islam, M. T., Hejnak, V., Vachova, P., Gaber, A., & Islam, T. (2021). Application of nanomaterials to ensure quality and nutritional safety of food. *Journal of Nanomaterials, 2021*(2), 1–19. https://doi.org/10.1155/2021/9336082

Hussain, M., Raja, N. I., Mashwani, Z.-U.-R., Muhammad, I., Muhammad, E., & Sumaira, A. (2019). Green synthesis and evaluation of silver nanoparticles for antimicrobial and biochemical profiling in Kinnow (*Citrus reticulata* L.) to enhance fruit quality and productivity under biotic stress. *IET Nanobiotechnology, 13*(3), 250–256. https://doi.org/10.1166/nnl.2018.2799

Ishkeh, S. R., Shirzad, H., Asghari, M., Alirezalu, A., Pateiro, M., & Lorenzo, J. M. (2021). Effect of chitosan nanoemulsion on enhancing the phytochemical contents, health promoting components, and shelf life of raspberry (*Rubus sanctus* Schreber). *Applied Sciences, 11*(5), 2224. https://doi.org/10.3390/app11052224

Jamdagni, P., Khatri, P., & Rana, J. S. (2018). Green synthesis of zinc oxide nanoparticles using flower extract of *Nyctanthes arbor-tristis* and their antifungal activity. *Journal of King Saud University – Science, 30*(2), 168–175. https://doi.org/10.1016/j.jksus.2016.10.002

Javed, R., Ain, N., Gul, A., Ahmad, M. A., Guo, W., Ao, Q., & Tian, S. (2022). Diverse biotechnological applications of multifunctional titanium dioxide nanoparticles: An up-to-date review. *IET Nanobiotechnology*, 1–19. https://doi.org/10.1049/nbt2.12085

Jung, J., Deng, Z., & Zhao, Y. (2019). A review of cellulose nanomaterials incorporated fruit coatings with improved barrier property and stability: Principles and applications. *Journal of Food Process Engineering, 43*(2), e13344. https://doi.org/10.1111/jfpe.13344

Kaewklin, P., Siripatrawan, U., Suwanagul, A., & Lee, Y. S. (2018). Active packaging from chitosan-titanium dioxide nanocomposite film for prolonging storage life of tomato fruit. *International Journal of Biological Macromolecules*, *112*, 523–529. https://doi.org/10.1016/j.ijbiomac.2018.01.124

Kamatyanatti, M., Singh, S. H., Sekhon, B. S., & Tripura, U. (2019). Nanotechnology: A novel technique in modern fruit production. *Think India Journal*, *22*(30), 426–440.

Kaphle, A., Navya, P. N., Umapathi, A., & Daima, H. K. (2018,). Nanomaterials for agriculture. food and environment: Applications, toxicity and regulation. *Environmental Chemistry Letters*, *16*(1), 43–58. https://doi.org/10.1007/s10311-017-0662-y

Khan, A. U., Malik, N., Khan, M., Cho, M. H., & Khan, M. M. (2018). Fungi-assisted silver nanoparticle synthesis and their applications. *Bioprocess and Biosystems Engineering*, *41*(1), 1–20. https://doi.org/10.1007/s00449-017-1846-3

Khezerlou, A., Tavassoli, M., Sani, M. A., Mohammadi, K., Ehsani, A., & McClements, D. J. (2021). Application of nanotechnology to improve the performance of biodegradable biopolymer-based packaging materials. *Polymers*, *13*(24), 4399. https://doi.org/10.3390/polym13244399

Khojah, E., Sami, R., Helal, M., Elhakem, A., Benajiba, N., Alkaltham, M. S., & Salamatullah, A. M. (2021). Postharvest physicochemical properties and fungal populations of treated cucumber with sodium tripolyphosphate/titanium dioxide nanoparticles during storage. *Coatings*, *11*(6), 613. https://doi.org/10.3390/coatings11060613

Kim, W., Han, T., Gwon, Y., Park, S., Kim, H., & Kim, J. (2022). Biodegradable and flexible nanoporous films for design and fabrication of active food packaging systems. *Nano Letters*, *22*(8), 3480–3487. https://doi.org/10.1021/acs.nanolett.2c00246

Kou, X., He, Y., Li, Y., Chen, X., Feng, Y., & Xue, Z. (2019). Effect of abscisic acid (ABA) and chitosan/nano-silica/sodium alginate composite film on the color development and quality of postharvest Chinese winter jujube (*Zizyphus jujuba* Mill. cv. Dongzao). *Food Chemistry*, *270*(1), 385–394. https://doi.org/10.1016/j.foodchem.2018.06.151

Kumar, D., Arora, S., Kumar, A., Abdullah, D. M., Danish, M., & Bahukhandi, K. D. (2020a). Green synthesis of silver nanoparticles using *Cucurbita pepo* leaves extract and its antimicrobial and antioxidant activities. In N. Siddiqui, S. Tauseef, R. Dobhal et al. (Eds.), *Advances in water pollution monitoring and control* (pp. 115–125). Springer. https://doi.org/10.1007/978-981-32-9956-6_13

Kumar, N., Kaur, P., Devgan, K., & Attkan, A. K. (2020b). Shelf-life prolongation of cherry tomato using magnesium hydroxide reinforced bio-nanocomposite and conventional plastic films. *Journal of Food Processing and Preservation*, *44*(4), e14379. https://doi.org/10.1111/jfpp.14379

Kumar, S., Mukherjee, A., & Dutta, J. (2020). Chitosan based nanocomposite films and coatings: Emerging antimicrobial food packaging alternatives. *Trends in Food Science and Technology*, *97*, 196–209. https://doi.org/10.1016/j.tifs.2020.01.002

Kumar, S., Ye, F., Dobretsov, S., & Dutta, J. (2019). Chitosan nanocomposite coatings for food, paints, and water treatment applications. *Applied Sciences*, *9*(12), 2409. https://doi.org/10.3390/app9122409

Kumar, S. K., & Krishnamoorti, R. (2010). Nanocomposites: Structure, phase behavior, and properties. *Annual Review of Chemical and Biomolecular Engineering*, *1*, 37–58. https://doi.org/10.1146/annurev-chembioeng-073009-100856

Larena, A., Tur, A., & Baranauskas, V. (2008). Classification of nanopolymers. *Journal of Physics: Conference Series*, *100*(1), 012023. https://doi.org/10.1088/1742-6596/100/1/012023

Lebre, F., Chatterjee, N., Costa, S., Fernández-de-Gortari, E., Lopes, C., Meneses, J., Ortiz, L., Ribeiro, A. R., Vilas-Boas, V., & Alfaro-Moreno, E. (2022). Nanosafety: An evolving concept to bring the safest possible nanomaterials to society and environment. *Nanomaterials*, *12*(11), 1810. https://doi.org/10.3390/nano12111810

Li, Y., Rokayya, S., Jia, F., Nie, X., Xu, J., Elhakem, A., Almatrafi, M., Benajiba, N., & Helal, M. (2021). Shelf-life, quality, safety evaluations of blueberry fruits coated with chitosan nano-material films. *Scientific Reports*, *11*(1), 55. https://doi.org/10.1038/s41598-020-80056-z

Liu, W., Zhang, M., & Bhandari, B. (2020a). Nanotechnology – A shelf-life extension strategy for fruits and vegetables. *Critical Reviews in Food Science and Nutrition*, *60*(10), 1706–1721. https://doi.org/10.1080/10408398.2019.1589415

Liu, X., Xu, Y., Zhan, X., Xie, W., Yang, X., Cui, S. W., & Xia, W. (2020b). Development and properties of new kojic acid and chitosan composite biodegradable films for active packaging materials. *International Journal of Biological Macromolecules*, *144*, 483–490. https://doi.org/10.1016/j.ijbiomac.2019.12.126

Liu, Y., Wang, S., Wang, Z., Yao, Q., Fang, S., Zhou, X., Yuan, X., & Xie, J. (2020c). The in-situ synthesis of silver nanoclusters inside a bacterial cellulose hydrogel for antibacterial applications. *Journal of Materials Chemistry. B*, *8*(22), 4846–4850. http://doi.org/10.1039/d0tb00073f

Luo, Y., Wang, Q., & Zhang, Y. (2020). Biopolymer-based nanotechnology approaches to deliver bioactive compounds for food applications: A perspective on the past, present, and future. *Journal of Agricultural and Food Chemistry*, *68*(46), 12993–13000. https://doi.org/10.1021/acs.jafc.0c00277

Mageswari, A., Srinivasan, R., Subramanian, P., Ramesh, N., & Gothandam, K. M. (2016). Nanomaterials: Classification, biological synthesis, and characterization. In S. Ranjan, N. Dasgupta, & E. Lichtfouse (Eds.), *Nanoscience in food and agriculture* (3rd ed., pp. 31–71). Springer International Publishing. https://doi.org/10.1007/978-3-319-48009-1_2

Majid, I., Ahmad Nayik, G., Mohammad Dar, S., & Nanda, V. (2018). Novel food packaging technologies: Innovations and future prospective. *Journal of the Saudi Society of Agricultural Sciences*, *17*(4), 454–462. http://doi.org/10.1016/j.jssas.2016.11.003

Malhotra, B., Keshwani, A., & Kharkwal, H. (2015). Natural polymer based cling films for food packaging. *International Journal of Pharmacy and Pharmaceutical Sciences*, *7*(4), 10–18.

Marboh, E. S., & Gupta, A. K. (2020). Advances in edible coatings and films for fresh fruits and vegetables. In K. Barman, S. Sharma, & M. W. Siddiqui (Eds.), *Emerging postharvest treatment of fruits and vegetables* (pp. 277–330). Apple Academic Press. https://doi.org/10.1201/9781351046312

Meena, M., Pilania, S., Pal, A., Mandhania, S., Bhushan, B., Kumar, S., Gohari, G., & Saharan, V. (2020). Cu chitosan nano-net improves keeping quality of tomato by modulating physio-biochemical responses. *Scientific Reports*, *10*(1), 21914. https://doi.org/10.1038/s41598-020-78924-9

Melo, N. F. C. B., de MendonçaSoares, B. L., Marques Diniz, K., Ferreira Leal, C., Canto, D., Flores, M. A. P., Henrique da Costa Tavares-Filho, J., Galembeck, A., Montenegro Stamford, T. L., Montenegro Stamford-Arnaud, T., & Montenegro Stamford, T. C. (2018). Effects of fungal chitosan nanoparticles as eco-friendly edible coatings on the quality of postharvest table grapes. *Postharvest Biology and Technology*, *139*, 56–66. https://doi.org/10.1016/j.postharvbio.2018.01.014

Melo, N. F. C. B., de Lima, M. C. A., Stamford, T. L. M., Galembeck, A., Miguel, A. P., Flores, M. A. P., de Campos Takaki, T. M., da Costa Medeiros, J. A., Stamford-Arnaud, T. M., & Stamford, T. C. M. (2020). Quality of postharvest strawberries: Comparative effect of fungal chitosan gel, nanoparticles and gel enriched with edible nanoparticle coatings. *International Journal of Food Studies*, *9*, 373–393. https://doi.org/10.1111/ijfs.14669

Metzger, C., Sanahuja, S., Behrends, L., Sängerlaub, S., Lindner, M., & Briesen, H. (2018). Efficiently extracted cellulose nanocrystals and starch nanoparticles and techno-functional properties of films made thereof. *Coatings*, *8*(4), 142. https://doi.org/10.3390/coatings8040142

Mohamed, S.A.A., El-Sakhawy, M., & El-Sakhawy, M. A.-M. (2020). Polysaccharides, protein, and lipid-based natural edible films in food packaging: A review. *Carbohydrate Polymers*, *238*, 116178. https://doi.org/10.1016/j.carbpol.2020.116178

Mokhena, T.C., John, M.J., Sibeko, M.A., Agbakoba, V.C., Mochane, M.J., Mtibe, A. T., Mokhothu, T. H., Motsoeneng, T.S., Phiri, M.M., Phiri, M.J., Hlangothi, P.S., & Mofokeng, T.G. (2020). Nanomaterials: types, synthesis and characterization. In M. Srivastava, N. Srivastava, P. Mishra, & V. Gupta (Eds.), *Nanomaterials in biofuels research, clean energy, production technologies* (pp. 115–141). Springer. https://doi.org/10.1007/978-981-13-9333-4_5

Moreno, M.A., Vallejo, A.M., Ballester, A.-R., Zampini, C., Isla, M.I., López-Rubio, A., & Fabra, M.J. (2020). Antifungal edible coatings containing Argentinian propolis extract and their application in raspberries. *Food Hydrocolloids*, *107*, 105973. https://doi.org/10.1016/j.foodhyd.2020.105973

Mortezaee, K., Najafi, M., Samadian, H., Barabadi, H., Azarnezhad, A., & Ahmadi, A. (2019). Redox interactions and genotoxicity of metal-based nanoparticles: A comprehensive review. *Chemico-Biological Interactions*, *312*, 108814. https://doi.org/10.1016/j.cbi.2019.108814

Mostafa, Y.S., Alamri, S.A., Alrumman, S.A., Hashem, M., & Baka, Z.A. (2021). Green synthesis of silver nanoparticles using pomegranate and orange peel extracts and their antifungal activity against *Alternaria solani*, the causal agent of early blight disease of tomato. *Plants*, *10*(11), 2363. https://doi.org/10.3390/plants10112363

Motelica, L., Ficai, D., Ficai, A., Trușcă, R. D., Ilie, C. I., Oprea, O. C., & Andronescu, E. (2020). Innovative antimicrobial chitosan/ZnO/Ag NPs/citronella essential oil nanocomposite – potential coating for grapes. *Foods*, *9*(12), 1801. https://doi.10.3390/foods9121801

Motol, R.K.C., Espineli, C.A., Tapit, C.M.V., & Tiangco, C.E. (2020). Synthesis and characterization of silver nanoparticles as a potential sensor for volatile organosulfides for visual detection of postharvest storage in garlic. *IOP Conference Series: Materials Science and Engineering*, *778*(1), 012002. https://doi.org/10.1088/1757-899X/778/1/012002

Nath, D., Santhosh, C.R., & Sarkar, P. (2022). Nanocomposites in food packaging: In K. Pal, A. Sarkar, N. Bandara, & V. Jegatheesan (Eds.), *Food, medical, and environmental-applications of nanomaterials: Micro and nano technologies* (pp. 167–120). Elsevier, Inc. https://doi.org/10.1016/B978-0-12-822858-6.00007-8

Ndwandwe, B. K., Malinga, S. P., Kayitesi, E., & Dlamini, B.C. (2021). Advances in green synthesis of selenium nanoparticles and their application in food packaging. *International Journal of Food Science and Technology*, *56*(6), 2640–2650. https://doi.org/10.1111/ijfs.14916

Neme, K., Nafady, A., Uddin, S., & Tola, Y.B. (2021). Application of nanotechnology in agriculture, postharvest loss reduction and food processing: Food security implication and challenges. *Heliyon*, *7*(12), e08539. https://doi.org/10.1016/j.heliyon.2021.e08539

Nxumalo, K.A., Aremu, A.O., & Fawole, O.A. (2021). Potentials of medicinal plant extracts as an alternative to synthetic chemicals in postharvest protection and preservation of horticultural crops: A review. *Sustainability*, *13*(11), 5897. https://doi.org/10.3390/su13115897

Oh, J.-W., Chun, S.C., & Chandrasekaran, M. (2019). Preparation and in vitro characterization of chitosan nanoparticles and their broad-spectrum antifungal action compared to antibacterial activities against phytopathogens of tomato. *Agronomy*, *9*(1), 21. https://doi:10.3390/agronomy9010021

Othman, S. H., Wane, B. M., Nordin, N., Noor Hasnan, N. Z., A Talib, R., & Karyadi, J. N. W. (2021). Physical, mechanical, and water vapor barrier properties of starch/cellulose nanofiber/thymol bio-nanocomposite films. *Polymers*, *13*(23), 4060. https://doi.org/10.3390/polym13234060

Pang, L. J., Adeel, M., Shakoor, N., Guo, K. R., Ma, D. F., Ahmad, M. A., Lu, G. Q., Zhao, M. H., Li, S. E., & Rui, Y. K. (2021). Engineered nanomaterials suppress the soft rot disease (*Rhizopus stolonifer*) and slow down the loss of nutrient in sweet potato. *Nanomaterials*, *11*(10), 2572. https://doi.org/10.3390/nano11102572

Pascuta, M. S., & Vodnar, D. C. (2022). Nanocarriers for sustainable active packaging: An overview during and post COVID-19. *Coatings*, *12*(1), 102. https://doi.org/10.3390/coatings12010102

Pathakoti, K., Manubolu, M., & Hwang, H. M. (2017). Nanostructures: Current uses and future applications in food science. *Journal of Food and Drug Analysis*, *25*(2), 245–253. https://doi.org/10.1016/j.jfda.2017.02.004

Patri, A. (2022). Nanotechnology: Over a decade of progress and innovation at FDA. *Polymers*, *21*(13), 3615. US Food and Drug Administration. Retrieved July 8, 2022, http://www.fda.gov/about-fda/nctr-research-focus-areas/nanotechnology; https://doi.org/10.3390/polym13213615

Ponni, P., Subramanian, K. S., Janavi, G. J., & Subramanian, J. (2020). Synthesis of nano-film from nanofibrillated cellulose of banana pseudostem (Musa spp.) to extend the shelf life of tomato. *BioResources*, *15*(2), 2882–2905. https://doi.org/10.15376/biores.15.2.2882-2905

Poonia, A., & Mishra, A. (2022). Edible nanocoatings: Potential food applications, challenges, and safety regulations. *Nutrition and Food Science*, *52*(3), 497–514. https://doi.org/10.1108/NFS-07-2021-0222

Pradhan, N., Singh, S., Ojha, N., Shrivastava, A., Barla, A., Rai, V., & Bose, S. (2015). Facets of nanotechnology as seen in food processing, packaging, and preservation industry. *BioMed Research International*, *2015*, 1–17. https://doi.org/10.1155/2015/365672

Prakash, A., Baskaran, R., Paramasivam, N., & Vadivel, V. (2018). Essential oil based nano-emulsions to improve the microbial quality of minimally processed fruits and vegetables: A review. *Food Research International*, *111*, 509–523. https://doi.org/10.1016/j.foodres.2018.05.066

Qadri, O. S., Younis, K., Srivastava, G., & Srivastava, A. K. (2019). Nanotechnology in packaging of fresh fruits and vegetables. In K. Barman, S. Sharma, & M. W. Siddiqui (Eds.), *E merging postharvest treatments of fruits and vegetables*. Apple Academic Press. https://doi.org/10.1201/9781351046312

Qiao, G., Xiao, Z., Ding, W., & Rok, A. (2019). Effect of chitosan/nano-titanium dioxide/thymol and tween films on ready-to-eat cantaloupe fruit quality. *Coatings*, *9*(12), 828. https://doi.org/10.3390/coatings9120828

Rai, M., Ingle, A. P., Paralikar, P., Anasane, N., Gade, R., & Ingle, P. (2018). Effective management of soft rot of ginger caused by *Pythium* spp. and *Fusarium* spp.: Emerging role of nanotechnology. *Applied Microbiology and Biotechnology*, *102*, 6827–6839. https://doi.org/10.1007/s00253-018-9145-8

Rajwade, J. M., Chikte, R. G., & Paknikar, K. M. (2020). Nanomaterials: New weapons in a crusade against phytopathogens. *Applied Microbiology and Biotechnology*, *104*(4), 1437–1461. https://doi.org/10.1007/s00253-019-10334-y

Rana, R. A., Siddiqui, M. N., Skalicky, M., Brestic, M., Hossain, A., Kayesh, E., Popov, M., Hejnak, V., Gupta, D. R., Mahmud, N. U., & Islam, T. (2021). Prospects of nanotechnology in improving the productivity and quality of horticultural crops. *Horticulturae*, *7*(10), 332. https://doi.org/10.3390/horticulturae7100332

Rashid, F., Ahmed, Z., Ameer, K., Amir, R. M., & Khattak, M. (2020). Optimization of polysaccharides based nanoemulsion using response surface methodology and application to improve postharvest storage of apple (*Malus domestica*). *Journal of Food Measurement and Characterization*, *14*(5), 2676–2688. https://doi.org/10.1007/s11694-020-00514-0

Rivero-Montejo, S. D. J., Vargas-Hernandez, M., & Torres-Pacheco, I. (2021). Nanoparticles as novel elicitors to improve bioactive compounds in plants. *Agriculture*, *11*(2), 134. https://doi.org/10.3390/agriculture11020134

Roberto, S. R., Youssef, K., Hashim, A. F., & Ippolito, A. (2019). Nanomaterials as alternative control means against postharvest diseases in fruit crops. *Nanomaterials*, *9*, 1752. http://doi:10.3390/nano9121752

Romanazzi, G., Feliziani, E., Baños, S. B., & Sivakumar, D. (2017). Shelf-life extension of fresh fruit and vegetables by chitosan treatment. *Critical Reviews in Food Science and Nutrition*, *57*(3), 579–601. https://doi.org/10.1080/10408398.2014.900474

Saedi, S., Shokri, M., Kim, J. T., & Shin, G. H. (2021). Semi-transparent regenerated cellulose/ZnONP nanocomposite film as a potential antimicrobial food packaging material. *Journal of Food Engineering*, *307*, O110665. https://doi.org/10.1016/j.jfoodeng.2021.110665

Salama, H. E., & Aziz, M. S. A. (2020). Optimized alginate and *Aloe vera* gel edible coating reinforced with *n*TiO$_2$ for the shelf-life extension of tomatoes. *International Journal of Biological Macromolecules*, *165*(Part B), 2693–2701. https://doi.org/10.1016/j.ijbiomac.2020.10.108

Salama, H. E., Aziz, M. S. A., & Alsehli, M. (2019). Carboxymethyl cellulose/sodium alginate/chitosan biguanidine hydrochloride ternary system for edible coatings. *International Journal of Biological Macromolecules*, *139*, 614–620. https://doi.org/10.1016/j.ijbiomac.2019.08.008

Sami, R., Almatrafi, M., Elhakem, A., Alharbi, M., Benajiba, N., & Helal, M. (2021a). Effect of Nano silicon dioxide coating films on the quality characteristics of fresh-cut cantaloupe. *Membranes*, *11*(2), 140. https://doi.org/10.3390/membranes11020140

Sami, R., Elhakem, A., Alharbi, M., Benajiba, N., Almatrafi, M., Abdelazez, A., & Helal, M. (2021b). Evaluation of antioxidant activities, oxidation enzymes, and quality of nano-coated button mushrooms (*Agaricus bisporus*) during storage. *Coatings*, *11*(2), 149. https://doi.org/10.3390/coatings11020149

Sani, M. A., Maleki, M., Eghbaljoo-Gharehgheshlaghi, H., Khezerlou, A., Mohammadian, E., Liu, Q., & Jafari, S. M. (2022). Titanium dioxide nanoparticles as multifunctional surface-active materials for smart/active nanocomposite packaging films. *Advances in Colloid and Interface Science*, *300*, 102593. https://doi.org/10.1016/j.cis.2021.102593

Santoscoy-Berber, L. S., Antunes-Ricardo, M., Gallegos-Granados, M. Z., García-Ramos, J. C., Pestryakov, A., Toledano-Magaña, Y., Bogdanchikova, N., & Chavez-Santoscoy, R. A. (2021). Treatment with Argovit® silver nanoparticles induce differentiated postharvest biosynthesis of compounds with pharmaceutical interest in carrot (*Daucus carota* L.). *Nanomaterials*, *11*(11), 3148. https://doi.org/10.3390/nano11113148

Satalkar, P., Elger, B. S., & Shaw, D. M. (2016). Defining nano, nanotechnology, and nanomedicine: Why should it matter? *Science and Engineering Ethics*, *22*(5), 1255–1276. https://doi.org/10.1007/s11948-015-9705-6

Shahvalizadeh, R., Ahmadi, R., Davandeh, I., Pezeshki, A., Seyed Moslemi, S. A. S., Karimi, S., Rahimi, M., Hamishehkar, H., & Mohammadi, M. (2021). Antimicrobial bio-nanocomposite films based on gelatin, tragacanth, and zinc oxide nanoparticles – microstructural, mechanical, thermo-physical, and barrier properties. *Food Chemistry*, *354*, 129492. https://doi.org/10.1016/j.foodchem.2021.129492

Shankar, S., & Rhim, J.-W. (2018). Preparation of sulfur nanoparticle-incorporated antimicrobial chitosan films. *Food Hydrocolloids*, *82*, 116-123. https://doi.org/10.1016/j.foodhyd.2018.03.054

Sharifan, H., Noori, A., Bagheri, M., & Moore, J. M. (2021). Postharvest spraying of zinc oxide nanoparticles enhances shelf-life qualities and zinc concentration of tomato fruits. *Crop and Pasture Science*, *73*(2), 22–31. https://doi.org/10.1071/CP21191

Sharma, C., Dhiman, R., Rokana, N., & Panwar, H. (2017). Nanotechnology: An untapped resource for food packaging. *Frontiers in Microbiology*, *8*(4), 1735. https://doi.org/10.3389/fmicb.2017.01735

Singh, A., & Prasad, S.M. (2017). Nanotechnology and its role in agro-ecosystem: A strategic perspective. *International Journal of Environmental Science and Technology*, *14*(10), 2277–2300. https://doi.org/10.1007/s13762-016-1062-8

Singh, P., & Mijakovic, I. (2022). Rowan berries: A potential source for green synthesis of extremely monodisperse gold and silver nanoparticles and their antimicrobial property. *Pharmaceutics*, *14*(1), 82. https://doi.org/10.3390/pharmaceutics14010082

Taha, I.M., Zaghlool, A., Nasr, A., Nagib, A., El Azab, I.H., Mersal, G.A.M., Ibrahim, M.M., & Fahmy, A. (2022). Impact of starch coating embedded with silver nanoparticles on strawberry storage time. *Polymers*, *14*(7), 1439. https://doi.org/10.3390/polym14071439

Tamimi, N., Mohammadi Nafchi, A., Hashemi-Moghaddam, H., & Baghaie, H. (2021). The effects of nano-zinc oxide morphology on functional and antibacterial properties of tapioca starch bionanocomposite. *Food Science and Nutrition*, *9*(8), 4497–4508. https://doi.org/10.1002/fsn3.2426

To, D.T., Van, P.T., Ngoc, L.S., Le, T.T., Phan, H.C.T., & Manh, T.D. (2022). Antimicroorganism activity of chitosan and nano-silicon dioxide mixture on damage causing strains isolated from postharvest guava. *Chemical Engineering Transactions*, *91*, 583–588. https://doi.org/10.3303/CET2291098

Upadhyay, S.K., Kumar, S., Rout, C., Vashistha, G., & Aggarwal, D. (2022). Risks and concerns of use of nanoparticles in agriculture. In V.D. Rajput, K.K. Verma, N. Sharma, & T. Minkina (Eds.), *The role of nanoparticles in plant nutrition under soil pollution* (pp. 371–394). Sustainable Plant Nutrition in a Changing World. Springer. https://doi.org/10.1007/978-3-030-97389-6_16

Valdés, A., Martínez, C., Garrigos, M.C., & Jimenez, A. (2021). Multilayer films based on poly (lactic acid)/gelatin supplemented with cellulose nanocrystals and antioxidant extract from almond shell by-product and its application on Hass avocado preservation. *Polymers*, *13*(21), 3615. https://doi.org/10.3390/polym13213615

Vimala Devi, P.S., Duraimurugan, P., Chandrika, K.S.V.P., Gayatri, B.R., & Prasad, R. D. (2019). Nanobiopesticides for crop protection. In K.A. Abd-Elsalam & R. Prasad (Eds.), *Nanobiotechnology applications in plant protection* (pp. 145–168). Nanotechnology in the Life Sciences, Springer. https://doi.org/10.1007/978-3-030-13296-5_8

Wantat, A., Seraypheap, K., & Rojsitthisak, P. (2022). Effect of chitosan coatings supplemented with chitosan-montmorillonite nanocomposites on postharvest quality of "Hom Thong" banana fruit. *Food Chemistry*, *374*(16), 131731. https://doi.org/10.1016/j.foodchem.2021.131731

Wardana, A.A., Suyatma, N.E., Muchtadi, T.R., & Yuliani, S. (2018). Influence of ZnO nanoparticles and stearic acid on physical, mechanical, and structural properties of cassava starch-based bionanocomposite edible films. *International Food Research Journal*, *25*(5), 1837–1844.

Widatalla, H.A., Yassin, L.F., Alrasheid, A.A., Rahman Ahmed, S.A., Widdatallah, M. O., Eltilib, S.H., & Mohamed, A.A. (2022). Green synthesis of silver nanoparticles using green tea leaf extract, characterization, and evaluation of antimicrobial activity. *Nanoscale Advances*, *4*(3), 911–915. https://doi.org/10.1039/d1na00509j

Winter, J., Nicolas, J., & Ruan, G. (2020). Hybrid nanoparticle composites. *Journal of Materials Chemistry. B*, *8*(22), 4713–4714. https://doi.org/10.1039/D0TB90071K

Xing, Y., Li, X., Guo, X., Li, W., Chen, J., Liu, Q., Xu, Q., Wang, Q., Yang, H., Shui, Y., & Bi, X. (2020a). Effects of different $TiO_2$ nanoparticles concentrations on the physical and antibacterial activities of chitosan-based coating film. *Nanomaterials*, *10*(7), 1365. https://doi.org/10.3390/nano10071365

Xing, Y., Xu, Q., Li, X., Chen, C., Ma, L., Li, S., Che, Z., & Lin, H. (2016). Chitosan-based coating with antimicrobial agents: Preparation, property, mechanism, and application effectiveness on fruits and vegetables. *International Journal of Polymer Science*, *2016*, 1–24. https://doi.org/10.1155/2016/4851730

Xing, Y., Yang, H., Guo, X., Bi, X., Liu, X., Xu, Q., Wang, Q., Li, W., Li, X., Shui, Y., Chen, C., & Zheng, Y. (2020b). Effect of chitosan/Nano-TiO$_2$ composite coatings on the postharvest quality and physicochemical characteristics of mango fruits. *Scientia Horticulturae, 263*, 109135. https://doi.org/10.1016/j.scienta.2019.109135

Xue, Y., Zhou, S., Fan, C., Du, Q., & Jin, P. (2019). Enhanced antifungal activities of eugenol-entrapped casein nanoparticles against anthracnose in postharvest fruits. *Nanomaterials, 9*(12), 1777. https://doi.org/10.3390/nano9121777

Yadav, S., Mehrotra, G. K., & Dutta, P. K. (2021). Chitosan based ZnO nanoparticles loaded gallic-acid films for active food packaging. *Food Chemistry, 334*, 127605. https://doi.org/10.1016/j.foodchem.2020.127605

Youssef, K., & Hashim, A. F. (2020). Inhibitory effect of clay/chitosan nanocomposite against *Penicillium digitatum* on citrus and its mode of action. *Jordan Journal of Biological Sciences, 13*(3), 349–355.

Yuan, G., Chen, X., & Li, D. (2016). Chitosan films and coatings containing essential oils: The antioxidant and antimicrobial activity, and application in food systems. *Food Research International, 89*(1), 117–128. https://doi.org/10.1016/j.foodres.2016.10.004

Zambrano-Zaragoza, M. L., Quintanar-Guerrero, D., Del Real, A., González-Reza, R. M., Cornejo-Villegas, M. A., & Gutiérrez-Cortez, E. (2020). Effect of nano-edible coating based on beeswax solid lipid nanoparticles on strawberry's preservation. *Coatings, 10*(3), 253. https://doi.org/10.3390/coatings10030253

Zare, M., Namratha, K., Ilyas, S., Sultana, A., Hezam, A., L, S., Surmeneva, M. A., Surmenev, R. A., Nayan, M. B., Ramakrishna, S., Mathur, S., & Byrappa, K. (2022). Emerging trends for ZnO nanoparticles and their applications in food packaging. *ACS Food Science and Technology, 2*(5), 763–781. https://doi.org/10.1021/acsfoodscitech.2c00043

Zhang, H., Li, R., & Liu, W. (2011). Effects of chitin and its derivative chitosan on postharvest decay of fruits: A review. *International Journal of Molecular Sciences, 12*(2), 917–934. https://doi.org/10.3390/ijms12020917

Zhao, J., Jiang, H., Huang, Q., Xu, J., Duan, M., Yu, S., Zhi, Z., Pang, J., & Wu, C. (2022). Carboxymethyl chitosan incorporated with gliadin/phlorotannin nanoparticles enables the formation of new active packaging films. *International Journal of Biological Macromolecules, 203*, 40–48. https://doi.org/10.1016/j.ijbiomac.2022.01.128

Ziv, C., & Fallik, E. (2021). Postharvest storage techniques and quality evaluation of fruits and vegetables for reducing food loss. *Agronomy, 11*(6), 1133. https://doi.org/10.3390/agronomy11061133

Zucchini, N. M., Florencio, C., Miranda, M., Borba, K. R., Oldoni, F. C. A., Oliveira Filho, J. G., Bonfim, N. S., Rodrigues, K. A., de Oliveira, R. M. D., Mitsuyuki, M. C., Hubinger, S. Z., Bresolin, J. D., & Ferreira, M. D. (2021). Effect of carnauba wax nanoemulsion coating on postharvest papaya quality. *Acta Horticulturae, 1325*, 199–206. https://doi.org/10.17660/ActaHortic.2021.1325.29

# 9 Integrating Innovative Technologies into Postharvest Fruit Storage Systems

*Ersin Çağlar*

## 9.1 INTRODUCTION

From the beginning, agriculture has been the most important source of livelihood and economy, and it is the first sector to play an important role in the development of the disciplines (Yarpuz-Bozdogan, 2018). Therefore, it has continued its existence sensitivity in every period. In the earliest times, agriculture and the discovery of fire gradually became the most important factors in forming the foundations of today's lifestyles with the discovery of agriculture and agricultural activities (Doss & Rayfield, 2021). After a while, people who discovered the land realized that they could not leave the fields that they planted, and thus their encounter with sedentary life began. In return for their planted crops, they also stepped into trade with the barter method and continued their economic activities by getting a share from the product (Erbay, 2013). With the inclusion of agriculture on the historical stage, commercial activities increased, markets were established, and social segments have begun to emerge with the benefits of settled life (Bielski et al., 2021). Agriculture, which emerged in this way in the historical process, constituted the first leg of development.

Today, basic needs such as water and land for production are gradually decreasing. Consequently, production is getting more difficult day by day. In addition to these basic needs, environmental factors adversely affect production opportunities (He et al., 2019). Technology is of great importance in increasing production in these adverse conditions. Therefore, it is necessary to correctly use technology to reduce the pressure on the environment while increasing production. For instance, the benefits of determining the actual need of water for plants through technology and performance of irrigation may be noted, or nutrients to be used in production with modern technological possibilities can be used at the required rate (DeLay et al., 2022).

The opportunities provided by the development of technology have gradually increased. Technology has influenced many sectors like health, education and agriculture that are important for human life with the opportunities that it provides. These opportunities have led to solutions to or have facilitated solutions to problems in many sectors. Agriculture is one of the sectors that greatly benefit from these

DOI: 10.1201/9781003452355-11

opportunities (Du et al., 2022). The Internet, IoT (Internet of Things), WSN (wireless sensor networks), cloud computing, and AI (artificial intelligence) are some of the most important opportunities of technology.

- **Internet**. In its simplest form, the Internet is a kind of communication system. By 2020, approximately 4.5 billion people were estimated to have access to the Internet (DeNardis, 2020; Schünemann et al., 2015).
- **IoT**. This is the network of any physical objects. These objects are called "things". Each object is embedded with sensors, software and other technologies to connect and exchange data (Koohang et al., 2022). With more than 7 billion connected IoT objects today, experts are expecting this number to grow to 10 billion by 2020 and to 22 billion by 2025 (Laghari et al., 2021). Figure 9.1 shows an IoT network connected to any kind of device.
- **WSN**. This consists of distributed sensors and one or more base station (sink nodes). These sensors monitor, in real-time, physical conditions, such as temperature, vibration or motion, and produce sensory data. Figure 9.2 illustrates a typical WSN architecture (Bala et al., 2018).

**FIGURE 9.1** Diagram of an IoT network.

- **Cloud computing**. Cloud computing is the on-demand delivery of IT resources over the Internet. You can access technology services, such as computing power, storage and databases instead of buying, owning and maintaining physical data centers and servers. In other words, cloud computing is the delivery of computing services. Figure 9.3 shows the cloud computing working principle (Alam, 2020; Srivastava & Khan, 2018).
- **AI**. Is demonstrated by machines, as opposed to the natural intelligence displayed by animals including humans. Simply, AI is the ability of a digital computer or computer-controlled robot to perform tasks commonly associated with intelligent beings (Zhang & Lu, 2021).

Apart from these opportunities, there are other opportunities through technology and benefits derived from these opportunities. But the most important opportunities provided by technology are those just described (Bukhman, 2021).

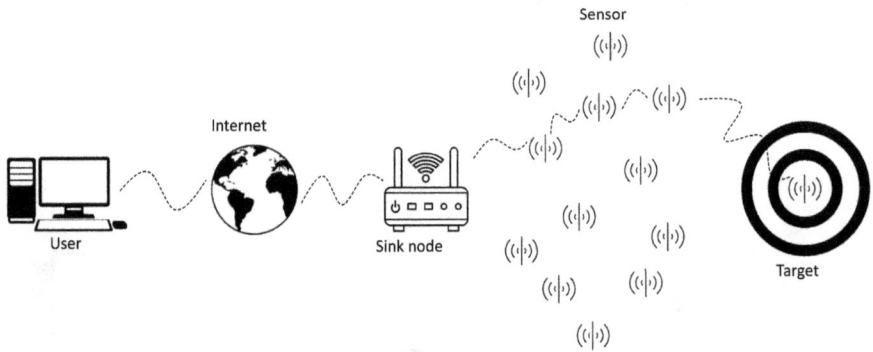

**FIGURE 9.2**   Typical WSN architecture.

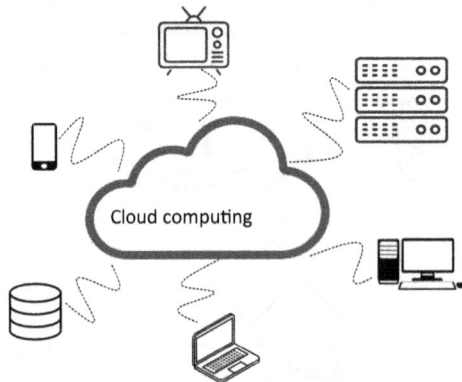

**FIGURE 9.3**   Diagram and working principle of cloud computing.

## 9.2  TECHNOLOGY AND AGRICULTURE

As mentioned, over the years, technology has revolutionized our world and daily lives with many opportunities that have moved all sectors into the digital era, whereas the ones that did not so move will not survive in a competitive environment. Thus technology is crucial for any sector (Habib et al., 2021). Any business sector must use all the opportunities technology offers, such as cloud computing and green computing (Vidmar et al., 2021; Trapp & Kanbach, 2021) as well as educational institutions (university, high/primary school, etc.). During the COVID-19 pandemic quarantine period, technology became more valuable since educational institutions used technology to offer their services to students (Valijonovna et al., 2022). Additionally, the health sector should also use technological opportunities for the provision of better treatment (Van der Wilt & Oortwijn, 2022).

Other than businesses, technology has a wide use in the entertainment sector and during sports activities, even in the Olympics. Nowadays, authorities are starting to discuss robots as referees for the World Cup 2022 (Dorer et al., 2022; Passault et al., 2022). Moreover, technology is vital in e-government to offer better services to citizens where it will change governance procedures in many aspects; at the same time, it changes the communication of all governments around the world (Kirsal & Caglar, 2019). In addition, a great many technological means are used in any field of agriculture such as plant protection, effective yield, postharvest, etc. (Sinha & Dhanalakshmi, 2022).

### 9.2.1  Effects of Technology in the Field of Agriculture

As mentioned previously, the world that we live in today offers us lots of facilities and charms, most of those charms run just because it's an advanced technology seems to be one of the most significant newborns of it. In recent years, technology has started to play a major role in daily life, extending our perceptions and ability to modify the environment around us (Du et al., 2022; Kembro & Norrman, 2022).

Among the many fields of endeavor, agriculture is the key field where technology is available. New models are being introduced daily for making operations productive and more effective. Technology contributes extensively to modern agriculture (Tiwari, 2022) by enabling farmers to manage their agricultural activities more efficiently. Technology has emerged as a blessing to humanity, allowing us to automate activities and save time and effort (Muangprathub et al., 2019). Around the year 2050, the world's population is expected to reach around 10 billion people (Sadigov, 2022). Agriculture will be all the more important in that we will need advanced agricultural methods to feed such a vast population (da Silveira et al., 2021).

### 9.2.2  Literature Review

Pursuant to the literature, various studies were performed in the field of education by combining PC games with Internet technology. These studies have been more successful than traditional methods. Researchers (Muangprathub et al., 2019) aimed to design and implement a WSN system for sensors in the crop field. In addition, users will be able to manage data via a smartphone and web applications. This proposed system can assist connected farmers anytime and anywhere and support mixed crop farming.

Besides WSN, IoT or AI technology is used in agriculture as well (Dagar et al., 2018). The proposed model is a simple IoT sensor architecture. This structure collects information and sends it to the server over the wi-fi network. There the server can perform actions depending on the information. Other research related to IoT and AI conducted by researchers (Chakshu, 2019) aimed to develop a highly cost-effective system for the food supply chain management system. It will provide status about the quality of these monitored foods and will also estimate the stock value/price based on the data received. On the other hand, researchers (Anoop et al., 2021) used smart storage to protect products by regulating storage parameters and maintaining environmental conditions along with the climate of the room used.

Apart from these technologies, cloud computing is also used in agriculture. Researchers (Lijin & Ankitha, 2022) introduced a modern technique to identify diseases on leaves. Moreover, this research also covers parameters for cloud computing, characteristics, deployment model, cloud service model, cloud benefits and cloud computing challenges in agriculture. Other research analyzes a smart irrigation system that is based on the Internet of Things and cloud-based architecture. The system is designed to measure soil moisture and humidity and then process this data in the cloud. With this system, farmers can use less water (Phasinam et al., 2022).

The possibilities provided by technology are not limited to these applications. Technology is also used in the detection and control of pests and diseases. Within the scope of smart agriculture, studies in the field of plant diseases have started (Aksoy et al., 2020). Wicaksono and Andryana (2020) used the Convolutional Neural Network (CNN) method for the detection of apple leaf disease where a total of 3151 image sets were used, including images of scabies, black rot, cedar rust disease and healthy leaves, using the PlantVillage dataset. While researchers achieved an average accuracy of 99.2% in the training data of the application, researchers achieved an accuracy of 94.9% in the testing process. Cruz et al. (2019) carried out a study on the detection with artificial intelligence of vine jaundice observed in grapefruit. In disease detection, researchers benefited from the change in leaf color. In the study, researchers labeled the grapevine images with the CNN method using the open-source PlantVillage dataset. Baranwal et al. (2019) GoogLeNet architecture in the CNN neural network set was used for apple leaf disease detection. Researchers performed the validation on a subset of the PlantVillage dataset, with the architecture containing 22 layers. Researchers used images of 1526 diseased and healthy leaves with black rot, scabies and rust disease as a dataset. As a result of the study, researchers achieved an average accuracy rate of 98.42%. Alruwaili et al. (2019), on the other hand, used the AlexNet architecture in CNN in the study for the detection of plant disease. In the study, researchers used the PlantVillage dataset, which contains a total of 54,306 images for 14 different plant species and 26 different diseases. As a result, researchers obtained 99.11% accuracy with the AlexNet model. In another study, Ferentinos (2018) used a dataset with 87,848 images and open source for training and testing in the model. Researchers developed using CNN for the diagnosis of plant diseases. This dataset includes 25 different plants and 58 different diseases observed in them. For the determination of tomato diseases, specially developed studies are extremely limited. In one of these studies, Mokhtar et al. (2015) concluded that it could detect with 99.83% accuracy due to the study on 800 healthy and diseased tomato leaves. As a result, disease detection based on artificial intelligence is an extremely current issue. In these studies, ready-made open-source libraries containing 26 disease samples belonging to 14 different plants, such as PlantVillage, were used.

## 9.3  PROPOSED METHODOLOGY

As previously mentioned, many technological opportunities are used in the field of agriculture. This chapter proposes a system related to postharvest that uses many technological opportunities like cloud computing, AI, WSN and IoT. This system is intended to increase postharvest productivity with the given technologies.

The system contains four phases to increase the efficiency of the postharvest system:

1. **IoT and WSN**. This is the first phase of the proposed system. The harvested products are placed in storage rooms that are full of sensors. These installed sensors have different features depending on the characteristics of the products together with the farmers' needs and requirements. Sensors can measure/follow several factors affecting product quality, such as temperature, relative humidity, oxygen concentration, carbon dioxide concentration, ethylene concentration, etc.
2. **AI**. With AI, the system can modify the temperature, relative humidity, oxygen and/or ethylene levels according to the requirements of the products. This modification is very important since farmers don't need to measure periodically and interfere. So, with the AI feature, farmers will save time and intervene early.

As mentioned, many of these technologies have been used in various aspects of agriculture. In some cases, some technologies have also been used together. Yet cloud computing and telephone coding are rarely used. A database can be created with cloud computing technology and can be accessed easily with a phone application. Unfortunately, this technology is not available in many parts of the world. For example, there has not been a study to create a software registry inventory (database) for disease detection in Turkey (Demir et al., 2021).

3. **Cloud computing**. All values measured by sensors will be recorded via cloud computing. These recorded values will be used in future evaluations and analyses so that more accurate ideas or solutions will be produced by utilizing the data recorded at each harvest season.
4. **Mobile computing**. Today, all computer technology is available on mobile phones. With an appropriate application, all data, measurements or camera recordings of the proposed system will be available on the mobile phone. In this way, farmer(s) will have access to everything more easily.

For future usage, these technologies may be used for detection in the storage area since the doors of the warehouse are not opened until the products in the storage are used. For this reason, any human intervention is impossible. However, some technological interventions can be made based on the observations to be made through sensors. Figure 9.4 illustrates the complete proposed model. As mentioned, the whole system process uses many innovative technologies in each phase. The first phase contains sensors and IoT devices in the storage area. The second phase includes AI technology. Cloud computing services are in the third phase. The last phase employs mobile technology.

In this system, any harvested product is kept in a storage area with the requested conditions based on the products. All measured data is saved in a cloud database with sensors. If the system detects a difference during the measurements, the atmosphere of the storage room is brought to the desired conditions with the AI feature. For example, if the temperature must be 5°C for a product, and the room atmosphere changes due to any condition, the atmosphere will be brought to the desired levels instantly with the AI feature. Cloud technology stores all measured data for any further analysis. Using these measured data, farmers will have an insight into their future storage. The last phase is the monitoring phase with mobile. Farmers can monitor the system all the time. If an unusual situation occurs, the farmer can also intervene in the system and change the conditions of the storage room.

Certainly, the Internet should be available for all the services to be used. The Internet connection speed should be high and wireless. Wireless provides Internet connectivity service to end users over a wireless communication network (Nazir et al., 2021). This study aims to minimize or completely solve the postharvest problems. These problems can be external and internal based on the commodity and environment, and, therefore, huge losses occur in products after harvest. External problems may be mechanical injury, parasitic diseases and internal factors. Postharvest losses are an average of 24–40% in developing countries and 2–20% in developed countries as a major source of waste (Chakshu, 2019).

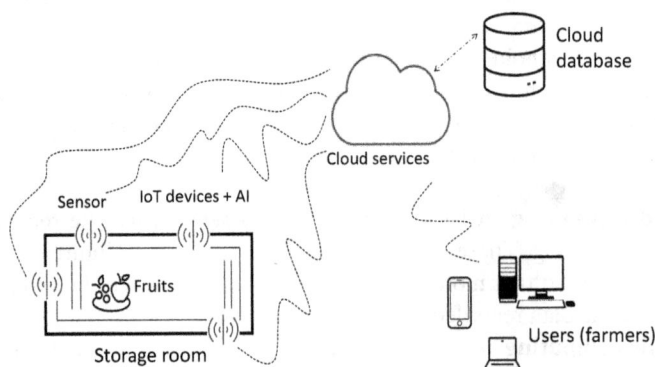

**FIGURE 9.4** Proposed model for the integration of IoT into postharvest fruit storage systems.

The lack of these technologies in postharvest handling causes 20–44% of the food supply.

These major problems are largely the result of an overall ineffective basic postharvest infrastructure in protecting crops, preventing damage from misuse, transportation, packaging and storage. High postharvest losses of products negatively affect food availability, food security and nutrition, as the producer sells less of the farm yield, and the net availability of these food commodities for consumption is reduced (Faqeerzada et al., 2018).

## 9.4 CONCLUSION

The contribution of technology to humanity is enormous. Today, the opportunities provided by technology are too important to be denied to any sector. The sectors that do not follow technological developments cannot survive in a competitive environment. Indisputably, technological developments have great benefits for everyone and every domain, whether business, education, health, agriculture, etc. Nowadays, agriculture benefits from many technological opportunities. As mentioned, cloud computing, AI, the Internet, IoT, mobile computing and WSN are used in agriculture to obtain more effective products.

This chapter proposed a model that contains almost all of today's newest technology. The aim is to minimize or eliminate postharvest problems with the existing technology. The proposed model consists of four phases. The first phase is related to IoT and WSN. This device analyzes and measures parameters and sends them to the cloud database. The second phase is related to AI. If any extraordinary situation occurs in the storage area, sensors will change the temperature, humidity or other parameters with AI technology. As mentioned, these technologies are used in many areas of agriculture. The following technological possibilities have not been used before (Demir et al., 2021). The third phase is related to the cloud database. Stored data will help farmers with further production. The last phase is related to the user or farmer, who uses a mobile phone to monitor and control all systems.

Technology is always difficult to catch up with. However, we must follow it. Only in this way can problems be minimized or solved.

## REFERENCES LIST

Aksoy, B., Halİs, H. D., & Salman, O. K. M. (2020). Identification of diseases in apple plants with artificial intelligence methods and comparison of the performance of artificial intelligence methods (in Turkish: Elma bitkisindeki hastalıkların yapay zekâ yöntemleri ile tespiti ve yapay zekâ yöntemlerinin performanslarının karşılaştırılması). *International Journal of Engineering and Innovative Research*, 2(3), 194–210. https://doi.org/10.47933/ijeir.772514

Alam, T. (2020). Cloud Computing and its role in the information technology. *IAIC Transactions on Sustainable Digital Innovation*, 1(2), 108–115. https://doi.org/10.34306/itsdi.v1i2.103

Alruwaili, M., Abd El-Ghany, S., & Shehab, A. (2019). An enhanced plant disease classifier model based on deep learning techniques. *International Journal of Advanced Technology and Engineering Exploration*, 9(1). https://doi.org/10.35940/ijcat.A1907.109119

Anoop, A., Thomas, M., & Sachin, K. (2021). IoT based smart warehousing using machine learning. In *Asian conference on innovation in technology (ASIANCON), 2021* (pp. 1–6). IEEE Publications. https://doi.org/10.1109/ASIANCON51346.2021.9544579

Bala, T., Bhatia, V., Kumawat, S., & Jaglan, V. (2018). A survey: Issues and challenges in wireless sensor network. *International Journal of Engineering and Technology*, 7(2), 53–55.

Baranwal, S., Khandelwal, S., & Arora, A. (2019). Deep learning convolutional neural network for apple leaves disease detection. In *Proceedings of the international conference on sustainable computing in science, technology and management (SUSCOM)*. Amity University. Retrieved February 26–28, 2019, from http://dx.doi.org/10.2139/ssrn.3351641

Bielski, S., Marks-Bielska, R., Zielińska-Chmielewska, A., Romaneckas, K., & Šarauskis, E. (2021). Importance of agriculture in creating energy security – a case study of Poland. *Energies*, 14(9), 2465. https://doi.org/10.3390/en14092465

Bukhman, I. (2021). *Technology for innovation: How to create new systems, develop existing systems and solve related problems* (p. 527). Springer Nature. https://doi.org/10.1007/978-981-16-1041-7

Chakshu, C. K. M. (2019). Postharvest crop management system using IoT and AI. *International Journal of Advance Research and Development*, 4(5), 42–44.

Cruz, A., Ampatzidis, Y., Pierro, R., Materazzi, A., Panattoni, A., De Bellis, L., & Luvisi, A. (2019). Detection of grapevine yellows symptoms in Vitis vinifera L. with artificial intelligence. *Computers and Electronics in Agriculture*, 157, 63–76. https://doi.org/10.1016/j.compag.2018.12.028

Dagar, R., Som, S., & Khatri, S. K. (2018). Smart farming–IoT in agriculture. In *International conference on inventive research in computing applications (ICIRCA), 2018* (pp. 1052–1056). IEEE Publications. https://doi.org/10.1109/ICIRCA.2018.8597264

da Silveira, F., Lermen, F. H., & Amaral, F. G. (2021). An overview of agriculture 4.0 development: Systematic review of descriptions, technologies, barriers, advantages, and disadvantages. *Computers and Electronics in Agriculture*, 189, 106405. https://doi.org/10.1016/j.compag.2021.106405

DeLay, N. D., Thompson, N. M., & Mintert, J. R. (2022). Precision agriculture technology adoption and technical efficiency. *Journal of Agricultural Economics*, 73(1), 195–219. https://doi.org/10.1111/1477-9552.12440

Demir, Ü., Nihal, K., & Uğurlu, B. (2021). Decision support model suggestion for the use of artificial intelligence in agriculture: Sample for tomato pest detection (in Turkish: Tarımda Yapay Zekâ Kullanımına Yönelik Karar Destek Modeli Önerisi: Domates Zararlısı Tespiti Örneği). *Lapseki Meslek Yüksekokulu Uygulamalı Araştırmalar Dergisi*, 2(4), 91–108.

DeNardis, L. (2020). *The Internet in everything*. Yale University Press. Retrieved July 21, 2022, from https://yalebooks.yale.edu/book/9780300233070/the-internet-in-everything/

Dorer, K., Giessler, M., Hochberg, U., Scharffenberg, M., Schillings, R., & Schnekenburger, F. (2022). Humanoid adult size champion 2021 sweaty. *Robot World Cup*, 352–359. https://doi.org/10.1007/978-3-030-98682-7_29

Doss, W., & Rayfield, J. (2021). Comparing Texas principal and agricultural education teacher perceptions of the importance of teaching activities in agricultural education programs. *Journal of Agricultural Education*, 62(1), 1–16. https://doi.org/10.5032/jae.2021.01001

Du, X., Wang, X., & Hatzenbuehler, P. (2022). Digital technology in agriculture: A review of issues, applications and methodologies. *China Agricultural Economic Review*, 15(1), 95–108. https://doi.org/10.1108/CAER-01-2022-0009

Erbay, R. (2013). The role of agriculture in economic development: An assessment of Turkey. In Turkish: Rolü, E. K. T. Türkiye üzerine bir değerlendirme. *Balkan Sosyal Bilimler Dergisi*, 2(4), 1–5.

Faqeerzada, M. A., Rahman, A., Joshi, R., Park, E., & Cho, B. K. (2018). Postharvest technologies for fruits and vegetables in South Asian countries: A review. *Korean Journal of Agricultural Science, 45*(3), 325–353. https://doi.org/10.7744/kjoas.20180050

Ferentinos, K. P. (2018). Deep learning models for plant disease detection and diagnosis. *Computers and Electronics in Agriculture, 145*, 311–318. https://doi.org/10.1016/j.compag.2018.01.009

Habib, M. N., Jamal, W., Khalil, U., & Khan, Z. (2021). Transforming universities in interactive digital platform: Case of city university of science and information technology. *Education and Information Technologies, 26*(1), 517–541. https://doi.org/10.1007/s10639-020-10237-w

He, G., Zhao, Y., Wang, L., Jiang, S., & Zhu, Y. (2019). China's food security challenge: Effects of food habit changes on requirements for arable land and water. *Journal of Cleaner Production, 229*, 739–750. https://doi.org/10.1016/j.jclepro.2019.05.053

Kembro, J., & Norrman, A. (2022). The transformation from manual to smart warehousing: An exploratory study with Swedish retailers. *International Journal of Logistics Management, 33*(5), 107–135. https://doi.org/10.1108/IJLM-11-2021-0525

Kirsal, Y., & Caglar, E. (2019). Analytical modelling and QoS evaluation of IoT applications in E-government. *Academic Conferences and Publishing Limited ECDG 2019 19th European Conference on Digital Government, 55*. https://doi.org/10.34190/ECDG.19.044

Koohang, A., Sargent, C. S., Nord, J. H., & Paliszkiewicz, J. (2022). Internet of things (IoT): From awareness to continued use. *International Journal of Information Management, 62*, 102442. https://doi.org/10.1016/j.ijinfomgt.2021.102442

Laghari, A. A., Wu, K., Laghari, R. A., Ali, M., & Khan, A. A. (2021). A review and state of art of Internet of things (IoT). *Archives of Computational Methods in Engineering*, 1–19. https://doi.org/10.1007/s11831-021-09622-6

Lijin, S., & Ankitha, P. (2022). The importance of cloud computing in agriculture. *Proceedings of the National Conference on Emerging Computer Applications 2022(NCECA 2022)*, 168–170. https://doi.org/10.5281/zenodo.6364831

Mokhtar, U., Bendary, N. E., Hassenian, A. E., Emary, E., Mahmoud, M. A., Hefny, H., & Tolba, M. F. (2015). SVM-based detection of tomato leaves diseases. In D. Filev et al. (Eds.), *Intelligent systems: Proceedings of the 7th IEEE international conference intelligent Systems IS, Vol. 2: Tools. Architectures, systems, applications, 2014* (pp. 641–652). Springer. https://doi.org/10.1007/978-3-319-11310-4_55

Muangprathub, J., Boonnam, N., Kajornkasirat, S., Lekbangpong, N., Wanichsombat, A., & Nillaor, P. (2019). IoT and agriculture data analysis for smart farm. *Computers and Electronics in Agriculture, 156*, 467–474. https://doi.org/10.1016/j.compag.2018.12.011

Nazir, R., Kumar, K., David, S., & Ali, M. (2021). Survey on wireless network security. *Archives of Computational Methods in Engineering*, 1–20. https://doi.org/10.1007/s11831-021-09631-5

Passault, G., Gaspard, C., & Ly, O. (2022). Robot soccer kit: Omni-wheel tracked soccer robots for education. In *IEEE international conference on autonomous robot systems and competitions (ICARSC), 2022* (pp. 34–39). IEEE Publications. https://doi.org/10.1109/ICARSC55462.2022.9784808

Phasinam, K., Kassanuk, T., Shinde, P. P., Thakar, C. M., Sharma, D. K., Mohiddin, M. K., & Rahmani, A. W. (2022). Application of IoT and cloud computing in automation of agriculture irrigation. *Journal of Food Quality, 2022*, 1–8. https://doi.org/10.1155/2022/8285969

Sadigov, R. (2022). Rapid growth of the world population and its socioeconomic results. *The Scientific World Journal, 2022*, 8110229. https://doi.org/10.1155/2022/8110229

Schünemann, H. J., Al-Ansary, L. A., Forland, F., Kersten, S., Komulainen, J., Kopp, I. B., Macbeth, F., Phillips, S. M., Robbins, C., van der Wees, P., Qaseem, A., & Board of Trustees of the Guidelines International Network. (2015). Guidelines international network: Principles for disclosure of interests and management of conflicts in guidelines. *Annals of Internal Medicine*, *163*(7), 548–553. https://doi.org/10.7326/M14-1885

Sinha, B. B., & Dhanalakshmi, R. (2022). Recent advancements and challenges of Internet of things in smart agriculture: A survey. *Future Generation Computer Systems*, *126*, 169–184. https://doi.org/10.1016/j.future.2021.08.006

Srivastava, P., & Khan, R. (2018). A review paper on cloud computing. *International Journal of Advanced Research in Computer Science and Software Engineering*, *8*(6), 17–20. https://doi.org/10.23956/ijarcsse.v8i6.711

Tiwari, S. P. (2022). Information and communication technology initiatives for knowledge sharing in agriculture. *Indian Journal of Agricultural Sciences*, *78*(9), 737–747. https://doi.org/10.48550/arXiv.2202.08649

Trapp, C. T. C., & Kanbach, D. K. (2021). Green entrepreneurship and business models: Deriving green technology business model archetypes. *Journal of Cleaner Production*, *297*, 126694. https://doi.org/10.1016/j.jclepro.2021.126694

Valijonovna, K. I., Rakhmatjonovich, T. D., & Mukhtoraliyevna, Z. S. (2022). Informational technology at education. *Spanish Journal of Innovation and Integrity*, *6*, 262–266.

Van der Wilt, G. J., Oortwijn, W., & Validate-HTA Consortium. (2022). Health technology assessment: A matter of facts and values. *International Journal of Technology Assessment in Health Care*, *38*(1), e53. https://doi.org/10.1017/S0266462322000101

Vidmar, D., Marolt, M., & Pucihar, A. (2021). Information technology for business sustainability: A literature review with automated content analysis. *Sustainability*, *13*(3), 1192. https://doi.org/10.3390/su13031192

Wicaksono, G., & Andryana, S. (2020). Aplikasi Pendeteksi penyakit pada daun tanaman apel dengan metode convolutional neural network. *Journal of Information Technology and Computer Science*, *5*(1), 9–16. https://doi.org/10.31328/jointecs.v5i1.1221

Yarpuz-Bozdogan, N. (2018). The importance of personal protective equipment in pesticide applications in agriculture. *Current Opinion in Environmental Science and Health*, *4*, 1–4.

Zhang, C., & Lu, Y. (2021). Study on artificial intelligence: The state of the art and future prospects. *Journal of Industrial Information Integration*, *23*, 100224. https://doi.org/10.1016/j.jii.2021.100224

# 10 Recent Technology and Advances in Fresh-Cut Products

*Sandra Horvitz, Cristina Arroqui, and Paloma Vírseda*

## 10.1 INTRODUCTION

Fresh fruit and vegetables are widely recognized as an important source of essential vitamins and minerals, dietary fiber and different bioactive compounds, like phenols, carotenoids, glucosinolates and tocopherols, that make them an important component of a healthy diet (Bhilwadikar et al., 2019; Botondi et al., 2021). In effect, the high prevalence of childhood obesity and different types of non-communicable, chronic diseases like Type II diabetes, obesity, cancer, stroke and heart diseases could be prevented with adequate consumption of fresh fruits and vegetables (Haß & Hartmann, 2018; Haynes-Maslow et al., 2013).

Despite their importance as a source of macro- and micronutrients and the multiple promotion campaigns developed worldwide (like the "5 a day"-type programs), the consumption of fresh fruits and vegetables for most individuals is still lower than the intake of a minimum of 400 g per day, recommended by the Food and Agriculture Organization of the United Nations (FAO) and the World Health Organization (WHO) (Saba et al., 2018). The main barriers to consumption include perishability, high prices, inconvenience, lack of time for preparation, dislike of the taste, unwillingness to replace other foods with fruits or vegetables, personal dietary choices and lack of information (Cohen et al., 1998).

In addition to other strategies aimed to promote an increase in fresh fruit and vegetable consumption, fresh-cut products arise as an alternative that could help in overcoming the inconvenience and the lack of time for preparation barriers. In fact, the market for these products exhibited an expansion in recent years with an increased offer of products in the main supermarket chains and an increased consumption due to the changes in consumers' alimentary habits (De Corato, 2020).

### 10.1.1 DEFINITION

Minimally processed fresh-cut fruits and vegetables (FCFV) are defined as "any fresh fruit or vegetable or any combination thereof that has been physically altered from its whole state after harvest (e.g., by chopping, dicing, peeling, ricing, shredding, slicing, or tearing), without any additional processing (such as blanching or cooking)" (USFDA, 2019). It is

DOI: 10.1201/9781003452355-12

generally accepted that FCFV involve 100% usable product, live tissues that remain in a fresh state and thus are fresh-like in character and quality (Wiley & Yildiz, 2017).

## 10.2 FRESH-CUT PREPARATION

Typical processing of fresh-cut products involves one or more of the following steps: washing, peeling, cutting/slicing/shredding, sanitizing, rinsing/spin drying, packing, storage, distribution and marketing (Figure 10.1). Modified atmosphere and vacuum are commonly used for packing purposes, and cold chain maintenance is essential from production to consumption. Regardless of some minor differences in the production steps within individual products, the fresh-cut industry faces the challenge of minimizing energy use, environmental pollution, food waste, and costs and at the same time maximizing the overall quality and convenience of FCFV (Yildiz, 2017).

## 10.3 ALTERATION OF FRESH-CUT PRODUCE

Traditional processing techniques like freezing, canning or drying aim to extend produce shelf life. On the contrary, minimal processing of fruits and vegetables usually promotes a faster alteration of the products (De Corato, 2020). In effect, the

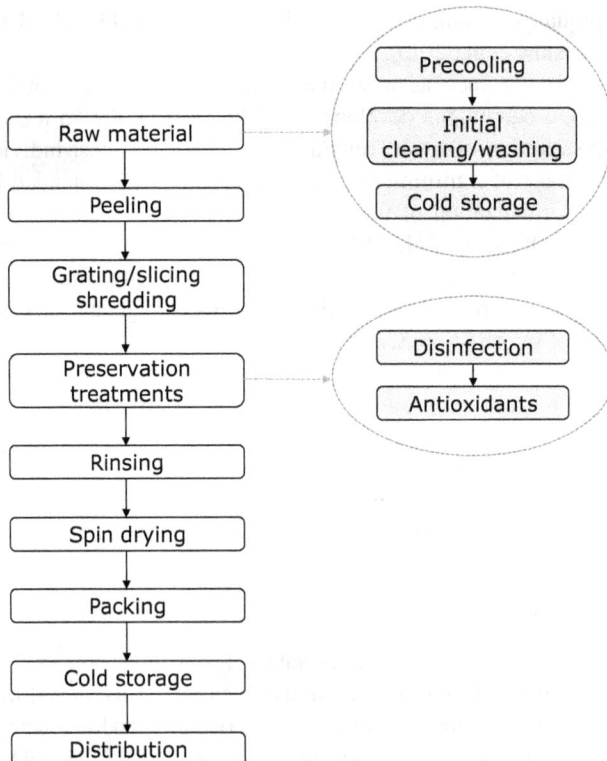

**FIGURE 10.1** General flowchart for fresh-cut products preparation.

mechanical damages inflicted during processing enhance physiological activity and biochemical changes, which in turn, may accelerate microbial growth and general quality loss, including changes in physicochemical (color, texture), sensory and nutritional attributes (Figure 10.2).

The prevailing type of alteration will be conditioned by a series of factors like the product considered, the variety or cultivar, maturity, and the growing, processing and storage conditions (Gross et al., 2016). In any case, FCFV have a short shelf life, with a consumption window of only a few days (Leneveu-Jenvrin et al., 2020), typically 4–7 for sliced vegetables and up to 10 days for sliced fruits (Botondi et al., 2021). Thus, the industry faces the challenge of extending the shelf life and ensuring produce safety, while complying with consumers' demands for convenient, fresh-like, natural FCFV, which at the same time are microbiologically safe and preserve a high nutritional and sensory quality (Qadri et al., 2015; Giannakourou & Tsironi, 2021).

### 10.3.1 Physiological and Biochemical Alterations

As previously mentioned, in response to the mechanical and physiological damage done to the tissues during FCFV preparation, a series of stress reactions are induced in the vegetal cells (Botondi et al., 2021). In effect, wounding enhances respiration rate, ethylene synthesis, susceptibility to microbial spoilage, enzymatic activity, water loss and the oxidation of organic acids and sugars, compounds responsible for FCFV flavor (Velderrain-Rodríguez et al., 2015). Consequently, ripening, softening and senescence are accelerated, and the shelf life of the produce is shortened when compared to whole products (Wiley & Yildiz, 2017).

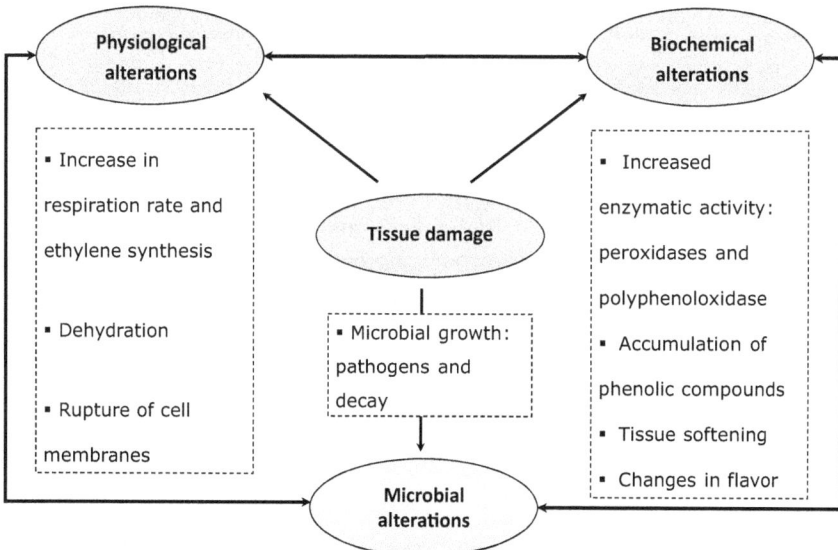

**FIGURE 10.2** Main alterations occurring in fresh-cut fruits and vegetables.

Color has a significant influence on consumers' perception of FCFV freshness and quality and depends on the presence of natural pigments (chlorophyll, carotenoids and anthocyanin) as well as on those resulting from enzymatic and non-enzymatic reactions (Miller et al., 2013). Color changes due to chlorophyll loss, formation of pigments and enzymatic browning, are among the main consequences of cutting-induced wounding in FCFV.

Enzymatic browning is originated when the phenolic compounds present in raw fruit and vegetables, like apples and potatoes, come into contact with oxygen and are oxidized by the polyphenol oxidase (PPO) enzyme (Kambhampati et al., 2019). PPO catalyzes the oxidation of phenolic compounds into colorless quinones which later polymerize in brown, melanin-type compounds. The phenylalanine ammonia-lyase (PAL) and peroxidases (POD) are also important enzymes responsible for enzymatic color changes in lettuce and in melons, watermelons and citrus fruits, respectively (Soliva-Fortuny & Martín-Belloso, 2003).

### 10.3.2 MICROBIAL ALTERATIONS

As mentioned, during FCFV processing, the raw material is subjected to different operations like peeling, cutting, shredding and/or slicing, which cause injuries and mechanical damages to the tissues (Almenar, 2021). Consequently, the natural compartmentalization of the cells is lost, substrates and enzymes may come in contact, and the internal fluids are exposed to the action of oxygen and light and constitute a source of nutrients for microorganisms' development (Qadri et al., 2015). At the same time, the cut surfaces provide an ideal environment for microorganisms' growth due to the irregularities where the microbes can remain attached and unavailable for sanitizers during the disinfection step (Feliziani et al., 2016; Raffo & Paoletti, 2022). Therefore, FCFV are more perishable than their fresh whole counterparts and can provide a favorable environment for the growth of both human pathogens and spoilage microorganisms (Siroli et al., 2015).

## 10.4 PRESERVATION METHODS

Postharvest washing is still the most widely used method for cleaning, disinfecting and extending the shelf life of fresh and FCFV (Gombas et al., 2017). However, in recent years there has been increasing evidence that this step may not be sufficient to ensure food safety and that it may even promote cross-contamination if the water becomes contaminated with microbial pathogens (Murray et al., 2017; Raffo & Paoletti, 2022). Thus, the aim of using sanitizers in the washing step is mainly for sanitization of the water and prevention of cross-contamination (Fan et al., 2021). Together with this, there is an increasing interest in alternative methods that could replace or supplement washing treatments (Meireles et al., 2016).

In the following sections, the main approaches developed to mitigate FCFV alterations are described. These include refrigeration, chlorine-alternative chemical compounds, physical and biological methods, and packaging (Figure 10.3).

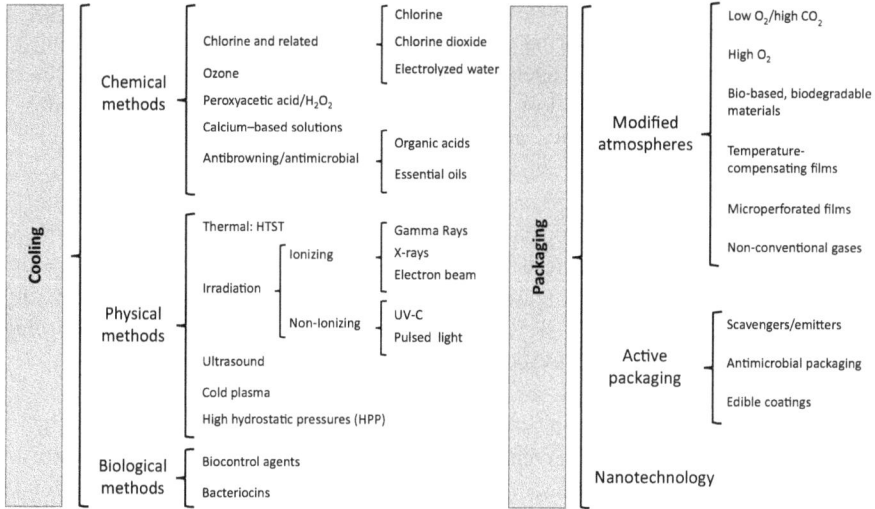

**FIGURE 10.3** Preservation methods used in fresh-cut fruits and vegetables.

### 10.4.1 REFRIGERATION

Temperature is the most important environmental factor affecting the postharvest life of both whole and fresh-cut fruit and vegetables, mainly through its influence, on one hand, on physiological activity and metabolic reactions (Rao, 2015) and, on the other hand, on the microbial growth of pathogenic and spoilage microorganisms (Abadias et al., 2012). Therefore, lowering produce temperature, reducing the time until optimal temperatures are reached and proper temperature management during processing and storage are key factors in sustaining FCFV quality. In fact, as storage temperature is directly related to respiration rate and biochemical reactions, processing temperatures of FCFV should be always maintained below 10 °C (Jideani et al., 2017; Yahia et al., 2019).

From what has been previously mentioned, it is evident that temperature control is essential during all the steps involved in the FCFV processing chain, from raw material reception until consumption by the final consumers, in particular:

- **Raw material harvest and reception**. Whenever possible, it is advisable to harvest early in the morning to avoid high ambient temperatures, especially during warm weather. For highly perishable products, with mid- or high respiration rates, it might be necessary to condition and carry out the precooling in the field, i.e., with hydrocooling, the most used system (Gil, 2021). If the produce is not precooled in the field, it should be promptly precooled once it arrives in the processing plant either by hydrocooling (cold water at <5 °C), cold forced air, or vacuum cooling. If processing will not be done immediately, produce should be kept in cold storage until use.

- **Processing**. During processing, it is important to maintain temperatures below 10, 4 and 5 °C during cutting/slicing/shredding, washing and storing the final product, respectively. In effect, keeping low temperatures reduces respiration rate, water loss, enzymatic activity and microbial growth, enhances produce quality, and extends shelf life (Kitinoja & Thompson, 2010).

The selection of the most appropriate cooling method is influenced by the physiological and morphological characteristics on individual products, like the porosity and the surface–volume relation (Figure 10.4) and is key to maximizing produce quality and shelf life (Garrido et al., 2015). In the case of FCFV, the most used precooling methods are hydrocooling and vacuum cooling, described here.

- **Hydrocooling**. This is a fast method in which the produce is cooled by contact with chilled water, either by immersion or by spraying chilled water on the surface of the product (Dar et al., 2020). Hydrocooling avoids water loss, and treatment duration will be a function of product size with small pieces requiring shorter times and vice versa. When using hydrocooling, it is critical to continuously control the microbiological quality of the water to avoid cross-contamination (Gil, 2021).
- **Vacuum cooling**. This method is based on rapid evaporation of part of the moisture of the product under vacuum. In effect, when pressure decreases, water boils at lower temperatures and evaporates, taking the energy for this process from the product, in the form of latent heat of evaporation (McDonald & Sun, 2000). This technique allows a rapid and uniform

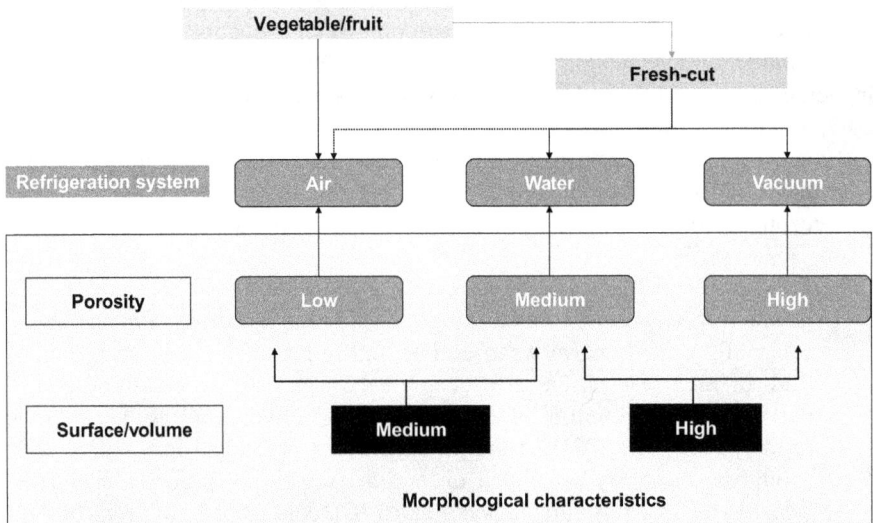

**FIGURE 10.4**  Precooling method selection as a function of produce characteristics.

cooling, and as no refrigerant medium is used, it is also considered safe and free from contaminants (He et al., 2013). On the other hand, the main disadvantages are the limitation to products with a high surface:volume ratio, and the water loss in the produce being cooled (Garrido et al., 2015). To compensate for water loss, produce is usually prewetted or sprayed with water prior to cooling (Ding et al., 2016).

## 10.4.2 Chemical Methods

### 10.4.2.1 Chlorine

Nowadays, chlorine is the most used disinfectant in aqueous solutions due to its availability, facility to apply and relatively low application cost (Ali et al., 2018). Chlorine-based sanitizers are used in concentrations of 50–200 mg $L^{-1}$ free chlorine and treatments of 1–5 min (Horvitz & Cantalejo, 2012). However, chlorine efficacy is dependent on free chlorine concentration in the wash solution, the pH and temperature of the water and the presence of organic matter. In addition, it may be corrosive to machinery and equipment and can produce unhealthy, toxic by-products that can adversely affect human and environmental safety (Goodburn & Wallace, 2013). This, together with the European Union (EU) restrictions on chlorine use, led to the development of alternative approaches for the decontamination of FCFV.

### 10.4.2.2 Chlorine Dioxide

Chlorine dioxide ($ClO_2$) can be produced by the reaction of sodium chlorite with chlorine gas or by the reaction of an acid with sodium chlorite (Ölmez & Kretzschmar, 2009). In the former, the aqueous form of $ClO_2$ is obtained while the latter gives the gaseous form of this disinfectant. Chlorine dioxide presents several advantages over chlorine. In effect, it has more oxidizing capacity, induces less potentially harmful by-products, has lower reactivity with organic matter, is less corrosive (Meireles et al., 2016), is stable in a wide pH range (6–10) and is effective against both vegetative cells and spores from foodborne pathogens (De Corato, 2020; Kambhampati et al., 2019). For the sanitation of processing equipment, a maximum of 200 mg $L^{-1}$ $ClO_2$ is admitted, while the maximum allowable for contact with whole produce is 3 mg $L^{-1}$. Moreover, after treatment with $ClO_2$, FCFV must be rinsed with potable water.

### 10.4.2.3 Electrolyzed Water

Electrolyzed water (EW) is obtained through the electrolysis of a diluted (0.1–0.2 %) sodium, potassium or magnesium chloride solution to produce electrolyzed acidic (pH 2–3) and basic (pH 10–13) water at the anode and cathode, respectively. A neutral electrolyzed solution (pH 6.5–7.5) is obtained by mixing these solutions (Raffo & Paoletti, 2022).

As a result of the process, oxygen and hydrogen ions and HOCl are produced in the anode, while the reaction on the cathode renders hydrogen and hydroxide ions. The antimicrobial properties of acidic EW derive from its low pH value, a high oxidation-reduction potential and the presence of chlorine-based reactants (De Corato, 2020).

This technology is considered an environmentally friendly approach as only a diluted NaCl water solution is needed (Raffo & Paoletti, 2022). In effect, the antimicrobial efficacy of neutral EW solutions (30 mg L$^{-1}$ free chlorine) was similar to that of conventional washing solutions containing 120 mg L$^{-1}$ free chlorine (Manzocco et al., 2015). Thus, washing with neutral EW could allow for the same antimicrobial efficacy while reducing chlorine concentration in the water. Finally, it is worth remarking that, in general, no deleterious effects were observed on the produce's sensory quality and nutrient content after treatment with EW, and thus it can be considered a chlorine substitute that could be used whenever chlorinated water use is allowed (Gil et al., 2015).

### 10.4.2.4   Ozone

Ozone (O$_3$) is produced as a gas and can be applied in gaseous or aqueous phases for the decontamination of fresh and FCFV produce, process water and food-contact surfaces (Botondi et al., 2021). It has a high oxidant capacity and can be used for the inactivation of a wide range of microorganisms, including viruses, bacteria and fungi, as well as for the degradation of pesticides and off-odors in storage rooms (Horvitz et al., 2021).

Once generated, ozone molecules decompose rapidly, and thus ozone must be generated continuously on-site and cannot be stored or transported. Its half-life is short, 20–30 minutes in water and up to 2–3 hours in air, breaking down thereafter, to molecular oxygen (Carletti et al., 2013). As ozone is efficient at low concentrations and short contact times, and it does not leave toxic residues on the treated produce, its use is considered an environmentally friendly technology (Horvitz & Cantalejo, 2015).

However, the effectiveness of ozone application can be influenced by a series of factors like the storage relative humidity and temperature, pH, gas concentration, duration of the treatment and the type and initial load of microorganisms present (Feliziani et al., 2016). In general, the best results were obtained under high relative humidity, low temperatures and neutral pH (6–8.5). For optimum efficiency, it is also essential that the gas is thoroughly and evenly distributed quickly. Otherwise, decomposition will occur before the O$_3$ is able to contact its target (Horvitz & Cantalejo, 2014).

### 10.4.2.5   Peroxyacetic Acid

Peroxyacetic acid (PA) is an oxidizing agent, synthesized by the reaction of hydrogen peroxide with either acetic acid or acetic anhydride in the presence of a catalyst like sulfuric acid (Bhilwadikar et al., 2019). The antimicrobial activity of PA is mainly due to the formation of reactive oxygen species (ROS), which damage microbial lipids and DNA and cause the denaturation of proteins and enzymes (Bhilwadikar et al., 2019). This sanitizer presents a series of advantages: it is effective against spoilage and pathogenic microorganisms in suspension, it is not corrosive, and it is not affected by temperature changes and the presence of organic matter (De Corato, 2020). PA decomposes to acetic acid, water and oxygen, leaving no toxic by-products after use (Ölmez & Kretzschmar, 2009). The use of PA is approved by the FDA as a food-grade sanitizer up to 80 ppm in wash water, which may not be enough to obtain a significant reduction in the microbial load of FCFV (Kambhampati et al., 2019).

### 10.4.2.6 Hydrogen Peroxide

Hydrogen peroxide ($H_2O_2$) is a strong oxidant with bactericidal and bacteriostatic effects, depending on the concentration, pH and temperature (Meireles et al., 2016). The antimicrobial activity is related to its oxidizing power and the generation of cytotoxic oxidizing species like hydroxyl free radicals, singlet oxygen species and hydrogen peroxides that attack cellular lipids, proteins and DNA (Ali et al., 2018). Hydrogen peroxide can be used in aqueous or vapor phase, either alone or combined with heat, which increases its antimicrobial activity (De Corato, 2020).

As hydrogen peroxide decomposes into oxygen and water in the presence of catalase, it does not form toxic residues (Van Haute et al., 2015). However, the main drawback is that in sensitive products like lettuce, berries and mushrooms, it may induce browning or bleaching of the tissues and degradation of nutraceutical compounds (Ölmez & Kretzschmar, 2009). To minimize this problem, residual $H_2O_2$ must be removed by either physical or chemical methods like rinsing with water immediately after treatment or by adding appropriate anti-browning agents like ascorbic acid (Ali et al., 2018).

### 10.4.2.7 Calcium-Based Solutions

Calcium-based treatments are used in fruits and vegetables mainly with the aim to preserve their firmness. Calcium ion can interact with pectin to form calcium pectate and thus maintain the plant cell wall integrity (Martin-Diana et al., 2005). For this purpose, different calcium salts like calcium lactate, calcium chloride, calcium phosphate, calcium propionate and calcium gluconate can be used (Ramos et al., 2013) in washing solutions in concentrations ranging from 0.5 to 3% and dipping times from 1 to 5 min (Kambhampati et al., 2019). The choice of the calcium source will be conditioned by different factors such as bioavailability and solubility, the flavor change it could induce in the treated produce and the interaction with food ingredients (De Corato, 2020). In this sense, calcium lactate is preferred to calcium chloride because it avoids the bitterness and off-flavors associated with the latter.

In addition to the use for texture improvement, different calcium salts have been studied for preventing decay and browning, for sanitation purposes, and for nutritional enrichment of fresh fruits and vegetables. Particularly, calcium lactate and calcium propionate proved to be effective sanitizer agents in comparison with chlorine in different FCFV, with the advantage of avoiding the formation of harmful by-products like chloramines and trihalomethanes (Martín-Diana et al., 2007). Calcium carbonate and calcium citrate are the calcium salts authorized to accomplish FCFV nutritional value enhancement (De Corato, 2020).

### 10.4.2.8 Organic Acids

Organic acids (such as lactic, citric, L-ascorbic, acetic, tartaric, sorbic and malic) are natural substances, classified as GRAS, and are used as preservatives in the FCFV industry due to their antimicrobial efficacy against fungi and psychrophilic and mesophilic bacteria, with the advantage of not forming toxic by-products (Raffo & Paoletti, 2022).

Their disinfectant activity is specific, and the main antimicrobial mechanism is through alterations in cell membranes permeability and nutrient transport system, acidification of the interior of the microbial cells and inhibition of metabolic reactions (Bhilwadikar et al., 2019). Relatively high concentrations are needed to achieve a rapid mortality of microorganisms, and as all the organic acids are non-selectively toxic, before use it is necessary to evaluate the risk of injury to the treated product. Additionally, there are other natural, organic acids, such as jasmonic and salicylic acids, which are not used as disinfecting agents but for their ability to induce plant defense mechanisms (Feliziani et al., 2016).

Besides the antimicrobial activity, L-ascorbic and citric acids also present antioxidant effects and have been used to prevent browning and oxidative reactions in FCFV as an alternative to sulfites, the traditional anti-browning agent, whose use on FCFV was banned by the FDA owing to their potential health risks (Dong & Wang, 2021).

Dipping peeled/cut FCFV in ascorbic or citric acid or in their mixture solutions could be effective in inhibiting browning reactions. Ascorbic acid (AsA) action on the polyphenols' system is complex and its efficacy is related to the acid concentration. In addition, the applied AsA is consumed during the reduction process, being irreversibly oxidized to dehydroascorbic acid. Thus, its anti-browning effect could be limited in time (Bobo-García et al., 2020). As regards to citric acid, it functions as an antioxidant by reducing the superficial pH value of several fresh-cut fruits (Soliva-Fortuny & Martín-Belloso, 2003) and by chelating the copper located in the active center of the PPO enzyme. However, in products very sensitive to browning, like peeled and sliced potatoes, the use of individual L-ascorbic and citric acids may not be effective enough, and they must be combined with other preservatives, antioxidants or modified atmospheres (Bobo-García et al., 2020).

### 10.4.2.9  Essential Oils

To harmonize consumers' demands for food products free of or with less use of synthetic additives, the trend in the food industry is to use natural preservatives (Negi, 2012; Kahramanoğlu et al., 2020). In this sense, plant extracts could be a good alternative to ensure food safety and to control browning of FCFV due to the large number of bioactive compounds (polyphenols, flavonoids terpenes and alkaloids) they contain (Matrose et al., 2021). However, many challenges, regarding plant material selection, extraction methods, lack of reproducibility of results and of experiments conducted on real food systems, currently prevent a widespread use of this type of compounds (Zhang et al., 2022).

Essential oils (EOs) include a mixture of aromatic, hydrophobic terpenes, alcohols, ketones, aldehydes and esters obtained from plant materials by distillation (González-Aguilar et al., 2008). These are volatile, naturally occurring compounds produced mainly by aromatic plants as secondary metabolites (Yousuf et al., 2021). Different essential oils (mainly terpenoids extracted from garlic, coriander, oregano, peppermint, thyme and cinnamon) have shown antimicrobial properties against food spoilage flora and pathogens. The mechanism of action was attributed to several targets in the microbial cells like disintegration of fungal hyphae, enhancement of cell membranes permeability and alteration of cell functional properties (Burt, 2004;

Pandey et al., 2017). Food properties like fat and protein content, enzymes, pH, and water activity can influence the effectiveness of EOs.

Different EOs such as vanillin, citral, hexanal, hexyl acetate, carvacrol, cinnamaldehyde, cinnamon, eugenol, thymol, eucalypthol and menthol could be considered natural food additives due to their antioxidant activity to improve the safety, shelf life, and quality of FCFV (Carocho et al., 2014; Patrignani et al., 2015). They can be used either alone or as conjugates into active packaging films under modified atmospheres. Despite their positive effects and the fact that most of the EOs are classified as GRAS, their use is challenging due to their high reactivity and volatility, intense aroma and the fact that certain EOs could be toxic when used in high concentrations (Yousuf et al., 2021).

### 10.4.3  Physical Methods

#### 10.4.3.1  Thermal Treatments

Short time heat treatments (STHT) are safe, environmentally friendly treatments applied to FCFV mostly as short-time hot water treatments with the aim to reduce surface microorganisms and prevent enzymatic browning (Giannakourou & Tsironi, 2021). In addition to hot water, heat treatments can be applied as hot vapor, hot air or hot water rinse brushing (Sivakumar & Fallik, 2013). Hot water dips are usually performed at 40–60 °C for a duration of only a few seconds to many minutes, depending on the plant organ, maturity, size and time of application. Microbial inactivation is a function of heating temperatures, treatment duration and heat resistance of the microorganisms (Fallik & Ilic, 2022). Thus, the appropriate selection of these variables (temperature and time of exposure) is essential to guarantee the produce quality during its shelf life (Nicola et al., 2022).

#### 10.4.3.2  Irradiation

*10.4.3.2.1 Ionizing Radiation: Gamma Ray, Electron Beam and X-ray*

These treatments produce high energy atoms, molecules and ions that react with water molecules, generating ROS and free radicals with high oxidizing and antimicrobial activities (Bhilwadikar et al., 2019). Irradiation provokes injuries to microorganisms' and insects' DNA directly and indirectly through the action of the ions and free radicals produced (Korkmaz & Polat, 2005). Due to its penetration ability, irradiation is effective against microorganisms located either on the surface or inside fresh and FCFV produce (Fan et al., 2021). It can be applied before or after packaging, leaving no harmful residues (Guerreiro et al., 2016) and is thus considered an environmentally friendly technology. Furthermore, it consumes less energy and does not cause heating of the irradiated produce, preserving the quality of heat-sensitive commodities.

Despite its advantages and the approval by regulatory agencies in several countries, this technology has not been widely implemented, mainly due to consumer concerns over irradiated products. The current uses of irradiation are mainly for phytosanitary purposes in order to meet quarantine requirements of disinfestation (Roberts & Follett, 2018).

*10.4.3.2.2. Non-ionizing Radiation: UV-C Light and Pulsed Light*

UV-C light treatments involve the exposure of produce to radiation in the range of 190–280 nm, for a specific time (Fallik & Ilic, 2022). From this range, the strongest microbiocidal effect of UV-C is in the range of 250–260 nm, where the DNA has its peak effectiveness of UV absorption (Ali et al., 2018). In effect, the germicidal effect of UV-C light results directly from its capability to damage microbial DNA and RNA and consequently prevent cell replication (Fan & Wang, 2022) and indirectly through the induction of resistance against human pathogens in different fruit and vegetables (De Corato, 2020).

UV light can be used in a continuous mode (UV lamps) or as pulses. The former presents high efficiency and short process times, while the latter is applied in short-duration (1–20 pulses per second), high-peak pulses of broad-spectrum light in the range of 200–1100 nm (Meireles et al., 2016). In general, it is accepted that the most important wavelength region for the bactericidal effects of pulsed light is the UV-C component (Fan & Wang, 2022). Pulsed light causes the destruction of microorganisms by a multitarget process of photochemical, photothermal and photophysical effects, resulting in more effective microbial inactivation than with continuous UV-light (Fan et al., 2021). In both types of application, the effectiveness may be limited by a low penetration capacity and shadowing effects. In addition to the disinfection of the fruit and vegetables surface, UV-C treatments have been also used to delay fruit ripening, reduce respiration rate and enzymatic browning by inhibition of the polyphenol oxidase, peroxidase and phenylalanine ammonia lyase enzymes activity (Hosseini et al., 2019).

### 10.4.3.3 Ultrasound

In the food industry, ultrasound is used with a power intensity in the 10–1000 W cm$^{-2}$ range at low frequency (20–100 kHz) to inactivate microorganisms and enzymes, extract bioactive compounds and clean surfaces (Ali et al., 2018). Ultrasound requires a liquid phase surrounding the solids to be treated, and it inactivates microorganisms through chemical (free radicals) and physical (pressure and heat gradients) mechanisms that provoke the disruption of cell walls and cellular death (Sango et al., 2014). This technology has limited practical application for FCFV produce decontamination when used alone. Thus, its effectiveness can be increased by combining it with other technologies as pressure, heat and washing with antimicrobial solutions (São José et al., 2014).

### 10.4.3.4 Cold Plasma

Cold plasma is produced at the ambient temperature under normal atmospheric pressure and refers to ionized gas consisting of a mixture of ions, free electrons, excited or non-excited atoms, radicals and photons (Pignata et al., 2017). Gas plasma is obtained by passing a high voltage through a gas phase, whether oxygen, helium, hydrogen and argon, the carrier gases most frequently used. Both the gas composition and the electric field strength used determine the composition of the plasma, and factors like the generating device, power input, mode and time of exposure, feeding gas, temperature and relative humidity affect the efficacy of the treatment (Murray et al., 2017).

Cold plasma may inactivate pathogenic and spoilage microorganisms on food surfaces, and it accomplishes its antimicrobial activity through the emission of UV light and the presence of reactive species, changes in the medium pH and the generation of oxidative compounds like $O_3$ and $H_2O_2$ (Niemira, 2012). For FCFV treatment, cold plasma can be applied directly, through washing with plasma-activated water in-package, which reduces the risk of postprocess contamination (Misra et al., 2019).

This technology presents a series of advantages like low temperature operation (<70 °C), short processing times, energy efficiency and antimicrobial efficacy together with low impact on the overall quality of the treated food and the environment (Bourke et al., 2018). However, quality may be affected if too long or too severe treatments are used (Fan & Wang, 2022). What is more, cold plasma efficacy at the industry level still needs to be validated.

### 10.4.3.5   High Hydrostatic Pressure

High hydrostatic pressure (HHP) is a non-thermal technology that entails submitting the produce to pressures in the range of 100–1000 MPa with the aim of inactivating microorganisms and enzymes, without causing degradation of color, flavor or nutrient compounds. Pressurization can be carried out at temperatures in the 0–100 °C range with a recommended duration of up to 20 min (Ali et al., 2018). Pressure is distributed uniformly irrespective of food composition, size and shape of the treated product.

Microbial inactivation after HHP treatment is a consequence of the damages in cell membranes and cellular integrity resulting in the interruption of the cellular functions responsible for reproduction and survival (Wang et al., 2016). In general, pressures in the range of 300–600 MPa are required to achieve reductions >1 log of pathogenic bacteria, but these pressures may impair the quality of fresh and fresh-cut fruits and vegetables (Rux et al., 2019).

As cell integrity and compartmentation of the FCFV may be lost after exposure to HHP, treated produce cannot be regarded as fresh, living organisms anymore, and therefore it is debatable whether HHP can be used for FCFV treatment or if HHP-treated products can be included under the FCFV denomination. Another barrier to HHP application is the high initial cost associated with the implementation of this technology.

### 10.4.4   Biological Methods

Bio-preservation technologies, consisting of the use of biological treatments for enhancing FCFV safety and extending shelf life, aim to limit pathogenic and spoilage microbial growth and to minimize sensory and nutritional quality losses of the treated food (Ananou et al., 2007; Ghanbari et al., 2013). Biological treatments use biocontrol agents (BCA), mainly yeasts, and spore-forming and lactic acid bacteria (LAB), naturally occurring on fruit surfaces, soil or rhizosphere, and/or its antimicrobial metabolites. These treatments are considered an eco-friendly and sustainable technique (Janisiewicz, 2013; Leneveu-Jenvrin et al., 2020).

The potential mechanisms of action for BCAs include production of volatile and antibiotic compounds, competition for nutrients or space, biofilm formation, synthesis of lytic enzymes and inhibition of spore germination as well as induction of host defenses. Frequently more than one mechanism is responsible for disease control (Spadaro & Droby, 2016; Wallace et al., 2018).

In the case of FCFV, bio-preservation is mainly accomplished using bacteriocins, which are antimicrobial peptides obtained through fermentation processes, and lactic acid bacteria, many of which are granted with the GRAS status and produce not only bacteriocins but also organic acids and oxidants with antimicrobial activity (Goodburn & Wallace, 2013). The main objective of using bio-preservation in FCFV is the inhibition of pathogenic and spoilage microflora microorganisms, together with a reduction in the use of chlorine as a sanitizer. In addition, the effects of BCA on quality attributes (color, vitamin C content, enzymatic and antioxidant activity) of different FCFV have been studied (Plaza et al., 2016; Siroli et al., 2016).

Despite the advantages of bio-preservation, many factors can affect its effectiveness and thus limit the extended use of these treatments, such as environmental conditions (temperature, pH), bacterial load and the interactions between native and competitive microorganisms (Bhilwadikar et al., 2019). Therefore, more than as an individual approach, biological preservation may be used in combination with other preservative methods to increase the safety and shelf life of FCFV (Fan et al., 2021; Siroli et al., 2015).

## 10.5  PRESERVATION BY PACKAGING

FCFV must be packaged to delay enzymatic, physical, chemical and microbiological changes. The use of proper packaging is essential for protecting FCFV and obtaining an optimal shelf life. In the following sections, the traditional and most innovative packaging currently used for FCFV is described.

### 10.5.1  Modified Atmosphere Packaging (MAP)

Modified atmosphere packaging (MAP) is probably the most widely used technology for packaging FCFV. MAP involves modifying the headspace gas composition, generally by decreasing oxygen concentration and increasing carbon dioxide, within the food package. The aim of MAP is to prevent microbial growth, inhibit oxidative browning and slow down the metabolism and respiration rate of FCFV produce, which in turn potentially extends shelf life (Giannakourou & Tsironi, 2021; Horev et al., 2012).

Gas composition inside the packages is determined by a series of factors: respiration rate and weight of product, storage temperature and relative humidity, and the permeability of the plastic films to oxygen and carbon dioxide (Bu et al., 2022). Low $O_2$ levels (1–5%) together with adequate $CO_2$ levels (3–20%) inhibits the growth of aerobic microorganisms, but special care must be taken to prevent the development of anaerobic conditions due to excessively low $O_2$ levels (Wilson et al., 2019). Moreover, the tolerance to high $CO_2$ concentrations varies among products, and sensitive products could be injured if $CO_2$ exceeds tolerable levels (Wang et al., 2015).

A different approach in MAP is the use of superatmospheric levels of $O_2$, $N_2$, and $CO_2$ concentrations. High $O_2$ concentrations effectively prevented anaerobic fermentation, off-flavors, and enzymatic browning. Still, the effectiveness of these atmospheres to control microbial growth, relies on the type of produce, the produce-microorganism combination, and the storage conditions (Kargwal et al., 2020). Finally, superatmospheric concentrations of either $CO_2$ or $N_2$ were generally less effective than high $O_2$ levels in enhancing FCFV quality and safety (Almenar, 2021).

Despite the beneficial effects associated with MAP, this technology has not always proven effective in enhancing the shelf life and safety of FCFV (Li et al., 2015). Therefore, different innovations in MAP, either used alone or combined with complementary technologies, have been studied and include the following:

- **Increased packaging sustainability**. This is accomplished by replacement of plastic films with bio-based, bio-degradable materials like polylactic acid (Jayeola et al., 2019; Mistriotis et al., 2016), and cellulose-based plastics (Ierna et al., 2017; Patanè et al., 2019). Nowadays, the main limitations for bio-based polymers are their high cost and poor technical performance (Wilson et al., 2019).
- **Temperature-compensating films**. Fluctuations in temperature occurring during transport and commercialization negatively affect the respiration rates and thus the quality of the packed FCFV (Jayeola et al., 2019). Different films capable of changing their permeability in response to temperature variations and allowing higher gas exchange at higher temperatures have been developed and are commercially available (Clarke, 2011).
- **Microperforated films**. These films have holes (30–300 µm diameter) that improve gas exchange through the film and thus, alleviate the too high/low concentrations of $CO_2/O_2$, respectively, and avoid anaerobiosis occurrence (Almenar, 2021). Furthermore, the number and size of the microholes can be adjusted to obtain different atmospheres, according to each product's requirement (González-Buesa & Salvador, 2022).
- **Non-conventional gas compositions**. Different gas mixtures, including argon, helium, xenon and nitrous and sulfur oxides, have been studied for FCFV packaging (Almenar, 2021). Positive results were found in preserving the overall quality of fresh-cut watercress (Pinela et al., 2016), rocket leaves (Baldassarre et al., 2015), red chard baby (Tomás-Callejas et al., 2011) and apples (Pardilla et al., 2015). The efficacy of these gases was attributed to their ability to lower the water activity of the packaged product (Caleb et al., 2013). Nevertheless, the high costs of these gases may limit their commercial viability (Wilson et al., 2019).

## 10.5.2 ACTIVE PACKAGING

Active packages provide functions other than protection and an inert barrier to the external environment by interacting with the packed produce and the environment. Active packages are intended to absorb or release bioactive compounds from or into the package environment. The selection of the active compound to be incorporated in the package depends on the specific requirements of the packaged product and can be divided in two groups:

1. **Scavengers/emitters**. These remove/absorb compounds that deteriorate food or release the compound of interest in the package headspace. This group includes oxygen (Cichello, 2015), carbon dioxide and ethylene scavengers (Kudachikar et al., 2011), moisture absorbers, carbon dioxide and ethanol emitters, and flavor releasing/absorbing systems. Sachets containing oxidizable iron and calcium hydroxide are used for removal of $O_2$ and $CO_2$, respectively (Lee, 2016). In the case of ethylene removers, they can adsorb, absorb or chemically alter ethylene and are generally placed in sachets inside the package (Awalgaonkar et al., 2020).

2. **Antimicrobial packaging**. This technology refers to packages containing active compounds with antimicrobial activity that are released gradually in the package headspace. They can be placed in sachets or pads inside the package or in the formulation of packaging films (Jung & Zhao, 2016; Almenar, 2021) or edible coatings. In the latter, the main component of the coating may be a polymer with antimicrobial properties (e.g., chitosan), or the antimicrobial agents are added in the film-forming solution (Giannakourou & Tsironi, 2021).

### 10.5.2.1  Edible Coatings

Edible coatings are a type of active packaging defined as a thin layer of edible material applied to the surface of food products, generally by dipping, spraying or brushing, with the aim to extend its shelf life (Velderrain-Rodríguez et al., 2015). Edible coatings are made from polysaccharides, proteins or lipids and slow down quality deterioration by forming a semipermeable barrier that limits moisture and solute migration, gas exchange, flavor loss, respiration and oxidative rates, as well as physiological disorders (Botondi et al., 2021). Furthermore, some coating materials like chitosan have antimicrobial properties and others, mainly polysaccharides, have been investigated as carriers for antimicrobials and antioxidants like essential oils, organic acids, bacteriocins and sulfites (Ali et al., 2018).

### 10.5.3  NANOTECHNOLOGY

Nanotechnology is defined as the production and use of materials with at least one external dimension in the size range of 1–100 nm (Cerqueira et al., 2018). Food nanotechnology is used to improve the functionalities of packaging materials regarding their mechanical, barrier, optical and antimicrobial properties (Fan et al., 2021). Other possible applications of nanotechnology in food packaging include the development of active packages with antimicrobial or antioxidant activities and nanosensors to detect food spoilage or contamination in smart packages (Peelman et al., 2013). Clay, metal and metal oxide nanoparticles such as silver, gold and zinc, and titanium oxides are among the most used nanomaterial in packaging (He et al., 2019).

In FCFV, nano Zn oxide, silver and $CaCO_3$ particles were successfully used to inhibit ethylene production in cut apples (Li et al., 2011), microbial growth in cut carrots (Becaro et al., 2016) and browning in cut yams (Luo et al., 2015), respectively.

Despite the mentioned advantages, information about the toxicity, risk of migration of nanomaterial to food products, and environmental impact is still very scarce and could condition consumers' acceptance and restrict an extended use of nanotechnology in food packaging.

## 10.6 FUTURE PERSPECTIVES

Despite the undeniable health benefits and convenience, FCFV are very sensitive to physiological, biochemical and microbial alterations that impair their quality and safety and shorten their shelf life. Furthermore, FCFV are normally consumed raw, increasing the risk for the occurrence of foodborne illnesses. Several preservation methods, together with appropriate packaging and cold chain maintenance, have been applied to guarantee safety and quality and extend the shelf life of FCFV.

Traditionally, chlorine and sulfites have been the most widely used sanitizer and antioxidant, respectively, in the fresh-cut industry. However, in recent years, their use has been restricted in many countries because they can adversely affect human and environmental safety. The need to enhance environmental sustainability, together with consumers' demands to minimize chemicals use due to concerns about environmental pollution, the possible contamination with agrochemicals residues and the inability to control diseases due to the appearance of tolerant strains of pathogens, has encouraged the search of alternative, eco-friendly approaches.

Alternative chemical preservatives, such as ozone, electrolyzed water, hydrogen peroxide, calcium-based solutions, organic acids, and essential oils, have been investigated to replace chlorine and sulfites and avoid the associated risks. In general, all of them were effective in controlling microbial growth and extending the shelf life of FCFV products. They do not form toxic by-products, and most of them are considered environmentally friendly. However, their effectiveness could be product-specific and be influenced by environmental factors, causing them in many cases to be insufficient for ensuring safety. Therefore, the combination of chemicals with other technologies, (e.g., MAP) is needed.

Similarly, physical techniques as thermal treatments, ultraviolet C light and ultrasound have low germicidal capacity when applied alone and need to be combined with other technologies. On the other hand, irradiation, high hydrostatic pressures and cold plasma have proven to be more effective as sanitizers than the former, but they are still not widely implemented due to the risk of negative effects on FCFV quality, lack of efficacy at the industry level, consumers' concerns over these technologies and the high initial cost associated with their implementation. Further research and technological optimization are needed for the application of these advanced strategies for each specific commodity.

Finally, innovative packages have also been developed aiming to extend FCFV shelf life, reduce food and packaging wastes and reduce the use of petroleum-based plastics. Among these, innovative MAP and active packages, which offer the possibility of incorporating active compounds, are expected to increase their presence in the FCFV packaging market.

One of the main challenges for the fresh-cut industry is the optimization of combined strategies for each specific product in order to ensure its safety and quality while complying with legal and market requirements. In this sense, any selected combination should be sustainable, cost-effective, feasible to apply at industrial level and safe for human health and the environment. Continuous collaboration between academia and the fresh-cut industry is essential to continue advancing in these areas.

## REFERENCES LIST

Abadias, M., Alegre, I., Oliveira, M., Altisent, R., & Viñas, I. (2012). Growth potential of Escherichia coli O157:H7 on fresh-cut fruits (melon and pineapple) and vegetables (carrot and escarole) stored under different conditions. *Food Control*, *27*(1), 37–44. https://doi.org/10.1016/j.foodcont.2012.02.032

Ali, A., Yeoh, W. K., Forney, C., & Siddiqui, M. W. (2018). Advances in postharvest technologies to extend the storage life of minimally processed fruits and vegetables. *Critical Reviews in Food Science and Nutrition*, *58*(15), 2632–2649. https://doi.org/10.1080/10408398.2017.1339180

Almenar, E. (2021). Recent developments in fresh-cut produce packaging. *Acta Horticulturae*, *1319*, 13–26. https://doi.org/10.17660/ActaHortic.2021.1319.2

Ananou, E., Maqueda, S., Martinez-Bueno, M., & Valdivia, M. (2007). Biopreservation, an ecological approach to improve safety and shelf-life of foods. In A. Méndez-Vilas (Ed.), *Communicating current research and educational topics and trends in applied microbiology* (pp. 475–486). Formatex.

Awalgaonkar, G., Beaudry, R., & Almenar, E. (2020). Ethylene-removing packaging: Basis for development and latest advances. *Comprehensive Reviews in Food Science and Food Safety*, *19*(6), 3980–4007. https://doi.org/10.1111/1541-4337.12636

Baldassarre, V., Navarro-Rico, J., Amodio, M. L., Artés-Hernández, F., & Colelli, G. (2015). Shelf-life of rocket leaves stored in argon enriched atmospheres. *Acta Horticulturae*, *1071*, 779–786. https://doi.org/10.17660/ActaHortic.2015.1071.103

Becaro, A. A., Puti, F. C., Panosso, A. R., Gern, J. C., Brandão, H. M., Correa, D. S., & Ferreira, M. D. (2016). Postharvest quality of fresh-cut carrots packaged in plastic films containing silver nanoparticles. *Food and Bioprocess Technology*, *9*(4), 637–649. https://doi.org/10.1007/s11947-015-1656-z

Bhilwadikar, T., Pounraj, S., Manivannan, S., Rastogi, N. K., & Negi, P. S. (2019). Decontamination of microorganisms and pesticides from fresh fruits and vegetables: A comprehensive review from common household processes to modern techniques. *Comprehensive Reviews in Food Science and Food Safety*, *18*(4), 1003–1038. https://doi.org/10.1111/1541-4337.12453

Bobo-García, G., Arroqui, C., Merino, G., & Vírseda, P. (2020). Antibrowning compounds for minimally processed potatoes: A review. *Food Reviews International*, *36*(5), 529–546. https://doi.org/10.1080/87559129.2019.1650761

Botondi, R., Barone, M., & Grasso, C. (2021). A review into the effectiveness of ozone technology for improving the safety and preserving the quality of fresh-cut fruits and vegetables. *Foods*, *10*(4), 748. https://doi.org/10.3390/foods10040748

Bourke, P., Ziuzina, D., Boehm, D., Cullen, P. J., & Keener, K. (2018). The potential of cold plasma for safe and sustainable food production. *Trends in Biotechnology*, *36*(6), 615–626. https://doi.org/10.1016/j.tibtech.2017.11.001

Bu, H., Hu, Y., & Dong, T. (2022). Changes in postharvest physiology, biochemistry, sensory properties and microbiological population of Allium mongolicum Regel regulated by adjusting the modified atmosphere inside the package during storage. *Journal of Food Processing and Preservation*, *46*(1), e16128. https://doi.org/10.1111/jfpp.16128

Burt, S. (2004). Essential oils: Their antibacterial properties and potential applications in foods – A review. *International Journal of Food Microbiology*, *94*(3), 223–253. https://doi.org/10.1016/j.ijfoodmicro.2004.03.022

Caleb, O. J., Mahajan, P. V., Al-Said, F. A.-J., & Opara, U. L. (2013). Modified atmosphere packaging technology of fresh and fresh-cut produce and the microbial consequences – a review. *Food and Bioprocess Technology*, *6*(2), 303–329. https://doi.org/10.1007/s11947-012-0932-4

Carletti, L., Botondi, R., Moscetti, R., Stella, E., Monarca, D., Cecchini, M., & Massantini, R. (2013). Use of ozone in sanitation and storage of fresh fruits and vegetables. *Journal of Food, Agriculture and Environment*, *11*(3–4), 585–589.

Carocho, M., Barreiro, M. F., Morales, P., & Ferreira, I. C. F. R. (2014). Adding molecules to food, pros and cons: A review on synthetic and natural food additives. *Comprehensive Reviews in Food Science and Food Safety*, *13*(4), 377–399. https://doi.org/10.1111/1541-4337.12065

Cerqueira, M. A., Vicente, A. A., & Pastrana, L. M. (2018). Chapter 1, Nanotechnology in food packaging: Opportunities and challenges. In M.A.P.R. Cerqueira, J.M. Lagaron, L.M. Pastrana Castro, & A.A.M. de Oliveira Soares Vicente (Eds.), *Nanomaterials for food packaging* (pp. 1–11). Elsevier. https://doi.org/10.1016/B978-0-323-51271-8.00001-2

Cichello, S. A. (2015). Oxygen absorbers in food preservation: A review. *Journal of Food Science and Technology*, *52*(4), 1889–1895. https://doi.org/10.1007/s13197-014-1265-2

Clarke, R. (2011). Breatheway® membrane technology and modified atmosphere packaging. In A.L. Brody, H. Zhuang, & J.H. Han (Eds.), *Modified atmosphere packaging for fresh-cut fruits and vegetables* (pp. 185–208). Blackwell Publishing Ltd. https://doi.org/10.1002/9780470959145.ch9

Cohen, N. L., Stoddard, A. M., Sarouhkhanians, S., & Sorensen, G. (1998). Barriers toward fruit and vegetable consumption in a multiethnic worksite population. *Journal of Nutrition Education*, *30*(6), 381–386. https://doi.org/10.1016/S0022-3182(98)70360-7

Dar, A. H., Bashir, O., Khan, S., Wahid, A., & Makroo, H. A. (2020). Fresh-cut products: Processing operations and equipments. In M.W. Siddiqui (Ed.), *Fresh-cut fruits and vegetables* (pp. 77–97). Academic Press.

De Corato, U. (2020). Improving the shelf-life and quality of fresh and minimally processed fruits and vegetables for a modern food industry: A comprehensive critical review from the traditional technologies into the most promising advancements. *Critical Reviews in Food Science and Nutrition*, *60*(6), 940–975. https://doi.org/10.1080/10408398.2018.1553025

Ding, T., Liu, F., Ling, J. G., Kang, M. L., Yu, J. F., & Ye, X. Q. (2016). Comparison of different cooling methods for extending shelf life of postharvest broccoli. *International Journal of Agricultural and Biological Engineering*, *9*(6), 178–185.

Dong, T., & Wang, Q. (2021). Browning of fresh-cut produce: Influencing factors and control technologies. *Acta Horticulturae*, *1319*, 47–58. https://doi.org/10.17660/ActaHortic.2021.1319.6

Fallik, E., & Ilic, Z. (2022). Chapter 20, Mitigating contamination of fresh and fresh-cut produce. In W.J. Florkowski, N.H. Banks, R.L. Shewfelt, & S.E. Prussia (Eds.), *Postharvest handling* (4th ed., pp. 621–649). Academic Press. https://doi.org/10.1016/B978-0-12-822845-6.00020-8

Fan, X., Mukhopadhyay, S., & Jin, T. (2021). Postharvest intervention technologies to enhance microbial safety of fresh and fresh-cut produce. *Acta Horticulturae*, *1319*, 27–36. https://doi.org/10.17660/ActaHortic.2021.1319.3

Fan, X., & Wang, W. (2022). Quality of fresh and fresh-cut produce impacted by nonthermal physical technologies intended to enhance microbial safety. *Critical Reviews in Food Science and Nutrition*, *62*(2), 362–382. https://doi.org/10.1080/10408398.2020.1816892

Feliziani, E., Lichter, A., Smilanick, J. L., & Ippolito, A. (2016). Disinfecting agents for controlling fruit and vegetable diseases after harvest. *Postharvest Biology and Technology*, *122*, 53–69. https://doi.org/10.1016/j.postharvbio.2016.04.016

Garrido, Y., Tudela, J. A., & Gil, M. I. (2015). Comparison of industrial precooling systems for minimally processed baby spinach. *Postharvest Biology and Technology, 102*, 1–8. https://doi.org/10.1016/j.postharvbio.2014.12.003

Ghanbari, M., Jami, M., Domig, K. J., & Kneifel, W. (2013). Seafood biopreservation by lactic acid bacteria – a review. *LWT – Food Science and Technology, 54*(2), 315–324. https://doi.org/10.1016/j.lwt.2013.05.039

Giannakourou, M. C., & Tsironi, T. N. (2021). Application of processing and packaging hurdles for fresh-cut fruits and vegetables preservation. *Foods, 10*(4), 830. https://doi.org/10.3390/foods10040830

Gil, M. I. (2021). Management of preharvest and postharvest factors related to quality and safety aspects of leafy vegetables. *Acta Horticulturae, 1319*, 1–12. https://doi.org/10.17660/ActaHortic.2021.1319.1

Gil, M. I., Gómez-López, V. M., Hung, Y. C., & Allende, A. (2015). Potential of electrolyzed water as an alternative disinfectant agent in the fresh-cut industry. *Food and Bioprocess Technology, 8*(6), 1336–1348. https://doi.org/10.1007/s11947-014-1444-1

Gombas, D., Luo, Y., Brennan, J., Shergill, G., Petran, R., Walsh, R., Hau, H., Khurana, K., Zomorodi, B., Rosen, J., Varley, R., & Deng, K. (2017). Guidelines to validate control of cross-contamination during washing of fresh-cut leafy vegetables. *Journal of Food Protection, 80*(2), 312–330. https://doi.org/10.4315/0362-028X.JFP-16-258

González-Aguilar, G. A., Ruiz-Cruz, S., Cruz-Valenzuela, R., Ayala-Zavala, J. F., De La Rosa, L. A., & Alvarez-Parrilla, E. (2008). New technologies to preserve quality of fresh-cut produce. In G.F. Gutiérrez-López, G.V. Barbosa-Cánovas, J. Welti-Chanes, & E. Parada-Arias (Eds.), *Food engineering* (pp. 105–115). Springer. https://doi.org/10.1007/978-0-387-75430-7_6

González-Buesa, J., & Salvador, M. L. (2022). A multi-physics approach for modeling gas exchange in microperforated films for modified atmosphere packaging of respiring products. *Food Packaging and Shelf Life, 31*, 100797. https://doi.org/10.1016/j.fpsl.2021.100797

Goodburn, C., & Wallace, C. A. (2013). The microbiological efficacy of decontamination methodologies for fresh produce: A review. *Food Control, 32*(2), 418–427. https://doi.org/10.1016/j.foodcont.2012.12.012

Gross, K. C., Wang, C. Y., & Saltveit, M. (Eds.). (2016). The commercial storage of fruits, vegetables, and florist and nursery stocks. In *Agriculture handbook* (Vol. 66). United States Department of Agriculture (United States Department of Agriculture).

Guerreiro, D., Madureira, J., Silva, T., Melo, R., Santos, P. M. P., Ferreira, A., Trigo, M.J., Falcão, A. N., Margaça, F. M. A., & Cabo Verde, S. (2016). Post-harvest treatment of cherry tomatoes by gamma radiation: Microbial and physicochemical parameters evaluation. *Innovative Food Science and Emerging Technologies, 36*, 1–9. https://doi.org/10.1016/j.ifset.2016.05.008

Haß, J., & Hartmann, M. (2018). What determines the fruit and vegetables intake of primary school children? An analysis of personal and social determinants. *Appetite, 120*, 82–91. https://doi.org/10.1016/j.appet.2017.08.017

Haynes-Maslow, L., Parsons, S. E., Wheeler, S. B., & Leone, L. A. (2013). A qualitative study of perceived barriers to fruit and vegetable consumption among low-income populations, North Carolina, 2011. *Preventing Chronic Disease, 10*, E34. https://doi.org/10.5888/pcd10.120206

He, S. Y., Zhang, G. C., Yu, Y. Q., Li, R. G., & Yang, Q. R. (2013). Effects of vacuum cooling on the enzymatic antioxidant system of cherry and inhibition of surface-borne pathogens. *International Journal of Refrigeration, 36*(8), 2387–2394. https://doi.org/10.1016/j.ijrefrig.2013.05.018

He, X., Deng, H., & Hwang, H. M. (2019). The current application of nanotechnology in food and agriculture. *Journal of Food and Drug Analysis, 27*(1), 1–21. https://doi.org/10.1016/j.jfda.2018.12.002

Horev, B., Sela, S., Vinokur, Y., Gorbatsevich, E., Pinto, R., & Rodov, V. (2012). The effects of active and passive modified atmosphere packaging on the survival of Salmonella enterica serotype Typhimurium on washed romaine lettuce leaves. *Food Research International, 45*(2), 1129–1132. https://doi.org/10.1016/j.foodres.2011.05.037

Horvitz, S., Arancibia, M., Arroqui, C., Chonata, E., & Vírseda, P. (2021). Effects of gaseous ozone on microbiological quality of Andean blackberries (*Rubus glaucus* Benth). *Foods, 10*(9), 2039. https://doi.org/10.3390/foods10092039

Horvitz, S., & Cantalejo, M. J. (2012). Effects of ozone and chlorine postharvest treatments on quality of fresh-cut red bell peppers. *International Journal of Food Science and Technology, 47*(9), 1935–1943. https://doi.org/10.1111/j.1365-2621.2012.03053.x

Horvitz, S., & Cantalejo, M. J. (2014). Application of ozone for the postharvest treatment of fruits and vegetables. *Critical Reviews in Food Science and Nutrition, 54*(3), 312–339. https://doi.org/10.1080/10408398.2011.584353

Horvitz, S., & Cantalejo, M. J. (2015). Effects of gaseous O3 and modified atmosphere packaging on the quality and shelf-life of partially dehydrated ready-to-eat pepper strips. *Food and Bioprocess Technology*, 8(8), 1800–1810. https://doi.org/10.1007/s11947-015-1537-5

Hosseini, F. S., Akhavan, H. R., Maghsoudi, H., Hajimohammadi-Farimani, R., & Balvardi, M. (2019). Effects of a rotational UV-C irradiation system and packaging on the shelf life of fresh pistachio. *Journal of the Science of Food and Agriculture, 99*(11), 5229–5238. https://doi.org/10.1002/jsfa.9763

Ierna, A., Rizzarelli, P., Malvuccio, A., & Rapisarda, M. (2017). Effect of different anti-browning agents on quality of minimally processed early potatoes packaged on a compostable film. *LWT – Food Science and Technology, 85*, 434–439. https://doi.org/10.1016/j.lwt.2017.03.043

Janisiewicz, W. J. (2013). Biological control of postharvest diseases: Hurdles, successes and prospects. *Acta Horticulturae, 1001*, 273–283. https://doi.org/10.17660/ActaHortic.2013.1001.31

Jayeola, V., Jeong, S., Almenar, E., Marks, B. P., Vorst, K. L., Brown, J. W., & Ryser, E. T. (2019). Predicting the growth of Listeria monocytogenes and Salmonella typhimurium in diced celery, onions, and tomatoes during simulated commercial transport, retail storage, and display. *Journal of Food Protection, 82*(2), 287–300. https://doi.org/10.4315/0362-028X.JFP-18-277

Jideani, A. I., Anyasi, T. A., Mchau, G. R., Udoro, E. O., & Onipe, O. O. (2017). Processing and preservation of fresh-cut fruit and vegetable products. In İ. Kahramanoğlu (Ed.), *Postharvest handling*. IntechOpen. https://doi.org/10.5772/intechopen.69763

Jung, J., & Zhao, Y. (2016). Chapter 18, antimicrobial packaging for fresh and minimally processed fruits and vegetables. In J. Barros-Velázquez (Ed.), *Antimicrobial food packaging* (pp. 243–256). Academic Press. https://doi.org/10.1016/B978-0-12-800723-5.00018-8

Kahramanoğlu, İ., Usanmaz, S., Okatan, V., & Chunpeng, W. (2020). Preserving postharvest storage quality of fresh-cut cactus pears by using different bio-materials. CABI. *Journal of Agricultural and Biological Science, 1*, 7. https://doi.org/10.1186/s43170-020-00008-5

Kambhampati, V., Suranjoy Singh, S., Ritesh, W., Soberly, M., Baby, Z., Baite, H., Mishra, S., & Pradhan, R. C. (2019). A review on postharvest management and advances in the minimal processing of fresh-cut fruits and vegetables. *Journal of Microbiology, Biotechnology and Food Sciences, 8*(5), 1178–1187. https://doi.org/10.15414/jmbfs.2019.8.5.1178-1187

Kargwal, R., Garg, M. K., Singh, V. K., Garg, R., & Kumar, N. (2020). Principles of modified atmosphere packaging for shelf life extension of fruits and vegetables: An overview of storage conditions. *International Journal of Chemical Studies, 8*(3), 2245–2252. https://doi.org/10.22271/chemi.2020.v8.i3af.9545

Kitinoja, L., & Thompson, J. (2010). Pre-cooling systems for small-scale producers. *Stewart Postharvest Review*, 6(2), 1–14. https://doi.org/10.2212/spr.2010.2.2

Korkmaz, M., & Polat, M. (2005). Irradiation of fresh fruit and vegetables. In W. Jongen (Ed.), *Improving the safety of fresh fruit and vegetables* (Vol. 13, pp. 387–428). Woodhead Publishing. https://doi.org/10.1533/9781845690243.3.387

Kudachikar, V. B., Kulkarni, S. G., & Prakash, M. N. (2011). Effect of modified atmosphere packaging on quality and shelf life of "robusta" banana (Musa sp.) stored at low temperature. *Journal of Food Science and Technology*, 48(3), 319–324. https://doi.org/10.1007/s13197-011-0238-y

Lee, D. S. (2016). Carbon dioxide absorbers for food packaging applications. *Trends in Food Science and Technology*, 57, 146–155. https://doi.org/10.1016/j.tifs.2016.09.014

Leneveu-Jenvrin, C., Charles, F., Barba, F. J., & Remize, F. (2020). Role of biological control agents and physical treatments in maintaining the quality of fresh and minimally processed fruit and vegetables. *Critical Reviews in Food Science and Nutrition*, 60(17), 2837–2855. https://doi.org/10.1080/10408398.2019.1664979

Li, J., Song, W., Barth, M. M., Zhuang, H., Zhang, W., Zhang, L., Wang, L., Lu, W., Wang, Z., Han, X., & Li, Q. (2015). Effect of modified atmosphere packaging (MAP) on the quality of sea buckthorn berry fruits during postharvest storage. *Journal of Food Quality*, 38(1), 13–20. https://doi.org/10.1111/jfq.12118

Li, X., Li, W., Jiang, Y., Ding, Y., Yun, J., Tang, Y., & Zhang, P. (2011). Effect of Nano-ZnO-coated active packaging on quality of fresh-cut "Fuji" apple. *International Journal of Food Science and Technology*, 46(9), 1947–1955. https://doi.org/10.1111/j.1365-2621.2011.02706.x

Luo, Z., Wang, Y., Jiang, L., & Xu, X. (2015). Effect of Nano-CaCO3-LDPE packaging on quality and browning of fresh-cut yam. *LWT – Food Science and Technology*, 60(2), 1155–1161. https://doi.org/10.1016/j.lwt.2014.09.021

Manzocco, L., Ignat, A., Anese, M., Bot, F., Calligaris, S., Valoppi, F., & Nicoli, M. C. (2015). Efficient management of the water resource in the fresh-cut industry: Current status and perspectives. *Trends in Food Science and Technology*, 46(2), 286–294. https://doi.org/10.1016/j.tifs.2015.09.003

Martín-Diana, A. B., Rico, D., Barry-Ryan, C., Frias, J. M., Mulcahy, J., & Henehan, G. T. M. (2005). Calcium lactate washing treatments for salad-cut iceberg lettuce: Effect of temperature and concentration on quality retention parameters. *Food Research International*, 38(7), 729–740. https://doi.org/10.1016/j.foodres.2005.02.005

Martín-Diana, A. B., Rico, D., Frías, J. M., Barat, J. M., Henehan, G. T. M., & Barry-Ryan, C. (2007). Calcium for extending the shelf life of fresh whole and minimally processed fruits and vegetables: A review. *Trends in Food Science and Technology*, 18(4), 210–218. https://doi.org/10.1016/j.tifs.2006.11.027

Matrose, N. A., Obikeze, K., Belay, Z. A., & Caleb, O. J. (2021). Plant extracts and other natural compounds as alternatives for post-harvest management of fruit fungal pathogens: A review. *Food Bioscience*, 41, 100840. https://doi.org/10.1016/j.fbio.2020.100840

McDonald, K., & Sun, D. W. (2000). Vacuum cooling technology for the food processing industry: A review. *Journal of Food Engineering*, 45(2), 55–65. https://doi.org/10.1016/S0260-8774(00)00041-8

Meireles, A., Giaouris, E., & Simões, M. (2016). Alternative disinfection methods to chlorine for use in the fresh-cut industry. *Food Research International*, 82, 71–85. https://doi.org/10.1016/j.foodres.2016.01.021

Miller, F. A., Silva, C. L. M., & Brandão, T. R. S. (2013). A review on Ozone-based treatments for fruit and vegetables preservation. *Food Engineering Reviews*, 5(2), 77–106. https://doi.org/10.1007/s12393-013-9064-5

Misra, N. N., Yepez, X., Xu, L., & Keener, K. (2019). In-package cold plasma technologies. *Journal of Food Engineering*, 244, 21–31. https://doi.org/10.1016/j.jfoodeng.2018.09.019

Mistriotis, A., Briassoulis, D., Giannoulis, A., & D'Aquino, S. (2016). Design of biodegradable bio-based equilibrium modified atmosphere packaging (EMAP) for fresh fruits and vegetables by using micro-perforated poly-lactic acid (PLA) films. *Postharvest Biology and Technology*, *111*, 380–389. https://doi.org/10.1016/j.postharvbio.2015.09.022

Murray, K., Wu, F., Shi, J., Jun Xue, S., & Warriner, K. (2017). Challenges in the microbiological food safety of fresh produce: Limitations of post-harvest washing and the need for alternative interventions. *Food Quality and Safety*, *1*(4), 289–301. https://doi.org/10.1093/fqsafe/fyx027

Negi, P. S. (2012). Plant extracts for the control of bacterial growth: Efficacy, stability and safety issues for food application. *International Journal of Food Microbiology*, *156*(1), 7–17. https://doi.org/10.1016/j.ijfoodmicro.2012.03.006

Nicola, S., Cocetta, G., Ferrante, A., & Ertani, A. (2022). Chapter 7, Fresh-cut produce quality: Implications for postharvest. In W. J. Florkowski, N. H. Banks, R. L. Shewfelt, & S. E. P russia (Eds.), *Postharvest handling* (4th ed., pp. 187–250). Academic Press. https://doi.org/10.1016/B978-0-12-822845-6.00007-5

Niemira, B. A. (2012). Cold plasma decontamination of foods. *Annual Review of Food Science and Technology*, *3*(1), 125–142. https://doi.org/10.1146/annurev-food-022811-101132

Ölmez, H., & Kretzschmar, U. (2009). Potential alternative disinfection methods for organic fresh-cut industry for minimizing water consumption and environmental impact. *LWT – Food Science and Technology*, *42*(3), 686–693. https://doi.org/10.1016/j.lwt.2008.08.001

Pandey, A. K., Kumar, P., Singh, P., Tripathi, N. N., & Bajpai, V. K. (2017). Essential oils: Sources of antimicrobials and food preservatives. *Frontiers in Microbiology*, *7*, 2161. https://doi.org/10.3389/fmicb.2016.02161

Pardilla, S., Mor-Mur, M., Vega, L. F., & Guri, S. (2015). Argon and high content of CO2: The future for fresh-cut apples packaged in MAP. *Acta Horticulturae*, *1071*, 731–737. https://doi.org/10.17660/ActaHortic.2015.1071.97

Patanè, C., Malvuccio, A., Saita, A., Rizzarelli, P., Siracusa, L., Rizzo, V., & Muratore, G. (2019). Nutritional changes during storage in fresh-cut long storage tomato as affected by biocompostable polylactide and cellulose based packaging. *LWT*, *101*, 618–624. https://doi.org/10.1016/j.lwt.2018.11.069

Patrignani, F., Siroli, L., Serrazanetti, D. I., Gardini, F., & Lanciotti, R. (2015). Innovative strategies based on the use of essential oils and their components to improve safety, shelf-life and quality of minimally processed fruits and vegetables. *Trends in Food Science and Technology*, *46*(2), 311–319. https://doi.org/10.1016/j.tifs.2015.03.009

Peelman, N., Ragaert, P., De Meulenaer, B., Adons, D., Peeters, R., Cardon, L., Van Impe, F. V., & Devlieghere, F. (2013). Application of Bioplastics for food packaging. *Trends in Food Science and Technology*, *32*(2), 128–141. https://doi.org/10.1016/j.tifs.2013.06.003

Pignata, C., D'Angelo, D., Fea, E., & Gilli, G. (2017). A review on microbiological decontamination of fresh produce with nonthermal plasma. *Journal of Applied Microbiology*, *122*(6), 1438–1455. https://doi.org/10.1111/jam.13412

Pinela, J., Barreira, J. C. M., Barros, L., Antonio, A. L., Carvalho, A. M., Oliveira, M. B. P. P., & Ferreira, I. C. F. R. (2016). Postharvest quality changes in fresh-cut watercress stored under conventional and inert gas-enriched modified atmosphere packaging. *Postharvest Biology and Technology*, *112*, 55–63. https://doi.org/10.1016/j.postharvbio.2015.10.004

Plaza, L., Altisent, R., Alegre, I., Viñas, I., & Abadias, M. (2016). Changes in the quality and antioxidant properties of fresh-cut melon treated with the biopreservative culture Pseudomonas graminis CPA-7 during refrigerated storage. *Postharvest Biology and Technology*, *111*, 25–30. https://doi.org/10.1016/j.postharvbio.2015.07.023

Qadri, O. S., Yousuf, B., & Srivastava, A. K. (2015). Fresh-cut fruits and vegetables: Critical factors influencing microbiology and novel approaches to prevent microbial risks – a review. *Cogent Food and Agriculture*, *1*(1), 1121606. https://doi.org/10.1080/2331193 2.2015.1121606

Raffo, A., & Paoletti, F. (2022). Fresh-cut vegetables processing: Environmental sustainability and food safety issues in a comprehensive perspective. *Frontiers in Sustainable Food Systems, 5*. https://doi.org/10.3389/fsufs.2021.681459

Ramos, B., Miller, F. A., Brandão, T. R. S., Teixeira, P., & Silva, C. L. M. (2013). Fresh fruits and vegetables – An overview on applied methodologies to improve its quality and safety. *Innovative Food Science and Emerging Technologies, 20*, 1–15. https://doi.org/10.1016/j.ifset.2013.07.002

Rao, C. G. (2015). *Engineering for storage of fruits and vegetables: Cold storage, controlled atmosphere storage, modified atmosphere storage.* Academic Press.

Roberts, P. B., & Follett, P. A. (2018). Chapter 9, Food irradiation for phytosanitary and quarantine treatment. In I.C.F.R. Ferreira, A.L. Antonio, & S.C. Verde (Eds.), *Food irradiation technologies* (pp. 169–182). Royal Society of Chemistry. https://doi.org/10.1039/9781788010252-00169

Rux, G., Gelewsky, R., Schlüter, O., & Herppich, W. B. (2019). High hydrostatic pressure effects on membrane-related quality parameters of fresh radish tubers. *Postharvest Biology and Technology, 151*, 1–9. https://doi.org/10.1016/j.postharvbio.2019.01.007

Saba, A., Moneta, E., Peparaio, M., Sinesio, F., Vassallo, M., & Paoletti, F. (2018). Towards a multi-dimensional concept of vegetable freshness from the consumer's perspective. *Food Quality and Preference, 66*, 1–12. https://doi.org/10.1016/j.foodqual.2017.12.008

Sango, D. M., Abela, D., McElhatton, A., & Valdramidis, V. P. (2014). Assisted ultrasound applications for the production of safe foods. *Journal of Applied Microbiology, 116*(5), 1067–1083. https://doi.org/10.1111/jam.12468

São José, d. AM, Andrade, M. C. D., Ramos, A. M., Vanetti, M. C. D., Stringheta, P. C., & Chaves, J. B. P. (2014). Decontamination by ultrasound application in fresh fruits and vegetables. *Food Control, 45*, 36–50. https://doi.org/10.1016/j.foodcont.2014.04.015

Siroli, L., Patrignani, F., Serrazanetti, D. I., Gardini, F., & Lanciotti, R. (2015). Innovative strategies based on the use of bio-control agents to improve the safety, shelf-life and quality of minimally processed fruits and vegetables. *Trends in Food Science and Technology, 46*(2), 302–310. https://doi.org/10.1016/j.tifs.2015.04.014

Siroli, L., Patrignani, F., Serrazanetti, D. I., Vannini, L., Salvetti, E., Torriani, S., Gardini, F., & Lanciotti, R. (2016). Use of a nisin-producing Lactococcus lactis strain, combined with natural antimicrobials, to improve the safety and shelf-life of minimally processed sliced apples. *Food Microbiology, 54*, 11–19. https://doi.org/10.1016/j.fm.2015.11.004

Sivakumar, D., & Fallik, E. (2013). Influence of heat treatments on quality retention of fresh and fresh-cut produce. *Food Reviews International, 29*(3), 294–320. https://doi.org/10.1080/87559129.2013.790048

Soliva-Fortuny, R. C., & Martín-Belloso, O. (2003). New advances in extending the shelf-life of fresh-cut fruits: A review. *Trends in Food Science and Technology, 14*(9), 341–353. https://doi.org/10.1016/S0924-2244(03)00054-2

Spadaro, D., & Droby, S. (2016). Development of biocontrol products for postharvest diseases of fruit: The importance of elucidating the mechanisms of action of yeast antagonists. *Trends in Food Science and Technology, 47*, 39–49. https://doi.org/10.1016/j.tifs.2015.11.003

Tomás-Callejas, A., Boluda, M., Robles, P. A., Artés, F., & Artés-Hernández, F. (2011). Innovative active modified atmosphere packaging improves overall quality of fresh-cut red chard baby leaves. *LWT – Food Science and Technology, 44*(6), 1422–1428. https://doi.org/10.1016/j.lwt.2011.01.020

US Food and Drug Administration. (2019). *Guide to minimize food safety hazards of fresh-cut produce: Draft guidance for industry.* Retrieved July 7, 2022, from http://www.fda.gov/media/117526/download

Van Haute, S., Tryland, I., Veys, A., & Sampers, I. (2015). Wash water disinfection of a full-scale leafy vegetables washing process with hydrogen peroxide and the use of a commercial metal ion mixture to improve disinfection efficiency. *Food Control, 50*, 173–183. https://doi.org/10.1016/j.foodcont.2014.08.028

Velderrain-Rodríguez, G. R., Quirós-Sauceda, A. E., González Aguilar, G. A., Siddiqui, M. W., & Ayala Zavala, J. F. (2015). Technologies in fresh-cut fruit and vegetables. In M. W. Siddiqui & M. S. Rahman (Eds.), *Minimally processed foods: Technologies for safety, quality, and convenience* (pp. 79–103). Springer International Publishing. https://doi.org/10.1007/978-3-319-10677-9_5

Wallace, R. L., Hirkala, D. L., & Nelson, L. M. (2018). Mechanisms of action of three isolates of Pseudomonas fluorescens active against postharvest grey mold decay of apple during commercial storage. *Biological Control, 117*, 13–20. https://doi.org/10.1016/j.biocontrol.2017.08.019

Wang, C. Y., Huang, H. W., Hsu, C. P., & Yang, B. B. (2016). Recent advances in food processing using high hydrostatic pressure technology. *Critical Reviews in Food Science and Nutrition, 56*(4), 527–540. https://doi.org/10.1080/10408398.2012.745479

Wang, Y., Bai, J., & Long, L. E. (2015). Quality and physiological responses of two late-season sweet cherry cultivars "Lapins" and "Skeena" to modified atmosphere packaging (MAP) during simulated long distance ocean shipping. *Postharvest Biology and Technology, 110*, 1–8. https://doi.org/10.1016/j.postharvbio.2015.07.009

Wiley, R. C., & Yildiz, F. (2017). Introduction to minimally processed refrigerated (MPR) fruits and vegetables. In F. Yildiz & R. C. Wiley (Eds.), *Minimally processed refrigerated fruits and vegetables* (pp. 3–15). Springer US. https://doi.org/10.1007/978-1-4939-7018-6

Wilson, M. D., Stanley, R. A., Eyles, A., & Ross, T. (2019). Innovative processes and technologies for modified atmosphere packaging of fresh and fresh-cut fruits and vegetables. *Critical Reviews in Food Science and Nutrition, 59*(3), 411–422. https://doi.org/10.1080/10408398.2017.1375892

Yahia, E. M., Fadanelli, L., Mattè, P., & Brecht, J. K. (2019). Chapter 13, controlled atmosphere storage. In E. M. Yahia (Ed.), *Postharvest technology of perishable horticultural commodities* (pp. 439–479). Woodhead Publishing. https://doi.org/10.1016/B978-0-12-813276-0.00013-4

Yildiz, F. (2017). Initial preparation, handling, and distribution of minimally processed refrigerated fruits and vegetables. In F. Yildiz & R. C. Wiley (Eds.), *Minimally processed refrigerated fruits and vegetables* (pp. 53–92). Springer. https://doi.org/10.1007/978-1-4615-2393-2_2

Yousuf, B., Wu, S., & Siddiqui, M. W. (2021). Incorporating essential oils or compounds derived thereof into edible coatings: Effect on quality and shelf life of fresh/fresh-cut produce. *Trends in Food Science and Technology, 108*, 245–257. https://doi.org/10.1016/j.tifs.2021.01.016

Zhang, X., Meng, W., Chen, Y., & Peng, Y. (2022). Browning inhibition of plant extracts on fresh-cut fruits and vegetables – a review. *Journal of Food Processing and Preservation, 46*(5), e16532. https://doi.org/10.1111/jfpp.16532

# 11 Ultraviolet and Blue-Light Illumination for Controlling Postharvest Decay and Preserving Storage Quality of Fresh Produce

*İbrahim Kahramanoğlu and Olga Panfilova*

## 11.1 INTRODUCTION

Fruits and vegetables constitute an important part of horticultural crops and have great importance in human nutrition. An important amount of fresh fruits and vegetables is being lost after harvest and does not reach final consumers. This loss is highly dependent on the crop types, production technologies and postharvest handling practices but is estimated to be around 30–50% of horticultural products (Gunders & Bloom, 2017; Kahramanoğlu, 2017). One of the most important causes of this loss is postharvest pathogens (Yahia, 2011). Moreover, the biggest share was devoted to postharvest pathogens, which may affect 10–30% of the crops (Jurick, 2022). The most important decay-causing phytopathogens of fruits and vegetables are alternaria rot (*Alternaria* spp.), anthracnose (*Colletotrichum* spp.), brown rot (*Monilia fructiocola*), blue mold (*Penicillium italicum*), green mold (*P. digitatum*) and gray mold (*Botrytis cinerea*) (Williamson et al., 2007; Yahia, 2011; Thomidis, 2014; Madbouly et al., 2020).

Fungi not only cause economic losses and may threaten human and animal life due to their harmful mycotoxins, i.e., aflatoxin by *Aspergillus* spp., fumonison by *Fusarium* spp., alternariol by *Alternaria* spp. and patulin by *Penicillium* spp. (Banani et al., 2016; Jurick, 2022). One of the most important ways for controlling postharvest pathogens of fruits and vegetables is the application of chemical fungicides. The most important active ingredients of the fungicides used for controlling postharvest fungi are azoxystrobin, difenoconazole, fludioxonil, imazalil, iprodione, propiconazole, pyrimethanil and thiabendazole (TBZ) (Palou, 2018; Jurick, 2022). However, the mis- and overuse of agrochemicals may cause resistance in pathogens and environmental problems and may produce risk for human health (Ruffo Roberto et al., 2019; Hosseini et al., 2020). In today's world, with the help of the Internet and public media, there is increasing

 DOI: 10.1201/9781003452355-13

public awareness about the harmful effects of agrochemicals and an important trend in reducing their use (El Khetabi et al., 2022). Therefore, a recent trend in postharvest applications aims to explore alternative technologies for agrochemicals, including light irradiation (Papoutsis et al., 2019; Kahramanoğlu et al., 2020).

The electromagnetic spectrum describes the range of frequencies of light with their wavelengths and photon energies. Most of that light is invisible to the human eye. The visible spectrum has a wavelength ranging from 400 nm (blue) to 700 nm (red). At the left side of the visible spectrum, UV light ranges from 100 nm to 400 nm. Less than 100 nm consists of X-rays, and gamma rays are less than 0.1 nm. On the right side of the visible spectrum is infrared radiation (700 nm to 1 cm), followed by microwave radiation (1 cm to 1 m) and finally the radio waves (>1 m). Most of this light spectrum is normally absorbed by atmospheric gases and do not reach the earth's surface. The visible light and a little of UV and infrared light are exempted, and they reach the earth's surface (Nikita et al., 2015). Light is essential for plants growth and development, due to its significant role in photosynthesis. Besides this well-known role of light, different spectra of light were also noted to have different impacts on plants or plant pathogens. One of the most important roles is the inhibition impact of UV and the blue-light on the fungal spores and production of reactive oxygen species (ROS) in cells of fungi, which damage and kill the fungi (Papoutsis et al., 2019). In addition to direct impact on the fungi, it is also well reported that abiotic stressors, including light irradiation, can induce adaptive responses in fresh produce (i.e., accumulation of ROS), which may result in pathogen control (Duarte-Sierra et al., 2020). This phenomenon is called hormesis, which is the beneficial effect of harmful substances or stress factors on living organisms when they are in small amounts/doses (Sakai, 2006). Details about the impacts of different light spectrum on the control of postharvest pathogens are discussed in this chapter.

## 11.2 HORMESIS MECHANISM

Hormesis is defined as the beneficial effects of harmful substances or stress factors on living organisms when they are in small amounts/doses (Sakai, 2006). It is a dose–response phenomenon where a low dose is associated with "stimulation", and a high dose is commonly characterized by "inhibition" (Kouda & Iki, 2010). Therefore, the beneficial impacts on the fruits stimulated by low doses of the specific stressors show a decreasing trend at higher doses (Duarte-Sierra et al., 2020). Hormesis has been reported to have several important beneficial impacts in harvested fresh produce, not limited to disease resistance but including delayed senescence and improvement in the accumulation of plant secondary metabolites (Figure 11.1). Stress-induced hormesis has been tested for various abiotic stressors, including UV light, heat, ozone, etc. (González-Estrada et al., 2021; Costa et al., 2022; Duarte-Sierra et al., 2022). These specific responses in fresh produce, i.e., disease resistance, can be induced by the accumulation of reactive oxygen species (ROS) (Duarte-Sierra et al., 2020) or by regulated gene expression (Podolec et al., 2021). In addition, the application of UV light has been noted to induce the defense compounds (several secondary metabolites) in fresh produce and to reduce/eliminate the attack of postharvest pathogens (Artés-Hernández et al., 2022).

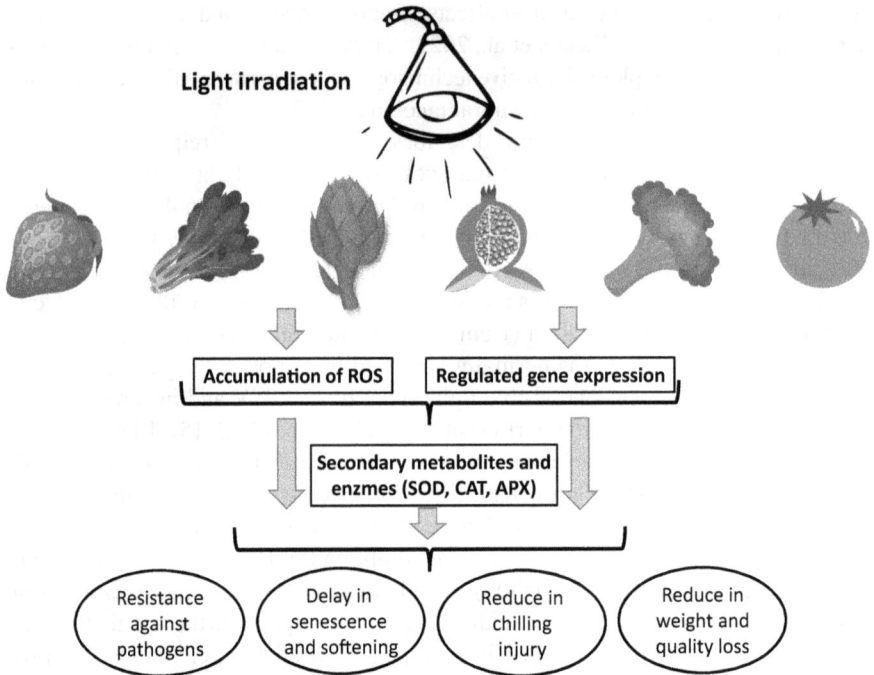

**FIGURE 11.1**   Postharvest hormesis induced by light irradiation [adapted from Zhang and Jiang (2019) and Duarte-Sierra et al. (2022)].

## 11.3   FERROPTOSIS MECHANISM (SUGGESTED METHOD FOR FUNGAL CELL DEATH)

Ferroptosis is a regulated, nonapoptotic type of iron-dependent cell death reported in mammalians. It is a new form of regulated cell death (RCD) (Stockwell et al., 2017). This is very similar to the mechanism of synthetic or biological fungicides, which also damage the fungal cell walls and cells (Zhu et al., 2019). Apoptosis is a type of programmed cell death, but ferroptosis is different, which is initiated by the inactivation of glutathione-dependent antioxidant defense and the subsequent iron-dependent accumulation of lipid reactive oxygen species (ROS) (Cao & Dixon, 2016). Therefore, ferroptosis requires two main biochemical processes: the accumulation of iron and lipid ROS. Tao et al. (2020) reported that ferroptosis can be regulated by numerous molecules through different mechanisms. Recent studies of Distéfano et al. (2017) suggested that stress factors, i.e., heat-induced ferroptosis-like cell death in plants. It is known and has been previously explained that light irradiation elevates ROS and triggers lipid peroxidation in fresh produce (Duarte-Sierra et al., 2020). UV light is a well-known trigger of apoptosis in different living organisms, i.e., plants (Nawkar et al., 2013), mammals (Lee et al., 2013, human pathogens (Narita et al., 2020)

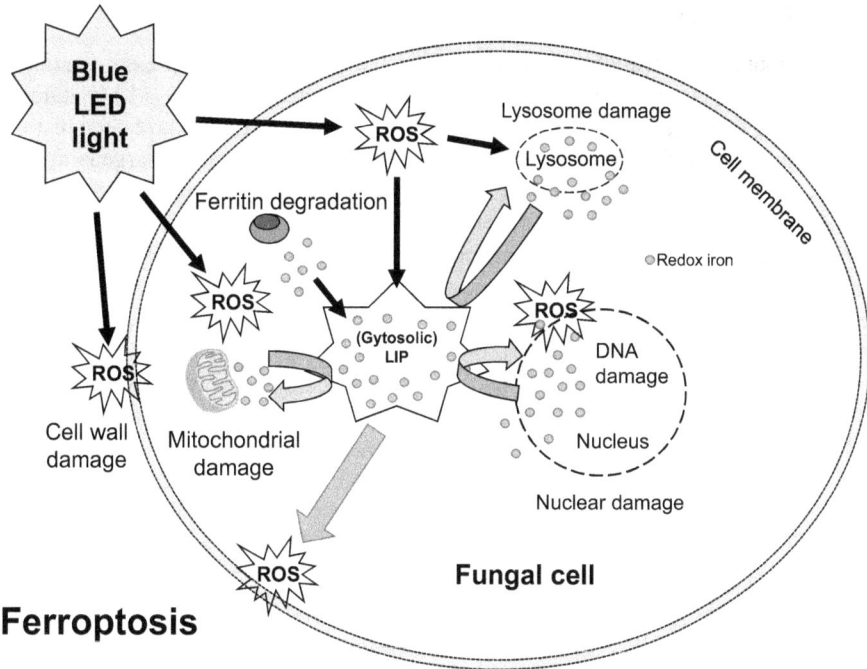

**FIGURE 11.2** Proposed mechanism of blue-light-induced ferroptosis in the fungal cell (Kahramanoğlu et al., 2020).

and algae (Bumbulis & Balog, 2013). The same phenomena can occur in fungal cells. In an earlier study, Fourtouni et al. (1998) reported that UV-B irradiation reduces the radial growth of *Alternaria solani*.

Similar reports exist for blue-light. For example, blue-light (470 nm) was reported to inhibit growth of *Penicillium digitatum* and *Fusarium graminearum* (De Lucca et al., 2012). In a different study, it was also noted that blue-light has significantly antifungal activity against *Scedosporium apiospermum*, *Scedosporium prolificans*, *Fusarium oxysporum* and *Fusarium solani* (Trzaska et al., 2017). The general mechanism of UV light and blue-light has not been fully understood. However, hormesis impacts light, and the impacts of ROS on the ferroptosis make it possible to come up with an idea that light may cause ferroptosis on fungal cells. In a recent study, Kahramanoğlu et al. (2020) suggested a mechanism involving the impacts of blue-light on the ferroptosis at the fungal cell level (Figure 11.2). It was recommended that the blue-light causes cell wall damage, which triggers ROS production. This ROS accumulation results in lipid peroxidation. This may result in damages to lysosomes. Since lysosomes have a significant role in controlling iron-related metabolic pathways, damages of the lysosomes may result in the redox iron and induce ferroptosis (Rizzollo et al., 2021) in the fungal cell (Kahramanoğlu et al., 2020).

## 11.4   COMMERCIAL LIGHT SOURCES

Several sources are available for UV and blue-light, including low-pressure mercury lamps, excimer lamps, light-emitting diodes (LEDs) and pulsed light (PL) (Natarajan et al., 2011; Bergman, 2021). Low-pressure mercury lamps (LPM) have been historically used for disinfecting water, air and surfaces. The LPM is a discharge lamp that works based on light emission from mercury (Hg) atoms. They are a very practical and effective way of producing germicidal ultraviolet (200–280 nm) (Bergman, 2021). Excimer lamps are quasimonochromatic sources that emit UV wavelengths in a wide range from 170 to 230 nm, dependent on the noble gas excimers present. These lamps are also useful in the UVC sterilization process (Bergman, 2021). PL is an emerging non-thermal food processing application that can sterilize food products by using white light. It involves the application of intense white light in the form of short pulses and a very high intensity, which is applied in a short period of time. High-energy pulses of light from a xenon light source are generally used for PL. The energy produced in 1 msec of pulsed light is nearly equivalent to the energy that would require 60 sec of continuous light. This technology can be used to control different type of pathogens including fungi, bacteria, parasites and viruses (Cacace & Palmieri, 2014; Mandal et al., 2020). The LED is a semiconductor diode capable of producing monochromatic light, which emits light when a specified voltage is applied across the two terminals. Different semiconductor materials are used for emitting different wavelengths of light, so that there are UV LEDs, IR LEDs and LED blue-lights available in the market (D'Souza et al., 2015).

## 11.5   UV LIGHT APPLICATION FOR CONTROLLING POSTHARVEST DECAY

Ultraviolet (UV) light has a wavelength range of 100–400 nm and is divided into three categories: UV-C (100–280 nm), UV-B (280–315 nm) and UV-A (315–400 nm). UV-A and UV-B pass through the atmosphere and reach the earth's surface, whereas UV-C is mostly absorbed by the ozone layer (Urban et al., 2016). UV light has been well studied for disinfecting air, water and surfaces, as well as for controlling postharvest pathogens (Koutchma et al., 2019). In addition, the use of UV for controlling postharvest pathogens in food storage was approved by the FDA (U.S. Food and Drug Administration) (FDA, 2000). The impacts of UV on microorganisms has been studied for a long period of time (Bintsis et al., 2000). Among the UV types, UV-C is the most studied and well-known due to its positive impacts on the prevention of fungal infections. The main reason is its absorption and low percentage of its existence on the earth. Moreover, it was reported that the relationship among UV light, plants and fungi is relatively complicated, and the impacts may significantly vary depending on the product type, fungi species and UV dose (Hideg et al., 2013).

As previously explained, UV-C may have two types of impact on fungi. One of them is the inhibition of the conidial germination and sporulation of fungi, such as colletotrichum on mango (Gunasegaran et al., 2018). This is a type of direct inhibition on the pathogen. In addition, as explained, another means of control is the hormesis mechanism. For example, according to Arcas et al. (2000), the application of UV-C (0.1 W m$^{-2}$) increases the flavonoids content in oranges and reduces pathogen growth by inducing defense system. There are several similar examples for this type of impact (see Table 11.1.). Moreover, the accumulation of phytoalexin and scoparone is the main means of plant-induced defense, which has been reported to reduce fungal growth in carrots (Mercier et al., 1993) and kumquats (Rodov et al., 1992), respectively. A similar mechanism was reported for tomato fruits, where the application of UV-C was noted to increase rishitin concentration and improve fruits' resistance to fungi (Charles et al., 2008). In addition to UV-C, the UV-B was also studied in postharvest. However, since plants are naturally exposed to UV-B, the research on UV-B is not as extensive as that on UV-C. In such a study, Ruiz et al. (2016) reported that UV-B (22,000 J m$^{-2}$ d$^{-1}$) treatment increases the accumulation of phenolic in the flavedo of lemon fruits and improves fruits' resistance against *P. digitatum*.

## 11.6 BLUE-LIGHT APPLICATION FOR CONTROLLING POSTHARVEST DECAY

Blue-light (400–500 nm) has shorter wavelengths and higher energy than other colors of the visible spectrum. It is normally absorbed by plant tissues and take important metabolic roles in living tissues. It is known to regulate the opening of stomata, therefore controlling water loss and carbon dioxide uptake (Lafuente & Alférez, 2015). In addition to the impacts on plants, blue-light was reported to have significant impacts on the prevention of fungal pathogens (see Table 11.2). Studies suggested that the blue-light increases the scoparone content in the flavedo of sweet oranges and therefore improves fruits' resistance to the infections of *P. digitatum* (Ballester & Lafuente, 2017). Studies with Satsuma mandarins also showed that the blue-light application improves the fruit's resistance against *P. italicum* (Yamaga et al., 2015). Further studies of Yamaga et al. (2015) reported that this positive impact of blue-light on Satsuma mandarins is due to the production of *Phytoalexin scoparone* in fruits.

## 11.7 LIGHT ILLUMINATION FOR PRESERVING STORAGE QUALITY OF FRESH PRODUCTS

Light illumination is a method of preserving the physical and biochemical properties of fruit and vegetable products during storage. In this case, in order to control the quality of fruits and vegetables during the storage and marketing of products, it is necessary to ensure uniform illumination of the product without glare and shadows (Gorodetskiy, 2016). This method is applicable not only for large vegetables and fruits but also for small berries, including strawberries and blueberries (Li et al., 2019; Zhirkova, 2021).

The use of a natural light source during storage is considered one of the ways to accelerate the ripening process of some vegetables, increase the formation of secondary metabolites, slow down the aging process of fruit and vegetable products. The development of fruit color after harvest is associated with the accumulation and formation of pigments in the light, such as lycopene, chlorophyll or anthocyanin (Xu

**TABLE 11.1**

**Impacts of Different UV Types on the Postharvest Pathogens in Different Crops**

| Crop | Fungi Species | UV Type | Application | Key Findings | References |
|---|---|---|---|---|---|
| Grapefruit | P. digitatum | UV-C | UV-C source (2.7 W m$^{-2}$) was placed 10 cm above the fruits | Accumulation of chitinase was observed and the growth of pathogen was reduced. | (Porat et al., 1999) |
| Kiwifruits | B. cinerea | UV-C | UV-C (between 2 and 6 kJ m$^{-2}$) combined with Meyerozyma guilliermondii | The combination was noted to induce defensive gene expression and total phenolics in kiwifruit, which induced disease resistance. | (Cheng et al., 2023) |
| Kumquats | P. digitatum | UV-C | UV-C source (0.2 × 103 J m$^{-2}$) was used | Accumulation of phytoalexin and scoparone was observed and the growth of pathogen was reduced. | (Rodov et al., 1992) |
| Lemon | P. digitatum | UV-B | UV-B (22,000 J m$^{-2}$ d$^{-1}$) source was placed 50 cm above the fruits | Increase in the accumulation of phenolics was observed in the flavedo. | (Ruiz et al., 2016) |
| Nectarine | Rhizopus stolonifer | UV-C | 3 kJ m$^{-2}$ UV-C | Increase in activities of antioxidant enzymes like superoxide dismutase (SOD), catalase (CAT) and ascorbate peroxidase (APX) and glutathione (GSH); and increase in disease resistance. | (Zhang et al., 2021) |
| Orange | P. digitatum | UV-C | UV-C source (0.1 W m$^{-2}$) was placed 65 cm above the fruits | Flavonoids content increased and pathogen growth was reduced. | (Arcas et al., 2000) |
| Pear | A. alternata | UV-C | Irradiation with low dose UV-C (0.36 kJ m$^{-2}$) | Increase in the contents of chitinase (CHI), SOD, peroxidase (POD), CAT, APX and phenylalanine ammonia-lyase (PAL) observed and an improvement in fungi resistance. | (Sun et al., 2022) |
| Potato tubers | Alternaria tenuissima | UV-C | UV-C between 10 and 20 kJ m$^{-2}$ combined with antagonistic yeast, Wickerhamomyces anomalus | Increase as observed in flavonoids and lignin in potato tubers together with the expression of defense-related genes, including polyphenol oxidase, peroxidase and β-1,3-glucanase, and thus an highest level of inhibition of pathogen was observed. | (Leng et al., 2022) |
| Strawberry | Botrytis cinerea | UV-C | 4.1 kJ m$^{-2}$ UV-C | Increases the activity of enzymes and proteins involved in defense against pathogens. | (Pombo et al., 2009; Forges et al. 2018) |
| Blueberry | Colletotrichum acutatum, C. gloeosporioides | UV-C | 2–4 kJ m$^{-2}$ UV-C | The total anthocyanin content increases by 20% after illumination and by 10% after storage. | (Perkins-Veazie et al., 2008; Wang et al. 2009a) |
| Tomato | Botrytis cinerea | UV-C | 3.7 kJ m$^{-2}$ UV-C at a distance of 30 cm from the radiation source for 3 m | Changes in the ultrastructure of the epidermal tissue of fruits, the physical barrier is formed that limits the invasion of Botrytis cinerea. | (Charles et al., 2008) |
| Grape | Botrytis cinerea | UV-B | 5.98–9.66 kJ m$^{-2}$ UV-B at a distance of 40 cm from the radiation source | The content of phenolics increased, and the antioxidant activity of berries rose. | (Ershov et al., 2018) |

**TABLE 11.2**

**Impacts of Blue-Light on the Postharvest Pathogens for Different Crops**

| Crop | Fungi Species | Application | Key Findings | References |
|---|---|---|---|---|
| Mandarin | *P. digitatum* and *P. italicum* | Blue-light at 465 nm with a photon flux of 80 $\mu$mol m$^{-2}$ s$^{-1}$ | Increase in the phytoalexin scoparone concentration, which results in the prevention of the microbial growth | (Yamaga & Nakamura, 2018) |
| Orange | *P. digitatum* | Blue-light (450 nm) at quantum fluxes between 210 and 630 $\mu$mol m$^{-2}$ s$^{-1}$ | Increase in scoparone at fruit flavedo and improvement in the fruit resistance to pathogen | (Ballester & Lafuente, 2017) |
| Sweet orange | *Geotrichum citri-aurantii* | Blue-light at 150 $\mu$mol m$^{-2}$ s$^{-1}$ | Increase in the production of lipid droplets (LDs) and ROs burst, which resulted in resistance against pathogen | (Du et al., 2023) |
| Tangerine | *P. digitatum* | Exposure of fruits to 410–540 nm blue-light at a fluency of 40 $\mu$mol m$^{-2}$ s$^{-1}$ | PLA2 gene expression and reduction in pathogen growth and development | (Alferez et al., 2012) |
| Lettuce | *B. cinerea* | 200 $\mu$mol m$^{-2}$ s$^{-1}$ | DPPH radical scavenging activity of lettuce seedlings increased. Increase of antioxidant capacity as well as the development of compact morphology. | (Kook et al., 2013) |
| Tomato | *B. cinerea* | 50–150 $\mu$mol m$^{-2}$ s$^{-1}$ | Increased production of osmoprotectants and antioxidative responses, enhanced accumulation of proline, including reactive oxygen species (ROS) scavenging enzymes. | (Kim et al., 2013) |
| Tomato | *P. digitatum, P. italicum* and *Phomopsis citri* | 40 $\mu$mol m$^{-2}$ s$^{-1}$ for 5–7 days | Decreased soft rot area and increased expression of 16 defense-related genes during the storage period. | (Poonia et al., 2022) |
| Strawberry | *B. cinerea* | 40 $\mu$mol m$^{-2}$ s$^{-1}$ for 12 days at 5°C. | Improvement in the total anthocyanin content, increase in the activities of glucose-6-phosphate, shikimate dehydrogenase, tyrosine ammonia-lyase, phenylalanine ammonia-lyase, cinnamate-4-hydroxylase, chalcone synthase, flavanone-3-$\beta$-hydroxylase, anthocyanin synthase and UDP-glycose flavonoid-3-O-glycosyltranferase | (Xu et al., 2014; Zhang et al., 2022) |

et al., 2014a, 2014b). Exposure to different lighting levels for 6–12 days at a temperature of 22–23°C on Brussels sprouts (*Brassica oleracea gemmifera*), broccoli (*Brassica oleracea*), tomato (*Solanum lycopersicum*) and red Chinese sand pears fruit (*Pyrus pyrifolia* Nakai) contributed to the improvement of fruit color (Büchert et al., 2011; Hasperué et al., 2016a, 2016b; Nájera et al., 2018). The illumination level of 20 W m$^{-2}$ at a temperature of 22°C contributed to a slowdown in weight loss for a period of 5 and 10 days in Brussels sprouts and broccoli, respectively (Hasperué et al., 2016a, 2016b). The antioxidant system (glutathione oxidase, catalase, ascorbate oxidase, superoxide dismutase and glutathione oxidase) of fruits and vegetables was significantly enhanced under the influence of light treatment (Xu et al., 2014a, 2014b). During storage in the light, the increase in antioxidant properties of such vegetables as tomatoes, Chinese cabbage (*Brassica* sp.), Peking cabbage (*Brassica napus* L.) and peas (*Pisum sativum* L.) is associated with the action of enzymes that activate DPPH free radicals, β-carotene and glucosinolates (Hee-Sun Kook, 2013; Johkan et al., 2010; Ki Lee et al., 2016; Wu et al., 2007). The increased activity of antioxidant enzymes reduces the development of citrus cancer (Ma et al., 2012).

The expression of genes such as MdMYB10 and MdUFGT under the influence of light contributes to the accumulation of vitamin C, soluble sugars and organic acids (Hasan et al., 2017a, 2017b). The influence of light on the production of secondary metabolites, which are important for the quality of crops during storage, has been proven (Dutta Gupta, 2017). The role of light in the induction of secondary metabolites in agricultural crops is associated with phenylalanine ammonia-lyase enzyme (Hasan et al., 2017a, 2017b). However, the effect of light on the formation of secondary metabolites has not yet been well studied. Light increases the content of phenolic compounds (i.e., flavonols, anthocyanins and ascorbic acid), carotenoids and GSH. (Wang et al., 2009b; Zhan et al., 2012a, 2012b). However, quantitative changes in the content of chlorophyll and carotenoids in fruits were observed during only 1 day of storage (Pérez-Ambrocio et al., 2018). Light affects the expression of fruit genes' encoding enzymes such as phenyl ammonium-lyase (PAL), chalconesynthase, flavanone-3-hydroxylase and others involved in various processes (Dutta Gupta, 2017). The intensity of the light level affects the content of anthocyanins in fruits during the postharvest period. Lighting red Chinese sand pears fruit for 10 days contributed to an increase in anthocyanins in fruits (Shi et al., 2015), while in grape berries (Gros Colman) this effect was observed after 72 hours. (Kataoka et al., 2003). In octoploid strawberry cultivars (*Fragaria ananassa*), anthocyanin biosynthesis is associated with activation of FaMYB10 gene expression. Besides, FaMYB10 accelerated anthocyanin synthesis of pelargonidin 3-glucoside and cyaniding 3-glucoside over 8 days (Kadomura-Ishikawa et al., 2013). Anthocyanin content increased by 21% in strawberries during storage (Kim et al., 2011). The light level temporarily slows down the aging process and prevents the formation of yellowing of some fruits and vegetables, such as cabbage, mandarins (*Citrus reticulate*), spinach leaves, (*Spinacia oleracea*), Chinese cabbage and kiwi (*Actinidia deliciosa*) (Liu et al., 2015; Lester et al., 2010; Zhan et al., 2012a, 2012b). The level of illumination reduces the number of pathogenic microorganisms in vegetables and fruits during the postharvest period. Photosensitizers or hotoreceptors absorb light, forming reactive oxygen species that react with biomolecules of microbial cells, causing destruction and prevention of spoilage (Luksiene & Zukauskas, 2009). The level of illumination is an effective means of slowing down the aging process of leaves for *Brassica oleracea* var. *sabellica*, broccoli, basil leaves, lettuce (Costa et al., 2013; Charles et al., 2018; Bárcena et al., 2019; Favre et al., 2018).

## 11.8 CONCLUSION

Current knowledge suggests that light illumination, both UV and blue-light, has potentially significant impacts on both the decay control and quality preservation of fruits and vegetables. It is clear that the light achieves this success with two main mechanisms, namely hormesis and ferroptosis. However, it is also clear that the product type, fungi species and light dose significantly impact the overall effect of the light. Therefore, it is highly important to conduct further studies on each product type.

## REFERENCES LIST

Alferez, F., Liao, H. L., & Burns, J. K. (2012). Blue light alters infection by *Penicillium digitatum* in tangerines. *Postharvest Biology and Technology*, *63*(1), 11–15. https://doi.org/10.1016/j.postharvbio.2011.08.001

Arcas, M. C., Botía, J. M., Ortuño, A. M., & Del Río, J. A. (2000). UV irradiation alters the levels of flavonoids involved in the defence mechanism of *Citrus aurantium* fruits against *Penicillium digitatum*. *European Journal of Plant Pathology*, *106*(7), 617–622. https://doi.org/10.1023/A:1008704102446

Artés-Hernández, F., Castillejo, N., & Martínez-Zamora, L. (2022). UV and visible spectrum led lighting as abiotic elicitors of bioactive compounds in sprouts, microgreens, and baby leaves – A comprehensive review including their mode of action. *Foods*, *11*(3), 265. https://doi.org/10.3390/foods11030265

Ballester, A. R., & Lafuente, M. T. (2017). LED Blue Light-induced changes in phenolics and ethylene in citrus fruit: Implication in elicited resistance against *Penicillium digitatum* infection. *Food Chemistry*, *218*, 575–583. https://doi.org/10.1016/j.foodchem.2016.09.089

Banani, H., Marcet-Houben, M., Ballester, A. R., Abbruscato, P., González-Candelas, L., Gabaldón, T., & Spadaro, D. (2016). Genome sequencing and secondary metabolism of the postharvest pathogen *Penicillium griseofulvum*. *BMC Genomics*, *17*(1), 19. https://doi.org/10.1186/s12864-015-2347-x

Bárcena, A., Gustavo Martínez, G., & Costa, L. (2019). Low intensity light treatment improves purple kale [*Brassica oleracea* var. *sabellica*] postharvest preservation at room temperature. *Heliyon*, *5*, 9. https://doi.org/10.1016/j.heliyon.2019.e02467

Bergman, R. S. (2021). Germicidal UV sources and systems†. *Photochemistry and Photobiology*, *97*(3), 466–470. https://doi.org/10.1111/php.13387

Bintsis, T., Litopoulou-Tzanetaki, E., & Robinson, R. K. (2000). Existing and potential applications of ultraviolet light in the food industry–a critical review. *Journal of the Science of Food and Agriculture*, *80*(6), 637–645. https://doi.org/10.1002/(SICI)1097-0010(20000501)80:6<637::AID-JSFA603>3.0.CO

Büchert, A. M., Gómez Lobato, M. E., Villarreal, N. M., Civello, P. M., & Martínez, G. A. (2011). Effect of visible light treatments on postharvest senescence of broccoli (*Brassica oleracea* L.). *Journal of the Science of Food and Agriculture*, *91*(2), 355–361. https://doi.org/10.1002/jsfa.4193

Bumbulis, M. J., & Balog, B. M. (2013). UV-C exposure induces an apoptosis-like process in Euglena gracilis. *ISRN Cell Biology*, *2013*, 1–6. https://doi.org/10.1155/2013/869216

Cacace, D., & Palmieri, L. (2014). High-intensity pulsed light technology. In *Emerging technologies for food processing* (pp. 239–258). Academic Press. https://doi.org/10.1016/B978-0-12-411479-1.00013-9

Cao, J. Y., & Dixon, S. J. (2016). Mechanisms of ferroptosis. *Cellular and Molecular Life Sciences*, *73*(11–12), 2195–2209. https://doi.org/10.1007/s00018-016-2194-1

Charles, F., Nilprapruck, P., Roux, D., & Sallanon, H. (2018). Visible light as a new tool to maintain fresh-cut lettuce post-harvest quality. *Postharvest Biology and Technology*, *135*, 51–56. https://doi.org/10.1016/j.postharvbio.2017.08.024

Charles, M. T., Benhamou, N., & Arul, J. (2008). Physiological basis of UV-C induced resistance to *Botrytis cinerea* in tomato fruit: III. *Postharvest Biology and Technology*, *47*(1), 27–40. https://doi.org/10.1016/j.postharvbio.2007.05.015

Cheng, L., Zhou, L., Li, D., Gao, Z., Teng, J., Nie, X., Guo, F., Wang, C., Wang, X., Li, S., & Li, X. (2023). Combining the biocontrol agent *Meyerozyma guilliermondii* with UV-C treatment to manage postharvest gray mold on kiwifruit. *Biological Control*, *180*, 105198. https://doi.org/10.1016/j.biocontrol.2023.105198

Costa, L., Millan Montano, Y., Carrión, C., Rolny, N., & Guiamet, J. J. (2013). Application of low intensity light pulses to delay postharvest senescence of *Ocimum basilicum* leaves. *Postharvest Biology and Technology*, *86*, 181–191. https://doi.org/10.1016/j.postharvbio.2013.06.017

De Lucca, A. J., Carter-Wientjes, C., Williams, K. A., & Bhatnagar, D. (2012). Blue light (470 nm) effectively inhibits bacterial and fungal growth. *Letters in Applied Microbiology*, *55*(6), 460–466. https://doi.org/10.1111/lam.12002

Distéfano, A. M., Martin, M. V., Córdoba, J. P., Bellido, A. M., D'Ippólito, S., Colman, S. L., Soto, D., Roldán, J. A., Bartoli, C. G., Zabaleta, E. J., Fiol, D. F., Stockwell, B. R., Dixon, S. J., & Pagnussat, G. C. (2017). Heat stress induces ferroptosis-like cell death in plants. *Journal of Cell Biology*, *216*(2), 463–476. https://doi.org/10.1083/jcb.201605110

Costa, D., Alviano Moreno, D. S., Alviano, C. S., & da Silva, A. J. R. (2022). Extension of Solanaceae food crops shelf life by the use of elicitors and sustainable practices during postharvest phase. *Food and Bioprocess Technology*, *15*(2), 249–274. https://doi.org/10.1007/s11947-021-02713-z

D'Souza, C., Yuk, H. G., Khoo, G. H., & Zhou, W. (2015). Application of light-emitting diodes in food production, postharvest preservation, and microbiological food safety. *Comprehensive Reviews in Food Science and Food Safety*, *14*(6), 719–740. https://doi.org/10.1111/1541-4337.12155

Du, Y., Sun, J., Tian, Z., Cheng, Y., & Long, C. A. (2023). Effect of blue light treatments on Geotrichum citri-aurantii and the corresponding physiological mechanisms of citrus. *Food Control*, *145*, 109468. https://doi.org/10.1016/j.foodcont.2022.109468

Duarte-Sierra, A., Tiznado-Hernández, M. E., & Jha, D. K. (2022). Postharvest hormesis in produce. *Current Opinion in Environmental Science and Health*, *29*. https://doi.org/10.1016/j.coesh.2022.100376

Duarte-Sierra, A., Tiznado-Hernández, M. E., Jha, D. K., Janmeja, N., & Arul, J. (2020). Abiotic stress hormesis: An approach to maintain quality, extend storability, and enhance phytochemicals on fresh produce during postharvest. *Comprehensive Reviews in Food Science and Food Safety*, *19*(6), 3659–3682. https://doi.org/10.1111/1541-4337.12628

Dutta Gupta, S. (2017). Light emitting diodes for agriculture: Smart lighting. *Light Emitting Diodes for Agriculture*, 1–334. https://doi.org/10.1007/978-981-10-5807-3

El Khetabi, A., Lahlali, R., Ezrari, S., Radouane, N., Lyousfi, N., Banani, H., Askarne, L., Tahiri, A., El Ghadraoui, L., Belmalha, S., & Barka, E. A. (2022). Role of plant extracts and essential oils in fighting against postharvest fruit pathogens and extending fruit shelf life: A review. *Trends in Food Science and Technology*, *120*, 402–417. https://doi.org/10.1016/j.tifs.2022.01.009

Ershov, B. G., Polikarpov, N. A., Fedotova, O. B., Goldstein, A. A., Kireev, S. G., Shashkovsky, S. G., Zhelayev, I. A., Tumashevich, K. A., & Seliverstov, A. F. (2018, September 26–28). The application of pulse ultraviolet technologies to increase the storage life of perishable foodstuffs. Radiation technologies in agriculture and food industry: Current state and prospects. In *Proceedings of the international research and practice conference* (pp. 262–266). Obninsk.

Favre, N., Bárcena, A., Bahima, J. V., Martínez, G., & Costa, L. (2018). Pulses of low intensity light as promising technology to delay postharvest senescence of broccoli. *Postharvest Biology and Technology*, *142*, 107–114. https://doi.org/10.1016/j.postharvbio.2017.11.006

FDA. (2000). Irradiation in the production, processing and handling of food: 21 CFR. *Federal Register*, *65*(179), 71056–71058.

Forges, M., Vàsquez, H., Charles, F., Sari, D. C., Urban, L., Lizzi, Y., Bardin, M., & Aarrouf, J. (2018). Impact of UV-C radiation on the sensitivity of three strawberry plant cultivars (*Fragaria* x *ananassa*) against *Botrytis cinerea*. *Scientia Horticulturae*, *240*, 603–613. https://doi.org/10.1016/j.scienta.2018.06.063

Fourtouni, A., Manetas, Y., & Christias, C. (1998). Effects of UV-B radiation on growth, pigmentation, and spore production in the phytopathogenic fungus *Alternaria solani*. *Canadian Journal of Botany*, *76*(12), 2093–2099. https://doi.org/10.1139/b98-170

González-Estrada, R. R., Blancas-Benitez, F. J., Aguirre-Güitrón, L., Hernandez-Montiel, L. G., Moreno-Hernández, C., Cortés-Rivera, H. J., Herrera-González, J. A. et al. (2021). Alternative management technologies for postharvest disease control. In *Food losses, sustainable postharvest and food technologies* (pp. 153–190). Academic Press. https://doi.org/10.1016/B978-0-12-821912-6.00008-0

Gorodetskiy, A. E. (2016). Smart electromechanical systems modules. In A. E. Gorodetskiy (Ed.), *Smart electromechanical systems* (pp. 7–15). Springer. https://doi.org/10.1007/978-3-319-27547-5_2

Gunasegaran, B., Ding, P., & Kadir, J. (2018). Morphological identification and in vitro evaluation of *Colletotrichum gloesporioides* in "Chok Anan" mango using UV-C irradiation. In A. Pacific (Ed.), *Acta Horticulturae III* (pp. 1213, 599–602). Symposium on Postharvest Research, Education and Extension: APS2014. https://doi.org/10.17660/ActaHortic.2018.1213.90

Gunders, D., & Bloom, J. (2017). *Wasted: How America is losing up to 40 percent of its food from farm to fork to landfill* (Vol. 26, pp. 1–26). Natural Resources Defense Council.

Hasan, M. M., Bashir, T., & Bae, H. (2017a). Use of ultrasonication technology for the increased production of plant secondary metabolites. *Molecules*, *22*(7). https://doi.org/10.3390/molecules22071046

Hasan, M. M., Bashir, T., Ghosh, R., Lee, S. K., & Bae, H. (2017b). An overview of LEDs' effects on the production of bioactive compounds and crop quality. *Molecules*, *22*(9), 1–12. https://doi.org/10.3390/molecules22091420

Hasperué, J. H., Guardianelli, L., Rodoni, L. M., Chaves, A. R., & Martínez, G. A. (2016a). Continuous white-blue LED light exposition delays postharvest senescence of broccoli. *LWT – Food Science and Technology*, *65*, 495–502. https://doi.org/10.1016/j.lwt.2015.08.041

Hasperué, J. H., Rodoni, L. M., Guardianelli, L. M., Chaves, A. R., & Martínez, G. A. (2016b). Use of LED light for Brussels sprouts postharvest conservation. *Scientia Horticulturae*, *213*, 281–286. https://doi.org/10.1016/j.scienta.2016.11.004

Hee-Sun Kook, K. K. (2013). The Effect of blue-light-emitting diodes on antioxidant properties and resistance to *Botrytis cinerea* in tomato. *Journal of Plant Pathology and Microbiology*, *4*(9). https://doi.org/10.4172/2157-7471.1000203

Hideg, E., Jansen, M. A., & Strid, A. (2013). UV-B exposure, ROS, and stress: Inseparable companions or loosely linked associates? *Trends in Plant Science, 18*(2), 107–115. https://doi.org/10.1016/j.tplants.2012.09.003

Hosseini, S., Amini, J., Saba, M. K., Karimi, K., & Pertot, I. (2020). Preharvest and postharvest application of garlic and rosemary essential oils for controlling anthracnose and quality assessment of strawberry fruit during cold storage. *Frontiers in Microbiology, 11*, 1855. https://doi.org/10.3389/fmicb.2020.01855

Johkan, M., Shoji, K., Goto, F., Hashida, S., & Yoshihara, T. (2010). Blue light-emitting diode light irradiation of seedlings improves seedling quality and growth after transplanting in red leaf lettuce. *HortScience, 45*(12), 1809–1814. https://doi.org/10.21273/HORTSCI.45.12.1809

Jurick, W. M. (2022). Biotechnology approaches to reduce antimicrobial resistant postharvest pathogens, mycotoxin contamination, and resulting product losses. *Current Opinion in Biotechnology, 78*, 102791. https://doi.org/10.1016/j.copbio.2022.102791

Kadomura-Ishikawa, Y., Miyawaki, K., Noji, S., & Takahashi, A. (2013). Phototropin 2 is involved in blue light-induced anthocyanin accumulation in *Fragaria* x *ananassa* fruits. *Journal of Plant Research, 126*(6), 847–857. https://doi.org/10.1007/s10265-013-0582-2

Kahramanoğlu, İ. (2017). Introductory chapter: Postharvest physiology and technology of horticultural crops. *Postharvest Handling, 1*(5). https://doi.org/10.5772/intechopen.69466

Kahramanoğlu, İ., Nisar, M. F., Chen, C., Usanmaz, S., Chen, J., & Wan, C. (2020). Light: An alternative method for physical control of postharvest rotting caused by fungi of citrus fruit. *Journal of Food Quality, 2020*, 1–12. https://doi.org/10.1155/2020/8821346

Kataoka, I., Sugiyama, A., & Beppu, K. (2003). Role of ultraviolet radiation in accumulation of anthocyanin in berries of "Gros Colman" grapes (*Vitis vinifera* L.). *Engei Gakkai Zasshi, 72*(1), 1–6. https://doi.org/10.2503/jjshs.72.1

Kim, B. S., Lee, H. O., Kim, J. Y., Kwon, K. H., Cha, H. S., & Kim, J. H. (2011). An effect of light emitting diode (LED) irradiation treatment on the amplification of functional components of immature strawberry. *Horticulture, Environment, and Biotechnology, 52*(1), 35–39. https://doi.org/10.1007/s13580-011-0189-2

Kim, K., Kook, H., Jang, Y., Lee, W., Kamala-Kannan, S., Chae, J., & Lee, K. (2013). The effect of blue-light-emitting diodes on antioxidant properties and resistance to *Botrytis cinerea* in tomato. *Journal of Plant Pathology and Microbiology, 4*(9). https://doi.org/10.4172/2157-7471.1000203

Kook, H. S., Park, S. H., Jang, Y. J., Lee, G. W., Kim, J. S., Kim, H. M., Oh, B.-T., Chae, J., & Lee, K. (2013). Blue LED (light-emitting diodes)-mediated growth promotion and control of Botrytis disease in lettuce. *Acta Agriculturae Scandinavica, Section B – Soil and Plant Science, 63*(3), 271–277. https://doi.org/10.1080/09064710.2012.756118

Kouda, K., & Iki, M. (2010). Beneficial effects of mild stress (hormetic effects): Dietary restriction and health. *Journal of Physiological Anthropology, 29*(4), 127–132. https://doi.org/10.2114/jpa2.29.127

Koutchma, T., Popović, V., & Green, A. (2019). Overview of ultraviolet (UV) LEDs technology for applications in food production. In *Ultraviolet LED technology for food applications* (pp. 1–23). Academic Press. https://doi.org/10.1016/B978-0-12-817794-5.00001-7

Lafuente, M. T., & Alférez, F. (2015). Effect of LED blue light on *Penicillium digitatum* and *Penicillium italicum* strains. *Photochemistry and Photobiology, 91*(6), 1412–1421. https://doi.org/10.1111/php.12519

Lee, C. W., Ko, H. H., Chai, C. Y., Chen, W. T., Lin, C. C., & Yen, F. L. (2013). Effect of *Artocarpus communis* extract on UVB irradiation-induced oxidative stress and inflammation in hairless mice. *International Journal of Molecular Sciences, 14*(2), 3860–3873. https://doi.org/10.3390/ijms14023860

Lee, M. K., Arasu, M. V., Park, S., Byeon, D. H., Chung, S. O., Park, S. U., Yong-, P., & Sun-, J. (2016). LED lights enhance metabolites and antioxidants in Chinese cabbage and kale. *Brazilian Archives of Biology and Technology*, *59*. https://doi. org/10.1590/1678-4324-2016150546

Leng, J., Dai, Y., Qiu, D., Zou, Y., & Wu, X. (2022). Utilization of the antagonistic yeast, *Wickerhamomyces anomalus*, combined with UV-C to manage postharvest rot of potato tubers caused by *Alternaria tenuissima. International Journal of Food Microbiology*, *377*, 109782. https://doi.org/10.1016/j.ijfoodmicro.2022.109782

Lester, G. E., Makus, D. J., & Hodges, D. M. (2010). Relationship between fresh-packaged spinach leaves exposed to continuous light or dark and bioactive contents: Effects of cultivar, leaf size, and storage duration. *Journal of Agricultural and Food Chemistry*, *58*(5), 2980–2987. https://doi.org/10.1021/jf903596v

Li, S., Luo, H., Hu, M., Zhang, M., Feng, J., Liu, Y., Dong, Q., & Liu, B. (2019). Optical non-destructive techniques for small berry fruits: A review. *Artificial Intelligence in Agriculture*, *2*, 85–98. https://doi.org/10.1016/j.aiia.2019.07.002

Liu, J. D., Goodspeed, D., Sheng, Z., Li, B., Yang, Y., Kliebenstein, D. J., & Braam, J. (2015). Keeping the rhythm: Light/dark cycles during postharvest storage preserve the tissue integrity and nutritional content of leafy plants. *BMC Plant Biology*, *15*(1), 92. https:// doi.org/10.1186/s12870-015-0474-9

Luksiene, Z., & Zukauskas, A. (2009). Prospects of photosensitization in control of pathogenic and harmful micro-organisms. *Journal of Applied Microbiology*, *107*(5), 1415–1424. https://doi.org/10.1111/j.1365-2672.2009.04341.x

Ma, G., Zhang, L., Kato, M., Yamawaki, K., Kiriiwa, Y., Yahata, M., Ikoma, Y., & Matsumoto, H. (2012). Effect of the combination of ethylene and red LED light irradiation on carotenoid accumulation and carotenogenic gene expression in the flavedo of citrus fruit journal of Agricultural and Food Chemistry, *60*, 197–201. https://doi.org/10.1016/j. postharvbio.2014.08.002

Madbouly, A. K., Abo Elyousr, K. A. M., & Ismail, I. M. (2020). Biocontrol of *Monilinia fructigena*, causal agent of brown rot of apple fruit, by using endophytic yeasts. *Biological Control*, *144*, 104239. https://doi.org/10.1016/j.biocontrol.2020.104239

Mandal, R., Mohammadi, X., Wiktor, A., Singh, A., & Pratap Singh, A. (2020). Applications of pulsed light decontamination technology in food processing: An overview. *Applied Sciences*, *10*(10), 3606. https://doi.org/10.3390/app10103606

Mercier, J., Arul, J., & Julien, C. (1993). Effect of UV-C on phytoalexin accumulation and resistance to *Botrytis cinerea* in stored carrots. *Journal of Phytopathology*, *139*(1), 17–25. https://doi.org/10.1111/j.1439-0434.1993.tb01397.x

Nájera, C., Guil-Guerrero, J. L., Enríquez, L. J., Álvaro, J. E., & Urrestarazu, M. (2018). LED-enhanced dietary and organoleptic qualities in postharvest tomato fruit. *Postharvest Biology and Technology*, *145*, 151–156. https://doi.org/10.1016/j. postharvbio.2018.07.008

Narita, K., Asano, K., Naito, K., Ohashi, H., Sasaki, M., Morimoto, Y., Igarashi, T., & Nakane, A. (2020). Ultraviolet C light with wavelength of 222 nm inactivates a wide spectrum of microbial pathogens. *Journal of Hospital Infection*, *105*(3), 459–467. https://doi. org/10.1016/j.jhin.2020.03.030

Natarajan, T. S., Natarajan, K., Bajaj, H. C., & Tayade, R. J. (2011). Energy efficient UV-LED source and TiO2 nanotube array-based reactor for photocatalytic application. *Industrial and Engineering Chemistry Research*, *50*(13), 7753–7762. https://doi. org/10.1021/ie200493k

Nawkar, G. M., Maibam, P., Park, J. H., Sahi, V. P., Lee, S. Y., & Kang, C. H. (2013). UV-induced cell death in plants. *International Journal of Molecular Sciences*, *14*(1), 1608–1628. https://doi.org/10.3390/ijms14011608

Nikita, P., Kevin, V., & Mateo, H. (2015). Electromagnetic radiation. *Chemistry LibreTexts*. Retrived April 23, 2023, from https://chem.libretexts.org/Bookshelves/Physical_ and_Theoretical_Chemistry_Textbook_Maps/Supplemental_Modules_(Physical_ and_Theoretical_Chemistry)/Spectroscopy/Fundamentals_of_Spectroscopy/ Electromagnetic_Radiation

Palou, L. (2018). Postharvest treatments with GRAS salts to control fresh fruit decay. *Horticulturae*, *4*(4), 46. https://doi.org/10.3390/horticulturae4040046

Papoutsis, K., Mathioudakis, M. M., Hasperué, J. H., & Ziogas, V. (2019). Non-chemical treatments for preventing the postharvest fungal rotting of citrus caused by Penicillium digitatum (green mold) and Penicillium italicum (blue mold). *Trends in Food Science and Technology*, *86*, 479–491. https://doi.org/10.1016/j.tifs.2019.02.053

Pérez-Ambrocio, A., Guerrero-Beltrán, J. A., Aparicio-Fernández, X., Ávila-Sosa, R., Hernández-Carranza, P., Cid-Pérez, S., & Ochoa-Velasco, C. E. (2018). Effect of blue and ultraviolet-C light irradiation on bioactive compounds and antioxidant capacity of habanero pepper (*Capsicum chinense*) during refrigeration storage. *Postharvest Biology and Technology*, *135*, 19–26. https://doi.org/10.1016/j.postharvbio.2017.08.023

Perkins-Veazie, P., Collins, J. K., & Howard, L. (2008). Blueberry fruit response to postharvest application of ultraviolet radiation. *Postharvest Biology and Technology*, *47*(3), 280–285. https://doi.org/10.1016/j.postharvbio.2007.08.002

Podolec, R., Lau, K., Wagnon, T. B., Hothorn, M., & Ulm, R. (2021). A constitutively monomeric UVR8 photoreceptor confers enhanced UV-B photomorphogenesis. *Proceedings of the National Academy of Sciences of the United States of America*, *118*(6), e2017284118. https://doi.org/10.1073/pnas.2017284118

Pombo, M. A., Dotto, M. C., Martínez, G. A., & Civello, P. M. (2009). UV-C irradiation delays strawberry fruit softening and modifies the expression of genes involved in cell wall degradation. *Postharvest Biology and Technology*, *51*(2), 141–148. https://doi.org/10.1016/j. postharvbio.2008.07.007

Poonia, A., Pandey, S., & Vasundhara. (2022). Application of light emitting diodes (LEDs) for food preservation, post-harvest losses and production of bioactive compounds: A review. *Food Production, Processing and Nutrition*, *4*(1), 8. https://doi.org/10.1186/ s43014-022-00086-0

Porat, R., Lers, A., Dori, S., Cohen, L., Weiss, B., Daus, A., Wilson, C.L., & Droby, S. (1999). Induction of chitinase and β-1, 3-endoglucanase proteins by UV irradiation and wounding in grapefruit peel tissue. *Phytoparasitica*, *27*(3), 233–238. https://doi.org/10.1007/BF02981463

Rizzollo, F., More, S., Vangheluwe, P., & Agostinis, P. (2021). The lysosome as a master regulator of iron metabolism. *Trends in Biochemical Sciences*, *46*(12), 960–975. https://doi. org/10.1016/j.tibs.2021.07.003

Rodov, V., Ben-Yehoshua, S., Kim, J. J., Shapiro, B., & Ittah, Y. (1992). Ultraviolet illumination induces scoparone production in kumquat and orange fruit and improves decay resistance. *Journal of the American Society for Horticultural Science*, *117*(5), 788–792. https://doi.org/10.21273/JASHS.117.5.788

Ruffo Roberto, S., Youssef, K., Hashim, A. F., & Ippolito, A. (2019). Nanomaterials as alternative control means against postharvest diseases in fruit crops. *Nanomaterials*, *9*(12), 1752. https://doi.org/10.3390/nano9121752

Ruiz, V. E., Interdonato, R., Cerioni, L., Albornoz, P., Ramallo, J., Prado, F. E., Hilal, M., & Rapisarda, V. A. (2016). Short-term UV-B exposure induces metabolic and anatomical changes in peel of harvested lemons contributing in fruit protection against green mold. *Journal of Photochemistry and Photobiology. B, Biology*, *159*, 59–65. https://doi. org/10.1016/j.jphotobiol.2016.03.016

Sakai, K. (2006). Biological responses to low dose radiation-hormesis and adaptive responses. *Yakugaku Zasshi: Journal of the Pharmaceutical Society of Japan*, *126*(10), 827–831. https://doi.org/10.1248/yakushi.126.827

Shi, Y., Wang, B. L., Shui, D. J., Cao, L. L., Wang, C., Yang, T., Wang, X. Y., & Ye, H. X. (2015). Effect of 1-methylcyclopropene on shelf life, visual quality and nutritional quality of netted melon. *Food Science and Technology International, 21*(3), 175–187. https://doi.org/10.1177/1082013214520786

Stockwell, B. R., Friedmann Angeli, J. P. F., Bayir, H., Bush, A. I., Conrad, M., Dixon, S. J., Fulda, S., Gascón, S., Hatzios, S. K., Kagan, V. E., Noel, K., Jiang, X., Linkermann, A., Murphy, M. E., Overholtzer, M., Oyagi, A., Pagnussat, G. C., Park, J., Ran, Q.,. . . Zhang, D. D. (2017). Ferroptosis: A regulated cell death nexus linking metabolism, redox biology, and disease. *Cell, 171*(2), 273–285. https://doi.org/10.1016/j.cell.2017.09.021

Sun, T., Ouyang, H., Sun, P., Zhang, W., Wang, Y., Cheng, S., & Chen, G. (2022). Postharvest UV-C irradiation inhibits blackhead disease by inducing disease resistance and reducing mycotoxin production in 'Korla' fragrant pear (Pyrus sinkiangensis). *International Journal of Food Microbiology, 362*, 109485. https://doi.org/10.1016/j.ijfoodmicro.2021.109485

Tao, N., Li, K., & Liu, J. (2020). Molecular mechanisms of ferroptosis and its role in pulmonary disease. *Oxidative Medicine and Cellular Longevity, 2020*, 9547127. https://doi.org/10.1155/2020/9547127

Thomidis, T. (2014). Fruit rots of pomegranate (cv. wonderful) in Greece. *Australasian Plant Pathology, 43*(5), 583–588. https://doi.org/10.1007/s13313-014-0300-0

Trzaska, W. J., Wrigley, H. E., Thwaite, J. E., & May, R. C. (2017). Species-specific antifungal activity of blue light. *Scientific Reports, 7*(1), 4605. https://doi.org/10.1038/s41598-017-05000-0

Urban, L., Charles, F., de Miranda, M. R. A., & Aarrouf, J. (2016). Understanding the physiological effects of UV-C light and exploiting its agronomic potential before and after harvest. *Plant Physiology and Biochemistry, 105*, 1–11. https://doi.org/10.1016/j.plaphy.2016.04.004

Wang, C. Y., Chen, C. T., & Wang, S. Y. (2009a). Changes of flavonoid content and antioxidant capacity in blueberries after illumination with UV-C. *Food Chemistry, 117*(3), 426–431. https://doi.org/10.1016/j.foodchem.2009.04.037

Wang, S. Y., Chen, C. T., & Wang, C. Y. (2009b). The influence of light and maturity on fruit quality and flavonoid content of red rasp-berries. *Food Chemistry, 112*(3), 676–684. https://doi.org/10.1016/j.foodchem.2008.06.032

Williamson, B., Tudzynski, B., Tudzynski, P., & Van Kan, J. A. (2007). *Botrytis cinerea*: The cause of grey mould disease. *Molecular Plant Pathology, 8*(5), 561–580. https://doi.org/10.1111/j.1364-3703.2007.00417.x

Wu, M. C., Hou, C. Y., Jiang, C. M., Wang, Y. T., Wang, C. Y., Chen, H. H., & Chang, H. M. (2007). A novel approach of LED light radiation improves the antioxidant activity of pea seedlings. *Food Chemistry, 101*(4), 1753–1758. https://doi.org/10.1016/j.foodchem.2006.02.010

Xu, F., Cao, S., Shi, L., Chen, W., Su, X., & Yang, Z. (2014). Blue light irradiation affects anthocyanin content and enzyme activities involved in postharvest strawberry fruit. *Journal of Agricultural and Food Chemistry, 62*(20), 4778–4783. https://doi.org/10.1021/jf501120u

Xu, F., Cao, S., Shi, L., Chen, W., Su, X., & Yang, Z. (2014a). Blue light irradiation affects anthocyanin content and enzyme activities involved in postharvest strawberry fruit. *Journal of Agricultural and Food Chemistry, 62*(20), 4778–4783. https://doi.org/10.1021/jf501120u

Xu, F., Shi, L., Chen, W., Cao, S., Su, X., & Yang, Z. (2014b). Effect of blue light treatment on fruit quality, antioxidant enzymes and radical-scavenging activity in strawberry fruit. *Scientia Horticulturae, 175*, 181–186. https://doi.org/10.1016/j.scienta.2014.06.012

Yahia, E. M. (Ed.). (2011). *Postharvest biology and technology of tropical and subtropical fruits* (p. 532). Woodhead Publishing Series in Food Science, Technology and Nutrition.

Yamaga, I., & Nakamura, S. (2018). Blue LED irradiation induces scoparone production in wounded satsuma mandarin 'Aoshima Unshu' and reduces fruit decay during long-term storage. *Horticulture Journal, 87*(4), 474–480. https://doi.org/10.2503/hortj.OKD-147

Yamaga, I., Takahashi, T., Ishii, K., Kato, M., & Kobayashi, Y. (2015). Antifungal effect of blue LED irradiation on the blue mold, *Penicillium italicum*, in satsuma mandarin fruits. *Horticultural Research, 14*(1), 83–87. https://doi.org/10.2503/hrj.14.83

Zhan, L., Hu, J., Li, Y., & Pang, L. (2012a). Combination of light exposure and low temperature in preserving quality and extending shelf life of fresh-cut broccoli (Brassica oleracea L.). *Postharvest Biology and Technology, 72*, 76–81. https://doi.org/10.1016/j.postharvbio.2012.05.001

Zhan, L., Li, Y., Hu, J., Pang, L., & Fan, H. (2012b). Browning inhibition and quality preservation of fresh-cut romaine lettuce exposed to high intensity light. *Innovative Food Science and Emerging Technologies, 14*, 70–76. https://doi.org/10.1016/j.ifset.2012.02.004

Zhang, W., & Jiang, W. (2019). UV treatment improved the quality of postharvest fruits and vegetables by inducing resistance. *Trends in Food Science and Technology, 92*, 71–80. https://doi.org/10.1016/j.tifs.2019.08.012

Zhang, W., Jiang, H., Cao, J., & Jiang, W. (2021). UV-C treatment controls brown rot in postharvest nectarine by regulating ROS metabolism and anthocyanin synthesis. *Postharvest Biology and Technology, 180*, 111613. https://doi.org/10.1016/j.postharvbio.2021.111613

Zhang, Y., Li, S., Deng, M., Gui, R., Liu, Y., Chen, X., Lin, Y., Li, M., Wang, Y., He, W., Chen, Q., Zhang, Y., Luo, Y., Wang, X., & Tang, H. (2022). Blue light combined with salicylic acid treatment maintained the postharvest quality of strawberry fruit during refrigerated storage. *Food Chemistry: X, 15*, 100384. https://doi.org/10.1016/j.fochx.2022.100384

Zhirkova, A.A., Balabanov, P.V., & Divin, A.G. (2021). Selecting lighting sources with optical-electronic control of fruit quality. *Vestnik Tambovskogo Gosudarstvennogo Tehnicheskogo Universiteta, 27*(4), 536–542. https://doi.org/10.17277/vestnik.2021.04.pp.536-542

Zhu, C., Lei, M., Andargie, M., Zeng, J., & Li, J. (2019). Antifungal activity and mechanism of action of tannic acid against *Penicillium digitatum*. *Physiological and Molecular Plant Pathology, 107*, 46–50. https://doi.org/10.1016/j.pmpp.2019.04.009

# Section III

Plant- and Animal-derived
Methods for Postharvest
Quality Preservation

# 12 Use of Chitosan in Postharvest Handling of Fruits and Vegetables

*Roghayeh Karimirad, Chunpeng (Craig) Wan, and İbrahim Kahramanoğlu*

## 12.1 INTRODUCTION

Horticultural crops are living organisms, even after being detached (harvested) from the crops, undergoing metabolic activities continually. During the process of storage and transportation, the characteristics of the products, such as nutrition, vitamins, carbohydrates, phytochemicals, flavor, and odor deteriorates due to respiration, transpiration, ethylene production, browning, rottenness and so on (Gatto et al., 2011). Therefore, the economic value of the fresh products decreases, and this causes an important economic damage to the farmers, exporter, importers, wholesalers and markets. To extend the postharvest life of products, some efficient actions including cold storage, controlled atmosphere storage, modified atmosphere packaging, coating application, fungicide application, heat treatments and irradiation have been practiced (Xanthopoulos et al., 2012; Castagna et al., 2013). In these applications, the edible treatments based on biopolymer is one of the effective and environmental techniques due to its unique features. Edible treatments could avoid water loss and flavor loss, prevent the penetration of $O_2$ gas into the product tissue (and reduce respiration) and/or prevent microorganism growth. Moreover, the edible materials are suitable for food safety standards (Mantilla et al., 2013). Nowadays, due to the increasing general concern about human health matters and ecological protection, there has been a raised interest in expanding natural biodegradable substances for the postharvest handling of fruit and vegetables. These would replace the commercial synthetic materials, mainly consisting of oxidized polyethylene (Dhall, 2013). Many substances, such as polysaccharides, proteins, lipids, essential oils, or different composites (Valencia-Chamorro et al., 2011) may be commercially applied as edible coatings (Santos et al., 2012; Vanzela et al., 2013).

Chitosan is one of the favorable biopolymers for economical application due to its biodegradability, biocompatibility and nontoxicity (Jianglian & Zhang, 2013). The U.S. Food and Drug Administration (FDA) has classified chitosan as generally recognized as safe (GRAS) (Luo & Wang, 2013). Chitosan is a biopolymer acquired from the deacetylation action of chitin from the outer skeleton of shellfish. It is the second most plentiful polysaccharide in nature after cellulose. This polysaccharide can be found in many sources, including crustaceans, mollusks, insects and fungi

DOI: 10.1201/9781003452355-15

(Jianglian & Zhang, 2013; Luo & Wang, 2013). It is well-known that the edible coatings produced from chitosan are very beneficial in maintaining the shelf life and postharvest quality of fruits and vegetables (Contreras-Oliva et al., 2012), and, in this regard, it is worth observing that chitosan demonstrated antifungal and antimicrobial characteristics. Therefore, its application as an edible coating substance aids in the prevention of microbial attack, and it might substitute for the application of synthetic fungicides (Elsabee & Abdou, 2013). This chapter attempts to brief the general application of various chitosan biopolymers based on the current situation of investigation and advances in the maintenance of fruits and vegetables. We expect that this chapter presents insights for investigators working on postharvest quality.

## 12.2　COMPOSITE COATING OF CHITOSAN

The deterioration of fruit corresponds to unfavorable physiological changes, especially loss of weight due to respiration, loss of tissue hardness and loss of resistance to microorganism damage. Such issues can happen during either storage, transportation or marketing of fresh produce, causing noticeable financial damage to producers, exporter, importers, wholesalers and markets. Edible coatings may be used on crops to improve appearance, decrease moisture loss and spoilage, delay ripening and extend postharvest life. Biopolymers such as chitosan could be the best solutions as a coating due to their film-forming features, biochemical characteristics, and natural, antimicrobial properties (Hossain & Iqbal, 2016). This polysaccharide coating protects against a broad kind of microbes, including algae, fungi and some bacteria. Chitosan had been successfully used to extend shelf life and control spoilage of many fruits. For example, biopolymer coatings applied to fresh banana fruits were reported to delay changes in total soluble solids (TSS) and in exterior color compared to control. The sample coated with 1% chitosan showed less weight loss and decreased darkening than the control sample. Researchers also reported that the occurrence and intensity of diseases were notably reduced by chitosan coating (Hossain & Iqbal, 2016). Other research has revealed that chitosan can prolong the shelf life and control the spoilage of many fruits, such as peaches, table grapes, strawberries, sweet cherries, mangos and apples (Petriccione et al., 2015; Jongsri et al., 2016; Cosme Silva et al., 2017). In another study, it was demonstrated that the chitosan application could remarkably extend the postharvest life of Santa Rosa plums. The edible coating creates a preservative barrier on the fruit surface, which preserves the firmness of treated fruits; delays water loss; causes changes in appearance, anthocyanin, titratable acidity, total antioxidant; and retards the respiration rate of fruits (Kumar et al., 2017).

Studies of Wang and Gao (2013) also reported that the intensity of spoilage in strawberries stored at either lower temperature was remarkably decreased and the postharvest life was prolonged by immersing samples in chitosan solutions (0.5, 1.0 and 1.5 g 100 mL$^{-1}$) for 5 min at 20°C. Their results revealed that chitosan could preserve high antioxidant enzyme activities. Antioxidant enzymes are effective free radical scavengers and efficiently repress reactive oxygen species in plant tissue and decrease oxidative stress caused by decay pathogens. Chitosan coatings can also prevent the increase of oxidative enzyme peroxidase and polyphenol oxidase activity.

An enhancement in antioxidant enzyme level and free radical scavenging ability, particularly at a lower temperature after 9 or 6 days of storage, would decrease the physiological change, increase the resistance of tissue against microbial attack and reduce the deterioration of fruits. Chitosan may also create a protective barrier on the sample's surface to reduce water loss due to transpiration. Therefore, the usage of chitosan coating could be a promising method in prolonging shelf life and preserving quality and in controlling the spoilage of strawberries (Wang & Gao, 2013).

Recently, the powder formulation of biopolymer has been broadly applied for the fresh fruit and vegetable industries as a postharvest treatment. However, before usage, the powder chitosan should be dissolved in a solution with an appropriate acid to be used for the postharvest handling of fresh fruits and vegetables. Before usage, the pH value of the chitosan solution must also be adjusted to a suitable pH range. These proceedings for providing chitosan solution should take multiple hours or even multiple days to complete, which is unsuitable for fresh produce posthar-vest handling, especially fresh crops treatment in the production area (Jiang et al., 2018a; Lin et al., 2019, 2020a). This problem can be resolved through the applica-tion of Kadozan (a commercial chitosan formula in liquid form). Kadozan can be diluted with aqua to acquire the different levels of Kadozan solutions without pH regulation or application of organic solvents (Jiang et al., 2018b). It was previously reported that the chitosan treatment increased the activities of reactive oxygen spe-cies scavenging enzymes (e.g., superoxide dismutase, catalase, ascorbate peroxidase and peroxidase), and the amounts of ascorbate and glutathione increased the scav-enging capacity of the DPPH radical and lowered the generation rate of oxygen and the amount of hydrogen peroxide and malondialdehyde in longan fruit pulp. These results revealed that chitosan treatment effectively delayed the breakdown of longan pulp, which was significantly related to the reduced reactive oxygen species level and membrane lipids peroxidation, and an increased integrity of cell membranes in fruit pulp. Additionally, these results also provide a suitable method for chitosan applica-tion with VKadozan: VKadozan + Water 1:500 for repressing fruit pulp breakdown and prolonging the quality of fresh longan fruits (Lin et al., 2020b).

Moreover, it is well-known that the ripe litchi fruit can quickly deteriorate and decay during storage, rendering a relatively short postharvest storability due to its susceptibility to spoilage, physiological change, browning, and microbial infections after harvest (Zhang et al., 2015). The short postharvest life and limited harvest time of litchi fruit severely limited its marketability. In previous literature, the influence of novel chitosan formulations such as Kadozana treatment on physiological features, the nutritional value, and the storage life of litchi fruit (Wuye) was studied. Kadozan treatment was reported to reduce browning and the disease index of fruit samples, to preserve a higher amount of commercially acceptable litchis and to exhibit a bet-ter storage life of fruits under ambient (25°C) temperatures. The ideal treatment for fruits is reported as 1:100 (VKadozan: VKadozan + Water) dilution, which might be a good technique for preserving the postharvest quality of fruits (Jiang et al., 2018).

The effects of various concentrations of chitosan, applied pre- or postharvest, on the storability of fresh raspberries was evaluated by Tezotto-Uliana et al. (2014). Researchers examined the 0, 0.5, 1.0 and 2.0% concentrations of chitosan. The post-harvest treatment was used instantly after harvest, as immersing the fruit samples into the solutions for a 5 min duration, and the preharvest treatment was performed by a hand-spray per week for 3 weeks' duration. In both tests, the fruit samples were stored at 0°C and 90% relative humidity. Pre- or postharvest chitosan at 1 or 2% were all effective in preserving titratable acidity and delaying respiration and ethylene biosynthesis and in preventing microbial attack and weight loss. The results showed that the application of chitosan (at 2% preharvest and 1 or 2% postharvest) were capable of preserving key quality parameters of raspberry fruits for 12 and 15 days, respectively (Tezotto-Uliana et al., 2014).

## 12.3 COMBINATION OF CHITOSAN AND ORGANIC COMPOUNDS

Chitosan is a polysaccharide that is recognized as nontoxic (Vimala, 2011), and it is mainly combined with other polymers such as poly-vinyl-alcohol, which is also biodegradable, and a nontoxic polymer. Therefore, considering that chitosan/poly-vinyl-alcohol can be safely applied for coating fresh products (Pereira et al., 2015), it is also possible to integrate chitosan composite into other compounds, namely natural agents or organic metals (Najafi et al., 2015). Using this composite as a coating for various fruit controls leads to many physiological changes such as respiration rate, and consequently prolongs storage life of fruits (Castello et al., 2010; Wang & Gao, 2013). So regarding this issue, chitosan/poly-vinyl-alcohol can be safely used for postharvest handling of fresh product (Pereira et al., 2015); for example, utilizing oxalic acid on fresh produce for anti-browning, due to its relation with the browning enzymes, polyphenol oxidase and peroxidase (Zheng et al., 2011). Lo'ay et al. (2017) reported that streak browning in bananas (cv. Williams) is related to destructive enzyme activities during storage life and that dipping fruits into chitosan/poly-vinyl-alcohol blended with oxalic acid provides anti-browning characteristics on the fruits. Researchers also presented that the chitosan/poly-vinyl-alcohol and oxalic acid (20 mM) remarkably reduce the fruit peel browning during storage life, reduce cell wall degradation and preserve the phenolic compounds from oxidation (Lo'ay et al., 2017).

Researchers also developed a carboxymethyl cellulose/chitosan bilayer coating that may be efficiently applied during the postharvest handling of fresh fruit, as this composite is stable, homogeneous and easy to prepare, and it does not need to utilize organic solvents, namely ammonia, and is biodegradable and safe. In addition, incorporating chitosan in the bilayer coating might also aid in preventing microbial attacks and decrease the requirement to apply fungicides (Poverenov et al., 2013). In a similar study, carboxymethyl cellulose/chitosan bilayer coating was reported to significantly increase skin glossiness of citrus fruits but was not very efficient in preventing weight loss. Therefore, it is still essential to test some other combinations for improving its water-vapor-barrier features, maybe by adding specific lipid ingredients (Mishra et al., 2010).

Kumari et al. (2015) investigated the impact of chitosan (2%) and salicylic acid (0.5 mM and 1.0 mM) for decreasing pericarp browning and protecting the sensory quality of litchi fruits (cv. Purbi). The fruits were treated with these applications alone or in a blend by the immersion method for 5 min. Then the fruits were dried in air and stored for 6 days at 4°C. Their data revealed that, among the tested treatments, 1.0 mM salicylic acid in a blend with 2% chitosan was more efficient than the control sample in decreasing pericarp browning (1.4-fold), spoilage loss (6.7-fold) and preserving higher anthocyanins (88%), ascorbic acid, flavonoids, phenolics (47%) and antioxidant ability (35%). Therefore, this composite can be applied to decrease pericarp browning and protect the quality of litchi fruit during storage life.

In other research, the impacts of using putrescine alone or in a blend with chitosan on preserving the quality of fresh table grapes were evaluated. The fruit samples were immersed in an aqueous solution comprising the various level of putrescine (0, 1 and 2 mM). Afterward, some fruits were coated with chitosan (1% w/v) and stored at 0–1°C and 90% RH for 60 days. Results demonstrated that the combination blend of 1 mM putrescine and 1% chitosan decreased the spoilage symptom, browning and grape cracking (Shiri et al., 2013).

## 12.4   COMBINATION OF CHITOSAN AND INORGANIC COMPOUNDS

Calcium can postpone senescence and ripening-associated processes such as susceptibility to microbial agents and nutritional loss. Calcium has also been presented to improve antioxidant ability of tissues, as well as preventing phys-iological changes at fruits during storage (Saure, 2014). Apart from this, cal-cium plays a significant role in defending against plant pathogens. This could be owing to the ability of calcium to prevent microbial pectolytic enzyme activities (Conway et al., 1994a). Even though calcium treatments express a positive influ-ence on fruit quality, extreme calcium levels could be causing adverse effects on the fruit skin, such as discoloration (Conway et al., 1994b) and unpleasant sensory flavor. Previous literature investigated the impact of a coating of calcium chloride ($CaCl_2$) blended with nanochitosan on the storage life of strawberries. Before being coated with 0.2 % nanochitosan, the fruits were immersed in vari-ous concentrations of $CaCl_2$ (1–4%). The results demonstrated that the coating of the fruits with 4% $CaCl_2$ and nanochitosan had remarkably protected fruit firmness, antioxidant levels and total phenolic content of the fruits. Both treat-ments effectively delayed the accumulation of malondialdehyde too. The sensory assessment also revealed a faint off-flavor, but no bitterness for the $CaCl_2$ (3%) and nanochitosan coated fruit after 15 days' storage at 4°C (Nguyen et al., 2020). In other research, strawberry fruits were treated with chitosan blended with cal-cium gluconate and stored at 10°C with a humidity of 70% for one week. No symptom of fungal spoilage was observed during the storage life for the samples coated with 1% chitosan +0.5% calcium gluconate or 1.5% chitosan. In opposi-tion to this, after 5 days of storage, 12.5% of the fruits coated with 1% chitosan without calcium salt were contaminated. Researchers reported that the addition of calcium to the biopolymer solution improved the nutrient value and firmness of the fruits (Munoz et al., 2008).

## 12.5   COMBINATION OF CHITOSAN BIOPOLYMER WITH ESSENTIAL OILS (EOS)

In addition to the inorganic and other organic materials, the chitosan biopolymers can also be combined with essential oils. Essential oils (EOs) are concentrated hydrophobic liquid (extracted from plants) containing volatile (easily evaporated) chemical compounds that retain the natural flavor of their source. The main methods for obtaining EOs are through distillation (via steam and/or water) or mechanical methods (i.e., cold press). Incorporation of chitosan with essential oils of cinnamaldehyde, carvacrol and trans-cinnamaldehyde were used on fresh blueberries. Results of that study showed that the chitosan coatings preserved the firmness of blueberry fruits and decreased pathogen growth on the fruit during storage (Sun et al., 2014).

EOs have almost no adverse influence on human health, and their benefit is much higher than synthetic chemical materials (Alves et al., 2017). However, as a general knowledge, it must be kept in mind that everything can be poison, even water, if consumed in high amounts. Therefore, the dose is the determinant of toxicity (Paracelsus, 1493–1541). For example, cinnamon EO is generally a yellow liquid with a special sweet odor (Shan et al., 2007). This oil is resistant to a large number of fungi and bacteria, and it is a highly effective, eco-friendly, natural antioxidant and a green preservative for food products. Nevertheless, its application in industry is restricted due to its volatility and irritating nature (Moreira et al., 2014).

Encapsulation techniques can increase solubility, reduce volatility and cover the flavors and disagreeable odors of oils. The untreated mangoes lost their nutrition and economic value after 14 days of storage, while fruit coated with 5 layers of chitosan and alginate (incorporated with cinnamon essential oil) still preserved their food and commercial value (Yin et al., 2019). In another study, the integration of *Ruta graveolens* L. EO provided excellent performance for achieving higher antimicrobial activity of biopolymer. *R. graveolens* ingredients, monoterpenes and 2-undecanone, 2-nonanone, have been identified as antifungal ingredients (Reddy & Al-Rajab, 2016; Kunicka-Styczynska & Gibka, 2010). The integration of the essential oil ingredients in the chitosan oil makes a better barrier, with antifungal, antimicrobial and antioxidant attributes, particularly against *Colletotrichum gloesporioides*, which is an important postharvest problem for guavas (Grande Tovar et al., 2019). Furthermore, chitosan in several examinations has been proved to prevent microbial decay in fresh fruits and vegetables (Elsabee & Abdou, 2013).

The potential advantages of any coatings for preserving fruit quality is highly dependent on the attributes and consistency of the forming emulsions of the coatings. The characteristics of the edible coating are related with molecular structure. Good elasticity is required in layers to inhibit cracking during storage and handling. The integration of *R. graveolens* L. EO in the chitosan biopolymer allowed for the acquisition of resistant emulsions of low viscosity and for securing adherence on the gooseberry fruit surfaces. It was observed that the integration of the EO improved the physicochemical properties of chitosan. Furthermore, microbiological analysis exhibited a reduction in the population of pathogens due to the antimicrobial and antifungal features of the ingredients. Chitosan + essential oil 1.5% presented the highest decrease in the colony forming unit (CFU) about the control. The chitosan +

*Ruta graveolens* L. oil coating applications were also observed to preserve the antioxidant capacity of gooseberries even after 12 days of treatment under storage.

Several investigations have been currently performed to encapsulate the EO for edible coating usage to reduce the odor and flavor of the coatings and to improve the acceptability of the coatings by consumers. The results of such studies suggested that the beneficial performance of chitosan as edible coatings on the physicochemical, antioxidant and microbial spoilage of *Physalis peruviana* are being significantly improved with the addition of *R. graveolens* L. EO (Gonzalez-Locarno et al., 2020).

Another important quality problem, mainly in most of the developing countries, is tomato rottenness. It is an essential challenge because it is a highly perishable fruit as a result of its high water content (Arah et al., 2015). Approximately 30% of newly harvested tomato fruits quickly deteriorate due to pathogen decays, mainly caused by fungi such as *Alternaria alternata*, *Botrytis cinerea* and *Rhizopus stolonifer* (Mohacsi-Farkas et al., 2014; Prusky, 2011). For instance, research using chitosan + lemongrass oils or thyme oil in a blend with propolis were reported to be effective in retarding the growth of *R. stolonifer* and maintaining the quality of fresh tomato fruits at 25°C storage (Athayde et al., 2016; Migliori et al., 2017). Some findings have introduced the influence of chitosan oil relying on nanoemulsions integrated with *Zatariamuti flora* and nutmeg seed oils oil in strawberries, with lemongrass oil in grape and thyme oil in avocadoes (Horison et al., 2019; Mohammadi et al., 2015a; Correa-Pacheco et al., 2017; Oh et al., 2017). In a different study, Robledo et al. (2018) reported a significantly reduction in the *B. cinerea* growth with the application of chitosan–thymol oil relying on nanoemulsion as the coating in cherry tomatoes. Regarding the severe antifungal activity of some *R. graveolens* ingredients, the effects of chitosan + EO to improve the firmness of fruits has been reported by their group to control *Colletotrichum gloesporioides* fungi decay in guavas (Grande Tovar et al., 2019), cape gooseberries for quality and microbial evaluation and papayas (Gonzalez-Locarno et al., 2020; Peralta-Ruiz et al., 2020).

Biopolymers such as chitosan combined with EOs to form edible coatings have been demonstrated to be efficient at prolonging the shelf life and storability of different horticultural products, such as sweet pepper (Xing et al., 2011) and table grapes (Sanchez-Gonzalez et al., 2011). A different study Vu et al. (2011) evaluated the antifungal effects of modified chitosan combined with peppermint oil and limonene oil on the fungal rot of strawberries during cold storage. Chitosan oil coatings did not improve remarkable changes in the physical and chemical quality of fruit during cold storage. Nevertheless, it was observed to have a little effect on the metabolic pattern of the sample in the presence of lemon oil, as reduced by some process induced in the respiration and color features. However, the combination of the chitosan with lemon oil caused an improvement in protection of the fruit against fungal rot development. According to the findings of that study, this treatment can be a replacement technique to prolong strawberry storage life. But lemon oil should be integrated at a lesser level in the film (lower than 1:3, chitosan: essential oil ratio) to keep its effect on the olfactory sensation (Perdones et al., 2012).

The results of the previous research revealed that the usage of chitosan blended from shrimp to form a composite severely prevented the mold contamination caused by *A. niger*, *B. cinerea*, *P. expansum* and *R. stolonifer* in table grapes during storage at 25°C and low temperatures. The influence of the coatings, including chitosan and *Mentha piperita* L. or *M. villosa* Huds, was the same in retarding the fungi growth and the incidence of the contaminated samples. This treatment did not adversely affect the physical and chemical or sensory features of table grapes (Conceicao Dantas Guerra et al., 2016).

Edible coatings acquired from chitosan and hydroxypropyl methylcellulose with and without bergamot oil are a suitable replacement for grape maintenance, improving fruit firmness. Chitosan coatings, including bergamot oil, generated the most efficient antimicrobial activity and indicated the greatest prevention of the respiration rates in terms of oxygen consumption and carbon dioxide production. Although the treatment did not appear to decrease the rate of grape browning, it prevented color expansion and improved the product's appearance as compared to untreated fruits (Sanchez-Gonzalez et al., 2011).

In that respect, in a previous study, the main efficiency was acquired with the application of chitosan nanoparticles as a composite integrated with *Cinnamomum zeylanicum* essential oil compared with the lone usage of essential oil for the control of *Phytophthora drechsleri* disease on cucumber (Mohammadi et al., 2015b). Other researchers have indicated that the application of essential oils and nanochitosan oil alone or in composition as a coating substance improves the firmness of fruits (Abdolahi et al., 2010; Eshghi et al., 2014).

Other research was conducted in vitro where Escherichia coli (wild-type) and *Penicillium digitatum* were cultured in favorable media and afterward subjected to 6 different EOs. Three compounds, cinnamaldehyde, carvacrol, and trans-cinnamaldehyde, possessed high antimicrobial activity and were chosen to study the shelf life of blueberries. Hereafter, 0.5% of each EO was added into a chitosan biopolymer and coated on freshly harvested blueberries. The chitosan coating considerably reduced bacteria, molds and yeasts on the fruits. Further concentrations tested indicated that the antimicrobial capacity remained even when lowering the carvacrol level to 0.1% and trans-cinnamaldehyde to 0.2%. Chitosan, carvacrol and trans-cinnamaldehyde also preserved fruit firmness (Sun et al., 2014).

Other research, relating to microbiological, physicochemical and sensory properties with a chitosan biopolymer alone or in combination with cinnamon EO in various levels, evaluated the postharvest life of slightly processed pineapples stored at 5°C for 15 days. The coatings applied in that study were effective in the preservation of pineapple in comparison to the control treatments. The combination of chitosan (2%) and cinnamon EO (0.5%) indicated the best results, 14.60% weight loss, demonstrating itself to be adequate in delaying microbial growth and extending the storage life of pineapple (Basaglia et al., 2021).

Studies of Xylia et al. (2021) also support this background information. It was reported that the effectiveness of chitosan is better when combined with *Origanum majorana* essential oil on the quality properties of fresh-cut lettuce. In the study, the chitosan was also incorporated with ascorbic acid and applied separately but was found to cause a less admissible crop (visual odor and flavor), whereas the combination of chitosan with marjoram oil was capable of maintaining the visual quality of the fruit samples and at the same time caused a pleasant aroma. The application of essential oil + ascorbic acid and Chitosan + ascorbic acid improved the total phenolics and antioxidant activity of the fruit samples on the 4 and 6 days of storage. Chitosan reduced molds and yeast counts in 6 days of storage, while essential oil, ascorbic acid, essential oil + chitosan and chitosan + ascorbic acid reduced molds and yeasts after 4 days of usage. The findings of that study indicated that the combination of ascorbic acid, essential oil and chitosan could be considered further as a replacement method for fresh-cut produce maintenance (Xylia et al., 2021).

A synergistic antifungal impact was reported when *Aloe vera* gel was integrated into chitosan oil (Vieira et al., 2016). The *A. vera* gel coating in several fruits, including mango (Shah & Hashmi, 2020), apple (Qi et al., 2011), orange (Rasouli et al., 2019) and strawberry (Sogvar et al., 2016), decreased transpiration, weight loss, microbial attack, softening and maintained other quality features that could extend the postharvest life of the fruits. Therefore, the blended impact of the coating raised resistance to pathogens and storage life in mango fruit (Shah & Hashmi, 2020; Ebrahimi & Rastegar, 2020) and in orange (Rasouli et al., 2019).

Recently, Khatri et al. (2020) also reported that *A. vera* gel, in combination with chitosan as an edible coating, helps to prolong the postharvest life of tomato fruits, together with their biochemical characteristics and antioxidative ability. Findings of Khatri et al. (2020) showed that the coating treatment provided a gradual rise in the total phenolic, as well as lycopene amounts, pectate lyase activity and a slow reduction in ascorbic acid amount and differentially induced antioxidative activities compared to the control sample. The blended *A. vera* gel and chitosan oil exhibited the best efficiency in retarding the ripening process and in prolonging the fruit storage life by up to 42 days. Similar studies have demonstrated that the blend of *A. vera* gel in combination with chitosan biopolymer as an edible coating improves the storability of bell peppers (Manoj et al., 2016) and blueberries (Vieira et al., 2016).

The study by Ehtesham Nia et al. (2021) showed the influence of preharvest foliar spraying (chitosan, 2.0% and 3.0%) and postharvest treatment (*Aloe vera* gel, 25% and 33%) to specify the quality of table grape (*Vitis vinifera*) during storage. Findings of the study exhibited that both treatments notably affected the storage life of table grapes. Furthermore, the chitosan and *A. vera* gel combinations reduced spoilage and decreased weight loss more than the other treatments. The applications remarkably decreased fruit spoilage, malondialdehyde content and polyphenol oxidase enzyme, also preserving the overall acceptance, total phenols, antioxidant capacity, anthocyanin and vitamin C content of the fruits. Based on their findings, these natural agents could be considered favorable replacements to prolong the marketable period of table grapes and keep down postharvest wastes.

## 12.6 INTEGRATED APPLICATION OF CHITOSAN BIOPOLYMER AND MODIFIED ATMOSPHERE PACKAGING

Modified atmosphere packaging (MAP) is a broadly applied technique for maintaining the storability of fresh horticultural products. MAP simply regulates the inner air composition of the packages (surrounding the fruits) and helps to reduce the respiration rate and senescence of the fruits and vegetables. But many economic polymeric films are restricted to use for atmosphere packaging applicability due to the low gas exchange of these films. Laser perforation is a new approach to supplying micropores that reduce gas permeability. Laser-induced micropore relying on modified atmosphere packaging with various pores and carbon-dot/chitosan with a CD level of 4.5% for the maintenance of fresh-cut cucumber was evaluated. According to their report, a micropore-based modified atmosphere (100 μm) can efficiently regulate the gas composition of samples during storage. Micropore relying on modified atmosphere packaging (100 μm) with four micropores and carbon-dot/chitosan efficiently preserved the lowest level of weight loss (4.1%), firmness (6.6 N), malondialdehyde concentration (2.94 μmol kg$^{-1}$), and prevented the degradation of flavor to the 15th day of storage. So micropore relying on modified atmosphere packaging, together with chitosan carbon-dot, can be applied as an efficient way for the maintenance of fresh-cut cucumber (Fan et al., 2021).

A previous study also assessed the influences of chitosan and modified atmosphere packaging on the postharvest quality and sensory index of table grapes (cv. Italia) during cold storage. The results indicated an essential effect of a high $CO_2$-modified atmosphere with chitosan alone on maintaining quality and sensory properties and retarding rot of table grape. The most efficient treatment in terms of the maintenance of quality, sensory and nutritional value was chitosan and high-$CO_2$-modified atmosphere packaging (Liguori et al., 2021).

The objective of the other research is to determine the influence of chitosan oil and modified atmosphere packaging on the postharvest quality and bioactive composition of pomegranate (cv. Hicaznar) fruit. Samples were subjected to chitosan treatment (0 or 1%) and packaged with or without modified atmosphere packaging bags. Chitosan, modified atmosphere packaging and chitosan + MAP treatments provided better husk color, titratable acidity and ascorbic acid content in comparison with the control treatment. During cold storage, chitosan alone was the most efficient treatment to control fungal rot and its uninterrupted effect during the shelf life period. The arils of chitosan-treated fruit were deep red and possessed the highest antioxidant capacity and total phenolic content. It was reported that control and chitosan-treated samples became unsalable after 6 months of cold storage and shelf life period, while modified atmosphere packaging and chitosan+ modified atmosphere packaging treated fruits were still marketable at the same time. Therefore, the consequence of that study is that the chitosan + MAP treatment is effective for controlling husk scald, rot and weight loss of fruit samples with preserving visual quality and primary red aril color severity for 6 months of cold storage (Candir et al., 2018).

The result of another work revealed that chitosan + limonene EOl and MAP could extend the postharvest quality of cucumber by preserving fruit quality features such as firmness, weight loss, color, pH and sensory characteristics and also decrease fruit rot by suppressing fungal growth. Chitosan and MAP function as semi-penetrable barriers to reduce water loss and respiration rate (Maleki et al., 2018). The clear consequence of these studies is that the combination of chitosan with MAP is an important practice for improving the postharvest storability of fruits and vegetables.

## 12.7   CONCLUSION

Chitosan, as an animal-derived, natural biodegradable compound has been proven to control several postharvest pathogens, delay senescence and maintain several quality parameters of fruits and vegetables. Studies revealed that chitosan has both a direct impact on the pathogens and an indirect impact by inducing the enzymatic defense system (polyphenol oxidase, glucanohydrolases, phenolics, specific phytoalexins, etc.) of the products. Chitosan also provides a barrier against gaseous and water vapor, reducing the respiration and transpiration of the fresh products. Studies also suggested that the combination of chitosan with organic or inorganic materials improves its efficacy. The combination of modified atmosphere packaging with chitosan is a promising method for the future of postharvest handling. Nowadays, the trend in the world is to control postharvest pathogens and reduce losses by avoiding the negative impacts of some synthetic tools on the environment and human health. Given that, chitosan is an important and safe tool for humanity.

## REFERENCES LIST

Abdolahi, A., Hassani, A., Ghosta, Y., Bernousi, I., & Meshkatalsadat, M. H. (2010). Study on the potential use of essential oils for decay control and quality preservation of Tabarzeh table grape. *Journal of Plant Protection Research*, *50*(1), 45–52. https://doi.org/10.2478/v10045-010-0008-2

Alves, M. M., Gonçalves, M. P., & Rocha, C. M. R. (2017). Effect of ferulic acid on the performance of soy protein isolate-based edible coatings applied to fresh-cut apples. *LWT*, *80*, 409–415. https://doi.org/10.1016/j.lwt.2017.03.013

Arah, I. K., Kumah, E. K., Anku, E. K., & Amaglo, H. (2015). An overview of postharvest losses in tomato production in Africa: Causes and possible prevention strategies. *Journal of Biology, Agriculture and Healthcare*, *5*, 78–88.

Athayde, A. J. A. A., de Oliveira, P. D. L., Guerra, I. C. D., da Conceição, M. L., de Lima, M. A. B., Arcanjo, N. M. O., Madruga, M. S., Berger, L. R. R., & de Souza, E. L. (2016). A coating composed of chitosan and *Cymbopogon citratus* (Dc. Ex Nees) essential oil to control Rhizopus soft rot and quality in tomato fruit stored at room temperature. *Journal of Horticultural Science and Biotechnology*, *91*(6), 582–591. https://doi.org/10.1080/14620316.2016.1193428

Basaglia, R. R., Pizato, S., Santiago, N. G., Maciel de Almeida, M. M., Pinedo, R. A., & Cortez-Vega, W. R. (2021). Effect of edible chitosan and cinnamon essential oil coatings on the shelf life of minimally processed pineapple (Smooth cayenne). *Food Bioscience*, *41*, 100966. https://doi.org/10.1016/j.fbio.2021.100966

Candir, E., Ozdemir, A. E., & Aksoy, M. C. (2018). Effects of chitosan coating and modified atmosphere packaging on postharvest quality and bioactive compounds of pomegranate fruit cv. "Hicaznar". *Scientia Horticulturae*, *235*(17), 235–243. https://doi.org/10.1016/j.scienta.2018.03.017

Castagna, A., Chiavaro, E., Dall'Asta, C., Rinaldi, M., Galaverna, G., & Ranieri, A. (2013). Effect of postharvest UV-B irradiation on nutraceutic quality and physical properties of tomato fruits. *Food Chemistry*, *137*(1–4), 151–158. https://doi.org/10.1016/j.foodchem.2012.09.095. https://doi:10.1016/j.foodchem.2012.09.095

Castelló, M. L., Fito, P. J., & Chiralt, A. (2010). Changes in respiration rate and physical properties of strawberries due to osmotic dehydration and storage. *Journal of Food Engineering*, *97*(1), 64–71. https://doi.org/10.1016/j.jfoodeng.2009.09.016

Guerra, I. C. D., de Oliveira, P. D. L., Santos, M. M. F., Lúcio, A. S. S. C., Tavares, J. F., Barbosa-Filho, J. M., Madruga, M. S., & de Souza, E. L. (2016). The effects of composite coatings containing chitosan and *Mentha* (*piperita* L. or x *Villosa* Huds) essential oil on postharvest mold occurrence and quality of table grape cv. Isabella. *Innovative Food Science and Emerging Technologies*, *34*, 112–121. https://doi.org/10.1016/j.ifset.2016.01.008

Contreras-Oliva, A., Rojas-Argudo, C., & Pérez-Gago, M. B. (2012). Effect of solid content composition of hydroxypropyl methylcellulose-lipid edible coatings on physicochemical and nutritional quality of "Oronules" mandarins. *Journal of the Science of Food and Agriculture, 92*(4), 794–802. https://doi.org/10.1002/jsfa.4649

Conway, W. S., Sams, C. E., & Kelman, A. (1994a). Enhancing the natural resistance of plant tissues to postharvest diseases through calcium applications. *HortScience, 29*(7), 751–754. https://doi.org/10.21273/HORTSCI.29.7.751

Conway, W. S., Sams, C. E., Wang, C. Y., & Abbott, J. A. (1994b). Additive effects of postharvest calcium and heat treatment on reducing decay and maintaining quality in apples. *Journal of the American Society for Horticultural Science, 119*(1), 49–53. https://doi.org/10.21273/JASHS.119.1.49

Correa-Pacheco, Z. N., Bautista-Banos, S., & Valle-Marquina, M. A. (2017). The effect of nanostructured chitosan and chitosan-thyme essential oil coatings on *Colletotrichum gloeosporioides* growth in vitro and on cv Hass avocado and fruit quality. *Journal of Phytopathology, 65*, 297–305. https://doi.org/10.1111/jph.12562

Cosme Silva, G. M., Silva, W. B., Medeiros, D. B., Salvador, A. R., Cordeiro, M. H. M., da Silva, N. M., Santana, D. B., & Mizobutsi, G. P. (2017). The chitosan affects severely the carbon metabolism in mango (*Mangifera indica* L. cv. Palmer) fruit during storage. *Food Chemistry, 237*, 372–378. https://doi.org/10.1016/j.foodchem.2017.05.123

Dhall, R. K. (2013). Advances in edible coatings for fresh fruits and vegetables: A review. *Critical Reviews in Food Science and Nutrition, 53*(5), 435–450. https://doi.org/10.1080/10408398.2010.541568

Ebrahimi, F., & Rastegar, S. (2020). Preservation of mango fruit with guar-based edible coatings enriched with *Spirulina platensis* and *Aloe vera* extract during storage at ambient temperature. *Scientia Horticulturae, 265*, 109258. https://doi.org/10.1016/j.scienta.2020.109258

Ehtesham Nia, A., Taghipour, S., & Siahmansour, S. (2021). Pre-harvest application of chitosan and postharvest *Aloe vera* gel coating enhances quality of table grape (*Vitis vinifera* L. cv. 'Yaghouti') during postharvest period. *Food Chemistry, 347*, 129012. https://doi.org/10.1016/j.foodchem.2021.129012

Elsabee, M. Z., & Abdou, E. S. (2013). Chitosan based edible films and coatings: A review. *Materials Science and Engineering. C, Materials for Biological Applications, 33*(4), 1819–1841. https://doi.org/10.1016/j.msec.2013.01.010

Eshghi, S., Hashemi, M., Mohammadi, A., Badii, F., Mohammadhoseini, Z., & Ahmadi, K. (2014). Effect of nanochitosan-based coating with and without copper loaded on physicochemical and bioactive components of fresh strawberry fruit (*Fragaria ananassa* Duchesne) during storage. *Food and Bioprocess Technology, 7*(8), 2397–2409. https://doi.org/10.1007/s11947-014-1281-2

Fan, K., Zhang, M., Guo, Ch., Dan, W., & Devahastin, S. (2021). Laser-induced microporous modified atmosphere packaging and chitosan Carbon-Dot coating as a novel combined preservation method for fresh-cut cucumber. *Food and Bioprocess Technology, 14*(5), 968–983. https://doi.org/10.1007/s11947-021-02617-y

Gatto, M. A., Ippolito, A., Linsalata, V., Cascarano, N. A., Nigro, F., Vanadia, S., & Di Venere, D. (2011). Activity of extracts from wild edible herbs against postharvest fungal diseases of fruit and vegetables. *Postharvest Biology and Technology, 61*(1), 72–82. https://doi.org/10.1016/j.postharvbio.2011.02.005

Gonzalez-Locarno, M., Maza Pautt, Y., Albis, A., Florez López, E., & Grande Tovar, C. D. (2020). Assessment of chitosan-rue (*Ruta graveolens* L.) essential oil-based coatings on refrigerated cape gooseberry (*Physalis peruviana* L.) quality. *Applied Sciences, 10*(8), 2684. https://doi.org/10.3390/app10082684

Grande Tovar, C. D., Delgado-Ospina, J., Navia Porras, D. P., Peralta-Ruiz, Y., Cordero, A. P., Castro, J. I., Chaur Valencia, M. N., Mina, J. H., & Chaves López, C. (2019). *Colletotrichum gloesporioides* inhibition in situ by chitosan-*Ruta graveolens* essential oil coatings: Effect on microbiological, physicochemical, and organoleptic properties of Guava (Psidium guajava L.) during room temperature storage. *Biomolecules, 9*(9), 399. https://doi.org/10.3390/biom9090399

Horison, R., Sulaiman, F. O., & Alfredo, D. (2019). Physical characteristics of nanoemulsion from chitosan/nutmeg seed oil and evaluation of its coating against microbial growth on strawberry. *Food Research, 3*, 821–827. https://doi.org/10.26656/fr.2017.3(6).159

Hossain, M. S., & Iqbal, A. (2016). Effect of shrimp chitosan coating on postharvest quality of banana (*Musa sapientum* L.) fruits. *International Food Research Journal, 23*, 277–283.

Jiang, X., Lin, H., Lin, M., Chen, Y., Wang, H., Lin, Y., Shi, J., & Lin, Y. (2018a). A novel chitosan formulation treatment induces disease resistance of harvested litchi fruit to *Peronophythora litchii* in association with ROS metabolism. *Food Chemistry, 266*, 299–308. https://doi.org/10.1016/j.foodchem.2018.06.010

Jiang, X., Lin, H., Shi, J., Neethirajan, S., Lin, Y., Chen, Y., Wang, H., & Lin, Y. (2018b). Effects of a novel chitosan formulation treatment on quality attributes and storage behavior of harvested litchi fruit. *Food Chemistry, 252*, 134–141. https://doi.org/10.1016/j.foodchem.2018.01.095

Jianglian, D., & Zhang, S. (2013). Application of chitosan based coating in fruit and vegetable preservation: A review. *Journal of Food Processing and Technology, 4*, 5. https://doi.org/10.4172/2157-7110.1000227

Jongsri, P., Wangsomboondee, T., Rojsitthisak, P., & Seraypheap, K. (2016). Effect of molecular weights of chitosan coating on postharvest quality and physicochemical characteristics of mango fruit. *LWT, 73*, 28–36. https://doi.org/10.1016/j.lwt.2016.05.038

Khatri, D., Panigrahi, J., Prajapati, A., & Bariya, H. (2020). Attributes of Aloe vera gel and chitosan treatments on the quality and biochemical traits of postharvest tomatoes. *Scientia Horticulturae, 259*, 108837. https://doi.org/10.1016/j.scienta.2019.108837

Kumar, P., Sethi, S., Sharma, R. R., Srivastav, M., & Varghese, E. (2017). Effect of chitosan coating on postharvest life and quality of plum during storage at low temperature. *Scientia Horticulturae, 226*, 104–109. https://doi.org/10.1016/j.scienta.2017.08.037

Kumari, P., Barman, K., Patel, V. B., Siddiqui, M. W., & Kole, B. (2015). Reducing postharvest pericarp browning and preserving health promoting compounds of litchi fruit by combination treatment of salicylic acid and chitosan. *Scientia Horticulturae, 197*, 555–563. https://doi.org/10.1016/j.scienta.2015.10.017

Kunicka-Styczynska, A., & Gibka, J. (2010). Antimicrobial Activity of Undecan-x-ones (x = 2–4). *Polish Journal of Microbiology, 59*(4), 301–306. https://doi.org/10.33073/pjm-2010-045

Liguori, G., Sortino, G., Gullo, G., & Inglese, P. (2021). Effects of modified atmosphere packaging and chitosan treatment on quality and sensorial parameters of minimally processed cv. "Italia" Table Grapes. *Agronomy, 11*(2), 328. https://doi.org/10.3390/agronomy11020328

Lin, Y., Chen, G., Lin, H., Lin, M., Wang, H., & Lin, Y. (2020a). Chitosan postharvest treatment suppresses the pulp breakdown development of longan fruit through regulating ROS metabolism. *International Journal of Biological Macromolecules, 165*(A), 601–608. https://doi.org/10.1016/j.ijbiomac.2020.09.194

Lin, Y., Li, N., Lin, H., Lin, M., Chen, Y., Wang, H., Ritenour, M. A., & Lin, Y. (2020b). Effects of chitosan treatment on the storability and quality properties of longan fruit during storage. *Food Chemistry, 306*, 125627. https://doi.org/10.1016/j.foodchem.2019.125627

Lin, Y., Lin, Y., Lin, Y., Lin, M., Chen, Y., Wang, H., & Lin, H. (2019). A novel chitosan alleviates pulp breakdown of harvested longan fruit by suppressing disassembly of cell wall polysaccharides. *Carbohydrate Polymers, 217*, 126–134. https://doi.org/10.1016/j.carbpol.2019.04.053

Lo'ay, A. A., & Dawood, H. D. (2017). Minimize browning incidence of banana by postharvest active chitosan/PVA Combines with oxalic acid treatment to during shelf-life. *Scientia Horticulturae, 226*, 208–215. https://doi.org/10.1016/j.scienta.2017.08.046

Luo, Y., & Wang, Q. (2013). Recent advances of chitosan and its derivatives for novel applications in food science. *Journal of Food Processing & Beverages, 1*, 13.

Maleki, G., Sedaghat, N., Woltering, E. J., Farhoodi, M., & Mohebbi, M. (2018). Chitosan-limonene coating in combination with modified atmosphere packaging preserve postharvest quality of cucumber during storage. *Journal of Food Measurement and Characterization, 12*(3), 1610–1621. https://doi.org/10.1007/s11694-018-9776-6

Manoj, H. G., Sreenivas, K. N., Shankarappa, T. H., & Krishna, H. C. (2016). Studies on chitosan and *Aloe vera* gel coatings on biochemical parameters and microbial population of bell Pepper (*Capsicum annuum* L.) under ambient condition. *International Journal of Current Microbiology and Applied Sciences, 5*(1), 399–405. https://doi.org/10.20546/ijcmas.2016.501.039

Mantilla, N., Castell-Perez, M. E., Gomes, C., & Moreira, R. G. (2013). Multilayered antimicrobial edible coating and its effect on quality and shelf-life of fresh-cut pineapple (*Ananas comosus*). *LWT – Food Science and Technology, 51*(1), 37–43. https://doi.org/10.1016/j.lwt.2012.10.010

Migliori, C. A., Salvati, L., Di Cesare, L. F., Lo Scalzo, R., & Parisi, M. (2017). Effects of preharvest applications of natural antimicrobial products on tomato fruit decay and quality during long-term storage. *Scientia Horticulturae, 222*, 193–202. https://doi.org/10.1016/j.scienta.2017.04.030

Mishra, B., Khatkar, B. S., Garg, M. K., & Wilson, L. A. (2010). Permeability of edible coatings. *Journal of Food Science and Technology, 47*(1), 109–113. https://doi.org/10.1007/s13197-010-0003-7

Mohacsi-Farkas, C., Nyirő-Fekete, B., Daood, H., Dalmadi, I., & Kiskó, G. (2014). Improving microbiological safety and maintaining sensory and nutritional quality of pre-cut tomato and carrot by gamma irradiation. *Radiation Physics and Chemistry, 99*, 79–85. https://doi.org/10.1016/j.radphyschem.2014.02.019

Mohammadi, A., Hashemi, M., & Hosseini, S. M. (2015b). Chitosan nanoparticles loaded with Cinnamomum zeylanicum essential oil enhance the shelf life of cucumber during cold storage. *Postharvest Biology and Technology, 110*, 203–213. https://doi.org/10.1016/j.postharvbio.2015.08.019

Mohammadi, A., Hashemi, M., & Hosseini, S. M. (2015a). Nanoencapsulation of *Zataria multiflora* essential oil preparation and characterization with enhanced antifungal activity for controlling *Botrytis cinerea*, the causal agent of gray mould disease. *Innovative Food Science and Emerging Technologies, 28*, 73–80. https://doi.org/10.1016/j.ifset.2014.12.011

Moreira, S. P., De Carvalho, W. M., Alexandrino, A. C., de Paula, H. C. B., Rodrigues, M. C. P., de Figueiredo, R. W., Maia, G. A., de Figueiredo, E. M. A. T., & Brasil, I. M. (2014). Freshness retention of minimally processed melon using different packages and multilayered edible coating containing microencapsulated essential oil. *International Journal of Food Science and Technology, 49*(10), 2192–2203. https://doi.org/10.1111/ijfs.12535

Muñoz, P. H., Almenar, E., Valle, V. D., Velez, D., & Gavara, R. (2008). Effect of chitosan coating combined with postharvest calcium treatment on strawberry (*Fragaria × ananassa*) quality during refrigerated storage. *Food Chemistry, 110*(2), 428–435. https://doi.org/10.1016/j.foodchem.2008.02.020

Najafi-Taher, R., Derakhshan, M. A., Faridi-Majidi, R., & Amani, A. (2015). Preparation of an ascorbic acid/PVA-chitosan electrospun mat: A core/shell transdermal delivery system. *RSC Advances, 5*(62), 50462–50469. https://doi.org/10.1039/C5RA03813H

Nguyen, V. T. B., Nguyen, D. H. H., & Nguyen, H. V. H. (2020). Combination effects of calcium chloride and Nano-chitosan on the postharvest quality of strawberry (*Fragaria* x *ananassa* Duch.). *Postharvest Biology and Technology, 162*, 111103. https://doi.org/10.1016/j. postharvbio.2019.111103

Oh, Y. A., Oh, Y. J., Song, A. Y., Won, J. S., Song, K. B., & Min, S. C. (2017). Comparison of effectiveness of edible coatings using emulsions containing lemongrass oil of different size droplets on grape berry safety and preservation. *LWT, 75*, 742–750. https://doi. org/10.1016/j.lwt.2016.10.033

Peralta-Ruiz, Y., Grande Tovar, C. G., Sinning-Mangonez, A., Bermont, D., Pérez Cordero, A., Paparella, A., & Chaves-López, C. (2020). Colletotrichum gloesporioides inhibition using chitosan-Ruta graveolens L. essential oil coatings: Studies in vitro and in situ on Carica papaya fruit. *International Journal of Food Microbiology, 326*, 108649. https:// doi.org/10.1016/j.ijfoodmicro.2020.108649

Perdones, A., Sánchez-González, L., Chiralt, A., & Vargas, M. (2012). Effect of chitosan–lemon essential oil coatings on storage-keeping quality of Strawberry. *Postharvest Biology and Technology, 70*, 32–41. https://doi.org/10.1016/j.postharvbio.2012.04.002

Pereira, V. A., de Arruda, I. N. Q., & Stefani, R. (2015). Active chitosan/PVA films with anthocyanins from *Brassica oleraceae* (Red Cabbage) as Time-temperature Indicators for application in intelligent food packaging. *Food Hydrocolloids, 43*, 180–188. https://doi. org/10.1016/j.foodhyd.2014.05.014

Petriccione, M., De Sanctis, F., Pasquariello, M. S., Mastrobuoni, F., Rega, P., Scortichini, M., & Mencarelli, F. (2015). The effect of chitosan coating on the quality and nutraceutical traits of sweet cherry during postharvest life. *Food and Bioprocess Technology, 8*(2), 394–408. https://doi.org/10.1007/s11947-014-1411-x

Poverenov, E., Danino, S., Horev, B., & Granit, R. (2013). Layer-by-layer electrostatic deposition of edible coating on fresh cut fruit model: Anticipated and unexpected effects of alginate–chitosan combination. *Food and Bioprocess Technology, 7*. https://doi. org/10.1007/s11947-013-1134-4

Prusky, D. (2011). Reduction of the incidence of postharvest quality losses, and future prospects. *Food Security, 3*(4), 463–474. https://doi.org/10.1007/s12571-011-0147-y

Qi, H., Hu, W., Jiang, A., Tian, M., & Li, Y. (2011). Extending shelf-life of Fresh-cut "Fuji" apples with chitosan-coatings. *Innovative Food Science and Emerging Technologies, 12*(1), 62–66. https://doi.org/10.1016/j.ifset.2010.11.001

Rasouli, M., Koushesh Saba, M., & Ramezanian, A. (2019). Inhibitory effect of salicylic acid and Aloe vera gel edible coating on microbial load and chilling injury of orange fruit. *Scientia Horticulturae, 247*, 27–34. https://doi.org/10.1016/j.scienta.2018.12.004

Reddy, D. N., & Al-Rajab, A. J. (2016). Chemical composition, antibacterial and antifungal activities of *Ruta graveolens L.* volatile oils. *Cogent Chemistry, 2*(1), 1220055. https:// doi.org/10.1080/23312009.2016.1220055

Robledo, N., Vera, P., López, L., Yazdani-Pedram, M., Tapia, C., & Abugoch, L. (2018). Thymol nanoemulsions incorporated in quinoa protein/chitosan edible films; antifungal effect in cherry tomatoes. *Food Chemistry, 246*, 211–219. https://doi.org/10.1016/j. foodchem.2017.11.032

Sanchez-Gonzalez, L., Pastor, C., Vargas, M., Chiralt, A., González-Martínez, C., & Cháfer, M. (2011). Effect of hydroxyl propyl methylcellulose and chitosan coatings with and without bergamot essential oil on quality and safety of cold-stored grapes. *Postharvest Biology and Technology, 60*(1), 57–63. https://doi.org/10.1016/j. postharvbio.2010.11.004

Santos, N. S. T., Athayde Aguiar, A. J., de Oliveira, C. E., Veríssimo de Sales, C., de Melo E Silva, S., Sousa da Silva, R., Stamford, T. C., & de Souza, E. L. (2012). Efficacy of the application of a coating composed of chitosan and *Origanum vulgare* L. essential oil to control *Rhizopus stolonifer* and Aspergillus niger in grapes (*Vitis labrusca* L.). *Food Microbiology, 32*(2), 345–353. https://doi.org/10.1016/j.fm.2012.07.014

Saure, M. C. (2014). Why calcium deficiency is not the cause of blossom-end rot in tomato and pepper fruit–a reappraisal. *Scientia Horticulturae, 174*, 151–154. https://doi.org/10.1016/j.scienta.2014.05.020

Shah, S., & Hashmi, M. S. (2020). Chitosan-Aloe vera gel coating delays postharvest decay of mango fruit. *Horticulture, Environment, and Biotechnology, 61*(2), 279–289. https://doi.org/10.1007/s13580-019-00224-7

Shan, B., Cai, Y. Z., Brooks, J. D., & Corke, H. (2007). Antibacterial properties and major bioactive components of Cinnamon Stick (*Cinnamomum burmannii*): Activity against foodborne pathogenic bacteria. *Journal of Agricultural and Food Chemistry, 55*(14), 5484–5490. https://doi.org/10.1021/jf070424d

Shiri, M. A., Ghasemnezhad, M., Bakhshi, D., & Sarikhani, H. (2013). Effect of postharvest putrescine application and chitosan coating on maintaining quality of table grape cv. "shahroudi" during long-term storage. *Journal of Food Processing and Preservation, 37*(5), 999–1007. https://doi.org/10.1111/j.1745-4549.2012.00735.x

Sogvar, O. B., Koushesh Saba, M., & Emamifar, A. (2016). *Aloe vera* and ascorbic acid coatings maintain postharvest quality and reduce microbial load of strawberry fruit. *Postharvest Biology and Technology, 114*, 29–35. https://doi.org/10.1016/j.postharvbio.2015.11.019

Sun, X., Narciso, J., Wang, Z., Ference, Ch., Bai, J., & Zhou, K. (2014). Effects of chitosan-essential oil coatings on safety and quality of fresh blueberries. *Journal of Food Science, 79*(5), M955–M960. https://doi.org/10.1111/1750-3841.12447

Tezotto-Uliana, J. V., Fargoni, G. P., Geerdink, G. M., & Kluge, R. A. (2014). Chitosan applications pre- or postharvest prolong raspberry shelf-life quality. *Postharvest Biology and Technology, 91*, 72–77. https://doi.org/10.1016/j.postharvbio.2013.12.023

Valencia-Chamorro, S. A., Palou, L., del Río, M. Á., & Pérez-Gago, M. B. (2011). Performance of hydroxypropyl methylcellulose (HPMC)-lipid edible coatings with antifungal food additives during cold storage of "Clemenules" mandarins. *LWT – Food Science and Technology, 44*(10), 2342–2348. https://doi.org/10.1016/j.lwt.2011.02.014

Vanzela, E. S. L., do Nascimento, P., Fontes, E. A. F., Mauro, M. A., & Kimura, M. (2013). Edible coatings from native and modified starches retain carotenoids in pumpkin during drying. *LWT – Food Science and Technology, 50*(2), 420–425. https://doi.org/10.1016/j.lwt.2012.09.003

Vieira, J. M., Flores-López, M. L., de Rodríguez, D. J., Sousa, M. C., Vicente, A. A., & Martins, J. T. (2016). Effect of chitosan-*Aloe vera* coating on postharvest quality of blueberry (*Vaccinium corymbosum*) fruit. *Postharvest Biology and Technology, 116*, 88–97. https://doi.org/10.1016/j.postharvbio.2016.01.011

Vimala, K. (2011). Fabrication of curcumin encapsulated chitosan-PVA silver nano-composite films for improved antimicrobial activity. *Journal of Biomaterials and Nanobiotechnology, 2*, 55–64. https://doi.org/10.4236/jbnb.2011.21008

Vu, K. D., Hollingsworth, R. G., Leroux, E., Salmieri, S., & Lacroix, M. (2011). Development of edible bioactive coating based on modified chitosan for increasing the shelf life of strawberries. *Food Research International, 44*(1), 198–203. https://doi.org/10.1016/j.foodres.2010.10.037

Wang, S. Y., & Gao, H. (2013). Effect of chitosan-based edible coating on antioxidants, antioxidant enzyme system, and postharvest fruit quality of strawberries (*Fragaria × ananassa* Duch.). *LWT – Food Science and Technology, 52*(2), 71–79. https://doi.org/10.1016/j.lwt.2012.05.003

Xanthopoulos, G., Koronaki, E. D., & Boudouvis, A. G. (2012). Mass transport analysis in perforation-mediated modified atmosphere packaging of strawberries. *Journal of Food Engineering*, *111*(2), 326–335. https://doi.org/10.1016/j.jfoodeng.2012.02.016

Xing, Y., Li, X., Xu, Q., Yun, J., Lu, Y., & Tang, Y. (2011). Effects of chitosan coating enriched with cinnamon oil on qualitative properties of sweet pepper (*Capsicum annuum* L.). *Food Chemistry*, *124*(4), 1443–1450. https://doi.org/10.1016/j.foodchem.2010.07.105

Xylia, P., Chrysargyris, A., & Tzortzakis, N. (2021). The combined and single effect of Marjoram essential oil, ascorbic acid, and chitosan on fresh-cut Lettuce preservation. *Foods*, *10*(3), 575. https://doi.org/10.3390/foods10030575

Yin, Ch., Huang, Ch., Wang, J., Liu, Y., Lu, P., & Huang, L. (2019). Effect of chitosan- and alginate-based coatings enriched with Cinnamon essential oil microcapsules to improve the postharvest quality of mangoes. *Materials*, *12*(13), 2039. https://doi.org/10.3390/ma12132039

Zhang, Z. K., Huber, D. J., Qu, H., Yun, Z., Wang, H., Huang, Z., Huang, H., & Jiang, Y. (2015). Enzymatic browning and antioxidant activities in harvested litchi fruit as influenced by apple polyphenols. *Food Chemistry*, *171*, 191–199. https://doi.org/10.1016/j.foodchem.2014.09.001

Zheng, X., Ye, L., Jiang, T., Jing, G., & Li, J. (2012). Limiting the deterioration of mango fruit during storage at room temperature by oxalate treatment. *Food Chemistry*, *130*(2), 279–285. https://doi.org/10.1016/j.foodchem.2011.07.035

# 13 Plant-Based Fixed Oil in Edible Coating Formulation for Postharvest Preservation of Horticultural Crops

*Bonga Lewis Ngcobo and Olaniyi Amos Fawole*

## 13.1 INTRODUCTION

The horticultural sector is currently under pressure to increase the production of fruits and vegetables to meet their fast-growing demands and to maintain their postharvest life. Food preservation is one of the major problems affecting the horticultural sector worldwide, and this contributes significantly to the postharvest losses of fruits and vegetables (Singh & Packirisamy, 2022). On the other side, there has been a rapid increase in the global demand for horticultural crops due to consumers' awareness of the health benefits of consuming fruit and vegetables (Galanakis et al., 2020).

Due to their highly perishable nature, in addition to cold storage technology, fruits and vegetables require other postharvest treatments to control decay and delay ripening and senescence. One of these postharvest treatments is edible coating applications. Edible coatings form a thin layer of edible substance applied on the surface of fresh produce, thereby providing it with protection (Riva et al., 2020). The application of edible coatings has been a reliable approach to improving the quality and extending the shelf life of fresh fruits and vegetables (Dhall, 2013; Ncama et al., 2018; Yousuf et al., 2018; Ngcobo & Bertling, 2021; Anjum et al., 2020; Fawole et al., 2020; Kawhena et al., 2020; Kumar et al., 2021). Edible coatings have the ability to reduce moisture loss (transpiration), gaseous exchange, respiration and oxidative reaction rates (Shinga & Fawole, 2023). In addition, edible coatings have a great potential to act as carriers of antimicrobial compounds such as essential oils and even microbial antagonists that can reduce the establishment of pathogens on the surfaces of fruits and vegetables and ultimately extend their shelf life (Riva et al., 2020; Blancas-Benitez et al., 2022; Kawhena et al., 2022).

According to Yousuf et al. (2021), novel approaches to formulating edible coatings help overcome various problems related to the effectiveness of coatings. Lipids, such as oils, are often incorporated into polysaccharide-based edible coatings to increase their hydrophobicity (Riva et al., 2020). A detailed review by Atarés and Chiralt

DOI: 10.1201/9781003452355-16

(2016) described the potential of essential oils as additives in films and coatings in food packaging. A few years later, Yousuf et al. (2021) compiled a comprehensive review article highlighting critical analysis to recapitulate the effects of using essential oil incorporated into coatings on fresh/fresh-cut fruits and vegetables. To our knowledge, there is no review on the impact of plant-based fixed oil on the potential performance or efficacy of edible coatings for postharvest preservation of horticultural crops. This review, therefore, raises an important awareness on the potential contribution of plant-based fixed oils in edible coatings in preserving horticultural crops. The unique properties of plant-based fixed oil, their composition and use in edible coatings and potential benefits in food preservation were discussed.

## 13.2   PLANT-BASED FIXED OIL: OVERVIEW

Plant-based fixed oils are nonvolatile at room temperature and are generally soluble in organic solvents. They are extracted or derived from plant species or parts and are generally a mixture of esters of fatty acids (Rosa et al., 2012, 2019; Kumar et al., 2016). Fixed oils (FO), also referred to as carrier oils, are pressed from the fatty portions of plant parts, such as the seeds, pulp, leaves, stems and flowers. The most popular FO include coconut, olive, argan and jojoba oil. The FO extracted from plant parts contains different types and concentrations of fatty acids; some oils contain either saturated, unsaturated or monosaturated fatty acids. Even though there is not enough data on the separation and quantification of fatty acids in FO from different plant species, it has been observed that gas chromatography–mass spectrometry (GC-MS) and high-performance liquid chromatography (HPLC) are the most widely used form of analysis to perform this duty. The separation methods of fatty acids in fixed oils vary. For instance, Elagbar et al. (2016) extracted and separated FO from berries of *Laurus nobilis* by supercritical $CO^2$, and the qualitative and quantitative information on the individual fatty acids of FO in *L. nobilis* berries was obtained by GC analysis. The authors demonstrated that the most represented fatty acids of FO were 12:0 (27.6%), 18:1 *n*-9 (27.1%), 18:2 *n*-6 (21.4%) and 16:0 (17.1%), the same study further detected the following unsaturated fatty acids by HPLC: 201.8 µg mg$^{-1}$ of 18:1 n-9, 127.5 µg mg$^{-1}$ of 18:2 *n*-6, and a minor amount of 18:3 *n*-3 (5.5 µg mg$^{-1}$) and 16:1 *n*-7. Results within the range of fatty acids composition were also reported by Castilho et al. (2005) on *L. nobilis*. Nickavar et al. (2003) also determined the total fatty acid composition of *Nigella sativa* seeds by GC and GC/MS, and the main fatty acids of the FO were linoleic acid (55.6%), oleic acid (23.4%) and palmitic acid (12.5%). Similar results were obtained by Marzouki et al. (2008), who determined fatty acid composition in the *Annona muricata* FO using organic solvent extraction and GC-FID. Further studies by de Morais et al. (2017) and Oz and Ulukanli (2012) confirmed that linoleic acid, oleic acid, lauric acid, stearic and palmitic (Figure 13.1) are the common fatty acids in the FO of various plant parts.

Fixed oils have not gained enormous attention in the agricultural sector; however, their use in pharmaceutical companies is common as they contain various therapeutic properties (Mahboubi et al., 2018; Rosa et al., 2019; Tubtimsri et al., 2021). Unlike essential oils, fixed oils are not commonly used in the agricultural sector as incorporations into edible coatings. Both these oils have, however, antimicrobial,

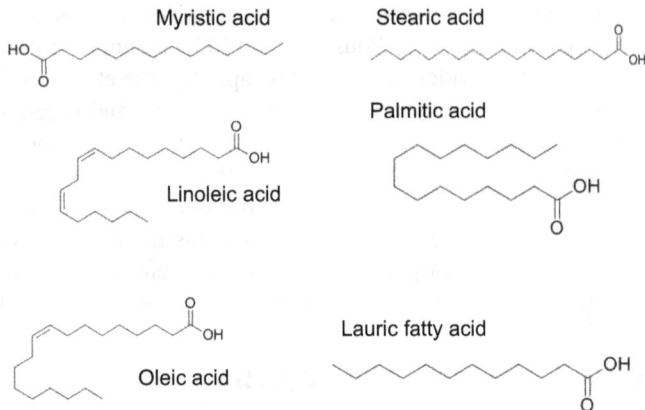

**FIGURE 13.1** Chemical structure of common fatty acids in fixed oils of various plant parts.

anti-browning, antioxidant properties and other active ingredients (Nobre et al., 2018; Ibiapina, 2021; Yousuf et al., 2021).

To achieve complete recovery of oils, a conventional method with nonpolar organic solvents such as *n*-hexane, petroleum ether, chloroform and ethyl acetate is used (Yeddes et al., 2012; Satriana et al., 2019). Alternatively, mechanical processes can be used (Piras et al., 2021). Most research on fixed oils suggests that this type of oil could prevent the occurrence of various diseases due to their phytochemical concentration, antibacterial, antifungal and nutraceutical properties in both plants and humans (Piras et al., 2013; Abushama et al., 2014; da Silva et al., 2020; Tubtimsri et al., 2021). Hence, this type of oil can be used in coating formulations to preserve fresh fruits and vegetables. In addition, FO does not evaporate, unlike essential oils, since they are nonvolatile. This is, therefore, an added characteristic that can improve the efficacy of edible coatings.

## 13.3 PLANT-BASED FIXED OILS IN EDIBLE COATINGS AND FILMS

The quality of horticultural commodities, particularly fruits and vegetables, is highly dependent on nutritional, organoleptic and microbiological properties that change dynamically from harvesting to marketing. Maintaining these properties on fruits and vegetables is still a work in progress in the scientific community. The use of edible coatings and films has been proven reliable in preserving various food products' nutritional, organoleptic and microbiological properties (Suhag et al., 2020). Some people confuse coatings with films. It is, therefore, worth mentioning that the two are entirely different; films are preformed independent structures that wrap or cover food after they are formed, whereas coatings are applied directly on the surface of the fruit, vegetable or other food items (Aguirre-Joya et al., 2018; Yousuf et al., 2018).

Most research on food preservation employing edible coating or film material uses various biomaterials alone or in combination with other biopolymers (Pop et al., 2020). Incorporating natural bioactive compounds and agents in the edible coating or film acts as a carrier and boosts the efficacy of coatings on the targeted food item as they release antioxidants, antimicrobials, vitamins, etc. (Riva et al., 2020; Sharma et al., 2019). The various bioactive compounds, among others, include essential and fixed oils. As mentioned earlier, essential oils have gained much attention in the agricultural sector (Yousuf et al., 2021), while plant-based fixed oil is still under-researched. Even though incorporating essential oils or the compounds derived from them into edible coatings or films preserves fruits and vegetables' phytochemical and physicochemical properties, some challenges include their high reactivity, volatility, and intense aroma. On the other side, plant-based fixed oils are nonvolatile and do not have an intense aroma; however, they can go rancid over time. Due to this and additional benefits, research on incorporating FO into edible coatings and films is slowly gaining popularity. Based on previous work, olive oil, coconut oil, and jojoba oil are useful for this purpose. Olive oil, in particular, has been used in composite edible coatings, starch-based edible coatings, and has been incorporated with other well-known edible coatings, including chitosan, gum Arabic, and waxes (Cruz et al., 2015; De León-Zapata et al., 2015; Dovale-Rosabal et al., 2015; Ramana et al., 2016; Ochoa-Reyes et al., 2021; Uthairatanakij et al., 2022). In some studies, plant-based fixed oils such as pomegranate seed and jojoba oil were used as plasticizers (Ochoa-Reyes et al., 2021; Teodosio et al., 2021b). An unusual fixed oil from tomatoes incorporated with gelatin preserved the bioactive compounds and quality of garambullo fruit (López-Palestina et al., 2018). Table 13.1 presents various plant-based fixed oil coating formulations for maintaining the quality of various crops. Even though the plant-based fixed oils are rich in certain fatty acids, no study has evaluated the contribution of specific fatty acids to coating formulation for the postharvest handling of horticultural crops.

## 13.4   MODE OF ACTION BEHIND THE SUCCESS OF FIXED OILS

Fixed oils work the same as essential oils, except that FO are nonvolatile and does not have an aroma. FO are slowly attracting interest as a preservative in the food sector due to their high content of bioactive compounds, with some extra antioxidant and antimicrobial activities (Nobre et al., 2018; Ibiapina, 2021; Yousuf et al., 2021). There are many plant-derived FO, each of which has a different biological activity and physicochemical properties. Recently, plant-based fixed oils such as pomegranate seed and jojoba oil were used as plasticizers. The latter assists in improving the flexibility of polymeric substances as well as overcoming the brittleness of edible coatings (Ochoa-Reyes et al., 2021; Teodosio etal., 2021a,b). Some of the FO, like coconut oil, have the ability to control the respiration and transpiration rate and bind the ethylene biosynthesis (Nasrin et al., 2020). Fixed oils are also a rich source of fatty acids, with lauric acid being one of the dominant ones. This fatty acid possesses antibacterial, antiviral and antifungal activities (Lieberman et al., 2006). Additionally, FO have been reported to contribute to the reduction of transpiration and respiration in fruit and vegetable as they are involved in closing the opening of stomata and lenticels (Bisen et al., 2012). Thus incorporating FO

**TABLE 13.1**

**Incorporation of Plant-Based Fixed Oils into Edible Coatings and Overall Effects on Various Crops**

| Coating in Which Fixed Oil Is Incorporated | Formulation Including Fixed Oil | Crop Investigated and Storage Regime | Key Findings | Reference |
|---|---|---|---|---|
| Whey protein | Coating solutions consist of 10% whey protein isolate (WPI) mixed with **olive oil** at concentrations of 0, 0.25, 0.5 and 1% | Coated and uncoated fresh-cut pineapples were packed in polypropylene (PP) boxes and stored at 4°C | Whey protein isolate + 1% olive oil significantly reduced weight loss and maintained the highest levels of ascorbic acid contents and total phenolic contents and delayed ripening. | (Uthairatanakij et al., 2022) |
| Chitosan | **Olive oil** was added at 2 and 4% (v/v) to both chitosan solutions at 1 and 2% (w/v) | Coated and uncoated homogeneous and uniform tomatoes were placed in plastic boxes and stored at 27 ± 1°C and 80% RH for 21 days. | Chitosan-olive oil emulsion delayed the ripening, maintained firmness, and contributed to extending the shelf life of tomatoes. | Dovale-Rosabal et al. (2015) |
| Beeswax | 100% coconut oil, **coconut oil** and beeswax mixture (9:1), coconut oil and beeswax mixture (4:1) | After coating, lemons were kept open in crates or modified atmospheric packages (MAP) and stored at 21 ± 2°C and 50 ± 5% RH for 18 days. | Coconut oil individually and in combination with beeswax (both formulations) coating, especially with MAP had a significant effect on retaining green color, reducing respiration, ethylene production and weight loss. | (Nasrin et al., 2020) |
| Candelilla wax, gum Arabic | More than 15 formulations of candelilla wax, gum Arabic, **jojoba oil**, and polyphenols from pomegranate were prepared. | Coated and uncoated pears were stored for 30 days at room temperature. | Camdelilla wax 3%, gum Arabic 4%, jojoba oil 0.15% and pomegranate polyphenols 0.015% reduced changes in pH, firmness, and weight loss and were effective in maintaining sensory quality in the pear fruit for 31 days. | (Cruz et al., 2015) |
| Sodium alginate | Composite edible coating of 2% sodium alginate and 0.2% **olive oil** with a combination of 1% ascorbic acid and 1% citric acid | Coated ber fruit were stored in food-grade storage bags of 42 μm thickness at 25 ± 2°C and 65% RH. | Findings revealed that the edible coating has the potential to control decaying incidence of ber fruit, extend its storage life and also improve its valuable nutritional characteristics. | (Ramana et al., 2016) |

| Coating type | Description | Conditions | Findings | Reference |
|---|---|---|---|---|
| Chlorella species | In the solutions of *Chlorella* sp. at concentrations of 0.5, 1.0, 1.5 and 2.0% (w/v) 0.3% (w/v) of **pomegranate seed oil (PSO)**, which was added as a plasticizer | Coated and uncoated *Spondias tuberosa* fruit were packed inside perforated polyethylene terephthalate (PET) trays and stored for 12 days at 14 ± 2°C and 85 ± 5% RH. | Coatings maintained vitamin C and phenolic compounds, reduced the mass loss and firmness and maintained the green color for at least 10 days. | (Teodosio et al., 2021a) |
| Sodium alginate | Sodium alginate [0–3.5% (w/w), 9 levels] was mixed with plasticizers (glycerol and sorbitol (0–20% (w/w), 6 levels) to make a solution. Surfactants (tween 40, tween 80, span 60, span 80, lecithin (0–5% (w/w), 11 levels), and **vegetable oils [sunflower oil, olive oil, rapeseed oil** (0–5% (w/w), 10 levels] were incorporated in the solution. | Strawberry | 1.25% sodium alginate, 2% glycerol, 0.2% sunflower oil, 1% span 80 and 0.2% tween 40 or tween 80 was the best formulation to obtain an effective coating for strawberry, a hydrophobic food surface. | (Senturk et al., 2018) |
| Beeswax | Beeswax was incorporated with **virgin coconut oil (VCO)**, oleic acid and cinnamaldehyde. | *Glomerella cingulate* in strawberry | The presence of beeswax, VCO, oleic acid and their mixtures slowed down the growth rate of *Glomerella cingulata* but could not completely inhibit the growth of the fungi. | Permana et al., 2021) |
| Gelatin (film) | **Tomato oil** was added to the coating solution in a proportion of 0%, 1% or 3% (v/v) | Garambullo fruit were stored at 5 ± 1°C and 95% RH for 0, 5, 10 or 15 days. | Gelatin-based coatings with tomato oil-3% was the most effective for preserving the bioactive compounds and quality of garambullo fruit. | (López-Palestina et al., 2018) |
| Starch-based coating | A solution of starch + glycerol (2:1, v/v%), starch + glycerol + **nigella sativa oil** (2:1:v/v: 600 ppm) and starch + glycerol + nigella sativa oil (2:1:v/v 600 ppm) was prepared, and the arils were separately immersed in each solution. | All treatment groups of pomegranate arils were stored at 4°C for 12 days. | Treatments/coatings greatly reduced softening of pomegranate arils, weight loss and browning, loss of vitamin C, loss of anthocyanin and finally delayed microbial decay. | Oz and Ulukanli (2012) |

*(Continued)*

**TABLE 13.1 (Continued)**

**Incorporation of Plant-Based Fixed Oils into Edible Coatings and Overall Effects on Various Crops**

| Coating in Which Fixed Oil Is Incorporated | Formulation Including Fixed Oil | Crop Investigated and Storage Regime | Key Findings | Reference |
|---|---|---|---|---|
| Arabic and xanthan gums | Three different biopolymers (pectin, Arabic and xanthan gums) were evaluated in mixtures with candelilla wax as the hydrophobic phase, **jojoba oil** as plasticizer and a crude extract of polyphenols as a source of bioactive compounds. | Green bell peppers were stored at $25 \pm 2$ °C for a maximum period of 10 days | Edible coatings applied to pepper maintained weight loss, appearance, color changes, pH, total soluble solids and texture. | (Ochoa-Reyes et al., 2021) |
| Gum Arabic | Candelilla wax, **jojoba oil** and fermented extract of tarbush was added into the Arabic gum edible coating and homogenized. | Apples were stored in marketing conditions for 8 weeks at $27 \pm 1$ °C. | Edible coating reduced weight loss and maintained water activity and firmness and extended the shelf life of apples. | (De León-Zapata et al., 2015) |
| *Chlorella* species | The 2% *Chlorella* sp. was mixed with 0.3% (w/v) of **pomegranate seed oil (PSO)**, which was added as a plasticizer. | Coated and uncoated umbu fruit with *Chlorella* enriched with PSO were packed inside perforated PET polyethylene terephthalate trays and then cold stored at $14 \pm 2$ °C and $85 \pm 5$% and then kept at $24 \pm 2$ °C, $85 \pm 5$% RH for three more days. | Coatings promoted postharvest shelf life of umbus for 12 days and maintained bioactive compounds (vitamin C and phenolic concentration) of the fruit. | (Teodosio et al., 2021b) |

*NB*: WVP: water vapor permeability; fixed oils used are highlighted in bold in edible coatings would always bring an added advantage. For instance, Senturk et al. (2018) revealed that surface tension could be decreased by fixed oils such as sunflower, olive and rapeseed oils, even at a low concentration, and could improve plasticity, flexibility and waxiness of coatings.

## 13.5 EXTRACTION AND INCORPORATION OF FIXED OILS INTO EDIBLE COATINGS

Edible coatings have unique functions/properties, including acting as carriers for bioactive and other functional ingredients (Riva et al., 2020). The use of fixed oils, particularly, has not gained much attention, and few studies have been reported. Extraction of these oils and other FO from plant parts is mainly performed using hexane solution followed by the GC-MS device for identifying and separating compounds present in them (Elagbar et al., 2016; Nobre et al., 2018; Satriana et al., 2019; Rezaei-Chiyaneh et al., 2021). Hexane is commonly used because it achieves a complete recovery of oils due to its nonpolar nature. Previous work has demonstrated that olive oil, coconut oil and jojoba oil play an increasing role in postharvest preservation (Cruz et al., 2015; De León-Zapata et al., 2015; Dovale-Rosabal et al., 2015; Ramana et al., 2016; Ochoa-Reyes et al., 2021; Uthairatanakij et al., 2022).

In terms of incorporating FO into edible coatings, many methods need to be investigated or are currently unknown. However, factors including FO concentration, temperature of composite mixture, the order in which FO is introduced into edible coating mixture and agitation methods play a role in the efficacy of edible coatings functionalized with FO. According to the literature, FO at 1% or even less is reported to improve the efficacy of edible coatings (Cruz et al., 2015; Ramana et al., 2016; Senturk et al., 2018; Teodosio et al., 2021a, 2021b; Uthairatanakij et al., 2022). Cruz et al. (2015) heated a mixture of specialized coating containing candelilla wax, gum Arabic, jojoba oil and pomegranate polyphenols at 80°C and homogenized it at 2500 rpm in an industrial homogenizer. Similarly, for alginate-based composite coating, Ramana et al. (2016) first suspended alginate (2.0% w/v) containing glycerol (0.75% v/v) in water at 80°C for 20 min, followed by the addition of olive oil (0.2% v/v) and stirred using a magnetic stirrer for 30 min.

## 13.6 CONCLUSION AND FUTURE PROSPECT

Edible coatings are an efficient alternative to chemical substances for preserving fruits and vegetables. This innovative and environmentally friendly approach can help to fulfill consumers' demand for fresh food items. Extensive studies have been carried out on using essential oils incorporated with coatings to preserve horticultural crops. However, plant-based fixed oils (nonvolatile oil) have not been extensively explored. Although the present study demonstrated that fixed oils could maintain quality and prolong the shelf life of fruits and vegetables, consumers still need to identify their preferences for specific combinations of horticultural crops and plant-based fixed oils.

# REFERENCES LIST

Abushama, M. F., Hilmi, Y. I., AbdAlgadir, H. M., Fadul, E., & Khalid, H. E. (2014). Lethality and antioxidant activity of some Sudanese medicinal plants' fixed oils. *European Journal of Medicinal Plants*, *4*(5), 563–570. https://doi.org/10.9734/EJMP/2014/7741

Aguirre-Joya, J. A., De Leon-Zapata, M. A., Alvarez-Perez, O. B., Torres-León, C., Nieto-Oropeza, D. E., Ventura-Sobrevilla, J. M., Aguilar, M. A., Ruelas-Chacón, X., Rojas, R., Ramos-Aguiñaga, M. E., & Aguilar, C. N. (2018). Basic and applied concepts of edible packaging for foods. In A. M. Grumezescu & A. M. Holban (Eds.), *Food packaging and preservation, handbook of food bioengineering* (pp. 1–61). Academic Press. https://doi.org/10.1016/B978-0-12-811516-9.00001-4

Anjum, M. A., Akram, H., Zaidi, M., & Ali, S. (2020). Effect of gum arabic and aloe vera gel based edible coatings in combination with plant extracts on postharvest quality and storability of 'Gola' guava fruits. *Scientia Horticulturae*, *271*, 109506. https://doi.org/10.1016/j.scienta.2020.109506

Atarés, L., & Chiralt, A. (2016). Essential oils as additives in biodegradable films and coatings for active food packaging. *Trends in Food Science and Technology*, *48*, 51–62. https://doi.org/10.1016/j.tifs.2015.12.001

Bisen, A., Pandey, S. K., & Patel, N. (2012). Effect of skin coatings on prolonging shelf life of kagzi lime fruits (*Citrus aurantifolia* Swingle). *Journal of Food Science and Technology*, *49*(6), 753–759. https://doi.org/10.1007/s13197-010-0214-y

Blancas-Benitez, F. J., Montaño-Leyva, B., Aguirre-Güitrón, L., Moreno-Hernández, C. L., Fonseca-Cantabrana, Á., Romero-Islas, L., & González-Estrada, R. R. (2022). Impact of edible coatings on quality of fruits: A review. *Food Control*, 139, 109063. https://doi.org/10.1016/j.foodcont.2022.109063

Castilho, P. C., Costa, M. D. C., Rodrigues, A., & Partidário, A. (2005). Characterization of laurel fruit oil from Madeira Island, Portugal. *Journal of the American Oil Chemists' Society*, *82*(12), 863–868. https://doi.org/10.1007/s11746-005-1156-4

Cruz, V., Rojas, R., Saucedo-Pompa, S., Martínez, D. G., Aguilera-Carbó, A. F., Alvarez, O. B., Rodríguez, R., Ruiz, J., & Aguilar, C. N. (2015). Improvement of shelf life and sensory quality of pears using a specialized edible coating. *Journal of Chemistry*, *2015*, 1–7. https://doi.org/10.1155/2015/138707

da Silva, M. A. M. P., Zehetmeyr, F. K., Pereira, K. M., Pacheco, B. S., Freitag, R. A., Pinto, N. B., Machado, R. H., Villarreal Villarreal, J. P., de Oliveira Hubner, S., Aires Berne, M. E., & da Silva Nascente, P. (2020). Ovicidal in vitro activity of the fixed oil of Helianthus annus L. and the essential oil of Cuminum cyminum L. against Fasciola hepatica (Linnaeus, 1758). *Experimental Parasitology*, *218*, 107984. https://doi.org/10.1016/j.exppara.2020.107984

de León-Zapata, M. A., Sáenz-Galindo, A., Rojas-Molina, R., Rodríguez-Herrera, R., Jasso-Cantú, D., & Aguilar, C. N. (2015). Edible candelilla wax coating with fermented extract of tarbush improves the shelf life and quality of apples. *Food Packaging and Shelf Life*, *3*, 70–75. https://doi.org/10.1016/j.fpsl.2015.01.001

de Morais, S. M., do Nascimento, J. E. T., Silva, A. A., Junior, J. E. R. H., Pinheiro, D. C. S. N., & de Oliveira, R. V. (2017). Fatty acid profile and anti-inflammatory activity of fixed plant oils. *Acta Scientiae Veterinariae*, *45*(1), 1–8. https://doi.org/10.22456/1679-9216.79403

Dhall, R. K. (2013). Advances in edible coatings for fresh fruits and vegetables: A review. *Critical Reviews in Food Science and Nutrition*, *53*(5), 435–450. https://doi.org/10.1080/10408398.2010.541568

Dovale-Rosabal, G., Casariego, A., Forbes-Hernandez, T. Y., & García, M. A. (2015). Effect of chitosan-olive oil emulsion coating on quality of tomatoes during storage at ambient conditions. *Journal of Berry Research*, *5*(4), 207–218. https://doi.org/10.3233/JBR-150103

Elagbar, Z. A., Naik, R. R., Shakya, A. K., & Bardaweel, S. K. (2016). Fatty acids analysis, antioxidant and biological activity of fixed oil of *Annona muricata* L. seeds. *Journal of Chemistry, 2016*, 1–6. https://doi.org/10.1155/2016/6948098

Fawole, O. A., Riva, S. C., & Opara, U. L. (2020). Efficacy of edible coatings in alleviating shrivel and maintaining quality of Japanese plum (*Prunus salicina* Lindl.) during export and shelf life conditions. *Agronomy, 10*(7), 1023. https://doi.org/10.3390/agronomy10071023

Galanakis, C. M., Aldawoud, T. M. S., Rizou, M., Rowan, N. J., & Ibrahim, S. A. (2020). Food ingredients and active compounds against the coronavirus disease (COVID-19) pandemic: A comprehensive review. *Foods, 9*(11), 1701. https://doi.org/10.3390/foods9111701

Ibiapina, A., Gualberto, L. D. S., Dias, B. B., Freitas, B. C. B., Martins, G. A. D. S., & Melo Filho, A. A. (2021). Essential and fixed oils from Amazonian fruits: Proprieties and applications. *Critical Reviews in Food Science and Nutrition*, 1–13. https://doi.org/10.1080/10408398.2021.1935702

Kawhena, T. G., Opara, U. L., & Fawole, O. A. (2022). Effects of gum arabic Coatings Enriched with lemongrass essential oil and pomegranate peel extract on quality maintenance of pomegranate whole fruit and arils. *Foods, 11*(4), 593. https://doi.org/10.3390/foods11040593

Kawhena, T. G., Tsige, A. A., Opara, U. L., & Fawole, O. A. (2020). Application of gum arabic and methyl cellulose coatings enriched with thyme oil to maintain quality and extend shelf life of "acco" pomegranate arils. *Plants, 9*(12), 1690. https://doi.org/10.3390/plants9121690

Kumar, A., Sharma, A., & Upadhyaya, K. C. (2016). Vegetable oil: Nutritional and industrial perspective. *Current Genomics, 17*(3), 230–240. https://doi.org/10.2174/1389202917666160202220107

Kumar, N., Pratibha, A. T., Neeraj, S. A., Petkoska, A. T., AL-Hilifi, S. A., & Fawole, O. A. (2021). Effect of chitosan-pullulan composite edible coating functionalized with pomegranate peel extract on the shelf life of mango (*Mangifera indica*). *Coatings, 11*(7), 764. https://doi.org/10.3390/coatings11070764

Lieberman, S., Enig, M. G., & Preuss, H. G. (2006). A review of monolaurin and lauric acid: Natural virucidal and bactericidal agents. *Alternative and Complementary Therapies, 12*(6), 310–314. https://doi.org/10.1089/act.2006.12.310

López-Palestina, C. U., Aguirre-Mancilla, C. L., Raya-Pérez, J. C., Ramírez-Pimentel, J. G., Gutiérrez-Tlahque, J., & Hernández-Fuentes, A. D. (2018). The effect of an edible coating with tomato oily extract on the physicochemical and antioxidant properties of garambullo (*Myrtillocactus geometrizans*) fruits. *Agronomy, 8*(11), 248. https://doi.org/10.3390/agronomy8110248

Mahboubi, M., Mohammad Taghizadeh Kashani, L. M. T., & Mahboubi, M. (2018). *Nigella sativa* fixed oil as alternative treatment in management of pain in arthritis rheumatoid. *Phytomedicine, 46*, 69–77. https://doi.org/10.1016/j.phymed.2018.04.018

Marzouki, H., Piras, A., Marongiu, B., Rosa, A., & Dessì, M. A. (2008). Extraction and separation of volatile and fixed oils from berries of *Laurus nobilis* L. by supercritical CO2. *Molecules, 13*(8), 1702–1711. https://doi.org/10.3390/molecules13081702

Nasrin, T. A. A., Rahman, M. A., Arfin, M. S., Islam, M. N., & Ullah, M. A. (2020). Effect of novel coconut oil and beeswax edible coating on postharvest quality of lemon at ambient storage. *Journal of Agriculture and Food Research, 2*, 100019. https://doi.org/10.1016/j.jafr.2019.100019

Ncama, K., Magwaza, L. S., Mditshwa, A., & Tesfay, S. Z. (2018). Plant-based edible coatings for managing postharvest quality of fresh horticultural produce: A review. *Food Packaging and Shelf Life, 16*, 157–167. https://doi.org/10.1016/j.fpsl.2018.03.011

Ngcobo, B.L., & Bertling, I. (2021). Combined effect of heat treatment and Moringa leaf extract (MLE) on colour development, quality and postharvest life of tomatoes. *Acta Horticulturae, 1306*, 323–328. https://doi.org/10.17660/ActaHortic.2021.1306.41

Nickavar, B., Mojab, F., Javidnia, K., & Amoli, M.A.R. (2003). Chemical composition of the fixed and volatile oils of Nigella sativa L. from Iran. *Zeitschrift Fur Naturforschung. C, Journal of Biosciences, 58*(9–10), 629–631. https://doi.org/10.1515/znc-2003-9-1004

Nobre, C.B., de Sousa, E.O., de Lima Silva, J.M.F., Melo Coutinho, H.D.M., & da Costa, J.G.M. (2018). Chemical composition and antibacterial activity of fixed oils of *Mauritia flexuosa* and *Orbignya speciosa* associated with Aminoglycosides. *European Journal of Integrative Medicine, 23*, 84–89. https://doi.org/10.1016/j.eujim.2018.09.009

Ochoa-Reyes, E., Martínez-Vazquez, G., Saucedo-Pompa, S., Montañez, J., Rojas-Molina, R., Leon-Zapata, M.A.D., Rodríguez-Herrera, R., & Aguilar, C.N. (2021). Improvement of shelf life quality of green bell peppers using edible coating formulations. *Journal of Microbiology, Biotechnology and Food Sciences, 2021*, 2448–2451.

Oz, A.T., & Ulukanli, Z. (2012). Application of edible starch-based coating including glycerol plus oleum nigella on arils from long-stored whole pomegranate fruits. *Journal of Food Processing and Preservation, 36*(1), 81–95. https://doi.org/10.1111/j.1745-4549.2011.00599.x

Permana, A.W., Sampers, I., & Van der Meeren, P. (2021). Influence of virgin coconut oil on the inhibitory effect of emulsion-based edible coatings containing cinnamaldehyde against the growth of *Colletotrichum gloeosporioides* (*Glomerella cingulata*). *Food Control, 121*, 107622. https://doi.org/10.1016/j.foodcont.2020.107622

Piras, A., Piras, C., Porcedda, S., & Rosa, A. (2021). Comparative evaluation of the composition of vegetable essential and fixed oils obtained by supercritical extraction and conventional techniques: A chemometric approach. *International Journal of Food Science and Technology, 56*(9), 4496–4505. https://doi.org/10.1111/ijfs.15098

Piras, A., Rosa, A., Marongiu, B., Porcedda, S., Falconieri, D., Dessì, M.A., Ozcelik, B., & Koca, U. (2013). Chemical composition and in vitro bioactivity of the volatile and fixed oils of Nigella sativa L. extracted by supercritical carbon dioxide. *Industrial Crops and Products, 46*, 317–323. https://doi.org/10.1016/j.indcrop.2013.02.013

Pop, O.L., Pop, C.R., Dufrechou, M., Vodnar, D.C., Socaci, S.A., Dulf, F.V., Minervini, F., & Suharoschi, R. (2020). Edible films and coatings functionalization by probiotic incorporation: A review. *Polymers, 12*(1). https://doi.org/10.3390/polym12010012

Ramana Rao, T.V., Baraiya, N.S., Vyas, P.B., & Patel, D.M. (2016). Composite coating of alginate-olive oil enriched with antioxidants enhances postharvest quality and shelf life of Ber fruit (Ziziphus mauritiana Lamk. Var. Gola). *Journal of Food Science and Technology, 53*(1), 748–756. https://doi.org/10.1007/s13197-015-2045-3

Rezaei-Chiyaneh, E., Battaglia, M.L., Sadeghpour, A., Shokrani, F., Nasab, A.D.M., Raza, M.A., & von Cossel, M. (2021). Optimizing intercropping systems of black cumin (Nigella sativa L.) and Fenugreek (*Trigonella foenum-Graecum* L.) through inoculation with bacteria and mycorrhizal fungi. *Advanced Sustainable Systems, 5*(9), 2000269. https://doi.org/10.1002/adsu.202000269

Riva, S.C., Opara, U.O., & Fawole, O.A. (2020). Recent developments on postharvest application of edible coatings on stone fruit: A review. *Scientia Horticulturae, 262*, 109–074. https://doi.org/10.1016/j.scienta.2019.109074

Rosa, A., Era, B., Masala, C., Nieddu, M., Scano, P., Fais, A., Porcedda, S., & Piras, A. (2019). Supercritical CO2 extraction of waste citrus seeds: Chemical composition, nutritional and biological properties of edible fixed oils. *European Journal of Lipid Science and Technology, 121*(7), 1800502. https://doi.org/10.1002/ejlt.201800502

Rosa, A., Rescigno, A., Piras, A., Atzeri, A., Scano, P., Porcedda, S., Zucca, P., & Assunta Dessì, M.A. (2012). Chemical composition and effect on intestinal Caco-2 cell viability and lipid profile of fixed oil from *Cynomorium coccineum* L. *Food and Chemical Toxicology, 50*(10), 3799–3807. https://doi.org/10.1016/j.fct.2012.07.003

Satriana, S., Supardan, M.D., Arpi, N., & Wan Mustapha, W.A. (2019). Development of methods used in the extraction of avocado oil. *European Journal of Lipid Science and Technology*, *121*(1), 1800210. https://doi.org/10.1002/ejlt.201800210

Senturk Parreidt, T., Schott, M., Schmid, M., & Müller, K. (2018). Effect of presence and concentration of plasticizers, vegetable oils, and surfactants on the properties of sodium-alginate-based edible coatings. *International Journal of Molecular Sciences*, *19*(3), 742. https://doi.org/10.3390/ijms19030742

Sharma, P., Shehin, V.P., Kaur, N., & Vyas, P. (2019). Application of edible coatings on fresh and minimally processed vegetables: A review. *International Journal of Vegetable Science*, *25*(3), 295–314. https://doi.org/10.1080/19315260.2018.1510863

Shinga, M.H., & Fawole, O.A. (2023). Opuntia ficus indica mucilage coatings regulate cell wall softening enzymes and delay the ripening of banana fruit stored at retail conditions. *International Journal of Biological Macromolecules*, *245*, 125550. https://doi.org/10.1016/j.ijbiomac.2023.125550

Singh, D.P., & Packirisamy, G. (2022). Biopolymer based edible coating for enhancing the shelf life of horticulture products. *Food Chemistry: Molecular Sciences*, *4*, 100085. https://doi.org/10.1016/j.fochms.2022.100085

Suhag, R., Kumar, N., Petkoska, A.T., & Upadhyay, A. (2020). Film formation and deposition methods of edible coating on food products: A review. *Food Research International*, *136*, 109582. https://doi.org/10.1016/j.foodres.2020.109582

Teodosio, A.E.M., Carlos Rocha Araújo, R.H., Figueiredo Lima Santos, B.G., Linné, J.A., da Silva Medeiros, M.L., Alves Onias, E., Alves de Morais, F., de Melo Silva, S., & de Lima, J.F. (2021a). Effects of edible coatings of Chlorella sp. containing pomegranate seed oil on quality of Spondias tuberosa fruit during cold storage. *Food Chemistry*, *338*, 127916. https://doi.org/10.1016/j.foodchem.2020.127916

Teodosio, A.E.M., Santos, B.G.F.L., Linné, J.A., de Lima Cruz, J.M.F., Onias, E.A., de Lima, J.F., & Araújo, R.H.C.R. (2021b). Preservation of *Spondias tuberosa* fruit with edible coatings based on chlorella sp. enriched with pomegranate seed oil during storage. *Food and Bioprocess Technology*, *14*(11), 2020–2031. https://doi.org/10.1007/s11947-021-02704-0

Tubtimsri, S., Limmatvapirat, C., Limsirichaikul, S., Akkaramongkolporn, P., Piriyaprasarth, S., Patomchaiviwat, V., & Limmatvapirat, S. (2021). Incorporation of fixed oils into spearmint oil-loaded nanoemulsions and their influence on characteristic and cytotoxic properties against human oral cancer cells. *Journal of Drug Delivery Science and Technology*, *63*, 102443. https://doi.org/10.1016/j.jddst.2021.102443

Uthairatanakij, A., Wote, Y.H., Laohakunjit, N., Jitareerat, P., & Kaisangsri, N. (2022). Whey protein incorporated with olive oil as novel edible coating for fresh cut pineapples. *Acta Horticulturae*, *1336*, 351–356. https://doi.org/10.17660/ActaHortic.2022.1336.46

Yeddes, N., Chérif, J.K., Jrad, A., Barth, D., & Trabelsi-Ayadi, M. (2012). Supercritical SC-CO2 and Soxhlet n-hexane extract of Tunisian *Opuntia ficus indica* seeds and fatty acids analysis. *Journal of Lipids*, *2012*, 1–6. https://doi.org/10.1155/2012/914693

Yousuf, B., Qadri, O.S., & Srivastava, A.K. (2018). Recent developments in shelf-life extension of fresh-cut fruits and vegetables by application of different edible coatings: A review. *LWT*, *89*, 198–209. https://doi.org/10.1016/j.lwt.2017.10.051

Yousuf, B., Wu, S., & Siddiqui, M.W. (2021). Incorporating essential oils or compounds derived thereof into edible coatings: Effect on quality and shelf life of fresh/fresh-cut produce. *Trends in Food Science and Technology*, *108*, 245–257. https://doi.org/10.1016/j.tifs.2021.01.016

# 14 Layer-by-Layer Application of Edible Coatings in Postharvest Handling of Horticultural Crops

*Mawande Hugh Shinga and Olaniyi Amos Fawole*

## 14.1 INTRODUCTION

Generally, horticultural crops are known for their perishability, and their economic value depreciates if they are not handled correctly after harvesting. Chemical treatment is one of the postharvest practices used to reduce food losses. Due to food regulations worldwide and the increased concern of consumers against chemicals, a lot of chemical-containing treatments in fruit and vegetables are prohibited for human health reasons. This has encouraged the food industries to innovate better techniques for retaining food quality and enhancing shelf life as the demand for fresh produce increases (Rico et al., 2007). The postharvest application of edible coatings is receiving increased interest from the food industry (Flores-López et al., 2016; Shinga & Fawole, 2023). Several researchers have studied edible coatings on different horticultural crops, including mango (Bambalele et al., 2021), citrus (Shinga et al., 2021), nut (Khoshnoudi-Nia & Sedaghat, 2019), pomegranate (Mwelase & Fawole, 2022) and spinach (Abedi et al., 2021).

Polysaccharide, protein and lipid-base edible coatings have been reported in many different foods (Sipahi et al., 2013). Lipid-based coating components include waxes, paraffin, acetoglycerides, organic fatty acids and resins. Lipid-based edible coatings are not only used in horticultural crops but are also applied to meat, poultry and seafood (Gennadios et al., 1997; Umaraw & Verma, 2017). They possess good moisture barrier characteristics for preventing food products from physiologically deteriorating due to their hydrophobic nature (Arnon-Rips & Poverenov, 2018). However, food with hydrophilic surfaces may possess poor mechanical properties and adhesion. Protein-based edible coatings possess a wide variety of physical and mechanical properties, therefore fitting the specific needs of different food products (Dhall, 2013; Umaraw & Verma, 2017). Protein-based components for edible coatings comprise gelatin, casein, whey protein, corn zein, wheat gluten, soy protein, mung bean protein and peanut protein. They normally show high gas permeability, good mechanical properties and low moisture barriers. In contrast, polysaccharide-based edible

DOI: 10.1201/9781003452355-17

coatings are highly available and therefore widely used, are normally water-soluble and generally have good mechanical properties (Bourtoom, 2008; Debeaufort et al., 1998; Umaraw & Verma, 2017). Polysaccharide-based edible coatings include cellulose derivatives, chitosan, starch, pectin, gums, algae-derived materials, etc. Combining two or more edible coatings from different types is necessary for improved properties. In this sense, the combined solution will have advantages. For instance, the poor mechanical strength of lipids can be improved by adding polysaccharides, while lipids can improve moisture permeability. Therefore, a layer-by-layer technique has been tested and reported as a promising alternative to this situation. This chapter presents a new insight into the application of layer-by-layer edible coating techniques to maximize the preservation of fruits and vegetables during storage and to prolong shelf life.

## 14.2  LAYER-BY-LAYER (LBL) EDIBLE COATING APPROACH: OVERVIEW

The common practice of fruit coating during postharvest storage has been used by industries to meet international trade standards (Gunaydin et al., 2017). Fruits are treated to reduce water loss and respiration rates and to enhance the visual quality of the commodity and also prolong shelf life (Gunaydin et al., 2017). Stand-alone edible coatings commonly possess advantages and some disadvantages. For instance, Arnon et al. (2014) found that carboxymethyl cellulose (CMC) coating imparted little gloss on citrus fruit surface, although CMC maintained high structural fruit integrity during storage. However, the chitosan coating as a stand-alone provided high glossiness to citrus fruit, but it was unstable and tended to peel off. CMC was first applied, then chitosan, for the reasons just mentioned. These coatings were applied separately because their composite formulation proved nonhomogeneous, yet chitosan on the CMC layer resulted in a homogeneous and stable coating (Arnon et al., 2014). This CMC-chitosan combination performed well in water-resistance, and its stability was maintained during cold storage throughout the shelf life, with no anaerobic conditions in the internal atmosphere of the fruit. This bilayer has been positively reported for use in the postharvest handling of fruit due to this ability and added that it is safe, relatively inexpensive and easy to prepare. The method of applying edible coatings on different fruits has been looked at closely. Soliva-Fortuny and Martín-Belloso (2003) suggested that the dipping technique does not allow the coating to stick to the hydrophilic surface of fresh-cut fruit. Therefore, LbL edible coatings became an alternative to overcome this problem. This technique also enables the edible coatings to bond physically and chemically to each other (Weiss et al., 2006). In this process, horticultural crops are subjected to different solutions, which comprise oppositely charged polyelectrolytes. This technique of LbL has been studied lately in papaya (Brasil et al., 2012), cantaloupe (Martiñon et al., 2014; Moreira et al., 2014), pineapple (Mantilla et al., 2013), watermelon (Sipahi et al., 2013) and citrus (Arnon et al., 2014).

The LbL technique can be limited by the lack of chemical affinity between selected molecules; therefore, molecule selection choice is important in the LbL method (Xia et al., 1999; Zhao et al., 1996). The overall fundamental concepts and mechanisms involved in the LbL technique have been reported in several articles (Ali et al., 2014; Arnon et al., 2014; Baldwin et al., 1999; Brasil et al., 2012; Daraei Garmakhany et al., 2014; Gunaydin et al., 2017; Hira et al., 2022; Kumar & Saini, 2021; Lara et al., 2020; Lu et al., 2021; Mantilla et al., 2013; Medeiros et al., 2012; Moreira et al., 2014; Sipahi et al., 2013; Souza et al., 2015; Treviño-Garza et al., 2015, 2017; Velickova et al., 2013; Yan et al., 2019). The compatibility of chemical properties of LbL edible coatings is significant since the nature of LbL edible coatings involves the method of alternating the sequential deposition of polyelectrolytes, polycations and polyanions on a charged surface (Decher et al., 1992). It allows a wide variety of materials to be explored (Crespilho et al., 2006), and film fabrication is done to preserve biomolecular activity under suitable conditions (Guo et al., 2004; Liu & Hu, 2005; Yang et al., 2006). To level up this approach, active and passive functional materials are created, including polymer anti-corrosion coatings (Dai et al., 2000; Melia et al., 2020; Percival et al., 2020), fire retardant coatings (Guin et al., 2015; Holder et al., 2017), nanoparticle electrocatalysts (Han et al., 2016; Yaqub et al., 2015) and ion-selective membrane (Harris et al., 2000). Among these materials of electroactive properties, several researchers conduct studies for LbL films or edible coatings for various applications.

LbL approach can be referred to as more than two substances that are used to form layers in food products. For instance, chitosan, pullulan and mucilage edible coatings can be used concurrently (Treviño-Garza et al., 2017). The combination of these edible coatings has been proven to be compatible. Chitosan has long been used as an edible coating, a cationic polymer derived from chitin. It is nontoxic, bio-functional and biocompatible, and it has bactericidal and fungicidal properties (Chung & Chen, 2008; Helander et al., 2001; Li et al., 2006), whereas pullulan is a non-ionic (neutral polysaccharide) polymer, water-soluble, and nontoxic. Leathers (2003) added that pullulan has an excellent film-forming agent. Mucilage possesses antibacterial, antifungal, and antioxidant properties (Escobedo-Lozano et al., 2015; Pandey & Mishra, 2010). It has been reported as an edible coating in various fruits (Benítez et al., 2015; Chauhan et al., 2014; Song et al., 2013). For example, the linseed mucilage has been studied and reported to consist of two fractions, acidic and neutral polysaccharides (Muralikrishna, 1987). Hence, it could be considered a promising application to prolong the shelf life of various food products. Several methods to form LbL materials involve an alternate deposition of various components and allow the formation of rationally designed composite materials with desired properties.

Percival et al. (2021) demonstrated the LbL deposition of the polyelectrolyte films, which was archived by dipping the nanoporous polyelectrolyte into solutions containing the desired cationic and anionic polymers. The cationic PEI (polyethyleneimine), the first polymer, coats the membrane during the dipping process. The following step is a rinse to remove weakly adhered excess polymers from the membrane on the surface of the food product. Then the same membrane is subjected to the second polymer solution, an anionic polymer [PAA (polyacrylic acid)], which then self-assembles onto the first layer of PEI, again followed by rinsing. In such studies, substances like amino acids could be added to encourage coating incorporation.

## 14.3 LAYER-BY-LAYER EDIBLE COATINGS APPLICATION IN HORTICULTURAL CROPS

The successful application of layer-by-layer edible coating lies in the principle of self-assembling electrostatic interactions within the polyelectrolyte. It is one of the best novel methods to improve the efficiency of barrier properties, increasing its shelf life over the single edible coating. Various authors have recently published their work on LbL-based coating, showing their importance in the postharvest handling of horticultural crops. For instance, Medeiros et al. (2012) used pectin and chitosan to form 5 layers of LbL coatings, and the combination of the antimicrobial and gas barrier properties of chitosan, with the low oxygen permeability of pectin layers, were efficient in the reduction of gas flow and on the extension of the shelf life of mangoes.

Chitosan and CMC LbL edible coatings were combined in the effect of strawberry quality and metabolite during storage in a study by (Yan et al., 2019) (Table 14.1). In this study, LbL edible coating resulted in a good response and extended the shelf life of strawberries. Brasil et al. (2012) prepared multilayer edible coatings based on polyelectrolyte interaction among opposite charges of chitosan, pectin and calcium chloride. The stability and uniform coating were formed and improved the microbiological and physicochemical quality of fresh-cut papaya (Table 14.1). Medeiros et al. (2012) conducted a study of the effect of polysaccharides (k-carrageenan and lysozyme) and a protein nanomultilayer coating on Rocha pear shelf life (Table 14.1). This study established the proof of concept of the use of nanolayered edible coatings for preserving food in postharvest handling. These coatings also showed enough mass transfer properties, oxygen permeability and adequate surface properties to be used as packaging materials for Rocha pears. This combination of coatings was also effective in improving the physicochemical quality of fresh-cut pears and therefore has a great potential for extending shelf life. Hira et al. (2022) researched layer-by-layer edible coatings on the shelf life and transcriptome of Kosui Japanese pear fruit (Table 14.1). The authors in this study prepared coatings from monolayer to pentalayer LbL, where monolayer was challenged by many constraints that hampered performance (Poverenov et al., 2014).

The application of chitosan and alginate-based tri-layer or penta-layer LbL edible coatings maintained physicochemical attributes more effectively than monolayer coatings. In pears, it is important to evaluate the marketability of crunchiness; this study revealed that chitosan and alginate-based LbL edible coating improved crunchiness compared to its counterpart. This LbL coating was also reported to prolong the shelf life of Kosui pear fruit. The successful use of this LbL edible coating might be because edible coatings act as physical barriers on food surfaces, preventing water vapor escape. The interaction between cationic (chitosan) and anionic (alginate) electrostatically resulted in high gas barrier properties as compared to monolayer coatings (Carneiro-da-Cunha et al., 2010; Poverenov et al., 2014).

In another research, transcriptome analysis suggested that LbL edible coatings induce changes in gene expression involved in glycolysis and the TCA cycle, which are key respiratory processes (Hira et al., 2022). According to the author, the LbL coating favored glycolysis as most genes were upregulated, while more TCA cycle genes were downregulated. The downregulation of genes was associated with the

# TABLE 14.1
## Effect of Layer-by-Layer Edible Coating on Horticultural Crops Quality Parameters During Postharvest Cold Storage

| Food Products | LbL Edible Coating Combination | Key Findings | References |
|---|---|---|---|
| Dragon fruit | Conventional chitosan and submicron chitosan | Double layer coating of 600 nm droplet size + 1% conventional chitosan improved fruit quality such as weight loss, antioxidant and respiration rate. | (Ali et al., 2014) |
| Fresh-cut cantaloupe | Chitosan and pectin | More effective in microbial systems, maintained physicochemical and sensory quality. | (Martiñon et al., 2014) |
| Fresh-cut pineapple | Mucilages, pullulan and chitosan | Improved physicochemical quality. Reduced ascorbic acid degradation. Controlled microbial effect. Preserve sensory qualities (color, odor, flavor, texture, overall acceptance and delayed sign of decay). | (Treviño-Garza et al., 2017) |
| Fresh-cut pineapple | Sodium alginate, pectin and calcium chloride | Slight difference in sensory quality, no effect on pH and Brix. Had significantly less juice leakage and texture. Reduced microbial growth. Extended shelf life to 15 days. | (Mantilla et al., 2013) |
| Strawberries | Pectin, pullulan and chitosan | Reduced weight loss, microbial growth and fruit softening and delayed color change and TSS content. Maintained AsA content. Conserved sensory quality (color, flavor, texture and acceptance). Shelf life increased by 9 days compared to control. | (Treviño-Garza et al., 2015) |
| Strawberries | Chitosan and beeswax | Reduced fungal infection, reduce weight loss and respiration rate, maintained firmness, color, TA, pH, TSS and sugars. Improved sensory quality. Prolonged shelf life. | (Velickova et al., 2013) |
| Strawberries | CMC and chitosan | Maintained firmness, yet little effect on TSS and TA. Delayed volatile decomposition. Decreased the resistance to environmental stress stimuli from exogenous coating. Delayed senescence. | (Yan et al., 2019) |
| Fresh-cut papaya | Polysaccharide-based (chitosan and pectin) | Improved microbiological and physicochemical quality and extended shelf life (by 7 days) compared to uncoated fruit. | (Brasil et al., 2012) |
| Mango | Hydroxypropyl cellulose and carnauba wax | Reduced decay and weight loss. No effect on TSS. Significantly higher TA. Delayed carboxylic acids accumulation, delayed ripening. Extended shelf life. Reduced softening. | (Baldwin et al., 1999) |
| Fresh-cut watermelon | Antimicrobial alginate-based (sodium-alginate, pectin, and calcium lactate) | Significantly reduced weight loss and preserved texture. Did not affect sensory quality (color, odor and flavor), The pH and Brix was also not affected. | (Sipahi et al., 2013) |
| Citrus fruit | CMC and chitosan | Prevented postharvest microbial spoilage and enhanced fruit glossiness and appearance but not very effective in weight loss. | (Amon et al., 2014) |

| Commodity | Coating | Findings | Reference |
|---|---|---|---|
| Plum | Hydroxypropyl methylcellulose and beeswax | Contains food additives, reduced brown rot caused by *M. fructicola* and physiological disorders. Delayed postharvest ripening process, reduced weight and firmness loss, and delayed color change. Improved gaseous exchange. Extended shelf life. | (Gunaydin et al., 2017) |
| Tomato | Whey protein isolate and xanthan gum | Successfully prolonged shelf life, exhibited low AsA degradation, total phenolic and flavonoid content, pH, TSS and TA compared to uncoated. Delayed weight loss, total sugars. Firmness and color were also sustained. | (Kumar & Saini, 2021) |
| Fresh-cut mango | Sodium alginate and chitosan | Microbiological and physicochemical qualities were improved. Shelf life was enhanced up to 8 days when compared to uncoated fruit. | (Souza et al., 2015) |
| Fresh-cut lotus root | Xantham gum + chitosan Guar gum + chitosan | Both LbL coatings showed higher functionality than a single layer, thus reducing the loss of water and preserved color. | (Lara et al., 2020) |
| Fresh-cut Rocha pear and whole Rocha pear | κ-carrageenan and lysozyme | As a result of this, uncoated fresh-cut pears and whole pears presented higher mass loss, higher TSS and lower TA when compared with coated fresh-cut pears and whole pears. Uncoated fresh-cut pears also showed a darker color. These findings suggest that these nanolayered coatings have a potential for enhancing shelf life. | (Lu et al., 2021) |
| Potato French fries | CMC, pectin and guar gum and xanthan gum | They reduced oil uptake, yet they are not recommended for some products and chips or for French fries. | (Daraei Garmakhany et al., 2014) |
| Kosui Japanese pear | Chitosan and alginate | Inhibited ethylene production, respiration and overall fruit ripening. Little effect on water vapor permeability; therefore, did not affect weight loss. Transcriptome analysis also showed that LbL coating is involved in up-regulation of glycolysis pathway genes and down-regulation of TCA cycle genes, which may lead to shelf life extension. | (Hira et al., 2022) |
| Plum | Chitosan and alginate | LbL coating effectively limited the gas exchange between fruit and atmosphere, inhibited the loss of water and reduced the respiration rate; thus a good effect on weight loss, maintaining firmness, TSS, titratable acid, ascorbic acid and phenolic content. LbL treatment further inhibited PAL activity significantly on the plums, which resulted in the inhibition of anthocyanin synthesis, therefore delaying color changes. LbL coating induced POD and SOD activities, hence restraining $O_2^-$ production. | (Li et al., 2022) |

*LbL:* layer-by-layer, CMC: carboxymethyl cellulose, TCA: tricarboxylic acid, TSS: total soluble solids, TA: titratable acids, AsA: ascorbic acids, PAL: phenylalanine ammonia lyase, POD: peroxidase, SOD: superoxide dismutase.

mitochondrion and peroxisome, which are the site for aerobic respiration and pho-
torespiration, respectively, which both require a high amount of oxygen (Hira et al.,
2022). Table 14.1. summarizes the efficacy of LbL edible coatings on various horticultural crops during postharvest storage.

## 14.4   LIMITATIONS OF LAYER-BY-LAYER EDIBLE COATINGS

The layer-by-layer edible coatings approach has been studied widely on different
food products. It has been reported to improve the quality parameters of fruits and
vegetables in postharvest handling. The main purpose of employing this technique
is to improve the quality parameters of agricultural products such as microbial quality, texture/firmness/appearance, weight loss, etc. Microbial attacks are one of the
significant reasons horticultural crops deteriorate. Edible coatings often contain antimicrobial properties in their composition, which might help cover microcracks and
wounds prone to microbial attacks. The interest in using LbL films/edible coatings
has increased recently. The significant input of LbL edible coatings is not question-
able as it brings diversity in the structure and components in the interaction of fresh
produce and surface coating(s) (An et al., 2018). LbL is based on polysaccharides as
the ideal building blocks for creating systems imitating the biochemical properties
of the in vivo cellular environment. These polymers have advantages over synthetic
polyelectrolytes; however, they may have difficulties at some point. As much as the
use of polysaccharides at the commercial level has increased with increasing demand,
wide variations in quality from one provider to another and from batch to batch can
be challenging (Crouzier et al., 2010). Besides, polysaccharides are mostly extracted
and purified from natural tissues and therefore depend on natural variations.

In LbL techniques, especially when using polysaccharide-based coatings only, the
molecular weight variations effect renders systematic studies; this concept is referred
to as polydispersity. Chemical modification can be an issue since organic solvents have
high hydration shells and poor solubility (Crouzier et al., 2010). However, the reactive
groups present on the different polysaccharides are limited in terms of variety. Since
polysaccharide-based coatings have longer persistence, their flexibility is lower than
that of synthetic polyelectrolytes. Also, low solubility, pH and ionic strength can only
be varied to a lower extent compared to synthetic. This may result in the buildup of
conditions being more limited due to their degrees of freedom. Escobedo-Lozano et al.
(2015) conducted a study on a mixed gel system where chitosan (CS) and acemannan
(AC) were combined. The author reported that the interaction established was hetero-
typic between these two polysaccharides, where AC chains interrupt the connectivity
of the CS network, resulting in maintained mechanical strength and antibacterial activity. However, the study also reflected on the detrimental effect of AC incorporation into
the CS hydrogels and concluded on the impracticality CS-AC mixture.

## 14.5   CONCLUSION AND FUTURE PROSPECTS

Overall, several studies have shown that layer-by-layer (LbL) edible coatings of two
or more components can improve edible coating properties by capitalizing on the
benefits from different materials. LbL edible coatings are currently mostly applied in

horticultural crops; however, there is literature on other food products. It was noted that the most used LbL coating materials are polysaccharides (mostly chitosan), probably due to their wide availability; however, LbL-based proteins and lipids can also be applied. This technique is usually formed through electrostatic interaction. Several studies also revealed the capability of LbL edible coating to prolong shelf life and enhance the physicochemical and phytochemical qualities, improve appearance, and inhibit microbial attack. Another key to adopting the LbL technique is understanding the mechanisms driving cell–multilayer film interactions.

## REFERENCES LIST

Abedi, A., Lakzadeh, L., & Amouheydari, M. (2021). Effect of an edible coating composed of whey protein concentrate and rosemary essential oil on the shelf life of fresh spinach. *Journal of Food Processing and Preservation*, *45*(4), e15284. https://doi.org/10.1111/jfpp.15284

Ali, A., Zahid, N., Manickam, S., Siddiqui, Y., & Alderson, P. G. (2014). Double layer coatings: A new technique for maintaining physico-chemical characteristics and antioxidant properties of dragon fruit during storage. *Food and Bioprocess Technology*, *7*(8), 2366–2374. https://doi.org/10.1007/s11947-013-1224-3

An, Q., Huang, T., & Shi, F. (2018). Covalent layer-by-layer films: Chemistry, design, and multidisciplinary applications. *Chemical Society Reviews*, *47*(13), 5061–5098. https://doi.org/10.1039/C7CS00406K

Arnon, H., Zaitsev, Y., Porat, R., & Poverenov, E. (2014). Effects of carboxymethyl cellulose and chitosan bilayer edible coating on postharvest quality of citrus fruit. *Postharvest Biology and Technology*, *87*, 21–26. https://doi.org/10.1016/j.postharvbio.2013.08.007

Arnon-Rips, H., & Poverenov, E. (2018). Improving food products' quality and storability by using Layer by Layer edible coatings. *Trends in Food Science and Technology*, *75*, 81–92. https://doi.org/10.1016/j.tifs.2018.03.003

Baldwin, E. A., Burns, J. K., Kazokas, W., Brecht, J. K., Hagenmaier, R. D., Bender, R. J., & Pesis, E. D. N. A. (1999). Effect of two edible coatings with different permeability characteristics on mango (*Mangifera indica* L.) ripening during storage. *Postharvest Biology and Technology*, *17*(3), 215–226. https://doi.org/10.1016/S0925-5214(99)00053-8

Bambalele, N. L., Mditshwa, A., Magwaza, L. S., & Tesfay, S. Z. (2021). The effect of gaseous ozone and Moringa leaf–carboxymethyl cellulose edible coating on antioxidant activity and biochemical properties of "Keitt" mango fruit. *Coatings*, *11*(11), 1406. https://doi.org/10.3390/coatings11111406

Benítez, S., Achaerandio, I., Pujolà, M., & Sepulcre, F. (2015). Aloe vera as an alternative to traditional edible coatings used in fresh-cut fruits: A case of study with kiwifruit slices. *LWT – Food Science and Technology*, *61*(1), 184–193. https://doi.org/10.1016/j.lwt.2014.11.036

Bourtoom, T. (2008). Edible films and coatings: Characteristics and properties. *International Food Research Journal*, *15*(3), 237–248.

Brasil, I. M., Gomes, C., Puerta-Gomez, A., Castell-Perez, M. E., & Moreira, R. G. (2012). Polysaccharide-based multilayered antimicrobial edible coating enhances quality of fresh-cut papaya. *LWT – Food Science and Technology*, *47*(1), 39–45. https://doi.org/10.1016/j.lwt.2012.01.005

Carneiro-da-Cunha, M. G., Cerqueira, M. A., Souza, B. W. S., Carvalho, S., Quintas, M. A. C., Teixeira, J. A., & Vicente, A. A. (2010). Physical and thermal properties of a chitosan/alginate nanolayered PET film. *Carbohydrate Polymers*, *82*(1), 153–159. https://doi.org/10.1016/j.carbpol.2010.04.043

Chauhan, C., Gupta, K. C., & Mukesh, A. (2014). Application of biodegradable Aloe vera gel to control post harvest decay and longer the shelf life of grapes. *International Journal of Current Microbiology and Applied Sciences, 3*(3), 632–642.

Chung, Y. C., & Chen, C. Y. (2008). Antibacterial characteristics and activity of acid-soluble chitosan. *Bioresource Technology, 99*(8), 2806–2814. https://doi.org/10.1016/j.biortech.2007.06.044

Crespilho, F. N., Zucolotto, V., Oliveira, Jr., O. N., & Nart, F. C. (2006). Electrochemistry of layer-by-layer films: A review. *International Journal of Electrochemical Science, 1*(5), 194–214. https://doi.org/10.1016/S1452-3981(23)17150-1

Crouzier, T., Boudou, T., & Picart, C. (2010). Polysaccharide-based polyelectrolyte multi-layers. *Current Opinion in Colloid and Interface Science, 15*(6), 417–426. https://doi.org/10.1016/j.cocis.2010.05.007

Dai, J., Sullivan, D. M., & Bruening, M. L. (2000). Ultrathin, layered polyamide and poly-imide coatings on aluminum. *Industrial and Engineering Chemistry Research, 39*(10), 3528–3535. https://doi.org/10.1021/ie000221d

Daraei Garmakhany, A., Mirzaei, H. O., Maghsudlo, Y., Kashaninejad, M., & Jafari, S. M. (2014). Production of low fat french-fries with single and multi-layer hydrocol-loid coatings. *Journal of Food Science and Technology, 51*(7), 1334–1341. https://doi.org/10.1007/s13197-012-0660-9

Debeaufort, F., Quezada-Gallo, J. A., & Voilley, A. (1998). Edible films and coatings: Tomorrow's packagings: A review. *Critical Reviews in Food Science and Nutrition, 38*(4), 299–313. https://doi.org/10.1080/10408699891274219

Decher, G., Hong, J.D., & Schmitt, J. (1992). Buildup of ultrathin multilayer films by a self-assembly process: III. Consecutively alternating adsorption of anionic and cationic poly-electrolytes on charged surfaces. *Thin Solid Films, 210–211*(2), 831–835. https://doi.org/10.1016/0040-6090(92)90417-A

Dhall, R. K. (2013). Advances in edible coatings for fresh fruits and vegetables: A review. *Critical Reviews in Food Science and Nutrition, 53*(5), 435–450. https://doi.org/10.1080/10408398.2010.541568

Escobedo-Lozano, A. Y., Domard, A., Velázquez, C. A., Goycoolea, F. M., & Argüelles-Monal, W. M. (2015). Physical properties and antibacterial activity of chitosan/acemannan mixed systems. *Carbohydrate Polymers, 115*, 707–714. https://doi.org/10.1016/j.carbpol.2014.07.064

Flores-López, M. L., Cerqueira, M. A., de Rodríguez, D. J., & Vicente, A. A. (2016). Perspectives on utilization of edible coatings and nano-laminate coatings for extension of postharvest storage of fruits and vegetables. *Food Engineering Reviews, 8*(3), 292–305. https://doi.org/10.1007/s12393-015-9135-x

Gennadios, A., Hanna, M. A., & Kurth, L. B. (1997). Application of edible coatings on meats, poultry and seafoods: A review. *LWT – Food Science and Technology, 30*(4), 337–350. https://doi.org/10.1006/fstl.1996.0202

Guin, T., Krecker, M., Milhorn, A., Hagen, D. A., Stevens, B., & Grunlan, J. C. (2015). Exceptional flame resistance and gas barrier with thick multilayer nanobrick wall thin films. *Advanced Materials Interfaces, 2*(11), 1500214. https://doi.org/10.1002/admi.201500214

Gunaydin, S., Karaca, H., Palou, L., de la Fuente, B., & Pérez-Gago, M. B. (2017). Effect of hydroxypropyl methylcellulose-beeswax composite edible coatings formulated with or without antifungal agents on physicochemical properties of plums during cold storage. *Journal of Food Quality, 2017*, 1–9. https://doi.org/10.1155/2017/8573549

Guo, M., Chen, J., Li, J., Nie, L., & Yao, S. (2004). Carbon nanotubes-based amperomet-ric cholesterol biosensor fabricated through layer-by-layer technique. *Electroanalysis, 16*(23), 1992–1998. https://doi.org/10.1002/elan.200403053

Han, C., Percival, S. J., & Zhang, B. (2016). Electrochemical characterization of ultra-thin cross-linked metal nanoparticle films. *Langmuir, 32*(35), 8783–8792. https://doi.org/10.1021/acs.langmuir.6b00710

Harris, J. J., Stair, J. L., & Bruening, M. L. (2000). Layered polyelectrolyte films as selective, ultrathin barriers for anion transport. *Chemistry of Materials*, *12*(7), 1941–1946. https://doi.org/10.1021/cm0001004

Helander, I. M., Nurmiaho-Lassila, E. L., Ahvenainen, R., Rhoades, J., & Roller, S. (2001). Chitosan disrupts the barrier properties of the outer membrane of Gram-negative bacteria. *International Journal of Food Microbiology*, *71*(2–3), 235–244. https://doi.org/10.1016/S0168-1605(01)00609-2

Hira, N., Mitalo, O. W., Okada, R., Sangawa, M., Masuda, K., Fujita, N., Ushijima, K., Akagi, T., & Kubo, Y. (2022). The effect of layer-by-layer edible coating on the shelf life and transcriptome of "Kosui" Japanese pear fruit. *Postharvest Biology and Technology*, *185*, 111787. https://doi.org/10.1016/j.postharvbio.2021.111787

Holder, K. M., Smith, R. J., & Grunlan, J. C. (2017). A review of flame retardant nanocoatings prepared using layer-by-layer assembly of polyelectrolytes. *Journal of Materials Science*, *52*(22), 12923–12959. https://doi.org/10.1007/s10853-017-1390-1

Khoshnoudi-Nia, S., & Sedaghat, N. (2019). Effect of active edible coating and temperature on quality properties of roasted pistachio nuts during storage. *Journal of Food Processing and Preservation*, *43*(10), e14121. https://doi.org/10.1111/jfpp.14121

Kumar, A., & Saini, C. S. (2021). Edible composite bi-layer coating based on whey protein isolate, xanthan gum and clove oil for prolonging shelf life of tomatoes. *Measurement: Food*, *2*, 100005. https://doi.org/10.1016/j.meafoo.2021.100005

Lara, G. R., Uemura, K., Khalid, N., Kobayashi, I., Takahashi, C., Nakajima, M., & Neves, M. A. (2020). Layer-by-layer electrostatic deposition of edible coatings for enhancing the storage stability of fresh-cut lotus root (Nelumbo nucifera). *Food and Bioprocess Technology*, *13*(4), 722–726. https://doi.org/10.1007/s11947-020-02410-3

Leathers, T.D. (2003). Biotechnological production and applications of pullulan. *Applied Microbiology and Biotechnology*, *62*(5–6), 468–473. https://doi.org/10.1007/s00253-003-1386-4

Li, B., Kennedy, J. F., Peng, J. L., Yie, X., & Xie, B. J. (2006). Preparation and performance evaluation of glucomannan–chitosan–nisin ternary antimicrobial blend film. *Carbohydrate Polymers*, *65*(4), 488–494. https://doi.org/10.1016/j.carbpol.2006.02.006

Li, H., Huang, Z., Addo, K. A., & Yu, Y. (2022). Evaluation of postharvest quality of plum (*Prunus salicina* L. cv. 'French') treated with layer-by-layer edible coating during storage. *Scientia Horticulturae*, *304*, 111310. https://doi.org/10.1016/j.scienta.2022.111310

Liu, H., & Hu, N. (2005). Comparative bioelectrochemical study of core– shell nanocluster films with ordinary layer-by-layer films containing heme proteins and CaCO3 nanoparticles. *Journal of Physical Chemistry. B*, *109*(20), 10464–10473. https://doi.org/10.1021/jp0505227

Lu, Z., Saldaña, M. D. A., Jin, Z., Sun, W., Gao, P., Bilige, M., & Sun, W. (2021). Layer-by-layer electrostatic self-assembled coatings based on flaxseed gum and chitosan for Mongolian cheese preservation. *Innovative Food Science and Emerging Technologies*, *73*, 102785. https://doi.org/10.1016/j.ifset.2021.102785

Mantilla, N., Castell-Perez, M. E., Gomes, C., & Moreira, R. G. (2013). Multilayered antimicrobial edible coating and its effect on quality and shelf-life of fresh-cut pineapple (*Ananas comosus*). *LWT – Food Science and Technology*, *51*(1), 37–43. https://doi.org/10.1016/j.lwt.2012.10.010

Martiñon, M. E., Moreira, R. G., Castell-Perez, M. E., & Gomes, C. (2014). Development of a multilayered antimicrobial edible coating for shelf-life extension of fresh-cut cantaloupe (*Cucumis melo* L.) stored at 4 C. *LWT – Food Science and Technology*, *56*(2), 341–350. https://doi.org/10.1016/j.lwt.2013.11.043

Medeiros, B. G. D. S., Pinheiro, A. C., Carneiro-da-Cunha, M. G., & Vicente, A. A. (2012). Development and characterization of a nanomultilayer coating of pectin and chitosan–Evaluation of its gas barrier properties and application on "Tommy Atkins" mangoes. *Journal of Food Engineering*, *110*(3), 457–464. https://doi.org/10.1016/j.jfoodeng.2011.12.021

Medeiros, B. G. D. S., Pinheiro, A. C., Teixeira, J. A., Vicente, A. A., & Carneiro-da-Cunha, M. G. (2012). Polysaccharide/protein nanomultilayer coatings: Construction, characterization and evaluation of their effect on 'Rocha' pear (*Pyrus communis* L.) shelf-life. *Food and Bioprocess Technology*, *5*(6), 2435–2445. https://doi.org/10.1007/s11947-010-0508-0

Melia, M. A., Percival, S. J., Qin, S., Barrick, E., Spoerke, E., Grunlan, J., & Schindelholz, E. J. (2020). Influence of Clay size on corrosion protection by Clay nanocomposite thin films. *Progress in Organic Coatings*, *140*, 105489. https://doi.org/10.1016/j.porgcoat.2019.105489

Moreira, S. P., de Carvalho, W. M., Alexandrino, A. C., de Paula, H. C. B., Rodrigues, M. D. C. P., de Figueiredo, R. W., Maia, G. A., de Figueiredo, E. M. A. T., & Brasil, I. M. (2014). Freshness retention of minimally processed melon using different packages and multilayered edible coating containing microencapsulated essential oil. *International Journal of Food Science and Technology*, *49*(10), 2192–2203. https://doi.org/10.1111/ijfs.12535

Muralikrishna, G., Salimath, P. V., & Tharanathan, R. N. (1987). Structural features of an arabinoxylan and a rhamno-galacturonan derived from linseed mucilage. *Carbohydrate Research*, *161*(2), 265–271. https://doi.org/10.1016/S0008-6215(00)90083-1

Mwelase, S., & Fawole, O. A. (2022). Effect of Chitosan-24-Epibrassinolide composite coating on the quality attributes of late-harvested pomegranate fruit under simulated commercial storage conditions. *Plants*, *11*(3), 351. https://doi.org/10.3390/plants11030351

Pandey, R., & Mishra, A. (2010). Antibacterial activities of crude extract of Aloe barbadensis to clinically isolated bacterial pathogens. *Applied Biochemistry and Biotechnology*, *160*(5), 1356–1361. https://doi.org/10.1007/s12010-009-8577-0

Percival, S. J., Melia, M. A., Alexander, C. L., Nelson, D. W., Schindelholz, E. J., & Spoerke, E. D. (2020). Nanoscale thin film corrosion barriers enabled by multilayer polymer clay nanocomposites. *Surface and Coatings Technology*, *383*, 125228. https://doi.org/10.1016/j.surfcoat.2019.125228

Percival, S. J., Russo, S., Priest, C., Hill, R. C., Ohlhausen, J. A., Small, L. J., Rempe, S. B., & Spoerke, E. D. (2021). Bio-inspired incorporation of phenylalanine enhances ionic selectivity in layer-by-layer deposited polyelectrolyte films. *Soft Matter*, *17*(26), 6315–6325. https://doi.org/10.1039/D1SM00134E

Poverenov, E., Danino, S., Horev, B., Granit, R., Vinokur, Y., & Rodov, V. (2014). Layer-by-layer electrostatic deposition of edible coating on fresh cut melon model: Anticipated and unexpected effects of alginate–chitosan combination. *Food and Bioprocess Technology*, *7*(5), 1424–1432. https://doi.org/10.1007/s11947-013-1134-4

Rico, D., Martín-Diana, A. B., Barat, J. M., & Barry-Ryan, C. (2007). Extending and measuring the quality of fresh-cut fruit and vegetables: A review. *Trends in Food Science and Technology*, *18*(7), 373–386. https://doi.org/10.1016/j.tifs.2007.03.011

Shinga, M. H., & Fawole, O. A. (2023). Opuntia ficus indica mucilage coatings regulate cell wall softening enzymes and delay the ripening of banana fruit stored at retail conditions. *International Journal of Biological Macromolecules*, *245*, 125550. https://doi.org/10.1016/j.ijbiomac.2023.125550

Shinga, M. H., Mditshwa, A., Magwaza, L. S., & Tesfay, S. Z. (2021). Evaluating carboxymethylcellulose encapsulating Moringa leaf extract as edible coatings for controlling rind pitting disorder in 'Marsh' grapefruit (*Citrus× paradisi*). *Acta Horticulturae*, (1306), 301–308. https://doi.org/10.17660/ActaHortic.2021.1306.38

Sipahi, R. E., Castell-Perez, M. E., Moreira, R. G., Gomes, C., & Castillo, A. (2013). Improved multilayered antimicrobial alginate-based edible coating extends the shelf life of fresh-cut watermelon (*Citrullus lanatus*). *LWT – Food Science and Technology*, *51*(1), 9–15. https://doi.org/10.1016/j.lwt.2012.11.013

Soliva-Fortuny, R. C., & Martín-Belloso, O. (2003). Microbiological and biochemical changes in minimally processed fresh-cut conference pears. *European Food Research and Technology*, *217*(1), 4–9. https://doi.org/10.1007/s00217-003-0701-8

Song, H. Y., Jo, W. S., Song, N. B., Min, S. C., & Song, K. B. (2013). Quality change of apple slices coated with Aloe vera gel during storage. *Journal of Food Science*, *78*(6), C817–C822. https://doi.org/10.1111/1750-3841.12141

Souza, M. P., Vaz, A. F. M., Cerqueira, M. A., Texeira, J. A., Vicente, A. A., & Carneiro-da-Cunha, M. G. (2015). Effect of an edible nanomultilayer coating by electrostatic self-assembly on the shelf life of fresh-cut mangoes. *Food and Bioprocess Technology*, *8*(3), 647–654. https://doi.org/10.1007/s11947-014-1436-1

Treviño-Garza, M. Z., García, S., del Socorro Flores-González, M., & Arévalo-Niño, K. (2015). Edible active coatings based on pectin, pullulan, and chitosan increase quality and shelf life of strawberries (*Fragaria ananassa*). *Journal of Food Science*, *80*(8), M1823–M1830. https://doi.org/10.1111/1750-3841.12938

Treviño-Garza, M. Z., García, S., Heredia, N., Alanís-Guzmán, M. G., & Arévalo-Niño, K. (2017). Layer-by-layer edible coatings based on mucilages, pullulan and chitosan and its effect on quality and preservation of fresh-cut pineapple (*Ananas comosus*). *Postharvest Biology and Technology*, *128*, 63–75. https://doi.org/10.1016/j.postharvbio.2017.01.007

Umaraw, P., & Verma, A. K. (2017). Comprehensive review on application of edible film on meat and meat products: An eco-friendly approach. *Critical Reviews in Food Science and Nutrition*, *57*(6), 1270–1279. https://doi.org/10.1080/10408398.2014.986563

Velickova, E., Winkelhausen, E., Kuzmanova, S., Alves, V. D., & Moldão-Martins, M. (2013). Impact of chitosan-beeswax edible coatings on the quality of fresh strawberries (*Fragaria ananassa* cv Camarosa) under commercial storage conditions. *LWT – Food Science and Technology*, *52*(2), 80–92. https://doi.org/10.1016/j.lwt.2013.02.004

Weiss, J., Takhistov, P., & McClements, D. J. (2006). Functional materials in food nanotechnology. *Journal of Food Science*, *71*(9), R107–R116. https://doi.org/10.1111/j.1750-3841.2006.00195.x

Xia, Y., Rogers, J. A., Paul, K. E., & Whitesides, G. M. (1999). Unconventional methods for fabricating and patterning nanostructures. *Chemical Reviews*, *99*(7), 1823–1848. https://doi.org/10.1021/cr980002q

Yan, J., Luo, Z., Ban, Z., Lu, H., Li, D., Yang, D., Aghdam, M. S., & Li, L. (2019). The effect of the layer-by-layer (LBL) edible coating on strawberry quality and metabolites during storage. *Postharvest Biology and Technology*, *147*, 29–38. https://doi.org/10.1016/j.postharvbio.2018.09.002

Yang, S., Li, Y., Jiang, X., Chen, Z., & Lin, X. (2006). Horseradish peroxidase biosensor based on layer-by-layer technique for the determination of phenolic compounds. *Sensors and Actuators. Part B*, *114*(2), 774–780. https://doi.org/10.1016/j.snb.2005.07.035

Yaqub, M., Imar, S., Laffir, F., Armstrong, G., & McCormac, T. (2015). Investigations into the electrochemical, surface, and electrocatalytic properties of the surface-immobilized polyoxometalate, TBA3K [SiW10O36 (PhPO) 2]. *ACS Applied Materials and Interfaces*, *7*(2), 1046–1056. https://doi.org/10.1021/am5017864

Zhao, X. M., Wilbur, J. L., & Whitesides, G. M. (1996). Using two-stage chemical amplification to determine the density of defects in self-assembled monolayers of alkanethiolates on gold. *Langmuir*, *12*(13), 3257–3264. https://doi.org/10.1021/la960044e

# 15 Recent Advances in Improvement of Postharvest Application of Edible Coatings on Fruit

*Olaniyi Amos Fawole*

## 15.1 INTRODUCTION

Several postharvest strategies have been employed to minimize quality losses that occur after harvest. Packaging is one of the most important ways by which the fruit's environment can be controlled. Paperboard makes up 48% of the packaging used in the food industry due to its protective properties (Opara & Mditshwa, 2013). It offers protection against mechanical injury, especially during transport and handling (Rolle, 2006), and is recyclable and low-cost (Opara & Mditshwa, 2013). Plastic is also another popular packaging material used due to its transparency, light weight and convenience. Plastic serves as a moisture, gas and contamination barrier. By reducing the amount of water diffusing out of the fruit, the firmness of fruit is maintained for an extended period. The regulation of gases also leads to the retarding of respiration and senescence and extending shelf life (Gonzalez & Tiznado, 1993). Despite the benefits provided by plastic, its use is becoming increasingly unpopular globally. Most plastics are single use, and their disposal in landfills leads to the release of harmful gases into the atmosphere (Jeevahan & Chandrasekaran, 2019). The use of plastic has also failed to prevent the occurrence of quality losses, despite the benefits it offers. Since 2017, the UN has been campaigning to reduce the impact of plastic pollution. The increasing awareness among consumers has also contributed toward making plastic an unfeasible packaging material to continue using in the effort to reduce postharvest losses.

The abundant water and nutrient content of fruit render them vulnerable to pests and pathogenic microorganisms such as *Penicillium* spp. and *Botrytis* spp. (Martinez et al., 2018). As microbial spoilage is one of the main factors reducing quality, the postharvest application of chemicals such as fungicides are also among the postharvest strategies employed. These chemicals reduce microbial loads and prevent decay on agricultural produce. Various modes of actions have been identified, including the inhibition of respiration, enzyme activity, cell division and elongation, as well as disruption of cell walls (Leroux, 1996). Despite the wide variety of products

DOI: 10.1201/9781003452355-18

available against specific pests, the continued use of pesticides and fungicides has led to microbial resistance and disruption of environmental ecosystems (Elshafie, 2015). Many of these chemicals also pose a toxic threat to human health due to bioaccumulation (Leroux, 1996). The need for environmental and consumer-friendly postharvest chemicals has birthed the development of edible coatings. Edible coatings are proteins, carbohydrates or fats directly formed on the food surface and are eaten as part of the food product (Jeevahan & Chandrasekaran, 2019). Edible coatings are often designed to combine the functions of both plastic packaging (to control the environment around the fruit) and chemicals (to reduce microbial decay). Edible coatings are generally regarded as safe (GRAS) and require no disposal as they are ingested with the fruit material. Plant-based edible coatings can (1) directly prevent the development of pathogens and/or (2) indirectly prevent pathogens by improving the fruits' tolerance (Wan et al., 2021). Despite the excitement surrounding edible coatings, its use has shown to impart off-flavors, discoloration and excessive glossiness in fruit (Zhong et al., 2014). Homogeneity and the formation of a stable coating has also proven to be a challenge. Also, edible coatings cannot be applied as a broad-spectrum solution, as different agricultural produce has different surface properties, affecting adhesion of the coating to the food surface. Although there are review papers on edible coatings as a technology to preserve fruit quality, according to author's knowledge, there is no review on the application techniques of edible coatings despite the important roles that coating application techniques play in the performance of the coatings. This review focuses on current knowledge on the postharvest application techniques of edible coatings on horticultural produce. Optimization of the various techniques are also reviewed, and future prospects on the application techniques of edible coatings are identified.

## 15.2  POSTHARVEST APPLICATION TECHNIQUES OF EDIBLE COATINGS: BENEFITS AND CHALLENGES

The transfer of a coating to a food product can be done using various methods. Edible coatings have been successfully applied to food products using dipping, spraying, brushing, panning, foaming, solvent casting, dripping and fluidized bed coating.

### 15.2.1  Dipping

Dipping is the process by which fruits are completely submerged in an aqueous coating medium for a specified amount of time. During dipping, the coating material adheres to the surface of the fruit. This method is especially recommended when applying viscous liquids (Andrade et al., 2013). Most studies on edible coatings have focused on dipping as it is a relatively simple process that requires very basic equipment. This application method is therefore easy to employ in a lab-scale environment. Despite the technological simplicity of the process, dipping poses challenges as the scale increases. Coating quantities increase significantly as the load of fruit increases because fruit must be completely submerged for the dipping to be effective. Coating potency is also lost during the process, and large volumes of coating must be replaced systematically to ensure that the coating maintains its active compounds

(Raghav et al., 2016). Coating contamination is also unavoidable with the dipping method, and undesirable materials such as stalks and leaves are often found in the residual coating. The industrial application of dipping is therefore limited by the high waste yield and loss of coating strength over time.

## 15.2.2 SPRAYING

Spraying, on the other hand, has become the most popular commercial technique used for the application of edible coatings. During spraying, the edible coating is forced through a nozzle to form small droplets in the form of a mist. These droplets are then transferred to the fruit surface. Spraying is typically used to apply thin layers of coating on the fruit surfaces as this method is suitable for less viscous solutions (Peretto et al., 2017). Spraying duration is often increased when increased coating deposition is required. This method has gained popularity due to its high through-put to fruit ratio. Spraying also produces minimal waste as the droplets are directly transferred to the fruit surface, and the coating also maintains its strength throughout the entire process. The volume of the coating used for this technique is minimal, and the risk of coating contamination is reduced, as opposed to dipping. Despite the benefits presented by spraying, an uneven coating is often observed (Andrade et al., 2013). As the sprayer is positioned overhead, only the surfaces of the fruit exposed to the nozzle are covered. The fruit also moves under the sprayer with a conveyer, which often leads to the removal of the thin coating. Nonuniform coverage often renders the edible coatings ineffective, and the desired results are not achieved within the fruit. The nozzles also cannot accommodate thick coatings, as blockages can occur.

## 15.2.3 DIPPING AND BRUSHING

Dripping is the method by which the coating material is applied to the fruit from above in the form of drops either directly to the fruit surface or to brushes, followed by drying, often with the aid of a fan (Zuhal et al., 2018). Brushing is especially popular in citrus packhouses, by which waxes are applied to fruit to reduce weight loss, provide a glossy appearance and act as biocontrol agent (Ladaniya, 2008). This method is viewed as the most economical process due to its high transfer efficiency and industrial compatibility. Dripping requires larger coating volumes than spraying as brushes are always required to be saturated, but offers all the other benefits of spraying. Due to the large droplet sizes, coating uniformity can only be achieved when the tumbling action over the saturated brushes is sufficient to allow even coverage.

## 15.2.4 PANNING

Panning is commonly used for coating nuts and sweets but has also been employed for processed fruits, such as raisins. The coated fruits are characterized by a smooth, consistent surface as a result of the polishing action in a rotating bowl, known as a pan. Panning is a slow process and the speed depends on the size of the center of the pan. The process consists of a closed stainless-steel pan possessing perforations on its side panels. A pump carries the coating to the spray guns positioned in the pan. The fruits are continuously rotated in the pan to ensure uniform coating. Coatings are then dried by means of forced air. The heat produced by the friction during rotation is removed by cold air (Andrade et al., 2012).

Despite the promising nature of edible coatings, a few drawbacks have limited their widespread application within the agricultural industry. Different fruits have different metabolisms and react differently to their storage environments after harvest. Given that edible coatings are custom designed to meet the needs of specific fruits, improvements or optimizations are necessary to ensure that the full potential of the edible coating is achieved. The vast majority of literature available on edible coatings has been successful in discussing the importance of the coating formulation and its effect on fruit quality. Recommendations on coating concentration (Daisy et al., 2020; Kubheka et al., 2020) and the addition of active compounds (Ebrahimi & Rastegar, 2020; Etemadipoor et al., 2020) are often given. There has been little consideration for the application method and its effect on coating performance within the literature. The steps of edible coating improvement are discussed next, and a closer look is taken at the technical aspects surrounding the application of coatings.

## 15.3 CONCENTRATION AND FORMULATION

The success of edible coatings has been attributed to the functional properties of the ingredients that make up the coating. Table 15.1 summarizes the ingredients generally used in the formulation of edible coatings as well as their functions. As different fruits have different physiological responses, edible coatings are formulated to meet the specific needs of the fruit. Stone fruits such as plums are susceptible to water loss, resulting in shriveling and reduction in quality. It is therefore necessary to formulate edible coatings to reduce the loss of water in plums, and lipids in the form of essential oils are often added to polysaccharide coatings to improve their hydrophobicity (Paladines et al., 2014). Other climacteric fruit such as bananas ripen very quickly, reducing their shelf life.

Edible coatings for bananas are therefore synthesized to offer a gas barrier to reduce metabolic process and retard ripening (Esyanti et al., 2019). When formulating edible coatings, it is important to understand that a balance must be maintained between coating components to ensure uniform and stable edible coatings. Ingredients with desired functions cannot be increased indefinitely, as consumer perception of fruit quality must also be considered. Examples of studies in which coating formulation was balanced with consumer acceptability included citrus (Rasouli et al., 2019), fresh-cut watermelon (Sipahi et al., 2013) and sweet cherries (Petriccione et al., 2015).

## 15.4 DRYING TECHNIQUES

The drying of edible coatings is essential in ensuring that a stable uniform coating forms around a fruit, providing it with the intended protection. Efficient drying of edible coatings is often overlooked due to the time it would take for coatings to properly set, especially in a commercial environment. The industry makes use of sprayer

**TABLE 15.1**
**Summary of Ingredients That Make Up Edible Coatings and Their Functions**

| Type | Examples | Function/Property | Reference |
|---|---|---|---|
| Fats | Beeswax, acetylated monoglycerides, fatty alcohols, fatty acids | Excellent moisture barriers<br>Brittle nature | (Khan et al., 2017)<br>(Andrade et al., 2012) |
| Proteins | Gelatin, casein, wheat gluten, zein, soy protein | Good mechanical properties<br>Low $O_2$ permeability due to strong intermolecular forces<br>Poor $H_2O$ properties | (Khan et al., 2017)<br>(Andrade et al., 2012) |
| Polysaccharides | Starch, starch derivatives, cellulose, pectin, alginate | Good mechanical properties<br>Low $O_2$ permeability due to strong intermolecular forces<br>Poor moisture barrier properties | (Khan et al., 2017)<br>(Andrade et al., 2012) |
| Polyols | Glycerol, polyethylene glycol | Acts as plasticizer | (Andrade et al., 2012) |
| Acids/bases | Acetic acid, lactic acid | Regulates the pH of coating medium | (Andrade et al., 2012) |
| Antimicrobials | Chitosan, essential oils, plant extracts | Reduces microbial biota | (Gol et al., 2013; Sánchez-González et al., 2011) |

systems to allow for large throughput of fruit volumes on packing lines, while ensuring thin layers of chemical treatments are applied to fruit surfaces to minimize the drying time. Two studies have reported on the effect of drying methods on edible chitosan films. Mayachiew et al. (2010) investigated the effect of ambient drying (30°C), hot air drying (40°C), vacuum drying and low-pressure superheated steam drying (LPSSD) (70–90°C, 10 kPA) on chitosan film properties. LPSSD-treated films had higher tensile strength and crystallinity. The study highlighted the significant effect of drying conditions on chitosan antimicrobial activity. Ambient drying produced the best antimicrobial films against *Staphylococcus aureus*, using the agar diffusion method. The higher activity was linked to antimicrobial preservation at low temperatures. Researchers also argued that lower temperatures aided the interaction between chitosan and the added active agent, limiting diffusion into the agar medium. Additionally, higher temperatures also favored intermolecular interaction, reducing the release of the antimicrobial agent. On the other hand, Thakhiew et al. (2010), investigated the effect of drying methods on the mechanical and physical properties of chitosan and confirmed that the increased drying temperature aided in the formation of crosslinks, yielding compact films and reducing swelling.

## 15.5   NANOTECHNOLOGY

Nanotechnology is a relatively new method capitalizing on materials of nanoscale (≤100 nm) to overcome the shortcomings presented by edible coatings (Flores-Lopez et al., 2016). This technology provides a medium through which active agents such as essential oils, antimicrobials and antioxidants can be transported in the edible coating. The main advantages introduced by nanotechnology to edible coatings are stable and biologically active systems, controlled release, improved mechanical properties, while maintaining coating appearance and transparency (Zambrano-Zaragoza et al., 2018). The value of nanotechnology can be attributed to the greater surface:volume ratio when compared to larger particles of the same chemical composition (Jeevahan & Chandrasekaran, 2019).

Both organic and inorganic nanocomposites in edible coatings have been reported in the literature. Common examples of organic components include polylactic acid (PLA), cellulose acetate phthalate (CAP), ethylcellulose, chitosan and alginate (Kumar et al., 2015). The most common nanoparticles used in edible coatings for fruit include nano-ZnO, nano-SiO$_2$, nano-TiO$_2$ and montmorillonite (Zambrano-Zaragoza et al., 2018). These inorganic components are GRAS (generally recognized as safe). The nature of these inorganic compounds makes them excellent antimicrobials, with ZnO offering great permeation barriers.

Esyanti et al. (2019) investigated the potential of nanoparticle-based chitosan edible coating on shelf life extension of banana. The reduction of the chitosan size was expected to improve its barrier properties to water and gases, while improving its antimicrobial effect. Chitosan nanoparticles were prepared by ionic gelation and applied to unripe green bananas using the spraying technique. After storage at 22°C for 6 days, the bananas treated with chitosan nanoparticles showed the biggest delay in ripening. The control fruit started turning yellow on the third day, chitosan-treated fruit between days 3–4, whereas the color change was only observed between days 4–5 in nanochitosan-treated fruit. The delay in ripening of the nanosystem was attributed to the disruption of the ethylene biosynthesis process by improved gas barrier properties. Nanomultilayer coatings of pectin and chitosan were investigated for their gas barrier properties in mangoes (Medeiros et al., 2012). After 45 days of storage at 4°C and 93% RH, it was concluded that the nanomultilayer system was effective at reducing mass loss and delaying ripening. The success of the coating was attributed to the increased gas permeability, reducing the flow of gas. Nanoparticles provide a means to improve the application of edible coatings and bioactive agents by increasing the surface:volume ratio. This technology is also commonly paired with the layer-by-layer application of edible coatings.

## 15.6　LAYER-BY-LAYER APPLICATION

Layer-by-layer application of edible coatings has generated great interest due to its ability to combine multiple edible coatings for the maintenance of fruit quality and the extension of shelf life. Layer-by-layer application is based on the principle of electrostatic interaction (Treviño-Garza et al., 2017). During this process, edible coatings consist of opposite charges applied to fruit materials sequentially, often involving a drying step between coatings (Brasil et al., 2012). This application technique addresses the problems often experienced with edible coatings, especially fresh-cut fruit. Adequate adhesion of the edible coatings to hydrophobic fruit surfaces limit the effectiveness of single coatings to maintain fruit quality. By combining electrostatic coatings, the adhesion properties of edible coatings are improved by layer-by-layer application, and the combined effect maintains and improves fruit quality (Yin et al., 2019). Chitosan, a cationic polysaccharide, has been a popular candidate coating for layer-by-layer application (Yin et al., 2019; Souza et al., 2015; Brasil et al., 2012) due to its bioactivity and antimicrobial properties (Velickova et al., 2015). Development of a single effective coating based on chitosan has been limited by its physical properties, such as its limited solubility and adhesion to fruit surfaces (Poverenov et al., 2014). Another drawback presented by chitosan edible coatings is their poor water barrier properties (Moreira et al., 2014). The benefit offered by layer-by layer-application is that anionic coatings such as sodium alginate (Mantilla et al., 2013) can be used to compensate for the shortcomings of chitosan.

## 15.7　NOVEL TECHNOLOGIES FOR APPLYING EDIBLE COATINGS

Recent research has reported the emergence of a novel coating application technology that has promising prospects for the future success of edible coatings. Initiating from the industrial spraying technique, electrospraying is a new technique that has shown great efficiency. Electrospraying originated from the paint industry (Khan et al., 2017) and has been shown to lend a big hand to the food industry. Electrospraying, like conventional spraying, applies a liquid coating to the fruit surface in the form of small droplets. The main difference between the two application methods is the way in which the droplets are formed. In a conventional sprayer, the liquid coating passes through an atomizer under high pressure. The combination of high pressure and the ejection through a small orifice ranging between 0.343 and 2.184 mm in diameter

(Mrak & Steward, 1962) causes the liquid coating to disperse into small droplets. During electrospraying, the liquid coating becomes charged by a high-voltage supply as it passes through the nozzle (Khan et al., 2017). The surface tension of the liquid is devastated by the charge and breaks into small droplets. The sheer stress provided by this action leads to the formation of a steady microscopic stream that disperses into droplets of uniform size (Rosell-Llompart et al., 2018). The charged droplets repel one another and are attracted to the grounded surface. The transfer efficiency of the coating to the fruit is increased by 80%, and droplets release their charge at deposition, as deposition is guided by electric field lines. Electrospraying provides the advantage of ultra-thin coatings on the fruit surface, maintaining the sensory aspect of the produce. Increased transfer efficiency and ultra-slim films lead to the reduction of coating waste. The small droplets are also able to move into extremely small cracks (Khan et al., 2017) in the fruit surface, ensuring that the quality is maintained for longer. Electrospraying, does, however, have higher running and maintenance costs as the high-voltage supply is vital to its success.

Peretto et al. (2017) investigated the potential of alginate to maintain strawberry quality by comparing the two application methods, including conventional spraying and electrospraying. Upon application, it was noted that electrospraying offered a higher transfer efficiency and more uniform coating compared with conventional spraying. In addition, electrosprayed strawberries demonstrated a 60% weight gain after coating compared to conventional spraying. The electrosprayed strawberries also exhibited higher firmness and lower decay incidence after the 13 day storage period. In another study by Cakmak et al. (2018) on dipping and electrospraying of apple slices with oil-water emulsions, the authors noted that the electrospraying method showed greater coating uptake and lower moisture loss compared to the dipped apple slices.

## 15.8  CONCLUSION AND FUTURE PROSPECTS

Horticultural produce is highly perishable by nature and is therefore highly susceptible to quality losses. Postharvest technology is therefore a vital tool in reducing losses and improving quality. These technologies are constantly challenged by consumers' awareness and their demand for sustainable practices. Edible coatings show great promise as a postharvest treatment for fresh fruit. Edible coatings have the potential to combine the functions of both plastic and pesticides to preserve fruit and extend their shelf life. The success of edible coatings is largely based on their application, and application techniques are vital in ensuring that coatings act at their optimal potential. Improvements in terms of coating formulation, drying, incorporation of nanoparticles and multilayer systems have positioned edible coatings as a strong candidate for overcoming these postharvest challenges. Future prospects for edible coatings include the expansion of nanoparticle systems and the layer-by-layer technique. The interaction of multilayers is not fully understood at this time, and its understanding should be improved. Electrospraying, a novel method that allows for higher transfer efficiencies and the formation of thinner, stable coatings is a field that should be further explored for real-life applications. Further research is required to determine the effect of other application parameters such as coating temperature and exposure time as these parameters effect the crosslinks that form, as well as the density and strength of the coating. It is also necessary to rationally develop a new generation of edible coatings that will consider the role of the application method in its success.

## REFERENCES LIST

Andrade, R. D., Skurtys, O., & Osorio, F. A. (2012). Atomizing spray systems for application of edible coatings. *Comprehensive Reviews in Food Science and Food Safety*, *11*(3), 323–337. https://doi.org/10.1111/j.1541-4337.2012.00186.x

Andrade, R. D., Skurtys, O., & Osorio, F. A. (2013). Drop impact behaviour on food using spray coating: Fundamentals and applications. *Food Research International*, *54*(1), 397–405. https://doi.org/10.1016/j.foodres.2013.07.042

Brasil, I. M., Gomes, C., Puerta-Gomez, A., Castell-Perez, M. E., & Moreira, R. G. (2012). Polysaccharide-based multilayered antimicrobial edible coating enhances quality of fresh-cut papaya. *LWT – Food Science and Technology*, *47*(1), 39–45. https://doi.org/10.1016/j.lwt.2012.01.005

Cakmak, H., Kumcuoglu, S., & Tavman, S. (2018). Production of edible coatings with twin-nozzle electrospraying equipment and the effects on shelf-life stability of fresh-cut apple slices. *Journal of Food Process Engineering*, *41*(1), e12627. https://doi.org/10.1111/jfpe.12627

Daisy, L. L., Nduko, J. M., Joseph, W. M., & Richard, S. M. (2020). Effect of edible gum arabic coating on the shelf life and quality of mangoes (*Mangifera indica*) during storage. *Journal of Food Science and Technology*, *57*(1), 79–85. https://doi.org/10.1007/s13197-019-04032-w

Ebrahimi, F., & Rastegar, S. (2020). Preservation of mango fruit with guar-based edible coatings enriched with *Spirulina platensis* and *Aloe vera* extract during storage at ambient temperature. *Scientia Horticulturae*, *265*, 109258. https://doi.org/10.1016/j.scienta.2020.109258

Elshafie, H. S., Mancini, E., Sakr, S., De Martino, L., Mattia, C. A., De Feo, V., & Camele, I. (2015). Antifungal activity of some constituents of *Origanum vulgare* L. essential oil against postharvest disease of peach fruit. *Journal of Medicinal Food*, *18*(8), 929–934. https://doi.org/10.1089/jmf.2014.0167

Esyanti, R. R., Zaskia, H., Amalia, A., & Nugrahapraja, H. (2019). Chitosan nanoparticle-based coating as post-harvest technology in banana. *Journal of Physics: Conference Series*, *1204*(1), 012109. https://doi.org/10.1088/1742-6596/1204/1/012109

Etemadipoor, R., Mirzaalian Dastjerdi, A. M., Ramezanian, A., & Ehteshami, S. (2020). Ameliorative effect of gum arabic, oleic acid and/or cinnamon essential oil on chilling injury and quality loss of guava fruit. *Scientia Horticulturae*, *266*, 109255. https://doi.org/10.1016/j.scienta.2020.109255

Flores-López, M. L., Cerqueira, M. A., de Rodríguez, D. J., & Vicente, A. A. (2016). Perspectives on utilization of edible coatings and nano-laminate coatings for extension of postharvest storage of fruits and vegetables. *Food Engineering Reviews*, *8*(3), 292–305. https://doi.org/10.1007/s12393-015-9135-x

Gol, N. B., Patel, P. R., & Rao, T. V. R. (2013). Improvement of quality and shelf-life of strawberries with edible coatings enriched with chitosan. *Postharvest Biology and Technology*, *85*, 185–195. https://doi.org/10.1016/j.postharvbio.2013.06.008

González, G., & Tiznado, M. (1993). Postharvest physiology of bell peppers stored in low density polyethylene bags. *LWT – Food Science and Technology*, *26*(5), 450–455. https://doi.org/10.1006/fstl.1993.1089

Jeevahan, J., & Chandrasekaran, M. (2019). Nanoedible films for food packaging: A review. *Journal of Materials Science*, *54*(19), 12290–12318. https://doi.org/10.1007/s10853-019-03742-y

Khan, M. K. I., Nazir, A., & Maan, A. A. (2017). Electrospraying: A novel technique for efficient coating of foods. *Food Engineering Reviews, 9*(2), 112–119. https://doi.org/10.1007/s12393-016-9150-6

Kubheka, S. F., Tesfay, S. Z., Mditshwa, A., & Magwaza, L. S. (2020). Evaluating the efficacy of edible coatings incorporated with Moringa Leaf extract on postharvest of "Maluma"avocado fruit quality and its biofungicidal effect. *Horticultural Science, 1*(aop), 1–6. https://doi.org/10.21273/HORTSCI14391-19

Kumar, V. D., Verma, P. R. P., & Singh, S. K. (2015). Development and evaluation of biodegradable polymeric nanoparticles for the effective delivery of quercetin using a quality by design approach. *LWT – Food Science and Technology, 61*(2), 330–338. https://doi.org/10.1016/j.lwt.2014.12.020

Ladaniya, M. S. (2008). Commercial fresh citrus cultivars and producing countries. In M. S. Ladaniya (Ed.), *Citrus fruit: Biology, technology and evaluation* (pp. 229–286). Academic Press. https://doi.org/10.1016/B978-012374130-1.50004-8

Leroux, P. (1996). Recent developments in the mode of action of fungicides. *Pesticide Science, 47*(2), 191–197. https://doi.org/10.1002/(SICI)1096-9063(199606)47:2<191::AID-PS415>3.0.CO

Mantilla, N., Castell-Perez, M. E., Gomes, C., & Moreira, R. G. (2013). Multilayered antimicrobial edible coating and its effect on quality and shelf-life of fresh-cut pineapple (*Ananas comosus*). *LWT – Food Science and Technology, 51*(1), 37–43. https://doi.org/10.1016/j.lwt.2012.10.010

Martinez, K., Ortiz, M., Albis, A., Gilma Gutiérrez Castañeda, C., Valencia, M. E., & Grande Tovar, C. D. (2018). The effect of edible chitosan coatings incorporated with Thymus capitatus essential oil on the shelf-life of strawberry (*Fragaria* x *ananassa*) during cold storage. *Biomolecules, 8*(4), 155. https://doi.org/10.3390/biom8040155

Mayachiew, P., Devahastin, S., Mackey, B. M., & Niranjan, K. (2010). Effects of drying methods and conditions on antimicrobial activity of edible chitosan films enriched with galangal extract. *Food Research International, 43*(1), 125–132. https://doi.org/10.1016/j.foodres.2009.09.006

Medeiros, B. G. D. S., Pinheiro, A. C., Carneiro-da-Cunha, M. G., & Vicente, A. A. (2012). Development and characterization of a nanomultilayer coating of pectin and chitosan–Evaluation of its gas barrier properties and application on "Tommy Atkins" mangoes. *Journal of Food Engineering, 110*(3), 457–464. https://doi.org/10.1016/j.jfoodeng.2011.12.021

Moreira, S. P., de Carvalho, W. M., Alexandrino, A. C., de Paula, H. C. B., Rodrigues, M. D. C. P., de Figueiredo, R. W., Maia, G. A., de Figueiredo, E. M. A. T., & Brasil, I. M. (2014). Freshness retention of minimally processed melon using different packages and multilayered edible coating containing microencapsulated essential oil. *International Journal of Food Science and Technology, 49*(10), 2192–2203. https://doi.org/10.1111/ijfs.12535

Mrak, E., & Steward, G. (Eds.). (1962). *Advances in food research* (2nd ed). Academic Press.

Opara, U. L., & Mditshwa, A. (2013). A review on the role of packaging in securing food system: Adding value to food products and reducing losses and waste. *African Journal of Agricultural Research, 8*(22), 2621–2630. https://doi.org/10.5897/AJAR2013.6931

Paladines, D., Valero, D., Valverde, J. M., Díaz-Mula, H. M., Serrano, M., & Martínez-Romero, D. (2014). The addition of rosehip oil improves the beneficial effect of Aloe vera gel on delaying ripening and maintaining postharvest quality of several stone fruit. *Postharvest Biology and Technology, 92*, 23–28. https://doi.org/10.1016/j.postharvbio.2014.01.014

Peretto, G., Du, W. X., Avena-Bustillos, R. J., Berrios, J. D. J., Sambo, P., & McHugh, T. H. (2017). Electrostatic and conventional spraying of alginate-based edible coating with natural antimicrobials for preserving fresh strawberry quality. *Food and Bioprocess Technology, 10*(1), 165–174. https://doi.org/10.1007/s11947-016-1808-9

Petriccione, M., De Sanctis, F., Pasquariello, M. S., Mastrobuoni, F., Rega, P., Scortichini, M., & Mencarelli, F. (2015). The effect of chitosan coating on the quality and nutraceutical traits of sweet cherry during postharvest life. *Food and Bioprocess Technology, 8*(2), 394–408. https://doi.org/10.1007/s11947-014-1411-x

Poverenov, E., Rutenberg, R., Danino, S., Horev, B., & Rodov, V. (2014). Gelatin-chitosan composite films and edible coatings to enhance the quality of food products: Layer-by-Layer vs. blended formulations. *Food and Bioprocess Technology, 7*(11), 3319–3327. https://doi.org/10.1007/s11947-014-1333-7

Raghav, P. K., Agarwal, N., & Saini, M. (2016). Edible coating of fruits and vegetables: A review. *International Journal of Scientific Research and Modern Education, 1*(1), 188–204.

Rasouli, M., Koushesh Saba, M. K., & Ramezanian, A. (2019). Inhibitory effect of salicylic acid and *Aloe vera* gel edible coating on microbial load and chilling injury of orange fruit. *Scientia Horticulturae, 247*, 27–34. https://doi.org/10.1016/j.scienta.2018.12.004

Rolle, R. S. (2006). Improving postharvest management and marketing in the Asia-Pacific region: Issues and challenges. *Postharvest Management of Fruit and Vegetables in the Asia-Pacific Region, 1*(1), 23–31.

Rosell-Llompart, J., Grifoll, J., & Loscertales, I. G. (2018). Electrosprays in the cone-jet mode: From Taylor cone formation to spray development. *Journal of Aerosol Science, 125*, 2–31. https://doi.org/10.1016/j.jaerosci.2018.04.008

Sánchez-González, L., Vargas, M., González-Martínez, C., Chiralt, A., & Cháfer, M. (2011). Use of essential oils in bioactive edible coatings: A review. *Food Engineering Reviews, 3*(1), 1–16. https://doi.org/10.1007/s12393-010-9031-3

Sipahi, R. E., Castell-Perez, M. E., Moreira, R. G., Gomes, C., & Castillo, A. (2013). Improved multilayered antimicrobial alginate-based edible coating extends the shelf life of fresh-cut watermelon (*Citrullus lanatus*). *LWT – Food Science and Technology, 51*(1), 9–15. https://doi.org/10.1016/j.lwt.2012.11.013

Souza, M. P., Vaz, A. F. M., Cerqueira, M. A., Texeira, J. A., Vicente, A. A., & Carneiro-da-Cunha, M. G. (2015). Effect of an edible nanomultilayer coating by electrostatic self-assembly on the shelf life of fresh-cut mangoes. *Food and Bioprocess Technology, 8*(3), 647–654. https://doi.org/10.1007/s11947-014-1436-1

Thakhiew, W., Devahastin, S., & Soponronnarit, S. (2010). Effects of drying methods and plasticizer concentration on some physical and mechanical properties of edible chitosan films. *Journal of Food Engineering, 99*(2), 216–224. https://doi.org/10.1016/j.jfoodeng.2010.02.025

Treviño-Garza, M. Z., García, S., Heredia, N., Alanís-Guzmán, M. G., & Arévalo-Niño, K. (2017). Layer-by-layer edible coatings based on mucilages, pullulan and chitosan and its effect on quality and preservation of fresh-cut pineapple (*Ananas comosus*). *Postharvest Biology and Technology, 128*, 63–75. https://doi.org/10.1016/j.postharvbio.2017.01.007

Velickova, E., Winkelhausen, E., Kuzmanova, S., Moldão-Martins, M., & Alves, V. D. (2015). Characterization of multilayered and composite edible films from chitosan and beeswax. *Food Science and Technology International, 21*(2), 83–93. https://doi.org/10.1177/1082013213511807

Wan, C., Kahramanoğlu, İ., & Okatan, V. (2021). Application of plant natural products for the management of postharvest diseases in fruits. *Folia Horticulturae, 33*(1), 203–215. https://doi.org/10.2478/fhort-2021-0016

Yin, C., Huang, C., Wang, J., Liu, Y., Lu, P., & Huang, L. (2019). Effect of chitosan- and alginate-based coatings enriched with cinnamon essential oil microcapsules to improve the postharvest quality of mangoes. *Materials, 12*(13), 2039. https://doi.org/10.3390/ma12132039

Zambrano-Zaragoza, M.L., González-Reza, R., Mendoza-Muñoz, N., Miranda-Linares, V., Bernal-Couoh, T.F., Mendoza-Elvira, S., & Quintanar-Guerrero, D. (2018). Nanosystems in edible coatings: A novel strategy for food preservation. *International Journal of Molecular Sciences, 19*(3), 705. https://doi.org/10.3390/ijms19030705

Zhong, Y., Cavender, G., & Zhao, Y. (2014). Investigation of different coating application methods on the performance of edible coatings on Mozzarella cheese. *LWT – Food Science and Technology, 56*(1), 1–8. https://doi.org/10.1016/j.lwt.2013.11.006

Zuhal, O., Yavuz, Y., & Kerse, S. (2018). Edible film and coating applications in fruits and vegetables. *Alınteri Zirai Bilimler Dergisi, 33*(2), 221–226. https://doi.org/10.28955/alinterizbd.368362

# Index

Note: Page numbers in *italics* indicate a figure and page numbers in **bold** indicate a table on the corresponding page.

For Product Safety Concerns and Information please contact our EU
representative  GPSR@taylorandfrancis.com
Taylor & Francis Verlag GmbH, Kaufingerstraße 24, 80331 München, Germany